Earth Science:
Physics and Chemistry of the Earth

Volume 2

EDITOR

Joseph L. Spradley, Ph.D.
Wheaton College, Illinois

Salem Press
A Division of EBSCO Publishing
Ipswich, Massachusetts   Hackensack, New Jersey

Cover Photo: New Zealand - North Island - Rotorua area Wai-O-Tapu (Sacred Waters)
© Barry Lewis/In Pictures/Corbis

Library of Congress Cataloging-in-Publication Data

Earth science : physics and chemistry of the earth / editor, Joseph L. Spradley.
    p. cm.
  Includes bibliographical references and index.
  ISBN 978-1-58765-989-8 (set) – ISBN 978-1-58765-973-7 (set (1 of 4)) – ISBN 978-1-58765-974-4 (vol. 1) – ISBN 978-1-58765-975-1 (vol. 2) 1. Earth sciences. I. Spradley, Joseph L.
  QE26.3.E27 2012
  550–dc23
                    2012004322

First Printing

PRINTED IN THE UNTED STATES OF AMERICA

CONTENTS

COMMON UNITS OF MEASURE

Notes: Common prefixes for metric units—which may apply in more cases than shown below—include giga- (1 billion times the unit), mega- (one million times), kilo- (1,000 times), hecto- (100 times), deka- (10 times), deci- (0.1 times, or one tenth), centi- (0.01, or one hundredth), milli- (0.001, or one thousandth), and micro- (0.0001, or one millionth).

UNIT	QUANTITY	SYMBOL	EQUIVALENTS
Acre	Area	ac	43,560 square feet 4,840 square yards 0.405 hectare
Ampere	Electric current	A *or* amp	1.00016502722949 international ampere 0.1 biot *or* abampere
Angstrom	Length	Å	0.1 nanometer 0.0000001 millimeter 0.000000004 inch
Astronomical unit	Length	AU	92,955,807 miles 149,597,871 kilometers (mean Earth-sun distance)
Barn	Area	b	10^{-28} meters squared (approx. cross-sectional area of 1 uranium nucleus)
Barrel (dry, for most produce)	Volume/capacity	bbl	7,056 cubic inches; 105 dry quarts; 3.281 bushels, struck measure
Barrel (liquid)	Volume/capacity	bbl	31 to 42 gallons
British thermal unit	Energy	Btu	1055.05585262 joule
Bushel (U.S., heaped)	Volume/capacity	bsh *or* bu	2,747.715 cubic inches 1.278 bushels, struck measure
Bushel (U.S., struck measure)	Volume/capacity	bsh *or* bu	2,150.42 cubic inches 35.238 liters
Candela	Luminous intensity	cd	1.09 hefner candle
Celsius	Temperature	C	1° centigrade
Centigram	Mass/weight	cg	0.15 grain
Centimeter	Length	cm	0.3937 inch
Centimeter, cubic	Volume/capacity	cm³	0.061 cubic inch
Centimeter, square	Area	cm²	0.155 square inch
Coulomb	Electric charge	C	1 ampere second

UNIT	QUANTITY	SYMBOL	EQUIVALENTS
Cup	Volume/capacity	C	250 milliliters 8 fluid ounces 0.5 liquid pint
Deciliter	Volume/capacity	dl	0.21 pint
Decimeter	Length	dm	3.937 inches
Decimeter, cubic	Volume/capacity	dm^3	61.024 cubic inches
Decimeter, square	Area	dm^2	15.5 square inches
Dekaliter	Volume/capacity	dal	2.642 gallons 1.135 pecks
Dekameter	Length	dam	32.808 feet
Dram	Mass/weight	dr *or* dr avdp	0.0625 ounce 27.344 grains 1.772 grams
Electron volt	Energy	eV	$1.5185847232839 \times 10^{-22}$ Btu $1.6021917 \times 10^{-19}$ joule
Fermi	Length	fm	1 femtometer 1.0×10^{-15} meters
Foot	Length	ft *or* '	12 inches 0.3048 meter 30.48 centimeters
Foot, cubic	Volume/capacity	ft^3	0.028 cubic meter 0.0370 cubic yard 1,728 cubic inches
Foot, square	Area	ft^2	929.030 square centimeters
Gallon (U.S.)	Volume/capacity	gal	231 cubic inches 3.785 liters 0.833 British gallon 128 U.S. fluid ounces
Giga-electron volt	Energy	GeV	$1.6021917 \times 10^{-10}$ joule
Gigahertz	Frequency	GHz	—
Gill	Volume/capacity	gi	7.219 cubic inches 4 fluid ounces 0.118 liter
Grain	Mass/weight	gr	0.037 dram 0.002083 ounce 0.0648 gram

UNIT	QUANTITY	SYMBOL	EQUIVALENTS
Gram	Mass/weight	g	15.432 grains 0.035 avoirdupois ounce
Hectare	Area	ha	2.471 acres
Hectoliter	Volume/capacity	hl	26.418 gallons 2.838 bushels
Hertz	Frequency	Hz	$1.08782775707767 \times 10^{-10}$ cesium atom frequency
Hour	Time	h	60 minutes 3,600 seconds
Inch	Length	in *or* "	2.54 centimeters
Inch, cubic	Volume/capacity	in^3	0.554 fluid ounce 4.433 fluid drams 16.387 cubic centimeters
Inch, square	Area	in^2	6.4516 square centimeters
Joule	Energy	J	$6.2414503832469 \times 1018$ electron volt
Joule per kelvin	Heat capacity	J/K	$7.24311216248908 \times 1022$ Boltzmann constant
Joule per second	Power	J/s	1 watt
Kelvin	Temperature	K	-272.15 degree Celsius
Kilo-electron volt	Energy	keV	$1.5185847232839 \times 10^{-19}$ joule
Kilogram	Mass/weight	kg	2.205 pounds
Kilogram per cubic meter	Mass/weight density	kg/m^3	$5.78036672001339 \times 10^{-4}$ ounces per cubic inch
Kilohertz	Frequency	kHz	—
Kiloliter	Volume/capacity	kl	—
Kilometer	Length	km	0.621 mile
Kilometer, square	Area	km^2	0.386 square mile 247.105 acres
Light-year (distance traveled by light in one Earth year)	Length/distance	lt-yr	5,878,499,814,275.88 miles 9.46×10^{12} kilometers
Liter	Volume/capacity	L	1.057 liquid quarts 0.908 dry quart 61.024 cubic inches
Mega-electron volt	Energy	MeV	—
Megahertz	Frequency	MHz	—

Unit	Quantity	Symbol	Equivalents
Meter	Length	m	39.37 inches
Meter, cubic	Volume/capacity	m³	1.308 cubic yards
Meter per second	Velocity	m/s	2.24 miles per hour 3.60 kilometers per hour
Meter per second per second	Acceleration	m/s²	12,960.00 kilometers per hour per hour 8,052.97 miles per hour per hour
Meter, square	Area	m²	1.196 square yards 10.764 square feet
Metric. See unit name			
Microgram	Mass/weight	mcg *or* μg	0.000001 gram
Microliter	Volume/capacity	μl	0.00027 fluid ounce
Micrometer	Length	μm	0.001 millimeter 0.00003937 inch
Mile (nautical international)	Length	mi	1.852 kilometers 1.151 statute miles 0.999 U.S. nautical mile
Mile (statute or land)	Length	mi	5,280 feet 1.609 kilometers
Mile, square	Area	mi²	258.999 hectares
Milligram	Mass/weight	mg	0.015 grain
Milliliter	Volume/capacity	ml	0.271 fluid dram 16.231 minims 0.061 cubic inch
Millimeter	Length	mm	0.03937 inch
Millimeter, square	Area	mm2	0.002 square inch
Minute	Time	m	60 seconds
Mole	Amount of substance	mol	6.02×10^{23} atoms or molecules of a given substance
Nanometer	Length	nm	1,000,000 fermis 10 angstroms 0.001 micrometer 0.00000003937 inch
Newton	Force	N	0.224808943099711 pound force 0.101971621297793 kilogram force 100,000 dynes
Newton-meter	Torque	N·m	0.7375621 foot-pound

UNIT	QUANTITY	SYMBOL	EQUIVALENTS
Ounce (avoirdupois)	Mass/weight	oz	28.350 grams 437.5 grains 0.911 troy or apothecaries' ounce
Ounce (troy)	Mass/weight	oz	31.103 grams 480 grains 1.097 avoirdupois ounces
Ounce (U.S., fluid or liquid)	Mass/weight	oz	1.805 cubic inch 29.574 milliliters 1.041 British fluid ounces
Parsec	Length	pc	30,856,775,876,793 kilometers 19,173,511,615,163 miles
Peck	Volume/capacity	pk	8.810 liters
Pint (dry)	Volume/capacity	pt	33.600 cubic inches 0.551 liter
Pint (liquid)	Volume/capacity	pt	28.875 cubic inches 0.473 liter
Pound (avoirdupois)	Mass/weight	lb	7,000 grains 1.215 troy or apothecaries' pounds 453.59237 grams
Pound (troy)	Mass/weight	lb	5,760 grains 0.823 avoirdupois pound 373.242 grams
Quart (British)	Volume/capacity	qt	69.354 cubic inches 1.032 U.S. dry quarts 1.201 U.S. liquid quarts
Quart (U.S., dry)	Volume/capacity	qt	67.201 cubic inches 1.101 liters 0.969 British quart
Quart (U.S., liquid)	Volume/capacity	qt	57.75 cubic inches 0.946 liter 0.833 British quart
Rod	Length	rd	5.029 meters 5.50 yards
Rod, square	Area	rd^2	25.293 square meters 30.25 square yards 0.00625 acre
Second	Time	s or sec	$\frac{1}{60}$ minute $\frac{1}{3,600}$ hour

Unit	Quantity	Symbol	Equivalents
Tablespoon	Volume/capacity	T or tb	3 teaspoons 4 fluid drams
Teaspoon	Volume/capacity	t or tsp	0.33 tablespoon 1.33 fluid drams
Ton (gross or long)	Mass/weight	t	2,240 pounds 1.12 net tons 1.016 metric tons
Ton (metric)	Mass/weight	t	1,000 kilograms 2,204.62 pounds 0.984 gross ton 1.102 net tons
Ton (net or short)	Mass/weight	t	2,000 pounds 0.893 gross ton 0.907 metric ton
Volt	Electric potential	V	1 joule per coulomb
Watt	Power	W	1 joule per second 0.001 kilowatt $2.84345136093995 \times 10^{-4}$ ton of refrigeration
Yard	Length	yd	0.9144 meter
Yard, cubic	Volume/capacity	yd^3	0.765 cubic meter
Yard, square	Area	yd^2	0.836 square meter
Yard, square	Area	yd^2	0.836 square meter

COMPLETE LIST OF CONTENTS

Volume 1

Volume 2

CATEGORY LIST OF CONTENTS

Earth Science:
Physics and Chemistry of the Earth

MAGNETIC REVERSALS

Investigation of the earth's magnetic field history, as recorded by diverse rock types, has disclosed that the magnetic field changes position relative to the surface of the earth. The information accumulated from the study of these reversals is used to explain many of the events that have occurred over the course of the earth's history, such as continental collisions.

PRINCIPAL TERMS

- **basalt:** dark-colored, fine-grained igneous rock frequently found beneath the sediment covering the ocean floor
- **detrital remanent magnetization (DRM):** sedimentary rock magnetization acquired by magnetic sediment grains aligning with the magnetic field
- **normal polarity:** orientation of the earth's magnetic field so that a compass needle points toward the Northern Hemisphere
- **polarity:** orientation of the earth's magnetic field relative to the earth
- **reverse polarity:** orientation of the earth's magnetic field so that a compass needle points toward the Southern Hemisphere
- **thermal remanent magnetization (TRM):** magnetization acquired as a magma's magnetic material becomes permanently magnetized

NORMAL AND REVERSE POLARITIES

Research into the history of the earth's magnetic field has revealed that the field has flipped polarity many times in the past. Presently, the field is oriented so that a compass needle points toward the Northern Hemisphere of the earth. This orientation is known as normal polarity. If a compass needle were to point toward the south, that would indicate a reverse polarity. The flipping, or reversal, of the field involves the exchange of pole positions from Northern Hemisphere to Southern Hemisphere, either normal to reverse or reverse to normal.

To determine whether the polarity change is a real field change or simply a modification in a rock's magnetic-recording mechanism, numerous rocks were analyzed to ascertain their magnetic characteristics and to determine whether these change over time. Only a very small percentage of the rocks studied,

including an igneous rock from Japan, displayed a self-reversing tendency. This finding persuaded geophysicists that self-reversing tendencies in rocks do not need to be considered in the study of the field's history. Therefore, geophysicists do not have to test every rock to determine whether it self-reverses.

THERMAL REMANENT MAGNETIZATION

Geologists must still verify that the polarity changes are real phenomena that are consistent from one region of the earth to another. They make use of the fact that when magnetic grains form in magma, they magnetically align themselves with the magnetic field present at that time. This type of rock recording of magnetic direction is known as thermal remanent magnetization (TRM). The best recorder of TRM is rock of basaltic composition.

The Hawaiian Islands are an example of basaltic rock formed from magma that has sporadically erupted from the Hawaiian volcanoes over a period of millions of years. The island of Hawaii is a large volcano that sits several kilometers below sea level and rises several kilometers above sea level. Measured from base to summit, Hawaii is the highest mountain in the world.

A detailed polarity history of the island is difficult to develop because volcanic eruptions are intermittent, with several thousands of years between eruptions. However, an overall appreciation of the field changes can be acquired by sampling the distinct layers located in the eroded sides of the volcanoes. In the laboratory, an "absolute" date for the rocks can be obtained using radioactive-dating procedures. Relative dates—the sequence of occurrence for the samples—can also be established, as sample A is from a layer that lies below sample B, and so on. Relative dating helps assure that the absolute dates are correct. If sample A has an absolute date of 120,000 years and sample B a date of 140,000 years, but sample B is located physically above sample A in the volcano, then something is wrong. Accurate dating is an important aspect in the establishment of the polarity time scale.

By using a magnetometer, the polarity and the field direction of the sample can be determined. If the field points down, the polarity is normal. A field that points up indicates a reversed polarity. Once enough data have been collected (several hundred samples), the sample polarities can be plotted against the sample date. In this manner, the polarity history can be determined for the past 4 million years.

The polarity scale shows that from 4 million to 3.3 million years ago, the field was reversed. This period is called the Gilbert reverse epoch (the major periods are named for scientists who have advanced the discipline of magnetism). The field was normal until 2.5 million years ago during the Gauss normal epoch, except for a brief period of reversed polarity around 3 million years ago, known as the Mammoth reverse event. The Matuyama reverse epoch continued until 700,000 years ago. This epoch contained two normal events: the Olduvai, around 2 million years ago; and the Jaramillo, about 1 million years ago. The Brunhes is the present-day normal epoch. Other normal or reverse events may have been present in these epochs.

DETRITAL REMANENT MAGNETIZATION

Geophysicists are compelled to find other methods that verify the validity of the polarity scale and to extend and add more detail to the existing scale. One technique utilizes the sediment layer covering most of the ocean basin. This sediment can record magnetic field direction by the mechanism of detrital remanent magnetization (DRM). Long sediment cores are obtained from various areas in the ocean, and the magnetic polarity of areas along the length of the individual cores is measured. Again, a pattern of polarity changes is evident.

Radioactivity cannot efficiently date the layers of the core. Fortunately, the sediment is laid down very slowly, and this rate is measurable. The rate is on the order of millimeters per 1,000 years; thus, a layer 10 millimeters from the surface was deposited approximately 15,000 years ago. Polarity is plotted against calculated age, and analysis shows that the sediment-based data correspond well with the land-based scale.

MAGNETIC SEAFLOOR STRIPES

In the 1950's, magnetometers were towed behind ships that sailed over the oceanic ridge to the south of Iceland. The data were plotted on a map of the research area, and something strange became evident: The recorded magnetic field varied over the area. The map revealed a striped pattern of weaker and stronger field intensities that was aligned parallel to the ridge, now known as magnetic seafloor stripes. Fred Vine and Drummond Matthews, working together, and Lawrence Morley, working alone, realized that polarity changes caused the stripes.

In the mid-1960's, a revolution in the earth sciences was occurring with the development of the theory of plate tectonics. Scientists theorized that the earth's surface rock was split into plates of thin but considerable area. These plates had boundaries that interacted in several possible ways: They could move together, or converge; they could move apart, or diverge; or they could slide past each other in an area known as a transform fault. At the diverging boundary, the motion should produce a breach between the plates, but none was found. Investigation disclosed that the volcanically active oceanic ridge was the diverging boundary and that basaltic magma quickly filled any gap. New plate material is formed at this diverging boundary, and the cooling magma records the magnetic field present at the time of cooling by thermal remanent magnetization. The cooled magma moves away parallel to the ridge as the plates diverge. The magnetic field of basaltic rock that recorded the earth's magnetic field during a period of the reversed polarity cancels some of the earth's present-day field. This cancellation produces an area of lower-intensity field parallel to the ridge. The rock recording normal polarity adds to the earth's field, resulting in a strong intensity stripe.

The last polarity change 700,000 years ago was represented by rock that was located many kilometers from the center of the ridge. By dividing that distance by 700,000 years, the rate at which the plate is forming can be calculated. That value is approximately 2-5 centimeters per year depending on which portion of the ridge is being measured. This rate is comparable to how fast human fingernails grow. As a result of this movement, North America, which is west of the ridge, is now about 25 meters farther west of Europe, which is east of the ridge, than when Christopher Columbus sailed in 1492.

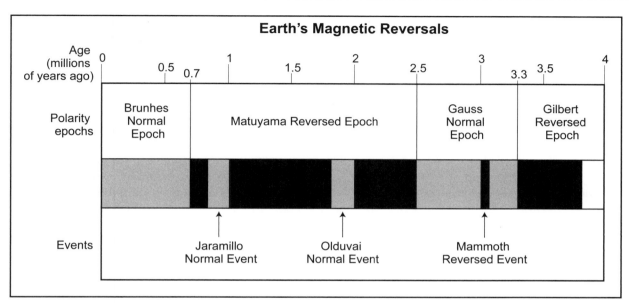

IDENTIFYING TRENDS

Other research has extended the polarity scale back hundreds of millions of years. This extension permits the identification of long-term trends depicting the manner in which the field has changed polarity. The field has remained very stable, with few polarity changes, several times in the past. In the Pennsylvanian period of the late Paleozoic era, for example, the field was predominantly in the reversed position; the field was normal for a long time in the Cretaceous period of the Late Mesozoic era. At other times, including the present Cenozoic era, the field has flipped many times.

ANALYZING ROCK SAMPLES

The procedure used to study magnetic reversals depends on the area of investigation. Land-based investigations are straightforward. The researcher chooses a likely site and conducts preliminary research to ascertain whether others have studied the area and whether the site will yield samples appropriate for study. If an area displays promise, the required rock samples are obtained. The rock should not be severely weathered, as that may alter the magnetic or radioactive components of the rock, which could lead to incorrect results. The sample is collected using a gasoline-powered drill with a tube-shaped, diamond-tipped drill bit. The resulting sample is a cylinder of rock still attached at its base to the original rock. A brass tube, the size of the drill

bit, is slipped over the cylinder. Brass is used because it is nonmagnetic and will not alter the sample's magnetic characteristics. Attached to the tube is a small platform on which is placed a Brunton compass, used to measure the orientation of the sample. The compass is very important, as sample orientation, the sampling site latitude and longitude, and the magnetic field direction of the sample are needed to calculate the sample's pole position. An orientation mark is made on the sample with a brass rod, and the sample is broken from the original rock. An identification number is assigned, which is carefully recorded along with all other pertinent information.

In the laboratory, the samples are prepared for the measurement of the rock's magnetic field direction by cutting them into lengths of 2.5 centimeters; thus, several small samples are obtained from each core, which can be used for dating purposes and for verifying the sample's polarity. Scientists do not rely on one measurement but make multiple assessments of a characteristic to ensure sample integrity. Because they are basalt samples, the rocks' magnetic directions are measured by a spinner magnetometer. A spinner works similarly to an electrical generator in that the sample is spun at high speed near coils of wire. The magnetic sample induces an electric current in the coils that is proportional to the strength of the sample's magnetic field. The rock's magnetic field direction is determined from these signals.

Computers perform the calculation of the final pole position using sample orientation, site latitude and longitude, and rock field direction. This pole position is plotted on a graph known as a stereonet, which is a two-dimensional representation of the earth's surface. Normal polarity poles plot in the Northern Hemisphere of the stereonet and those of reversed polarity in the Southern Hemisphere.

The basalt is also analyzed to determine its age. The amount of a suitable radioactive element (the parent isotope) and the amount of the element into which the radioactive element decays (the daughter isotope) are measured. The sample's age is calculated from these measurements.

OBTAINING THE SEDIMENT POLARITY SCALE

The sediment polarity scale is more difficult to obtain, because the cores come from the ocean bottom—3 kilometers below sea level, in some places—using a coring device that is dropped from a ship. Back in the laboratory, samples are taken along the length of the core. Their position on the core is measured, as their distance from the top of the core determines the age of the sample. The original core is not oriented as it is taken from the sea floor, because the scientist is not interested in the sample's pole position. The scientist is interested in the sample's polarity, and an unoriented core will yield that information.

The sediment has a weak magnetic field, so a superconducting magnetometer is used for the measurements. When some materials are cooled close to absolute zero, near the boiling point of liquid helium, they have no resistance to electrical current. Superconducting magnetometers that can detect small magnetic fields, such as those of sediment, employ these materials. Again, the polarity of the sediment sample is determined from the magnetometer readings.

DETECTING SEAFLOOR STRIPES

The detection of the seafloor stripes is a simple but tedious endeavor that requires the towing of a magnetometer "fish" several hundred meters behind a ship along parallel tracks across the area of interest. The magnetometer is towed to prevent the detection of ship-related magnetic fields. The readings—signal strength and ship's position—are plotted on a map and the stripe patterns are observed.

GEOLOGIC APPLICATIONS

The fact that field reversals are a worldwide occurrence and instantaneous from a geological perspective led to the establishment of the field of magnetostratigraphy for correlating rock layers from various continents. In piecing together the earth's history as revealed in the rocks, it is necessary to know what is occurring around the world at approximately the same time. To accomplish this feat, ways have been developed to correlate rocks in one area with rocks in another area. One technique involves using index fossils, which are fossils that are widespread and common but for which the organisms lived for only a short time. The problem is that index fossils are limited in distribution and cannot be used for correlating between continents. Magnetostratigraphy bypasses this difficulty.

Magnetic reversals require that the strength of the magnetic field decrease to near zero and then build back with reversed polarity. This change in polarity can take anywhere from 2,000 to 20,000 years. The magnetic field strength of the past has been both greater and weaker than it is at present. The magnetic field of the earth shields living organisms from damaging cosmic radiation. Periods of reduced magnetic field strength allow increased exposure of life-forms to radiation. Periods of mass extinctions or accelerated genetic changes may accompany these periods of weakened magnetic field strength.

The fact that the earth's magnetic field has reversed many times in the past is important because it verifies the theory of plate tectonics. Before the plates rupture, the rock layers along the potential rift area dome upward; to relieve the pressure, the dome splits in a three-armed rift. Two of the arms expand in length and join arms from other domed areas. These joined arms form the boundary between the two plates; the third arm fails to enlarge but forms a wedge-shaped basin that fills with sediment. Over time, organic material in the sediment is converted to petroleum. These sources of petroleum were unknown until the development of the theory of plate tectonics. Understanding plate tectonics also increases the understanding of earthquakes and their origin, which could lead to their prediction and even to their control.

Stephen J. Shulik

FURTHER READING

Butler, Robert F. *Paleomagnetism: Magnetic Domains to Geologic Terranes.* Boston: Blackwell Scientific Publications, 1992. Butler's exploration of the earth's magnetic fields begins with basic descriptions of what paleomagnetism is and how it occurs. The book is filled with illustrations to back up difficult concepts covered in the text.

Campbell, Wallace H. *Earth Magnetism: A Guided Tour through Magnetic Fields.* Burlington, MA: Harcourt/ Academic Press, 2001. Provides a history of research in magnetism followed by fundamental physics concepts. Later chapters discuss more advanced topics in geomagnetism. The text is written in a manner that a person with some geology or physics background could follow without instruction. Well suited for undergraduate students.

Cox, Allan, ed. *Plate Tectonics and Geomagnetic Reversals.* San Francisco: W. H. Freeman, 1973. Cox was a leader in establishing the magnetic polarity scale for the past 4 million years. In this book, he provides fascinating introductions to chapters that are composed of seminal papers concerning magnetic reversals and their contribution to the development of the theory of plate tectonics. The papers are advanced for the average reader, but have many graphs, diagrams, and figures that merit attention. The introductions are good, as they reveal the human side of scientists in their quest for knowledge.

Garland, G. D. *Introduction to Geophysics.* 2d ed. London: W. B. Saunders, 1979. Used as a text for introductory geophysics, this book contains (in sections 21.4 through 21.6) very readable material on magnetic reversals, magnetic anomalies, and the magnetic character of the oceans and continents. The many figures and graphs are of interest to the general reader.

Glen, William. *The Road to Jaramillo: Critical Years of the Revolution in Earth Science.* Stanford, Calif.: Stanford University Press, 1982. A history of the plate tectonics revolution of the mid-1950's to the mid-1960's. Chapter 6 is devoted to the evolution of the magnetic polarity scale. Chapter 7 deals with magnetic seafloor stripes and their interpretation. Other chapters are also important, such as Chapter 5, with its discussion of the participants in the earth science revolution. Best suited for students of geology; the writing is very technical.

Jacobs, John. *Deep Interior of the Earth.* London: Chapman and Hall, 1992. This introductory geophysics textbook is formidable for the average student because there is considerable mathematics in some chapters, but it does cover many useful topics such as geomagnetism, the earth's present magnetic field, auroras, and the magnetosphere. It contains a minimum of equations but many figures and graphs.

Kennett, J. P., ed. *Magnetic Stratigraphy of Sediments. Benchmark Papers in Geology*, Vol. 54. Stroudsburg, Pa.: Dowden, Hutchinson, and Ross, 1980. A memorial edition to Norman Watkins, who was intimately involved with the research aimed at the development of plate tectonics theory via the study of magnetization of sediments. A collection of seminal papers concerned with the magnetic stratigraphy of sediments, which relies on the fact of magnetic reversals. As a collection, the reading is uneven, as some papers are very involved, while others are less technical. The editor provides short essays at the beginning of each chapter. Graphs, tables, figures, and diagrams abound.

_____*Marine Geology.* Englewood Cliffs, N.J.: Prentice-Hall, 1982. Pages 71-78 are devoted to magnetostratigraphy and magnetic reversals. The text is a veritable all-you-need-to-know book about marine geology: plate tectonics, oceanic structure, sediments, margins, and history. Various dating methods are also described. Very readable, with no mathematics and some figures.

Lapedes, Daniel N., ed. *McGraw-Hill Encyclopedia of Geological Sciences.* New York: McGraw-Hill, 1978. Pages 704-708, under the heading "Rock Magnetism," provide a concise description of many aspects associated with rock magnetism: the process of rock magnetization, the present field, magnetic reversals, secular variation, and apparent polar wandering, among others. Very readable, with no mathematics and a fair number of graphs, tables, and figures.

Merrill, Ronald T. *Our Magnetic Earth: The Science of Geomagnetism.* Chicago: University of Chicago Press, 2010. This book dives right into magnetic fields and reversals. Later chapters discuss the effects the magnetic field has on animals, climate, and plate tectonics. A clearly written text for nonmathematical readers.

Merrill, R. T., and M. W. McElhinney. *The Magentic Field of the Earth: Paleomagnetism, the Core, and the Deep Mantle.* San Diego: Academic Press, 1998. The authors cover the basic material associated with the earth's magnetic field. Chapters deal with the origin of the magnetic field, as well as the origin of secular variation and field reversals. A strong background in mathematics is recommended. Bibliography and index. Numerous tables, figures, and mathematical equations.

Monastersky, Richard. "The Flap over Magnetic Flips." *Science News* 143 (June 1993): 378-380. In an article written for the general public, the author reviews recent data and hypothesis as to what happens during the time that the magnetic field is reversing polarity. The simple dipole field becomes more complex, but the field configuration returns to similar configuration each time it reverses.

Motz, Lloyd, ed. *Rediscovery of the Earth.* New York: Van Nostrand Reinhold, 1979. As a collection of articles for the nonscientist by renowned scientists in their respective fields, this text makes very interesting reading and is augmented with many colorful illustrations. The chapter titled "The Earth's Magnetic Field and Its Variations" was written by Takesi Nagata, who has written hundreds of articles on diverse aspects of geophysics besides the earth's magnetic field.

Plummer, Charles C., and Diane Carlson. *Physical Geology.* 12th ed. Boston: McGraw-Hill, 2007. A college-level introductory geology textbook that is clearly written and wonderfully illustrated. An excellent sourcebook of basic information on geologic terminology and fundamentals of geologic processes. An excellent glossary.

See also: Continental Drift; Earth's Magnetic Field; Geobiomagnetism; Magnetic Stratigraphy; Milankovitch Hypothesis; Mountain Building; Plate Tectonics; Polar Wander; Radioactive Decay; Rock Magnetism.

MAGNETIC STRATIGRAPHY

The earth's magnetic field has fluctuated between a polarity like that of today's field ("normal") and one completely opposite ("reversed") thousands of times in the last 600 million years. The magnetic minerals in erupting lavas and in settling sediments align with the prevailing field at the time the rock forms, thereby recording the earth's polarity history. The pattern of polarity changes in a thick sequence of rock and can be matched from area to area, providing scientists with a powerful tool of correlation.

PRINCIPAL TERMS

- **correlation:** matching the sequence of events (distinctive layers, fossils, magnetic polarity intervals) between two stratigraphic sections
- **Curie point:** the temperature at which a magnetic mineral locks in its magnetization
- **magnetic domain:** a region within a mineral with a single direction of magnetization; mineral grains smaller than about 100 microns contain only one domain, while larger grains can contain several domains
- **magnetic polarity time scale:** the geologic history of the changes in the earth's magnetic polarity
- **magnetic remanence:** the ability of the magnetic minerals in a rock to "lock in" the magnetic field of the earth prevailing at the time of their formation
- **paleomagnetism:** the study of the ancient magnetic field of the earth, as recorded by magnetic minerals in rocks
- **radiometric dating:** the estimation of the numerical age of a rock by measuring the decay of radioactive minerals, such as uranium, rubidium, or potassium
- **stratigraphy:** the study and interpretation of geologic history from layered rock sequences (usually sedimentary)

MAGNETIC REMANENCE

A compass shows that the earth's magnetic field lines point toward the North Pole, but 800,000 years ago, a compass needle would have pointed to the South Pole. The earth's magnetic field has apparently changed polarity thousands of times in the geologic past and an excellent record of its history extends back over the last 150 million years. This history is recorded in the magnetic minerals of rocks that were deposited or erupted in the geologic past.

Several minerals common in the earth's crust are known to be magnetic, but the most important are the iron oxides magnetite and hematite. Magnetite contains three atoms of iron and four of oxygen; hematite contains two atoms of iron and three of oxygen. When a magma cools, the magnetic domains (areas within a crystal that have the same magnetic direction) within crystals align with the field at that time and lock in that direction as the rest of the rock crystallizes. This process is known as thermal remanent magnetization (TRM). Since TRM is formed by cooling, it is found only in igneous and metamorphic rocks. The only igneous rocks that are commonly layered and capable of stratigraphic study are lava flows. Stacked sequences of lava flows were the source of the first discovery that the earth's magnetic field had reversed. The temperature at which this magnetization is locked in is known as the Curie point (named after French physicist Pierre Curie). For magnetite, the Curie point is about 578 degrees Celsius, but for hematite, it can be as high as 650 degrees Celsius. The actual Curie point varies with the iron and titanium content in the mineral.

When rocks with magnetic minerals are eroded, the magnetic grains become sedimentary particles that are transported by wind and water. As these particles settle, they, too, align with the prevailing field. As the rest of the sediment is hardened into rock, the sedimentary rock records the direction of the field at the time it was formed. This is known as detrital (or depositional) remanence magnetization (DRM). Since most stratigraphic sequences are sedimentary, most magnetic stratigraphy is concerned with the DRM of sediments.

After a rock is formed, it is possible for water seeping through it to oxidize the iron and precipitate new minerals (particularly hematite and iron hydroxides, such as goethite). Since these new minerals are formed by chemical activity, the magnetic field they lock in is known as chemical remanent magnetization (CRM). These minerals lock in a magnetic field that records the time of chemical alteration rather than the time of the formation of the rock. This magnetization is usually a secondary, "overprinted" one that obscures the original magnetization, which is the most interesting to the paleomagnetist.

MAGNETIC POLARITY TIME SCALE

Thick sequences of lava flows or layers of sediments that span long periods of time record the changes in the earth's magnetic field through that time interval. By sampling many levels through such a sequence, the paleomagnetist can determine the magnetic sequence, or magnetic stratigraphy, of that local section. Under the right conditions, the magnetic pattern of a section is distinctive. It can be matched to the pattern in a number of other sections of approximately the same age, and these sections can be correlated by the polarity changes. If the pattern is long and distinctive enough and its numerical age can be estimated (usually by radiometric dating), then it is possible to match the pattern to the worldwide magnetic polarity time scale and to estimate an even more precise age.

The worldwide magnetic polarity time scale was first developed in the 1960's, when a group of scientists found that all lava flows with potassium-argon dates less than about 700,000 years were normally magnetized (like the earth's present field), and those older than 700,000 years were usually reversely magnetized (opposite the earth's present field). They began to seek out more and more lava flows around the world, sampling them for both their magnetism and their potassium-argon age. In about five years of sampling, they found a consistent pattern: All rocks of the same age had the same magnetic polarity, no matter where they were located. This immediately suggested that their magnetic properties were caused by worldwide magnetic field reversals rather than by local peculiarities of the rocks themselves.

SEAFLOOR SPREADING

Continuous sequences of lava flows that could be dated, however, were not available for time periods older than about 13 million years. What was needed was a terrestrial process that continuously recorded the earth's magnetic field behavior and could be dated. Such a process was discovered in the early 1960's at the same time that magnetic polarity reversals were documented. The crust of the ocean floor is constantly pulling apart, and the gap is filled by magma from the mantle below. When the magma cools, it locks in the magnetic polarity prevailing at

the time. Continual seafloor spreading pulls apart this newly cooled crustal material and carries it away from the mid-ocean ridge, causing new magma to fill in the rift, to cool, and to lock in a new polarity. This process of cooling, magnetization, and spreading acts as a "tape recorder" that produces a magnetic record of the present field at the center of the ridge and progressively older fields away from the ridge crest. In a few places in the ocean basin, this ocean-floor "tape recording" goes back about 150 million years.

The steady spreading of oceanic plates provides the only continuous record of the changes in the earth's magnetic field between 13 million and about 150 million years ago. In 1968, the first attempt was made to construct a magnetic polarity time scale. Using the known rates of spreading of several mid-ocean ridges, scientists extrapolated several oceanic spreading records back to about 100 million years and placed tentative dates on all the polarity events that were recorded. Since 1968, many attempts have been made to date this polarity time scale more precisely. Ironically, most of the new dates have shown that the original 1968 extrapolation was remarkably good, and new versions of the time scale differ very little from the first version. This proves the assumption that seafloor spreading is a relatively steady, constant process.

EVALUATION AS A CORRELATION TOOL

Magnetic stratigraphy has proved to be one of the most powerful tools of correlation and dating available. It has many features that other methods of correlation do not. Unlike correlation by distinctive rock units or by the changes in fossils through time, magnetic polarity changes happen on a worldwide basis and can be recorded in any type of rock (lavas or marine or nonmarine sediments) formed at that time. No rock type is formed worldwide, and fossils are restricted by the environments in which they lived. Thus, rocks formed in both the oceans and land can be directly correlated by magnetic stratigraphy, even though the rock sequences are different and they do not share the same fossils.

Another unique feature of magnetic stratigraphy is that polarity changes take place within about 4,000-5,000 years, which is considered instantaneous in a geological sense for any event that occurred more than about a million years ago. Thus, a polarity zone boundary represents a worldwide, geologically instantaneous "time plane" that can be used as a very precise marker wherever it is found. By contrast, the changes in fossil assemblages in a stratigraphic section can seldom resolve events down to a few thousand years, and radiometric dates typically have analytical errors that are anywhere from hundreds of thousands to millions of years.

The major limitation of magnetic stratigraphy is that most magnetic patterns are not unique. When paleomagnetists sample a rock, they get only normal or reversed polarity, not a numerical age. To date a sequence, some other form of dating must be used to place the magnetic pattern on the magnetic polarity time scale. For example, a sequence of "normal-reversed-normal" is not unique by itself; it has occurred many times in the geologic past. If, however, a distinctive set of fossils or a radiometric date can constrain that pattern to a certain period in earth history, then there may be only one part of the magnetic polarity time scale that matches that pattern at that particular point in time. This match gives a more precise age estimate than does the fossil or radiometric date alone.

ROCK SAMPLING AND MEASUREMENT METHODS

Paleomagnetists study ancient magnetic fields by sampling rocks of the proper age and rock type. If it is a lava or other very hard rock, they use a portable drill that collects a short core about 2 centimeters in diameter. Lavas tend to be strongly magnetized compared to other rock types. If the rock to be sampled is a softer sedimentary rock that might break up while drilling, then simple chisels and scrapers are used to extract a hand sample. Sediments tend to have magnetizations that are weaker than those of lavas by a factor of one hundred to one thousand. In addition, only fine-grained sediments (siltstones, claystones, fine sandstones, and limestones) record a remanence; coarse sandstones have more than one magnetic domain within each magnetic grain, which cancel one another out. In both cases, the direction of the present earth's magnetic field is marked on the sample, so it can be compared with the direction recorded in the rock.

The samples are then measured in a device called a magnetometer, which determines the direction and intensity of the field recorded by the sample (its natural remanent magnetization [NMR]. Some magnetometers are portable, but they are only suitable for measuring strongly magnetized lavas. Most

labs now use a superconducting cryogenic magnetometer. Its sensing area is kept at 4 degrees Celsius above absolute zero (−269 degrees Celsius) so that it is superconducting, or has almost no resistance to electrical current. When a sample is lowered into the sensing area, even weak magnetic fields in the sample cause changes in electrical current, which are then converted into a magnetic signal.

Typically, the field direction found in the sample (NRM) is a composite of several different magnetic fields. For example, if the rock were deposited during a period of reversed polarity, it may still have a young magnetic overprint acquired during the normal polarity that is seen today. The interaction of these two directions may give an NRM that is neither normal nor reversed, but some intermediate direction. To get rid of unwanted overprinting, the samples must be treated with high temperatures (thermal demagnetization) or high external magnetic fields (alternating field demagnetization), which destroys the less stable (and presumably young overprinted) component of the magnetization. After each treatment at progressively higher temperatures or progressively higher applied fields, the sample is measured again. Interpreting the change of direction and strength of the magnetic component during this stepwise demagnetization enables the paleomagnetist to decide which magnetic mineral is the carrier of the magnetic remanence and also which temperature or field is best for magnetically "cleaning" samples.

After magnetic cleaning, each sample produces a direction that presumably represents the field direction at the time the rock was formed. This remanence is known as the primary, or characteristic, remanence. Because several samples are taken of each lava flow or of each sedimentary bed, the directions of all of the samples from a given site are averaged to omit random "noise." The more tightly all the directions from a site cluster, the more reliable they are likely to be. There are statistical methods that measure this clustering and allow the paleomagnetist to determine the quality of the data. Data that cluster poorly or give nonsensical results can be rejected.

ROLE IN UNDERSTANDING GEOLOGIC HISTORY

Magnetic stratigraphy has become one of the most powerful tools of dating geologic events. It is critical to understanding geologic history and provides a much greater understanding of certain aspects of the geological past than was previously possible. For example, there has been great controversy over how fast evolution takes place or when mass extinctions occurred. By more precisely dating the sequences in which these events are recorded, scientists can determine rates of evolution much more precisely or determine a much more accurate date for the timing of a mass extinction, which may, in turn, allow the determination of the causes of these events and resolve many long-standing controversies. Magnetic stratigraphy has been used to date the long history of evolution of fossil mammals and dinosaurs in the terrestrial environment and the details of the evolution of the world ocean in marine sections. In many marine sections, the use of magnetic stratigraphy has allowed precise dating of climatic changes, particularly the glacial-interglacial fluctuation of the last ice age. This precise dating, in turn, has allowed scientists to determine that the glacial-interglacial cycles were controlled by changes in the earth's orbital motions, and they thus deciphered the cause of the ice ages. A better understanding of how some of these events (climate change, ice ages, mass extinctions) occurred in the past will help scientists to decide whether such events are likely to happen again in the near future.

Donald R. Prothero

FURTHER READING

Bebout, Gray E. *Subduction Top to Bottom.* Washington, D.C.: American Geophysical Union, 1996. Bebout's book gives clear definitions and explanations of subduction, folding, faults, and orogeny. Illustrations and maps help to clarify some difficult concepts.

Boggs, Sam, Jr. *Principles of Sedimentology and Stratigraphy.* Columbus, Ohio: Charles E. Merrill, 1986. A college-level textbook on stratigraphy and sedimentology, which devotes a chapter (Chapter 15) to magnetic stratigraphy. This chapter, although brief and not concerned with the practical aspects of magnetic stratigraphy, does give one of the few up-to-date accounts available in any stratigraphy textbook.

Brookfield, Michael E. *Principles of Stratigraphy.* Hoboken, N.J.: Wiley-Blackwell, 2004. Written for undergraduate students, this text provides an overview of the principles and applications of stratigraphy. Includes stratigraphic techniques and case studies. Organized into three sections; beginning with foundational material, followed by data collection and research topics, and completed with interpretations and analysis.

Brush, Stephen G. *Nebulous Earth: The Origin of the Solar System and the Core of the Earth from Laplace to Jeffreys.* Cambridge: Cambridge University Press, 1996. Brush's book, volume 1 in the *History of Modern Planetary Physics* series, contains useful information on the nebular hypothesis, the origin of the solar system, and the earth's core. Includes a bibliography and index.

Cox, Allan, ed. *Plate Tectonics and Geomagnetic Reversals.* San Francisco: W. H. Freeman, 1973. A collection of the classic papers that led to the plate tectonics revolution, edited by a man who was responsible for the paleomagnetic data that propelled it. It includes many of the pioneering papers that first described the reversals of the earth's magnetic field as well as the discovery of the magnetic polarity time scale and seafloor spreading. One of its best features is the editorial introductions, which place the papers in historical context.

Cox, Allan, and R. B. Hart. *Plate Tectonics: How It Works.* Palo Alto, Calif.: Blackwell Scientific, 1986. A college-level textbook that explains many facets of plate tectonics, with examples and problem sets. Several chapters give an excellent discussion of paleomagnetism.

Glen, William. *The Road to Jaramollo: Critical Years of the Revolution in Earth Sciences.* Stanford, Calif.: Stanford University Press, 1982. A history of the plate tectonics revolution, recounting the important individuals and their discoveries that led to the discovery of continental drift and seafloor spreading. The development of the magnetic polarity time scale was a key part of this revolution, and the rivalry between various labs in discovering and dating the magnetic reversals is described in detail.

Gurnis, Michael, et al., eds. *The Core-Mantle Boundary Region.* Washington, D.C.: American Geophysical Union, 1998. This collection of articles is one volume of the American Geophysical Union's *Geodynamics* series. Although intended for the specialist, the essays contain plenty of information suitable for the careful college-level reader. Bibliography.

Hamblin, William K., and Eric H. Christiansen. *Earth's Dynamic Systems.* 10th ed. Upper Saddle River, N.J.: Prentice Hall, 2003. This geology textbook offers an integrated view of the earth's interior not common in books of this type. The text is well organized into four easily accessible parts. The illustrations, diagrams, and charts are superb. Includes a glossary and laboratory guide. Suitable for high school readers.

Kearey, Philip, Keith A. Klepeis, and Frederick J. Vine. *Global Tectonics.* 3rd ed. Cambridge, Mass.: Wiley-Blackwell, 2009. This college text gives the reader a solid understanding of the history of global tectonics, along with current processes and activities. The book is filled with colorful illustrations and maps.

Kenneth, J. P., ed. *Magnetic Stratigraphy of Sediments.* Stroudsburg, Pa.: Dowden, Hutchinson, and Ross, 1980. An anthology of classic papers on magnetic stratigraphy. Most of the papers deal with magnetic stratigraphy of marine sediments and their application to paleo-oceanographic problems, but some also cover terrestrial magnetic stratigraphy. The editor wrote introductions that place all the papers in historical context.

Lindsay, E. H., et al. "Mammalian Chronology and the Magnetic Polarity Time Scale." In *Cenozoic Mammals of North America*, edited by M. O. Woodburn. Berkeley: University of California Press, 1987. This chapter contains one of the best reviews of the practical aspects of magnetic stratigraphy as applied to terrestrial sections. Although it is written on the professional level, it assumes little or no background in rock magnetism.

McElhinny, M. W. *Paleomagnetism and Plate Tectonics.* New York: Cambridge University Press, 1973. One of the most popular college textbooks on paleomagnetism. Although some of the text is outdated, it contains an excellent discussion of the field as it was at that time.

Merril, Ronald T., and Philip L. McFadden. "The Use of Magnetic Field Excursions in Stratigraphy." *Quaternary Research.* 63 (2005): 232-237. This article addresses the pros and cons of using magnetic field excursions. Well suited for researchers in the field.

Ogg, James G., Gabi Ogg, and Felix M. Gradstein. *The Concise Geologic Time Scale.* New York: Cambridge University Press, 2008. This book is a complete overview of the geological time scale, including stratigraphy topics such as chronostratigraphy and magnetic stratigraphy. The book is organized by geological time periods. Also contains a reference appendix, the geological time scale table, and indexing.

Prothero, D. R. *Interpreting the Stratigraphic Record.* New York: W. H. Freeman, 1989. A college-level textbook on stratigraphy, containing a chapter on magnetic stratigraphy. The discussion on magnetic stratigraphy is considerably clearer, more thorough, and more up-to-date than that by Boggs.

_____. "Mammals and Magnetostratigraphy." *Journal of Geological Education* 36 (1988): 227. A nontechnical article detailing the practical aspects of terrestrial magnetostratigraphy. It also reviews the progress on the terrestrial, fossil-mammal-bearing record up to the time of the article.

Tarling, D. H. *Palaeomagnetism: Principles and Applications in Geology, Geophysics, and Archaeology.* London: Chapman and Hall, 1983. One of the best books available on paleomagnetism. It discusses magnetic stratigraphy on a much more general level than does McElhinny. A good first resource in reading about the subject.

See also: Earth's Magnetic Field; Geobiomagnetism; Magnetic Reversals; Polar Wander; Rock Magnetism; Volcanism.

MANTLE DYNAMICS AND CONVECTION

Mantle dynamics is the study of the motion of the earth's mantle, which is primarily generated by convection. Convection within the mantle causes the transfer of heat from one region of the earth to another. This convection facilitates the movement of the lithospheric plates of the earth, resulting in mountain building, earthquakes, volcanism, and the evolution of continents and ocean basins. Understanding mantle dynamics and convection is a major component of the framework for explaining how the earth developed, how it works, and why it is constantly changing.

PRINCIPAL TERMS

- **asthenosphere:** the weak zone directly below the lithosphere, from 10 to 200 kilometers below the earth's surface, believed to consist of soft material that yields to viscous flow
- **convection cell:** a pattern of movement of mantle material in which the central area is uprising and the outer area is downflowing because of density changes produced by heat variations
- **core-mantle boundary:** the seismic discontinuity 2,890 kilometers below the earth's surface that separates the mantle from the outer core
- **lithosphere:** the relatively rigid outer zone of the earth, which includes the continental crust, the oceanic crust, and the part of the upper mantle lying above the weaker asthenosphere
- **lower mantle:** the seismic region of the earth between 670 and 2,890 kilometers below the surface, consisting of the DN and DO layers
- **mantle plume:** a vertical cylindrical distribution of material in the mantle within which abnormal amounts of heat are conducted upward to form a hot spot at the earth's surface
- **seismic tomography:** a processing technique for constructing a cross-sectional image of a slice of the subsurface from seismic data
- **upper mantle:** the part of the mantle that lies above a depth of about 670 kilometers, consisting of the B layer and the C layer

CHEMICAL AND MECHANICAL PROPERTIES OF THE MANTLE

The earth's interior consists of a series of shells of different compositions and mechanical properties. Based on chemical composition, the outermost layer is the crust, consisting of both continental and oceanic crust. The next major compositional layer of the earth is the mantle, which is approximately 2,890 kilometers thick and constitutes about 82 percent of the earth's volume and 68 percent of its mass.

By studying fragments of the mantle that have been brought to the surface by volcanic eruptions, earth scientists have deduced that the mantle is chemically composed of silicate rocks containing primarily silicon, oxygen, iron, and magnesium.

Based upon physical, or mechanical, properties, the solid, strong, rigid outer layer of the earth is termed the lithosphere ("rock sphere"). The lithosphere includes the crust and the uppermost part of the mantle. The earth's lithosphere varies greatly in thickness, from as little as 10 kilometers in some oceanic regions to as much as 300 kilometers in some continental areas. The earth's lithosphere is broken up into a series of large fragments, or rigid plates. Seven major plates and a number of smaller ones have been distinguished, and they grind and scrape against one another as they move independently, similar to chunks of ice in water. Much of the earth's dynamic activity occurs along plate boundaries, and the global distribution of associated tectonic phenomena, particularly earthquakes and volcanism, delineates the boundaries of the plates.

Within the upper mantle, there is a major zone where the temperature and pressure are such that part of the material melts, or nearly melts. Thus the rocks in this region of the earth lose much of their strength, becoming soft and plastic like, so that they can slowly flow as a viscous liquid. This zone of easily deformed mantle is termed the asthenosphere ("weak sphere"). Seismic velocities are about 6 percent lower in the asthenosphere than in the lithosphere. Although there is no fundamental change in chemical composition between the two regions, the lithosphere and the asthenosphere are mechanically distinct.

The lithosphere rides over the plastic, partly molten asthenosphere. As the lithosphere moves, the continents split, and the large plates drift thousands of kilometers across the earth's surface. All the major structural features of the earth are the result of a system of moving lithospheric plates. Movement in the plate tectonic system is driven by the loss of internal heat energy, primarily from within the mantle. This heat-driven internal movement is responsible for the creation of ocean basins and continents, as well as deformations of the earth's solid outer layers that generate earthquakes, mountain belts, and volcanic activity at the plate boundaries. The primary source of heat within the mantle appears to be the radioactive decay of uranium, thorium, and potassium.

Below the asthenosphere, the rock becomes stronger and more rigid. The higher pressure below the asthenosphere offsets the effect of higher temperatures, making the rock stronger than in the overlying asthenosphere. The layer from the base of the earth's crust to a depth of approximately 670 kilometers is designated the upper mantle, which includes the asthenosphere and the lower part of the lithosphere. Earth scientists have had difficulty unraveling all the layers in the upper mantle. One model of the structure of the upper mantle designates a B layer, nearly 400 kilometers thick and of fairly uniform composition, and a C layer, between 200 and 300 kilometers thick, in which the chemical composition appears to be quite variable.

The transition from the upper to the lower mantle is quite gradual, with the depth of the separating, or transition, boundary usually thought to be between 600 and 670 kilometers. One of the fundamental questions about the seismic discontinuity that separates the upper mantle from the lower mantle is whether it is a barrier to lower mantle convection. One theory is that this boundary temporarily prevents the penetration of mantle material but ultimately will allow passage. As seismic wave velocities appear to be very steady throughout the DN region, a layer between 670 and 2,700 kilometers deep, the lower mantle is assumed to have a less complex structure than the upper mantle. Variations of properties inside the DN region appear to be predominantly caused by the effects of simple compression. However, inside the DO region, from 2,700 kilometers deep down to the outer core, seismic velocity falls continuously, indicating some continuous changes in physical properties and chemical composition that could produce deep mantle convection.

EVIDENCE FOR MANTLE CONVECTION

Better understanding of plate tectonics from analysis and interpretation of vast amounts of seismic data, as well as the application of improved observational and experimental techniques to the study of the properties of mantle materials, has confirmed the existence of mantle convection. Apparent episodic material exchanges between the upper and lower mantle, the genesis of plumelike upwellings, and the ultimate fate of subducted slabs within the earth are all aspects of the mantle's dynamic convection system. One of the most significant questions in mantle dynamics is whether mantle convection is isolated in the upper mantle or involves the whole mantle.

The advent of seismic tomography in the mid-1980's, coupled with new laboratory and computational capabilities in the 1990's, has profoundly impacted the understanding of mantle convection. Global seismic tomography has helped to resolve the parameters that characterize mantle dynamics, particularly a viscosity increase from the upper to the lower mantle, an endothermic phase transition at the upper mantle-lower mantle boundary, heat flow across the core-mantle boundary (CMB), and the effect of the motion of rigid surface plates on convection patterns. Using seismic tomography, images of three-dimensional variations in seismic velocities have revealed the deep structure of the underlying surficial plates. Patterns of high and low seismic velocity have been identified throughout the mantle, with the strongest variations found in the upper 300 kilometers.

Unexpected features, such as deep roots of high-velocity material extending 300 to 400 kilometers below continental cratons, have greatly enhanced the understanding of plate tectonics and continental formation. Deep-seated upwellings under the ocean ridges and beneath major volcanic hot spots are indicated by low-velocity regions and by deflections of transition-zone discontinuities. While some tomographic images show descending lithospheric slabs that are apparently blocked at the upper mantle-lower mantle boundary, others show lithospheric slabs sinking nearly to the CMB. Numerical models indicate that reasonable simulations of subducted lithospheric slabs and plumes will penetrate the upper mantle-lower mantle boundary, although they may be temporarily blocked. Laboratory models of descending sheets interacting with contrasting density and viscous interfaces support this conclusion. Some slabs of subducting lithosphere have been seismically imaged as high-velocity tabular downwellings extending throughout the mantle.

Studies conducted in 2004 indicate that large lithospheric slabs could be subducted to a depth of 2,900 kilometers, in support of whole-mantle convection.

UPPER MANTLE DYNAMICS AND CONVECTION

Because of its ability to flow, the asthenosphere figures prominently in the dynamic theories on the causes of vertical motion observed at the earth's surface, such as postglacial rebound. Periodic compensatory adjustments that take place in the interior of the earth in response to changing mass distributions at the surface that arise from erosion, sedimentation, glaciation and deglaciation, and volcanism are thought to occur through flow in the asthenosphere.

Likewise, the asthenosphere plays a prominent role in models of the large horizontal movements of the lithosphere as observed in continental drift and plate tectonics. As it slowly churns in large convection cells, the asthenosphere is the lubricating layer over which the plates glide. Thermal convection in the asthenosphere is thought to be a fundamental force in driving the tectonic plates. According to this scenario, hot mantle material rises at the mid-oceanic spreading ridges (divergent boundaries), where it escapes as magma, cools, and generates new oceanic crust. The sea floor moves in conveyor-belt fashion, ultimately to be destroyed at convergent plate boundaries, where it is subducted, or carried down, into the asthenosphere and eventually remelted. The rest of the hot mantle material spreads out sideways beneath the lithosphere, slowly cooling in the process. As it flows outward, it drags the overlying lithosphere outward with it, thus continuing to open the ridges. When the hot mantle material cools, the flowing material becomes dense enough to complete the convection cycle by sinking back deeper into the mantle, and tomographic images indicate that this is happening under subduction zones at converging boundaries.

Some debate continues as to whether convection is confined to the upper mantle in a thin asthenosphere or whether it occurs throughout the mantle. The convection cells need not be confined to the asthenosphere. Seismic velocity data indicate that oceanic lithosphere can be subducted to depths of approximately 700 kilometers. Thus convection cells

may operate at least down to those depths. In addition, in some places in the asthenosphere, the temperature may reach the rock-melting temperature and produce magma, thus giving the asthenosphere another dynamical role as the source region of many types of igneous rocks.

LOWER MANTLE DYNAMICS AND CONVECTION

A serious ongoing debate among earth scientists about the size of convection cells in the mantle began in the early 1980's. Because of the different chemical compositions of the upper and lower mantle, geochemists have argued that the upper and lower mantles must have isolated convection cells with virtually no mixing between them. Thus slabs of the lithosphere that sink below the surface at the edge of tectonic plates should stay within the upper mantle, with their material being recycled there. In addition, geochemical evidence also exists for distinct reservoirs in the mantle, emphasizing the importance of plume flows from internal boundary layers as distinct from large-scale flows associated with the oceanic plates.

Opposing this view, many geophysicists have maintained that convection involves the entire mantle. In the late 1990's, numerous three-dimensional seismic tomographic studies mapped seismic speeds in the earth's mantle, and the interpretations provide strong evidence that the lithosphere is sinking well into the lower mantle. The tomographic images of seismic wave speeds at different depths are a rough indication of the temperature distribution in the mantle. Waves travel more quickly in regions that are colder, and more slowly in hotter regions. In numerous locations beneath the earth, tomographic images show cold regions to be a continuous function of depth far into the lower mantle, suggesting the descent of some slabs of oceanic lithosphere into the lower mantle at the edge of tectonic plates. One interpretation of the tomographic images shows that the mantle's heterogeneity is dominated by large-scale structures, which support a mantle convection system dominated by large flow patterns. Evidence is not conclusive as to whether the mantle is only a layered convective system, but it does suggest that significant material transport occurs across the upper mantle-lower mantle boundary.

In addition, increasing geophysical evidence supports the conjecture that the earth's core interacts with the surrounding mantle. Images from seismic tomography reveal that the lowermost 200 to 400 kilometers of the mantle is one of the most heterogeneous regions of the earth. Above the CMB, seismic images indicate the presence of two laterally variable seismic discontinuities, one at 130 to 400 kilometers and another at 5 to 50 kilometers above the CMB. According to the tomographic images, both regions have anisotropic properties, and the complexities of physical and dynamic processes are as sophisticated as those present in the lithosphere and shallow asthenosphere, which supports the idea of a dynamic lower mantle involved in convection.

In the late 1990's, the discovery that the DO layer O is associated with dramatically reduced seismic velocities at 5 to 50 kilometers above the CMB changed the persisting idea that the deep region of the lower mantle was solid. From tomographic images, some of the most unusual anomalies seen in the lower mantle are thin patches, less than 40 kilometers thick, in which seismic velocities are locally reduced by 10 percent or more. Such ultra-low-velocity zones are not seen anywhere else in the mantle. Explaining their presence requires massive local melting within the lowermost mantle, meaning that this region is most likely convecting. The significant heterogeneity, which likely involves locally hot and partially molten zones near the CMB, is indicative of the dynamical behavior of the DO layer. Some earth scientists reason that many of the volcanic plumes associated with hot spots at the earth's surface, such as the Hawaiian island chain, represent upwelling jets of hot rock in the mantle that are preferentially lined up above the ultra-low-velocity molten patches in the DO layer.

INTEGRATED MANTLE CONVECTION

In the late 1990's, many earth scientists believed that it was necessary to make a compromise between whole-mantle convection and isolated-mantle convection. With regard to the existing geochemical evidence, the main requirement is that there exist chemically distinct reservoirs that do not have to be totally confined to the lower mantle. Three suggested models have emerged. Some earth scientists support a model of the mantle that contains isolated, discontinuous volumes of material dispersed throughout. Other earth scientists suggest that a reasonable model is one in which the mass transport in the mantle does not occur in a steady, continuous fashion but rather in an intermittent, nonsteady

state. Still others believe that lithospheric slabs descending into the mantle lose geochemically monitored elements in the upper mantle as they sink to lower depths. The best model may contain aspects of all three of these alternatives.

Laboratory experiments in the mid-1990's have shown that the oxides of the earth's deep mantle react vigorously when placed in contact with liquid iron alloys, thought to exist in the outer core, at the high pressures and temperatures at the CMB. These experiments suggest that the rocky mantle is slowly dissolving, over geological time spans, into the liquid metal of the outer core. The slow dissolution appears to be related to a fundamental change in the bonding character of oxygen at high pressures. Whereas oxygen forms insulating compounds at low pressures, it can become a metal-alloying component at high pressures. Thus, when coupled with seismic tomography, experimental and theoretical investigations of high pressures point to the CMB as perhaps being the most chemically active region in the earth's interior. Numerical models of the mantle suggest that the earth may have undergone a transition from layered to whole-mantle convection caused by a combination of secular cooling and a decrease in heat production in the mantle from radioactive decay.

The products of the chemical reactions at the CMB, where insulating oxides meet metallic alloys, may well explain the seismologically observed heterogeneity of the DO layer in the mantle. In addition, piles of oceanic crust that have settled toward the bottom of the mantle may further contribute to the heterogeneity of the region. The possible occurrence of varying amounts of metal alloys at the base of the mantle is particularly important because metal conducts heat much more readily than insulating oxides do. Consequently, heat may be emerging from the CMB in a spatially variable manner that determines the pattern of convection throughout the earth's mantle.

SIGNIFICANCE

Many of the geophysical and geological phenomena of the earth's crust are consequences of the dynamics and thermal convection within the underlying mantle. The major features of mantle convection are deduced from seismic data, laboratory investigations, and computational modeling. The emerging picture is that the tectonic plates of the lithosphere are the most active component of mantle convection, whereas mantle plumes are an important secondary component. Direct consequences of mantle dynamics and convection include the relative motions of the lithospheric plates, the spreading of the sea floor and formation of new crust, volcanism in its various tectonic settings, much of the earth's seismic activity, and the majority of the observed heat flow through the earth's surface.

Unraveling the complexities of the mantle continues to be a challenge for seismology, but the results play a crucial role in answering questions regarding the composition, dynamics, and evolution of the earth. Better understanding of the present-day mantle is providing a much more complete picture of the evolution and interaction of the earth's thermal and tectonic regimes. Plausible arguments indicate that the mantle was episodically layered in Precambrian times and that plate tectonics would not have worked when the mantle was more than about 50 degrees Celsius hotter than at present. Whether plumes were more or less important in the past is being studied. Based on the present model of the mantle's dynamic and convection patterns, plumes do not offer an alternative to plate tectonics because they are derived from a different thermal boundary layer (the DO layer).

Insight into the dynamic interactions in the lower part of the mantle near the CMB is very important for better understanding past geological phenomena of significant magnitude. In particular, there is evidence for periods of massive volcanic eruptions (superplumes) that were hundreds of times greater than anything the earth has experienced in recent geological time. Models of the deep mantle based on three-dimensional tomographic images and laboratory observations indicate that superplume events could be the surface manifestation of fluid-dynamical instabilities triggered at the CMB. Such models have generated the first glimpses of how such massive instabilities are initiated deep inside the earth.

Alvin K. Benson

FURTHER READING

Brown, G. C., and A. E. Mussett. *The Inaccessible Earth.* 2d ed. New York: Chapman and Hall, 1993. Provides an understanding of mantle convection and how the lithospheric plates are an essential part of it. Contains insights concerning whether the mantle's D layer is a barrier to whole-mantle convection. Basic groundwork for general readers is included in a series of notes at the end of the book.

Davies, Geoffrey F. *Mantle Convection for Geologists.* New York: Cambridge University Press, 2011. Begins with strong foundational material upon which to build convection concepts. Although the title implies technical writing, the author intended the text be for anyone studying geological processes or university level students.

Hamblin, W. K., and E. H. Christiansen. *Earth's Dynamic Systems.* 10th ed. Upper Saddle River, N.J.: Prentice Hall, 2003. Very readable description of the earth's two major dynamic systems: the hydrologic system and the tectonic system. Basic explanations of mantle composition, dynamics, and convection. The text is well organized into four easily accessible parts. Many excellent color photos and illustrations.

Jackson, I., ed. *The Earth's Mantle.* Cambridge, England: Cambridge University Press, 1998. Comprehensive overview of the composition, structure, and evolution of the mantle layer. Reviews the evolution of the earth. Draws on perspectives from isotope geochemistry, cosmochemistry, fluid dynamics, petrology, seismology, geodynamics, and mineral and rock physics. Written for more advanced readers.

Jeanloz, R., and B. Romanowicz. "Geophysical Dynamics at the Center of the Earth." *Physics Today* 50 (August 1997): 22. Description of what occurs in the earth's core and at the core-mantle boundary. Based on geophysical observations and laboratory and computational models. Explanation of interactions at the DO layer in the lower mantle.

Lay, T., and Q. Williams. "Dynamics of Earth's Interior." *Geotimes* 43 (November 1998): 26. Excellent overview of the present understanding of the thermal, chemical, and dynamical state of the earth's deep interior. Includes colored cross sections of the earth's interior at various depths generated from seismic data.

Lillie, R. J. *Whole Earth Geophysics.* Upper Saddle River, N.J.: Prentice Hall, 1999. Introductory book that illustrates how different types of geophysical observations, especially seismic, have helped determine the earth's gross structure and composition. Good explanation of the theory of plate tectonics and the mantle's role in it.

Lowrie, William. *Fundamentals of Geophysics.* 2d ed. New York: Cambridge University Press, 2007. Excellent overview of geophysics topics written for the student with a strong physics background. Lowrie presents the mathematics at a level understood by mid-level university students.

Schubert, Gerald, Donald L. Turcotte, and Peter Olson. *Mantle Convection in the Earth and Planets.* New York: Cambridge University Press, 2001. Comprehensive text on thermodynamics of the earth. Contains high-level mathematical equations and solutions, requiring the reader to have a strong background in physics or mathematics. The authors further extrapolate concepts to other planets in our solar system. This text makes important contributions to the graduate's studies.

Strahler, A. N. *Plate Tectonics.* Cambridge, Mass.: GeoBooks, 1998. A textbook of basic principles and important plate tectonic data. Generally descriptive presentation, with supportive quantitative data. Basic explanations of mantle convection and its role in plate tectonics.

See also: Continental Drift; Earthquakes; Earth's Core; Earth's Differentiation; Earth's Mantle; Heat Sources and Heat Flow; Lithospheric Plates; Mountain Building; Plate Motions; Plate Tectonics; Plumes and Megaplumes; Seismic Tomography; Tectonic Plate Margins; Volcanism.

MASS EXTINCTION THEORIES

Mass extinctions—those periods in Earth's history in which large numbers of species rapidly become extinct—result from environmental changes caused by factors including volcanic activity, extraterrestrial impacts, shifting continents, and, most recently, destructive human activity. Although devastating to life on Earth, mass extinctions also lead to periods of evolutionary expansion, including major shifts in the types of animals on the planet.

PRINCIPAL TERMS

- **background extinction rate:** the rate at which species become extinct given the absence of any extraordinary environmental phenomena
- **biodiversity:** a measure of the diversity of life forms in a certain area, which may be limited to a certain biome or may include the entire biosphere of the planet
- **extinction:** the disappearance of a species from the biota of the earth, as occurs when the last representative of a species dies without leaving offspring
- **fossil record:** the history of life on Earth as interpreted from the fossilized remains of extinct life forms
- **global warming/cooling:** the increase or decrease in the average temperature of Earth, as caused by either naturally occurring geochemical processes or by the activity of life forms on Earth
- **K-T boundary:** the layer of sediment deposited at the end of the Cretaceous period that contains minerals that may have arrived on Earth from a meteorite impact
- **mass extinction:** a period in which the extinction rate for species rises above the background extinction rate predicted for species under relatively constant conditions
- **press/pulse model:** the theory that extinctions result from the slow and constant buildup of environmental pressures set into motion by major environmental changes
- **radioisotope dating:** the process of using the decay of radioactive isotopes found in mineral deposits to obtain the date at which nearby geological items were deposited in the surrounding soil
- **sixth extinction:** a theoretical extinction predicted to occur sometime in the future and potentially caused at least partially by human activity on Earth's surface
- **supercontinents:** large landmasses containing two or more of Earth's continents in a single landmass

DEFINING MASS EXTINCTIONS

Mass extinctions, also called major extinctions, are periods in the earth's past in which large numbers of species became extinct in a relatively short period. These periods contrast with the background extinction rate, which is the rate at which species go extinct in an absence of any catastrophic environmental change. To be considered a mass extinction, according to paleontologists, the event must lead to the extinction of a relatively large number of species and must be a global or geographically widespread phenomenon, affecting species from distantly related evolutionary groups.

Paleontologists have identified five mass extinctions in Earth's history: Triassic-Jurassic, Permian-Triassic, Devonian-Carboniferous, Ordovician-Silurian, and Cretaceous-Tertiary. Each of the five was a major turning point for life on Earth, and each coincided with environmental catastrophes so severe that they have left clear signs in the geological record—signs that are used to demarcate the shift between geologic periods. In addition to the five major extinctions, there have been other periods (for example, the end of the Ice Age in North America) in which extinction rates rose far above the background level, but where the resulting extinction affected a narrower geographical area or number of species.

The best-known and most-studied mass extinction resulted in the disappearance of most dinosaur species and marks the transition from the Cretaceous period to the Tertiary. This is also the most recent mass extinction, having occurred approximately 65 million years ago. Paleontologists estimate that the Cretaceous-Tertiary event led to the extinction of a minimum of 62 percent of Earth's species and 11 percent of the families of species. The extinction of entire families is significant because it means that there are no representatives of that larger group to repopulate the world with representative species in subsequent generations.

The Permian-Triassic extinction event, which occurred approximately 250 million years ago and marked the beginning of the age of dinosaurs, was the most massive extinction episode in Earth's history. Some scientists have estimated that the extinction resulted in the loss of more than 90 percent of the species on Earth and of more than one-half of Earth's families of organisms.

EVIDENCE FOR MASS EXTINCTIONS

Evidence for mass extinctions is derived from the fossil record, which is the history of life on Earth as revealed from the fossilized remains of extinct organisms. The fossil record has been created by dating and organizing fossils into a temporal progression that can be used to measure the amount of time that a certain species appeared on Earth.

Extinction rates vary among organisms. Mammal species typically emerge and become extinct within approximately one million years, while there are examples of reptile species that have existed virtually unchanged for up to ten million years. Taking this into account, scientists estimate that between 10 and 100 million species become extinct each year. This is considered the background extinction level for the earth.

In the nineteenth century, pioneering researchers discovered that, at certain levels of the geologic strata, many or most of the fossil species present at one level were missing from subsequent layers. At first, paleontologists thought that these "gaps" in the fossil record were caused by incomplete knowledge. Scientific leaders of the nineteenth century, such as Charles Darwin and Charles Lyell, believed that extinction was a gradual process brought about by local, immediate pressures; the two scientists could not conceive of conditions that would cause a mass extinction. However, paleontologists discovered that gaps in the fossil record occurred around the world at similar points in the geological strata. Standard evolutionary theory could not account for the seemingly global disappearance of species.

In the mid-twentieth century, paleontologists began using decayed radioactive isotopes to determine the age of geological samples. Radioactive isotopes occur naturally in the earth's mineral surface and, because the substances are chemically unstable, they decay, giving rise to another substance called the decay product. When scientists find a fossil in the earth, they can examine the proportion of isotopes to decay products in nearby sediment and are thereby able to determine the age of the sample.

Radioisotope dating strengthened the case for mass extinctions, as paleontologists refined their temporal understandings of the fossil record. With this new evidence, paleontologists realized that extinction sometimes occurred relatively quickly. Paleontologist Norman D. Newell consolidated evidence for mass extinctions in his 1967 paper "Revolutions in the History of Life," after which most paleontologists turned their attention toward the goal of discovering how and why mass extinctions occur.

CAUSES OF MASS EXTINCTIONS

Biologists sometimes identify extinctions as having either external or internal causes. Internal causes arise from within living organisms, such as the development of a fatal bacterium or virus that causes the extinction of one or more species. Alternatively, the origin of one species can, in some cases, lead to the extinction of other species. The evolution of the human species has led to the extinction of hundreds of other species relatively rapidly, and these extinctions would be considered internally caused. However, most internal extinction events affect only a few species and, therefore, have little potential to cause mass extinctions on their own.

Paleontologists believe that the five major extinctions were largely the result of external causes, which are changes in the physical environment brought about by geochemical or extraterrestrial events. These environmental catastrophes can lead to widespread extinction, as many species are left without the evolutionary adaptations that would enable them to adjust to changing conditions.

Volcanism has been one of the most important forces of environmental change, and scientists have discovered evidence of increased volcanic activity closely associated with mass extinctions. Lava emanating from volcanoes leaves a layer of material in the sediment that geologists refer to as flood basalt, and this characteristic combination of minerals has been found in samples preceding the Cretaceous-Tertiary, Triassic-Jurassic, and Permian-Triassic extinctions.

In addition to devastating lava flows, volcanic activity can bring about global changes in temperature, as massive amounts of ash and gas are ejected by this activity into the atmosphere. Volcanoes are therefore naturally occurring sources of both global warming and global cooling. In addition, geologists have discovered that there can be "supervolcanoes," which grow much larger than standard volcanoes and have a correspondingly greater potential to wreak environmental havoc. Few of these supervolcanoes exist on the modern Earth, but geologists have found evidence suggesting that supervolcanoes may have been more common in Earth's distant past.

Researcher Henrik Svenson of the University of Oslo in Norway has discovered evidence of a supervolcano that erupted in the area that is now Siberia and that may have been a major factor in the Permian-Triassic extinction. Svenson's research, published in 2009, indicates that this supervolcano may have been active for more than 200,000 years,

resulting in numerous large eruptions that covered the surrounding environment with lava and that spewed toxic gasses into the air. In the longer term, this volcano might have led to widespread changes in temperature, thereby bringing about environmental collapse.

Volcanic activity also is related to the movement of continents, a phenomenon that scientists refer to as plate tectonics. The continents move when pressure is exerted from deep within the earth's surface, driven by activity in the earth's core. The continents have, in the distant past, been joined into "supercontinents," which are defined as the combination of

Dr. Walter Alvarez of the University of California at Berkeley and Dr. Premoli Silva in Italy, examining the place where the top Cretaceous and base of the Tertiary layer are mixed in an old submarine landslide. In the early 1980s Alvarez discovered a worldwide layer of clay with unusually high iridium content and postulated that it had been deposited by a large asteroid impact which caused the extinction of the dinosaurs. At the time this was a controversial theory (scientists followed Darwin's theory that extinction, like evolution, was a gradual process, unlikely to be caused by a single catastrophe) but now it is considered the most plausible explanation of the dinosaurs' abrupt demise. (Gianni Tortoli/Photo Researchers, Inc.)

two or more continents. At the end of the Permian period, when the largest mass extinction in Earth's history took place, most of the earth's continents were united in a landmass known as Pangaea. The joining of continents necessitates a wide range of environmental changes, including increased volcanic activity, reduced coastal habitat for colonization, and climatic changes caused by shifting ocean and wind currents.

Species living on supercontinents may be more vulnerable to internal extinction pressures, such as the spread of disease or the introduction of a new predator. As the continents join, land bridges are created between them and species begin to move between both areas. In these conditions, a predator from one continent can prove devastating to prey animals on another continent because the prey species lack the evolutionary adaptations to avoid predation. Species diversity is also lower on supercontinents, because the same species can occupy available niches across a larger area. Paleontologists believe that the formation of Pangaea led to a decline in the number of species and, therefore, left the biota more vulnerable to extinction when the environment changed because of external factors.

Climate change is one of the most important causes of mass extinction, and paleontologists have found evidence of major climate changes associated with each of the five major extinctions. Geologists now know that the planet alternates between periods of relatively mild climate and periods in which large portions of the earth are covered in glaciers, sometimes called ice ages. Paleontologists have found evidence of a massive, global ice age that occurred around the time of the Ordovician-Silurian extinction, which occurred more than 440 million years ago. Geologists have also noted that global warming was occurring during other extinction events and most likely played a leading role in the extinction event. Global warming and global cooling can be affected by the shifting of continents and by more rapid catastrophic events such as volcanism.

In the 1970s, father and son researchers Luis and Walter Alvarez found evidence suggesting that the Cretaceous-Tertiary extinction may have been partially caused by the impact on Earth of a large meteor originating from elsewhere in the galaxy. The team found a layer of iridium and other rare elements deposited in the soil at what paleontologists have named

the K-T boundary, which is the layer that marks the end of the Cretaceous. Whereas iridium is rare on Earth, evidence suggests that it frequently occurs in meteors originating elsewhere in the solar system. With the publication of the Alvarezes' theory in 1980 came controversy and much debate in the paleontological community. The Alvarezes based part of their claim on the existence of the Chicxulub Crater in the Yucatán Peninsula of Mexico. Paleontologists believe that the crater resulted from an impact that occurred approximately 250,000 to 300,000 years before the Cretaceous-Tertiary extinction.

The Chicxulub impact would have created an explosion thousands of times more powerful than the detonation of the global military arsenal, immediately vaporizing plants and animals for miles surrounding the impact site. In addition, the aftereffects would have included earthquakes, volcanic eruptions, and massive storms reaching miles inland from the coast. Finally, the impact may have spewed a dense cloud of particulate matter into the atmosphere, eventually leading to long-lasting climate change that may have led to the loss of many plant species and a collapse of the food chain for many other species.

Paleontologists have found evidence linking extraterrestrial impact events with four of the five major extinctions. However, geologists also have found significant evidence for impact events that did not result in a mass extinction, leading to questions regarding exactly which conditions must be present for a meteor impact to facilitate a mass extinction.

COMBINATION OF CAUSES

Paleontologists now believe that major extinctions result from a special set of circumstances in which environmental conditions shift rapidly because of many interconnected, causal events in a relatively short time. As geological research continues, paleontologists are recognizing that each of the extinctions in Earth's history seems to be connected with more than one potential agent of environmental change.

In a 2006 article in *Geological Society of America*, researchers presented the press/pulse model of extinction, which proposes that mass extinctions result from a combination of gradual factors (press) exacerbated by some major event (pulse) that results in a chain reaction of environmental collapse. The "press" part of the model can involve any variety of environmental change, slowly putting pressure on the biosphere. During the Permian-Triassic extinction, the press may have involved environmental changes resulting from the unification of the earth's continents. This gradual pressure then received a "pulse" from volcanism or an impact event, or both, thereby accelerating the mass extinction event.

In 2008, geologist Shanan Peters published the results of a comprehensive research study indicating that, in each of the five major extinction events, changing sea level was a major determining factor. For example, during the Cretaceous-Tertiary extinction event, it appears that the global sea level fell by a significant margin. This sea level "regression" may have had a devastating effect on marine life and would also have led to global warming because of changes in ocean and wind currents. Peters believes that all the major extinction events were, at minimum, partially caused by falling sea levels, like those that occurred in the Cretaceous, setting the stage for a mass extinction event with influence from a major environmental pulse.

In each case, it appears that mass extinctions can occur only when the biota of the earth has been experiencing a steady increase in gradual extinction pressures, facilitating lower species diversity and greater vulnerability to environmental change. Meteor impacts, volcanic eruptions, and other major catastrophes then push environmental change into overdrive. The oceans swell or shrink, the temperature rises or falls, and the food chains collapse.

Luis and Walter Alvarez, and a number of other researchers, have theorized that extinction events may occur on periodic cycle, perhaps linked with cosmic cycles in the solar system that lead to periods in which Earth is more vulnerable to meteor impacts. Attempts to develop a cyclic theory of extinction have met with some success, but the exact timeline of extinction events is still a matter of considerable debate among paleontologists. The periodicity of extinctions remains an area of speculation, as paleontologists attempt to understand the factors that might contribute to an extinction cycle.

THE SIXTH EXTINCTION

Some scientists have speculated that the human species will have sufficient impact on the earth to usher in a mass extinction event, sometimes referred to as the sixth extinction. If this occurs, it may be the first known mass extinction to have been caused by internal, rather than external, factors.

Human activity poses a threat to life on Earth in several different ways. First, humans kill animals directly, either for food or for sport, as well as unintentionally, as occurs in collisions between automobiles and wildlife. Additionally, humans remove vast amounts of vegetation, leading to loss of habitat and a collapse of local food chains. Many scientists also believe that humans are causing significant global climate change, both through industrial activity and through the removal of vegetation that acts to regulate climate. Biologists estimate that human activity, taken as a whole, has increased the extinction rate to seven times the background level, resulting in the loss of between seventy and seven hundred species each year.

MASS EXTINCTIONS IN EVOLUTION

Each of the earth's major extinction events resulted in the loss of a minimum of 60 percent of species around the world. In thousands of years, the environmental conditions contributing to the event inevitably begin to dissipate, however, and the world is left with a relatively small number of species occupying a vast area. Large and highly specialized species are typically the first to succumb to extinction pressures, followed by species with restrictive needs for certain types of habitats. In the wake of the event, most of the species that survive are small, opportunistic species that are able to survive in a variety of habitats. The relatively hardy "pest" species that remain after the event proceed to populate the globe, giving rise to new species as they radiate to fill available niches in the surrounding environment. As a result, mass extinction events also represent periods in which the dominant species of the earth shift from one group to another.

To give one of many examples, the Cretaceous-Tertiary extinction resulted in the loss of the terrestrial dinosaurs, but the direct ancestors of the dinosaurs, the birds, survived the extinction and spread around the world, occupying many of the environmental niches that were formerly occupied by prehistoric reptiles and dinosaurs. Similarly, the only mammal species that lived during the age of the dinosaurs were small, scavenger species. In the wake of the dinosaurs, the mammals were free to radiate, growing into a variety of species and occupying all habitats on Earth. While extinctions bring an end to many types of organisms, they also enable new types of life to emerge and flourish in the new environment.

Micah L. Issitt

FURTHER READING

Adams, Jonathan S. *Species Richness: Patterns in the Diversity of Life.* New York: Springer, 2009. Chapter 3 of this book examines extinction events from a modern perspective informed by studies drawing from the life sciences and geology. It also contains information about the history of research into the nature of mass extinctions.

Jablonski, David. "Mass Extinctions and Macroevolution." *Paleobiology* 31 (June 2005): 192-210. This article from one of the pioneers in extinction research discusses the role of mass extinctions in the evolution of new species and examines some of the current research into the timing and causes of mass extinctions.

Lieberman, Bruce M., and Roger Kaesler. *Prehistoric Life: Evolution and the Fossil Record.* Hoboken, N.J.: Wiley-Blackwell, 2010. A review of the literature, discoveries, and major concepts in paleontology. Contains sections on mass extinctions and discussions of extinction theories.

Peters, Shanan E. "Environmental Determinants of Extinction Selectivity in the Fossil Record." *Nature* 454 (2008): 626-629. Discusses evidence suggesting that changes in ocean levels are closely linked to, and potentially important causal factors behind, extinction events. Also discusses the relationship of ocean levels to other possible causes of extinction.

Raup, David. *Extinction: Bad Genes or Bad Luck?* New York: W. W. Norton, 1992. An overview of the debates and theories surrounding mass extinctions, presented by one of the pioneering scientists in the field. Raup discusses many potential contributing factors to extinctions and also discusses the potential benefits of extinction to the overall diversity of global ecosystems.

Ridley, Mark. *Evolution.* Hoboken, N.J.: Wiley-Blackwell, 2004. An engaging introductory text that presents many of the major issues in current research into evolution. The section "Macroevolution" examines the major extinctions and their role in evolution.

See also: Asteroid Impact Craters; Climate Change: Causes; Earth Tides; Glaciation and Azolla Event; Milankovich Hypothesis; Plate Motions; Plate Tectonics; Plumes and Megaplumes; Radioactive Decay; Radiocarbon Dating; Volcanism.

MASS SPECTROMETRY

Mass spectrometry is the technique used for determining particle abundances by their mass and charge characteristics in an evacuated electromagnetic field. Its principal uses in the earth sciences are in determining the isotope ratios of light, stable substances and in measuring the isotopic abundances of radioactive and radiogenic substances.

PRINCIPAL TERMS

- **absolute date or age:** the numerical timing of a geologic event, as contrasted with relative, or stratigraphic, timing
- **geochronology:** the study of the absolute ages of geologic samples and events
- **half-life:** the time required for a radioactive isotope to decay by one-half of its original weight
- **ions:** atoms or molecules that have too few or too many electrons for neutrality and are therefore electrically charged
- **isochron:** a line connecting points representing samples of equal age on a radioactive isotope (parent) versus radiogenic isotope (daughter) diagram
- **isotope:** a species of an element having the same number of protons but a different number of neutrons and therefore a different atomic weight
- **nuclide:** any observable association of protons and neutrons
- **radioactive decay:** a natural process whereby an unstable, or radioactive, isotope transforms into a stable, or radiogenic, isotope

EARLY MASS SPECTROMETRY

The years 1896 to 1906 are sometimes referred to as a "golden decade" of physics because this time period saw critical discoveries and experiments that resulted in the quantitative analysis of charged particles by mass. One hundred years ago, experiments with cathode rays led to their identification as streams of electrons by German physicists Eugen Goldstein and Wilhelm Wien and British physicist Joseph John Thomson. Using the earliest application of mass analysis, Thomson identified the two isotopes of neon, neon-20 and neon-22. This work was followed in 1918 and 1919 by Canadian-American physicist A. J. Dempster and British chemist F. W. Aston, who designed mass spectrographs which were used in succeeding years to determine most of the naturally occurring isotopes of the periodic table.

In 1896, French physicist Antoine-Henri Becquerel presented his discovery of the phenomenon of radioactivity to the scientific community in Paris. This finding was followed rapidly by the seminal work of French chemist Marie Curie in radioactivity, a term she coined. Her discovery of the intensely radioactive elements radium and plutonium led Ernest Rutherford to delimit three kinds of radioactivity—alpha, beta, and gamma—and, in 1910, led English radiochemist Frederick Soddy to formulate a theory of radioactive decay. Soddy later proposed the probability of isotopes, the existence of which was demonstrated on early mass spectrographs and mass spectrometers.

GEOCHRONOLOGIC TIME MEASUREMENT

Rutherford and Soddy's theory of the time dependence of radioactive decay, followed by breakthroughs in instrumentation for the measurement of these unstable species and their radiogenic "daughter" nuclides, caught the attention of early geochronologists and had a revolutionary effect on the study of geology. In 1904, Rutherford proposed that geologic time might be measured by the breakdown of uranium in uranium-bearing minerals, and a few years later, American radiochemist Bertram Boltwood announced the "absolute" ages of three samples of uranium minerals. The ages, which approximated half a billion years, indicated that at least some earth materials were much older than had been thought—an idea developed by British geologist Arthur Holmes in his classic *The Age of the Earth* (1913). Holmes's early time scale for the earth and his enthusiasm for the developing study of radioactive decay were not met with instant acceptance by most contemporary geologists, but eventually absolute ages would become the prime quantitative components in the field of geology.

After the early study of the isotopes of uranium came the discovery of other unstable isotopes and the formulation of the radioactive decay schemes that have become the workhorses of geochronology. The theory of the radioactive decay of the parent, or unstable, nuclide (or the growth of the daughter, or stable, nuclide) developed in the early 1900's has not changed; it is still the basis for geochronologists' measurement of time. This field is one of the arenas for the use of mass spectrometry.

STABLE ISOTOPE FRACTIONATION

The other use of mass spectrometry in the earth sciences results from isotopes' potential to fractionate, or change their relative abundance proportions, during geological processes, for physicochemical reasons other than radioactive decay and radiogenic buildup. Fractionation not resulting from radioactive decay (stable isotope fractionation) comes about because the thermodynamic properties

Researcher preparing samples for mass spectrometry. Mass spectrometry uses strong magnetic and electric fields to separate the components of a sample by mass and charge. The individual components can then be identified. (Simon Fraser/Photo Researchers, Inc.)

of molecules depend on the mass of the atoms from which they are made. The total energy of a molecule can be described in terms of the electronic interactions of its atoms and the other energetic components of these atoms, such as their rotation, vibration, and translation. Molecules that contain in their molecular configurations different isotopes will have differing energies, because of the different energy components (usually vibrational and rotational) that are mass-dependent. The total energy of molecules also decreases with decreasing temperature; at zero kelvin, or absolute zero, this energy has a finite value known as its zero-point energy. The vibrational component of energy, the most important factor in fractionation, is inversely proportional to the square root of its mass. A molecule with the heavier of two isotopes will have a lower vibrational energy and thus a lower zero-point energy than a similar but lighter isotope molecule. Other factors being equal, the chemical bonds of a molecule with lighter isotopic composition will be more easily broken than those of the heavier isotope analogue, and the heavier molecule thus will be less reactive chemically.

Geologic processes that result in stable isotopic fractionation are the redistribution of isotopes as a function of isotopic exchange; nonthermodynamic (kinetic) processes that depend on the amounts of the species present during a reaction; and a range of strictly physical processes, including diffusion, evaporation, condensation, adsorption, desorption, crystallization, and melting. Physical conditions such as these undoubtedly were much more intense during preaccretion events, such as star formation, than during more typical "geologic" processes, such as sedimentation or volcanism; consequently, fractionation effects are observable in primitive materials, such as some components of relatively unprocessed meteorites. These materials show extremely interesting stable isotope fractionation effects even among the heaviest elements.

ELEMENTS OF MASS SPECTROMETERS

As commonly used, the term "mass spectrometer" refers to an instrument in which beams of ionized isotopes are separated magnetically. The earlier, more qualitative mass spectrograph focused ion beams onto a photographic plate. Mass spectrometers have three common elements: a source component, wherein elemental species are ionized so that they can be accelerated electrically; an analyzer section, where isotopic species are separated in a magnetic field by their mass-charge ratio; and a collector assembly, where the ion beams are quantitatively measured.

MAGNETIC-SECTOR MASS SPECTROMETRY

The most common instrument in geologic use is the magnetic-sector machine, in which a uniform magnetic field is bound in a region, or sector, commonly by a stainless steel tube that can be evacuated to very low pressures to prevent sample contamination. The source region may consist of a solid source; a purified and spiked sample of a heavy element is introduced in the solid state onto a filament of purified metal such as tantalum or rhenium, and the filament is heated electrically until a sufficient percentage of the element is vaporized and ionized for efficient measurement. Alternatively, the source may be a gas. In this case, the desired (commonly light) elements in a gaseous state are introduced into an evacuated region and bombarded with electrons to produce a sufficient percentage of ionized species for acceleration into the analyzer section of the instrument. The ionized species are accelerated electrically through a series of slits, onto which variable electric potentials can be applied for the purpose of acceleration and focusing, so that a well-defined, focused beam of the element or its gaseous compound is beamed into the analyzer tube.

The analyzer sector, which is commonly constructed so that the lowest pressure possible can be maintained and the least number of contaminant species will be struck by the focused beam, is bent at angles of 90 or 120 degrees as they pass through a magnetic field capable of efficiently separating the ion beams by their mass-to-charge ratios. (Where the charges are uniform, as is usual in earth science research, the separation is, as desired, only by mass.)

The collector assembly commonly consists of a Faraday cup; the separated isotope beams enter, hit the metal cup, and impart unit charges to the cup as the atoms are neutralized. The resulting direct current is exceedingly low and, in many instruments, must be converted to an alternating current so that the intensity of the signal can be increased for measurement with a strip-chart recorder or, more commonly, for digital readout. Accelerating voltage in the source assembly is adjusted with the magnetic

field in the analyzer sector (commonly monitored and controlled with a very precise gaussmeter) so that a beam of a unique charge-to-mass ratio (or of a unique mass, for ions of the same charge) passes through a final slit into the collector. Ions entering the collector are neutralized by electrons that flow from the ground to the metal collector cup, across a resistor whose voltage difference is amplified and measured with a digital or analogue voltmeter. These data are exhibited as a strip-chart readout or, more commonly, as digital output that is computer-collected and reduced for analysis. The collection of large numbers of highly precise isotopic ratios in computer-reduced digital form has made possible the modern use of mass spectrometry in the earth sciences and the determination of isotopic parameters that would not otherwise have been obtainable.

ADDITIONAL TYPES OF MASS SPECTROMETRY

Many advances have been made, so it is now possible to obtain extremely precise ratio measurements of tiny pieces of material in a relatively short time. Ion probe mass spectrometers allow these measurements on *in situ* samples in thin sections that, concomitantly, can be studied petrologically. Ion probe mass spectrometry involves the combination of a microbeam probe (using ions, rather than the lighter electrons, as "bullets" for ionization) and a magnetic-sector mass spectrometer. Accelerator mass spectrometry employs the use of a particle accelerator or cyclotron as the mass analyzer; it is useful primarily to make high-abundance measurements for cosmogenic nuclides such as carbon-14. Accelerator mass spectrometry makes possible the precise measurement of cosmogenic nuclides on tiny samples.

The developing field of resonance ionization mass spectrometry holds much promise in earth science studies because of its potentially high ionization efficiency and, therefore, sensitivity. Other possible mass spectrometric practices may include high-accuracy isotope dilution analysis utilizing a plasma ion source, and ion cyclotron resonance (Fourier transform) mass spectrometry. The super-machine of the future may combine some or even all of these potential advances.

PRINCIPAL MECHANICS OF SPECTROMETRY

Although some modern methods of determining absolute time do not involve isotopes, most do, and the standard method for their quantitative measurement is by mass spectrometry. Because the various radioactive nuclides useful in geochronology are also varied in their chemical characteristics, several instruments and techniques are involved. The principal mechanics of spectrometry, however, are mainly the same. The standard method involves placing the purified samples of the materials in question as solids on purified metal filaments and inserting the loaded filaments into a solid-source mass spectrometer. Evacuated to very low pressures, the spectrometer source regions are made so that the metal filaments can be heated to the point that elements such as rubidium and strontium ionize. The charged, ionized sample is accelerated through a series of collimating slits into the high-vacuum analyzing tube, where it encounters a controlled electromagnetic field. The beams of ions are separated by charge-mass ratios into beams of separated isotopes. As the charge of the elements is the same for each atom, however, the ions in this case are separated on the basis of mass only. Specific isotopic beams, controlled by the magnetic field, are channeled through more collimating slits to the collector part of the spectrometer. Commonly, a Faraday cup is used to analyze the number of atoms of each isotope by conversion of each atomic impact into a unit of charge, which is subsequently amplified, often with a vibrating reed electrometer. A digital readout is then produced. The actual output is isotope ratio measurements, which are converted by a mathematical program to the required parameters for determining time.

Scientists determine the age graphically, with the use of an isochron diagram, in which isotope ratios collected in the spectrometer are used as coordinates. A line known as an isochron connects points representing samples of equal ages. An isochron has an age value indicated by its slope on the figure; a horizontal isochron has a zero-age value, while successively greater positive slopes have increasingly greater ages given in terms of the slope and the half-life of the parent isotope. A single sample from a mineral or rock is represented by only one point in the diagram. Therefore, for an isochron to be drawn, an estimate of the sample's initial isotopic composition would be necessary. Ages calculated this way are termed "model ages."

EQUILIBRIUM AND NONEQUILIBRIUM FRACTIONATION

Stable isotope fractionation, or the enrichment of one isotope relative to another in a chemical or physical process, also has earth science applications. The two processes of this sort are equilibrium fractionation, which is useful in determining geologic paleotemperatures, and kinetic (nonequilibrium) fractionation, which is useful in establishing biologically mediated geochemical processes, such as the bacterial utilization of sulfur.

Isotopic fractionation in these processes is measured by the fractionation factor α, defined as A/B, where A is the ratio of the heavy to the light isotope in molecule A, and B is that ratio for molecule B. Although α may be calculated theoretically, in geologic use it is derived mainly from empirical data. This factor, which is largely dependent on the vibrational energies of the molecules involved, is a function of temperature; thus, it is a measure of ambient geologic processes.

ESTABLISHING PALEOTEMPERATURES OF ANCIENT SEAWATER

Many earth science applications of stable isotope fractionation are in use, but perhaps the best-known example is the use of oxygen isotope ratios to establish paleotemperatures of ancient seawater. Surface seawater, in at least partial equilibrium with the atmosphere, contains oxygen with a characteristic isotopic composition. This composition is provided by the ratio of the most abundant species: oxygen-18 and oxygen-16. Marine plants and animals, such as foraminifera, that build their hard parts out of components dissolved in seawater, such as calcium, carbon, and oxygen (as in calcium carbonate), utilize oxygen that is isotopically characteristic of the seawater. Although this process also depends on other, incompletely understood, factors, it is primarily a function of water temperature. Therefore, the ratio of oxygen-18 to oxygen-16 in the foraminifera is a measure of the water temperature. Because the calcium carbonate does not readily reequilibrate with ambient water after it is precipitated, it retains its characteristic isotopic composition after sedimentary burial for many millions of years. Isotopic data collected from foraminifera recovered from deep-sea cores are therefore used to record water temperatures (and consequently, climate) of the geologic past. More than any other paleothermometry device, this application has been extremely useful in providing a record of global temperature changes, especially of the past glacial periods, for use in constructing and testing quantified models of the causes of climate change.

For this application, the sample is introduced into the source region of the mass spectrometer as a gas, commonly carbon dioxide. Ionization of the gas may be accomplished by bombardment of the molecules with electrons. The positively charged ions created are accelerated through collimating slits into the analyzer section of the spectrometer. In this type of gaseous analysis, use is made of the double-focusing mass spectrometer, in which the isotopic composition of the sample is determined relative to that of the standard in iterative, alternating measurements.

VALUE TO PALEONTOLOGY

The revolution in the earth sciences, largely a result of the plate tectonics paradigm which was introduced in the early 1960's, was preceded by an even more important revolution, one that received little fanfare. In the 1940's and 1950's, earth science began to be significantly influenced by quantitative investigations which may be considered to have provided the quantitative foundation for the plate tectonic revolution. Of these studies, none was more significant than the use of mass spectrometry for determining absolute ages of minerals and rocks and, later, for paleothermometry. Absolute-age determinations gave a firm basis for paleontology and established not only the earth's antiquity but also a quantitative sequencing for its rocks and sediments—the geologic time scale.

E. Julius Dasch

FURTHER READING

Barker, James. *Mass Spectrometry.* 2d ed. New York: Wiley, 1999. A college text concerning the field of mass spectrometry and its protocol and applications. There is a fair amount of analytical chemistry involved, so the reader without a scientific background may have a difficult time. Bibliographical references and index included.

Chapman, John Roberts. *Practical Organic Mass Spectrometry: A Guide for Chemical and Biochemical Analysis.* 2d ed. Chichester, N.Y.: Wiley, 1995. A good account of mass spectrometry and its use in the analysis of organic compounds. The book offers an interesting look into the relationships among organic compounds, biochemical structures, and spectrometry. Two appendices, index, and bibliography.

de Hoffmann, Edmond, and Vincent Stroobant. *Mass Spectrometry: Principles and Applications.* 3rd ed. New York: Wiley-Interscience, 2007. A good resource for advanced undergraduate and graduate level students. As the title states, this text covers the principles and applications of mass spectrometry.

Duckworth, H. E. *Mass Spectrometry.* Cambridge, England: Cambridge University Press, 1958. An older, very technical but informative treatise on the basic principles of mass spectrometry. Suitable for college-level readers.

Faure, Gunter. *Isotopes: Principles and Applications.* 3rd ed. New York: John Wiley & Sons, 2004. Originally titled *Principles of Isotope Geology*, this is an excellent, though technical, introduction to the use of radioactive and stable isotopes in geology. It includes an introductory treatment of mass spectrometric principles and techniques. The work is well illustrated and indexed. Suitable for college-level readers.

Levin, Harold L. *The Earth Through Time.* 9th ed. Philadelphia: Saunders College Publishing, 2009. Chapter 5, "Time and Geology," reviews the geologic time scale and then turns to techniques for determining absolute age. It offers a history of early attempts at geochronology, an overview of radiometric dating, and a discussion of the principal dating methods. There is a simple description and diagram of a mass spectrometer. Easy to read and suitable for high school students.

Santamaria-Fernandez, Rebeca. "Precise and Traceable Carbon Isotope Ratio Measurements by Multicollector ICP-MS: What Next??" *Analytical & Bioanalytical Chemistry* 397 (2010): 973-978. This article reviews a new method of mass spectrometry used to measure carbon ratios. Provides a good description of mass spectrometry methodology for the intermediate or beginner MS analyzer.

Smith, David G., ed. *The Cambridge Encyclopedia of Earth Sciences.* Cambridge, England: Cambridge University Press, 1981. Organized as a compilation of high-quality and authoritative scientific articles rather than a typical encyclopedia. Chapter 8 in this clearly written text contains a section on mass spectrometry and its uses. Other analytical techniques are also discussed. Includes a table of trace element abundances in common rocks and minerals. For general audiences.

Watson, J. Throck, and O. David Sparkman. *Introduction to Mass Spectrometry: Instrumentation, Applications, and Strategies for Data Interpretation*. New York: Wiley, 2007. This is an extremely detailed, comprehensive publication on mass spectrometry. Provides historical reference and up-to-date applications. An associated website with exercise materials is available to students.

See also: Earth's Age; Earth's Oldest Rocks; Electron Microprobes; Electron Microscopy; Experimental Petrology; Geologic and Topographic Maps; Infrared Spectra; Neutron Activation Analysis; Petrographic Microscopes; Potassium-Argon Dating; Radioactive Decay; Radiocarbon Dating; Rubidium-Strontium Dating; Samarium-Neodymium Dating; Uranium-Thorium-Lead Dating; X-ray Fluorescence; X-ray Powder Diffraction.

METAMORPHISM AND CRUSTAL THICKENING

Metamorphism produces a wide array of Earth's rocks that form under circumstances related to high pressure and heat. Tectonic activity, particularly when it results in crustal thickening, is a driving factor in producing metamorphic rocks. Uplifting at faults has brought many metamorphic rocks to the surface, providing scientists the opportunity to make inferences about Earth's interior.

PRINCIPAL TERMS

- **diagenesis:** change that can occur in sedimentary rocks under 200 degrees Celsius and 300 megapascals; a distinct process from metamorphism
- **facies:** groupings of metamorphic rocks by chemical and mineral compositions that are typically found within specific pairings of temperature and pressure conditions
- **foliated:** having distinct layers that occur following the rotation of minerals (such as mica or chlorite) as a rock undergoes stress or strain on one side
- **grade:** an indicator of the relative pressure and temperature conditions under which a metamorphic rock forms; described as low-grade or high-grade
- **igneous rock:** a product of lava or magma that has cooled and solidified
- **magma intrusion:** the entrance of hot magma into a rock
- **metamorphosis:** a physical and chemical transformation of a rock into a different rock under the influence of high temperature and pressure
- **metasomatism:** a process that adds chemicals from surrounding rocks, often carried by water, causing a drastic chemical or physical change in the contacted metamorphic rock
- **petrology:** the scientific study of rocks
- **protolith:** the originating rock of a metamorphic rock; can be igneous, sedimentary, or metamorphic
- **sedimentary rock:** a product of the deposition of organic material and minerals at Earth's surface and within bodies of water

METAMORPHIC ROCKS

Petrology, the scientific study of rocks, recognizes three main types of rocks: sedimentary, igneous, and metamorphic. Sedimentary rocks are formed from the deposition of organic material and minerals at Earth's surface and within bodies of water. Igneous rocks are the products of magma and lava that have cooled and solidified. Metamorphic rocks are rocks (of any type) that have transformed physically, chemically, or both due to the influence of high temperature and pressure. Metamorphic rocks make up a large part of Earth's crust; common types include gneiss, marble, quartzite, schist, and slate.

The first form of a metamorphic rock is called the protolith, and it can be igneous, sedimentary, or even an older metamorphic rock. When a protolith is subjected to a temperature greater than about 200 degrees Celsius (C), or 392 degrees Fahrenheit (F), and a pressure above about 300 megapascals (MPa)—which equals 43,500 pounds per square inch—that protolith can transform in many ways. Metamorphism is a distinct process from diagenesis, which describes changes that can occur in sedimentary rocks under 200 degrees C and 300 MPa. Likewise, at temperatures and pressures high enough that a metamorphic rock begins to melt, it is no longer considered a metamorphic process; it is igneous.

Metamorphic rocks contain index minerals that are found only at the high temperatures and pressures of metamorphism; these minerals include andalusite, kyanite, sillimanite, and staurolite. Some other minerals can also be found in metamorphic rocks, including amphiboles, feldspars, micas, olivines, pyroxenes, and quartz. These minerals are stable at relatively high temperatures and pressures, and the points at which they become unstable can be used to deduce information about the conditions under which the metamorphic rock formed.

Three general types of classification can be applied to metamorphic rocks: facies, grades, and textures. Facies are groupings of metamorphic rocks by chemical and mineral compositions that are typically found within specific pairings of temperature and pressure conditions. For example, the lowest-grade facies is the zeolite facies; it describes metamorphic rocks at the lowest temperature and pressure conditions that can occur without entering diagenesis. This generally applies to certain sediments that undergo burial metamorphism. The highest grade is called the eclogite facies, which occurs at pressures around 1.2 gigapascals (174,000 pounds per square inch) and temperatures exceeding 600 degrees Celsius (1112 degrees Fahrenheit).

Grades describe relative pressure and temperature conditions under which a metamorphic rock formed. For example, metamorphic rocks are considered low grade between about 200 and 320 degrees Celsius (392 and 608 degrees Fahrenheit) and under about 600 MPa (87,000 pounds per square inch), while high grade refers to rocks formed at hotter, higher-pressure conditions. Low-grade metamorphic rocks tend to have an abundance of hydrous minerals, whereas high-grade rocks do not, because hydrous volatiles such as water and carbon dioxide evaporate as the temperature rises.

Texture provides even more details. In general, metamorphic rocks can be foliated or nonfoliated. Foliated rocks have distinct layers that develop through the rotations of minerals (such as mica or chlorite) as a rock undergoes stress or strain on one side. Nonfoliated rocks, which are less common, undergo stress uniformly on all sides, so layering does not occur. Foliated rocks are further divided into classes of slates, phyllites, schists, and gneisses, while nonfoliated rocks can be granoblastic or hornfelsic. Metamorphic rock textures also can be described as idioblastic or xenoblastic. Idioblastic rocks are bounded by their crystal faces, while xenoblastic rocks do not show their crystal faces.

Metamorphism occurs because the application of heat and pressure causes the ions and atoms in rocks to reorganize themselves, which alters the crystal structure. A variety of conditions provide the necessary heat and pressure for a metamorphic rock to form. Being located deep in the earth is enough to cause change. Tectonic processes also contribute, as pieces of crust are pushed under one another, causing increased friction, heat, and pressure. The intrusion of magma into a rock also causes metamorphism.

Tectonic activity has caused uplifting and erosion to reveal many metamorphic rocks at Earth's surface. This process gives scientists easy access to study the rocks and to make inferences about Earth's interior.

TYPES OF METAMORPHISM

Metamorphism varies widely in its process and its results, and it can be classified through a variety of types. These types are discussed here.

Regional metamorphism refers to changes through a wide region, such as the lower continental crust, brought on by an orogenic event (tectonic activity leading to severe deformation of Earth's crust). The lower crust is affected by its depth and by tectonic processes such as continental collisions, which result in uplifting and subduction of crust (processes full of friction, heat, and pressure). Most rock formed by regional metamorphism is highly foliated; slate, schist, and gneiss are common occurrences.

Contact metamorphism is caused by the intrusion of magma. Hot magma cools into an igneous rock, forming a region called the contact metamorphism aureole; metamorphism occurs locally due to heat, which lessens farther from the aureole. Contact metamorphism typically produces nonfoliated hornfels, and it is also common to find ore minerals around the contact zone. This process, called metasomatism, adds chemicals from surrounding rocks, often carried by water, and causes a drastic chemical or physical change in the contacted metamorphic rock.

Dynamic metamorphism, also called cataclasis, occurs at areas of moderate to high strain, such as fault zones. The rock undergoes metamorphosis due to mechanical deformation, which itself is caused by shearing and sliding along a fault. In simpler terms, the rock is crushed or shattered. This process does not involve much temperature change.

Hydrothermal metamorphism occurs under conditions of high temperature and moderate pressure, and in the presence of a hydrothermal fluid such as magma and hot groundwater or hot ocean water. Hydrothermal metamorphism generally occurs at the surface, often resulting in rich ore deposits. One area that exhibits hydrothermal metamorphism is Yellowstone National Park in the northwestern United States.

Shock or impact metamorphism is the result of shock waves—ultra-high-pressure conditions caused by a comet or meteorite impact, or a huge volcanic eruption. This type of metamorphism forms minerals like silicon dioxide polymorphs (such as coesite and stishovite), which are stable at high pressures. Shock metamorphism leaves tell-tale textural signs such as planar fractures, shock lamellae, and shatter cones. The effects of shock metamorphism have been found at every identified impact site on Earth.

Burial metamorphism crosses somewhat with diagenesis, but the process "grades up" to metamorphosis as pressure and temperature rise. Sedimentary rocks buried several hundred meters below the surface came from this type of metamorphism.

Prograde metamorphism occurs as temperature and pressure increase, and water and carbon dioxide are lost. Retrograde metamorphism is the opposite process. It is less common because the volatiles (water and carbon dioxide) must be present.

METAMORPHIC PROCESSES

Metamorphism can occur through a variety of processes, such as recrystallization, neocrystallization, phase change, pressure solution, and plastic deformation. The aforementioned metasomatism is also an important contributor.

Recrystallization occurs at about one-half the melting point of a rock (noted by degrees on the Kelvin scale). Particles change size and shape, but their identities, as atoms and ions pack together to form new crystal structures. Limestone recrystallizes into marble, for example, and small calcite crystals in sedimentary limestone transform into larger particles in metamorphic marble. Another example is clay, which can recrystallize to muscovite mica. Neocrystallization is the formation of new crystals

Skarn metamorphic rock. Skarn rocks are formed when igneous or sedimentary rocks are chemically altered by changes in heat and pressure conditions. They are largely composed of calcium, magnesium, and iron silicate minerals. The pink skarn in this sample is a breccia—a rock with fragments of one material (dark specks) cemented within a matrix of a different material. These fragments were probably incorporated during the high-intensity phases of metamorphosis when the rock was highly mobile. This sample measures 200 millimeters across. (Dirk Wiersma/Photo Researchers, Inc.)

not found in the protolith, a very slow process that involves the diffusion of atoms through solid crystal. Phase change also refers to the formation of new minerals, but these have the same formula as the protolith.

Pressure solution occurs when a rock is subjected to high pressure on one side in the presence of hot water. The rock's minerals partly dissolve and diffuse through the water to precipitate elsewhere. This process contributes to cleavage, a planar rock foliation. Plastic deformation occurs when pressure causes a metamorphic rock to shear and bend but not break. The temperature has to be high enough to avoid brittle fractures but low enough to avoid the diffusion of crystals.

Metasomatism alters a rock by introducing new chemicals from the surrounding environment— chemicals that are often delivered by a fluid. One way fluid delivery occurs is through the break down of hydrous minerals under high temperature and pressure in the lower crust. This break down releases fluid to the upper crust, where it interacts with and alters rocks.

TECTONIC PROCESSES AND
METAMORPHISM

Tectonic activity is a major factor of metamorphism, particularly in relation to crustal thickening, which can occur through crustal shortening or when one crust rides over another at a convergent boundary (causing the formation of mountains, among other results). Thickening crust becomes warmer and thus weaker. The lower portion of the crust is even warmer than the upper portion because of depth, so the lower portion becomes more plastic; this leads, ultimately, to collapse and the creation of a rift under the growing mountain. These steps all contribute to changing temperature and pressure conditions, thus causing metamorphosis in the rocks in a given region.

Tectonic activity can also lead to the stoppage of metamorphism. Crust that gets uplifted at faults is subjected to weathering and erosion, which cools the rock and returns it to a sedimentary state. All three types of rocks—sedimentary, igneous, and metamorphic—are closely connected, as demonstrated by the rock cycle. As the equilibrium of a rock's environment changes with time, the rock transforms into one of the other types.

To understand a simplified rock cycle, one can picture an igneous rock. Under the influence of weathering and erosion, that rock can break down into sediments, which undergo compaction and cementation into a sedimentary rock. This sedimentary rock then transforms under heat and pressure to become a metamorphic rock, which melts into magma. This new form then cools into an igneous rock, completing the cycle. The cycle is continuous and interacts with tectonic activity and the water cycle to affect nearly all aspects of life on Earth.

Rachel Leah Blumenthal

FURTHER READING

Blakey, Ronald C., Wolfgang Frisch, and Martin Meschede. *Plate Tectonics: Continental Drift and Mountain Building.* New York: Springer, 2011. This book provides an introduction to plate tectonics, covering Earth's early history to the present day. Topics covered include subduction zones, mid-ocean ridges, and the formation of mountains.

Chilingar, George V., et al. *Evolution of Earth and Its Climate: Birth, Life, and Death of Earth.* Boston: Elsevier, 2011. This text provides ample information on Earth's formation, composition, and the differentiation of its layers. It also explores the formation of the moon and its influence on Earth.

Dickey, John S. *On the Rocks: Earth Science for Everyone.* New York: Wiley, 1996. Covering topics from stardust to Earth's formation to Earth's planetary neighbors, this book is ideal for Earth science novices. Easy to understand but comprehensive, *On the Rocks* is an enjoyable overview of the history of the earth.

Lutgens, Frederick K., and Edward J. Tarbuck. *Earth Science.* 13th ed. Upper Saddle River, N.J.: Prentice Hall/Pearson, 2012. Originally published in 1976, *Earth Science* is an updated introductory textbook geared to undergraduates, including those without a science background. Covers geology, astronomy, and other Earth science topics.

Monroe, James S., and Reed Wicander. *The Changing Earth: Exploring Geology and Evolution.* 5th ed. Belmont, Calif.: Brooks/Cole, Cengage Learning, 2009. This textbook offers a solid introduction to geology in an easily readable format, supplemented with relevant photographs, diagrams, and real-world examples. The newest edition has been condensed from previous editions.

Williams, Linda D. *Earth Science Demystified.* New York: McGraw-Hill, 2004. A quick and easy self-teaching guide aimed at readers without formal science training. Covers the basic range of Earth science topics and includes summaries and questions.

Wu, Chun-Chieh. *Solid Earth.* Vol. 26 in *Advances in Geosciences.* London: World Scientific, 2011. This book highlights recent research papers in seismology, planetary exploration, the solar system, and other topics relevant to Earth science.

See also: Earthquakes; Earth's Interior Structure; Earth's Mantle; Earth's Oldest Rocks; Experimental Petrology; Experimental Rock Deformation; Fluid Inclusions; Geodynamics; Heat Sources and Heat Flow; Mantle Dynamics and Convection; Mountain Building; Petrographic Microscopes; Phase Changes; Plate Motions; Plate Tectonics; Radioactive Decay; Radiocarbon Dating; Relative Dating of Strata; Rock Magnetism; Tectonic Plate Margins; Volcanism.

MILANKOVITCH HYPOTHESIS

Milankovitch cycles refer to various periodic changes in Earth's orbital revolution and axial rotation—cycles that cause corresponding changes in the amount of solar radiation the earth receives. The Milankovitch hypothesis seeks to correlate these cycles with long-term climate changes on Earth, including the ice ages that have occurred about every 100,000 years during the last one or two million years.

PRINCIPAL TERMS

- **aphelion:** Earth's greatest distance from the sun in its elliptical orbit
- **celestial equator:** intersection of Earth's equatorial plane with the celestial sphere
- **eccentricity:** the degree of elongation of an elliptical orbit from a circular orbit with zero eccentricity
- **ecliptic plane:** the plane of Earth's orbit around the sun
- **equinox:** the points on the celestial sphere where the sun appears to cross the celestial equator, moving northward at the vernal equinox and southward at the autumnal equinox; corresponds to equal hours of night and day
- **greenhouse gases:** atmospheric gases such as water vapor, carbon dioxide, and methane that trap heat by absorption of solar radiation and reemission of longer wavelengths that cannot escape from the atmosphere
- **insolation:** the amount of incident solar radiation on a unit area of the earth's surface at any given latitude
- **nutation:** the periodic change in the angle of tilt of Earth's rotational axis that results in a bobbing of Earth's axis during precession
- **obliquity:** the angle of tilt of Earth's rotational axis (about 23.5 degrees) from a perpendicular to its orbital plane (its ecliptic)
- **perihelion:** the closest point of Earth from the sun on its elliptical orbit of the sun (located at one focal point of the ellipse)
- **precession of the equinoxes:** the 26,000-year wobble of the earth's axis that causes the axis to slowly shift along a conical path away from its current direction toward the North Star Polaris
- **solstice:** the farthest points on the celestial sphere of the sun's apparent path above and below the celestial equator, where the sun stands at midsummer and midwinter

ATTEMPTS TO EXPLAIN THE ICE AGES

The Milankovitch hypothesis attempts to explain the periodic occurrence of ice ages on Earth through the last one million years by calculating the combined effects of various periodic Earth motions on Earth's long-term climate patterns. Milutin Milanković, a Serbian geophysicist, engineer, and professor of applied mathematics at the University of Belgrade, worked on these calculations while interned during World War I. His interest in this topic began with a 1912 scientific paper, "Contribution to the Mathematical Theory of Climate." In 1913, he published *The Schedule of Sun Radiation on the Earth's Surface,* which was followed by another scientific work, "About the Issue of the Astronomical Theory of Ice Ages" (1914). After the war, Milanković published a monograph, "Mathematical Theory of Thermal Phenomena Caused by Solar Radiation" (1920), which was based on his scientific calculations and established the Milankovitch cycles.

Milanković's ideas were first suggested by the French mathematician Joseph Adhemar in 1842. Adhemar recognized that the earth's elliptical orbit brings it closer to the sun (perihelion) in winter (on January 3) and farther from the sun (aphelion) in summer (July 4) in the Northern Hemisphere. The opposite occurs in the Southern Hemisphere, with the earth farther from the sun in winter, so that Antarctic winters are colder than Arctic winters. Kepler's laws show that the earth moves more slowly at aphelion, so Antarctic winters are several days longer than Antarctic summers. Adhemar believed this would cause more cooling and would produce a larger Antarctic ice sheet; he made this prediction before the ice sheet was discovered.

Adhemar also recognized that the precession of the equinoxes slowly shifts the axis of Earth's rotation and the corresponding seasons. The axis leans toward the sun in summer and away from the sun in winter. Precession reverses this effect during half of the precession cycle, and thus might cause the ice ages in the Northern Hemisphere when the axis is leaning away from the sun at aphelion.

Adhemar's ideas were developed further by Scottish scientist James Croll, who taught himself physics and astronomy. After corresponding with British geologist Sir Charles Lyell on the ice ages and their possible connection with orbital variations, Croll devoted himself to research on this topic and published his results in the books *Climate and Time, in Their Geological Relations* in 1875 and *Climate and Cosmology* in 1885. Croll related the 26,000-year precession of the equinoxes to a series of ice ages. He followed Adhemar in suggesting that when winter occurs near aphelion it will be colder for a longer period of time. Croll was the first to suggest that the resulting ice sheet will reflect more sunlight and remain frozen longer, producing colder temperatures for a "positive feedback" that reinforces the severity of an ice age even during summers. However, his calculation that the last ice age ended some 80,000 years ago was contradicted by new evidence at the end of the nineteenth century, suggesting a more recent end to the last ice age. Croll's ideas fell out of favor until they were revived by Milanković.

MILANKOVITCH CYCLES

Milanković calculated the changes of solar radiation on Earth caused by various periodic motions of Earth, including the 26,000-year axial precession of the equinoxes, the 112,000-year orbital precession of Earth's elliptical orbit, the 41,000-year nutation (nodding) of Earth's rotational axis, and changes in the eccentricity (elongation) of Earth's orbit during a cycle of about 100,000 years.

The precession of the equinoxes was discovered about 150 B.C.E. by Greek astronomer and mathematician Hipparchus, who determined the precession to be a slow shift of the stars, in which the equinoxes shifted along the celestial equator by about one degree every 72 years, or 360 degrees in about 26,000 years. When Polish astronomer Nicolaus Copernicus proposed in his monumental work *On the Revolutions of the Heavenly Spheres* (1543) that the earth rotates on its axis and revolves around the sun, he recognized that precession of the equinoxes was caused by a 26,000-year wobble of the earth's axis, which traces a 23.5-degree cone around a perpendicular to the earth's orbital plane. English scientist-mathematician Sir Isaac Newton showed in the late seventeenth century that this wobble results from gravitational forces of the sun and moon acting on the equatorial bulge of the earth. Newton attributed this bulge to centrifugal forces caused by Earth's rotation,

which resulted in its flattened spherical shape (oblate spheroid).

Although the earth's axis precesses once every 26,000 years relative to the stars, its axis precesses faster relative to the perihelion of the earth's orbit. This is true because the orbit itself is precessing about once every 112,000 years due to a slight inclination of Earth's orbit relative to the orbits of Jupiter and Saturn and due to gravitational forces acting to reduce this inclination. The result of this rotation of Earth's orbit to the ecliptic plane is that the 26,000-year axial precession relative to the stars is only about 21,000 years relative to the earth's orbit and its perihelion and aphelion points.

Because the greatest seasonal differences between summer and winter occur when the earth is closest to the sun (perihelion), at the summer solstice, and farthest from the sun (aphelion), at the winter solstice, it is this 21,000-year cycle that should have the greatest climate effect. The coldest and longest winters should occur when the earth is farther from the sun, near aphelion every 21,000 years. The increasing ice sheet on Earth would then reflect the sun's radiation during the following summers, slowing any melting of ice and helping to offset the larger amount of radiation that Earth receives when it is closer to the sun at perihelion.

The tilt of Earth's axis also varies in a 41,000-year nutation period, bobbing between 22.1 and 24.5 degrees of obliquity. The nutation of Earth's axis was discovered by English astronomer James Bradley in 1728 and was later shown to be caused, primarily, by the changing gravitational forces of the sun and moon on the earth's equatorial bulge. Increasing obliquity increases the amount of solar radiation in summers, when Earth's axis leans farther toward the sun and decreases it in winters, when the axis leans farther from the sun. This effect is amplified at the higher latitudes, where there are greater amounts of land that experience faster temperature changes than the oceans, in which mixing and cooling can occur. In a similar fashion, decreasing obliquity decreases solar radiation intensity in summers and increases it in winters. Even though winters are warmer, it is thought that the cooler summers melt less of the previous winters' ice build-up, contributing to the onset of an ice age.

Studies have shown that tilts of more than 24.5 degrees could raise temperatures enough to prevent the emergence of higher forms of animal life. Studies also have shown that the earth's unusually large moon stabilizes the tilt of the earth in a narrow range in contrast to the tilt of Mars, which varies widely and prevents the long-term stability of seasons needed for the development of living organisms. The current 23.5-degree obliquity of Earth is near the middle of the nutation range and is moving toward decreasing obliquity and its associated cooling trend; however, this may be offset by global warming, which is caused by human contributions to greenhouse gases in the atmosphere.

The eccentricity (elongation) of the earth's elliptical orbit has several periodic variations, mainly due to the gravitational forces of Jupiter and Saturn on the earth. The longest of these variations is 413,000 years, but several shorter variations combine to give a stronger periodic variation of about 100,000 years. The major axis of Earth's elliptical orbit remains nearly constant, which stabilizes the length of the year; but the minor axis varies, with an associated change in the distance between the focal points on the major axis, and thus the distance between the earth and the sun located at one focus.

The elongation of the earth's orbit varies from near circularity (lowest eccentricity of 0.005) to mildly elongated (highest eccentricity of 0.058) and has an average eccentricity of 0.028. Earth's orbit is currently in the lower end of the range, with an eccentricity of 0.017. Solar radiation received by the earth at perihelion is 6.8 percent more than radiation received at aphelion. The maximum variation in radiation received by the earth is 23 percent, when its orbit is at maximum elongation.

CONFIRMATION OF THE MILANKOVITCH HYPOTHESIS

Milanković combined the various Milankovitch cycles to predict that temperatures on Earth would oscillate with a variety of different periods. The shorter term periods, caused by the combined effect of axial and orbital precessions (21,000 years), and the changing axial tilt (obliquity) caused by nutation (41,000 years), would be superposed on the longer variations, caused by changes in the eccentricity of the earth's orbit (about 100,000 years). Milanković believed that decreasing obliquity, with its cooler summers at aphelion, would have the strongest influence on increasing glaciation, especially at higher northern latitudes (which have larger landmasses), and thus would cause ice ages approximately every 41,000 years. This led Milanković to recognize that ice ages would not alternate between the Northern and Southern Hemispheres but rather would be dominated by the higher-latitude landmasses and occur simultaneously in both hemispheres. Information about previous

ice ages was limited at the time, so no clear confirmation of this hypothesis was possible.

The development of radiocarbon dating in the 1950s appeared to conflict with the Milankovitch hypothesis, and for a time the hypothesis was abandoned. More recent research has shown that the longer 100,000-year variations in the elliptical shape of the earth's orbit have a stronger effect than nutation, and that ice ages have occurred about every 100,000 years during the last couple of million years.

A more complete and accurate record of ice ages began to emerge in the 1970s from deep-ocean sediment cores and Antarctic ice cores, which together revealed past global sea levels and temperatures. These data were the basis for an influential paper by J. D. Hays, J. Imbrie, and N. J. Shackleton ("Variations in the Earth's Orbit: Pacemaker of the Ice Ages," 1976) that clearly revealed the three Milankovitch cycles. This revival of the Milankovitch hypothesis differed in the relative strengths of the cycles, with the measured peaks separated by about 23,000 years, 42,000 years, and 100,000 years, corresponding to approximately 10, 25, and 50 percent of the climatic variance caused by precession, obliquity, and eccentricity.

PROBLEMS, PROPOSALS, AND PREDICTIONS

The revival of the Milankovitch hypothesis in the 1970s had strong support in the observed climatic cycles, with their periodicities closely matching each of the Milankovitch cycles. All the cycles were evident in temperature variations, although the 100,000-year eccentricity cycle dominated and gave the best correlation with ice-age glaciations of the last 1 million years. This success, however, was not without problems, although these problems did result in new research and theories.

One problem immediately evident was that the 100,000-year eccentricity cycle had the weakest theoretical influence on solar radiation on Earth, yet it appeared to have the strongest climatic influence, as seen in the data from core samples of ocean sediments and Antarctic ice. Another problem was a clear shift from the dominant influence of the 41,000-year obliquity cycle to the recent dominance of the 100,000-year cycle, beginning about 1 million years ago.

In 1994, Richard A. Muller, at the University of California, Berkeley, proposed a new theory to resolve these problems. He focused on another orbital motion not considered in the original Milankovitch hypothesis: the changing inclination of Earth's orbit relative to the orbit of Jupiter. This orbital inclination varies over a few degrees with a 100,000-year period similar to that of the earth's orbital eccentricity, both of which result mainly from the gravitational influence of Jupiter and Saturn.

In 1983, the Infrared Astronomical Satellite detected a thin ring of dust and debris around the sun approximately in the plane of Jupiter's orbit. This ring apparently resulted from collisions between asteroids of the Themis and Koronis families. The earth still passes through it, on January 9 and July 9 of each year, an event accompanied by an increase in radar-detected meteors. Muller postulated that about every 100,000 years, when Earth's orbit aligns with this ring of debris, the ring itself gathers enough dust by accretion in the atmosphere to cool Earth's climate and reinforce the weak eccentricity cycle. Muller also proposed that if asteroid collisions occurred about every 1 million years, the resulting increase in dust would explain the shift from the 41,000-year cycle to the 100,000-year cycle.

Muller's proposal has not been widely accepted, and other problems remain unresolved. The Milankovitch cycles continue to have wide support as the best explanation for the ice ages. Summer radiation received in northern latitudes (calculated at 65 degrees north) appears to be the strongest driver of the 100,000-year ice-age cycle. Extrapolating the Milankovitch cycles into the future has led to various predictions of long-term temperature trends and ice-age glaciations. A widely quoted 1980 study by John Imbrie and John Z. Imbrie concluded that orbital forces upon Earth's climate will continue the long-term cooling trend that began 6,000 years ago for another 23,000 years unless the trend is reversed by greenhouse gases. A 2009 study, led by Darrell Kaufman of Northern Arizona University, indicated an Arctic cooling trend during the last 2,000 years that should continue for another 4,000 years, but this cooling trend appears to have been reversed by global warming.

Joseph L. Spradley

FURTHER READING

Fagan, Brian M., ed. *The Complete Ice Age: How Climate Change Shaped the World*. London: Thames and Hudson, 2009. This well-illustrated book is a collection of articles on the causes and effects of the ice ages, including their effect on animal and human life.

Huybers, P. J. "Early Pleistocene Glacial Cycles and the Integrated Summer Insolation Forcing." *Science* 313 (July 2006): 508–511. An interesting discussion of the difference between the 21,000-year precession cycle and the 41,000-year obliquity cycle, and how summer insolation strengthens the latter.

Imbrie, John, and Katherine Palmer Imbrie. *Ice Ages: Solving the Mystery*. 2d ed. Cambridge, Mass.: Harvard University Press, 1986. A pioneer in climate studies and his daughter wrote this comprehensive and readable book on the history, geology, astronomy, and climatology of the ice ages, with special attention to the Milankovitch hypothesis.

Macdougall, Doug. *Frozen Earth: The Once and Future Story of Ice Ages*. Berkeley: University of California Press, 2006. Biographies of four ice-age scientists, including James Croll and Milutin Milanković, are interwoven with the evidence for widespread glaciations and the causes and effects of ice ages.

Milanković, Milutin. *Canon of Insolation and the Ice-Age Problem*. 1941. Reprint. Belgrade: Agency for Textbooks, 1998. An English translation of one of the original works establishing the Milankovitch hypothesis, with a new biographical essay and index.

Muller, Richard A., and Gordon J. MacDonald. *Ice Ages and Astronomical Causes: Data, Spectral Analysis, and Mechanisms*. London: Springer, 2000. A highly technical book on long-term temperature analysis with many graphs and equations. Includes a helpful introduction with a description of Muller's orbital inclination theory.

Raymo, Maureen E., and Peter Huybers. "Unlocking the Mysteries of the Ice Ages." *Nature* 451 (January 2008): 284-285. A good review of Milankovitch cycles and their correlation with ice-age data. Includes a discussion of problems that remain and the need for a more complete theory.

Roe, Gerard. "In Defense of Milankovitch." *Geophysical Research Letters* 33, no. 24 (December 2006). Argues that the correlation between global ice volume and the Milankovitch cycles can be improved by focusing on the time rate of change of global ice volume.

See also: Climate Change: Causes; Deep-Earth Drilling Projects; Earth's Mantle; Geodynamics; Geologic and Topographic Maps; Glaciation and Azolla Effect; Jupiter's Effect on the Earth.

MOUNTAIN BUILDING

A diversity of geological and geodynamic factors contributes to the formation of the earth's mountains. Plate tectonics—the movement of massive stone plates beneath Earth's outer crust—is central to this process. As these plates contact and then pull away from each other, the outer crust is pushed outward, forming mountains. In some cases this push is dramatic, while in others it is more gradual and "gentle." Many mountains are remnants of much larger, ancient mountains that have since been eroded by geodynamic forces.

PRINCIPAL TERMS

- **convergent boundary:** the geodynamic process in which two tectonic plates directly collide
- **divergent boundary:** the geodynamic process whereby two tectonic plates separate and create a space for magma to fill
- **folding:** the effect of the collision of two plates pushing the earth's outer crust outward
- **lithosphere:** the outermost geological layer of the earth that consists of massive plates of rock
- **orogenesis:** the process of mountain building
- **plate tectonics:** the theoretical concept in which the earth's lithosphere, comprising continental and smaller plates, is in constant motion
- **transform boundary:** the geodynamic process in which two tectonic plates collide in a side-swiping motion

Basic Principles

Mountains are formed in a variety of ways, but most scientists believe that it is primarily done through plate tectonics. The theory of plate tectonics, which has been overwhelmingly accepted in the scientific community, states that beneath the earth's outer crust lies a layer of seven massive rock plates called the lithosphere. These continental plates (the North American, African, South American, Eurasian, Antarctic, Australian, and Pacific), along with other smaller plates, are constantly in motion and moving in different directions.

The tectonic plates occasionally collide with one another at plate boundaries. In the case of convergent boundaries, the force of these direct collisions causes the boundaries to fold; the upward fold pushes outward in the form of mountains and mountain ranges. When plates move away from one another (to create divergent boundaries), parts of the lithosphere break away and fall into the superheated asthenosphere beneath, causing hot magma to flow into the empty spaces. The magma then pushes

against the outer crust, creating cone- and dome-shaped mountains. The mountain becomes an active volcano if lava is pushed through the surface of the domes. When plates sideswipe each other (to create transform boundaries), the shearing effect produces mountainous deformations on the surface.

Mountains are also formed on the earth's surface through erosion rather than plate tectonics. In this manner, wind, ice floe, and running water carve lower-level mountains into the earth's crust. These elements are also responsible for sculpting the shape of mountains through millions of years.

Background and History

Throughout history, humans have been awed by the grandeur and powerful visage of mountains. Ancient Tibetans, from 20,000 years B.C.E., believed that gods resided atop the Himalaya. Ancient Greeks later shared that sentiment, believing that their gods resided atop Mount Olympus.

In the early nineteenth century, scientists began to explore the nature of higher elevations and mountain ranges. In 1829, French geologist Jean-Baptise Élie de Beaumont theorized that mountains formed as the earth cooled and contracted. In 1842, American scientist Henry Darwin Rogers argued that mountains were the result of wavelike undulations caused by molten rock flowing beneath Earth's crust.

In 1912, German climatologist Alfred Wegener moved the field of geology to an entirely new level. Wegener noticed that fossils of certain species of plants and animals were unearthed thousands of miles apart, even across oceans. He expanded upon the observations of others, who believed that each continent seemed to fit together with other continents like pieces of a puzzle. Wegener further argued, in developing his theory of continental drift, that the continents were once connected and that they had later broken apart, and were now drifting away from one another. Continental drift theory remained relevant until the 1960s, when advances in seismological and volcanic studies seemed to show the outlines of tectonic plate boundaries. Plate tectonics has since been overwhelmingly embraced by the scientific community, with clear applications in orogenesis, the formation of mountains on the earth's surface.

CONVERGENT BOUNDARIES AND OROGENESIS

The concept of plate tectonics is central to the process of orogenesis. The shape and configuration of mountains and mountain ranges depend on how tectonic plates converge. For example, the Himalayan range, which includes the world's tallest mountain (Mount Everest), was formed at a convergent boundary of the Eurasian and Indian plates, which first collided 25 million years ago. The force

by which the plates collided caused a crumpling effect, pushing rock outward in the form of mountain peaks. The collision is ongoing, which means that the Himalayas continue to form and grow.

Scientists frequently use the Himalayan example to study other mountain ranges. In one study, scientists attempting to understand the orogenesis of the Caledonide mountain range in Scandinavia have noticed similarities between that range and the composition of the Himalayas. This comparison has led them to argue that the Caledonides were formed by a convergent boundary plate collision. Such information is useful for scientists who are surveying this range for geothermal energy and certain minerals within its bedrock.

Plate boundary. This fissure marks the boundary between the North American and the European tectonic plates. The two plates are moving apart, creating cracks in rock at the Earth's surface. The boundary, known as the mid-Atlantic ridge, divides Iceland and is effectively splitting it apart. Photographed in Thingvellir, Iceland. (Daniel Sambraus/Photo Researchers, Inc.)

DIVERGENT BOUNDARIES AND OROGENESIS

In the second form of orogenesis, known as divergent boundaries, plates move from one another and leave an opening between them. This opening is filled with magma from the earth's inner regions; the magma pushes outward, creating mountains. One of the best-documented examples of divergent boundary orogenesis is the Mid-Atlantic Ridge, which rests on the floor of the Atlantic Ocean, spreading from the Arctic regions to beyond the Cape of Good Hope off the southern coast of South Africa.

The divergent boundary of the Mid-Atlantic Ridge provides scientists with a rare ability to monitor the spread of the two plates in question (the North American and Eurasian plates). Scientists have placed a number of observation stations in Iceland, which lies along the boundary of the two plates. In addition to the volcanic activity that occurs frequently from this divergence, seismic activity enables scientists to use a wide range of scientific technologies (including global positioning systems) to monitor the continuing rate of divergence and the continuing formation of mountains.

TRANSFORM BOUNDARIES AND OROGENESIS

The third form of orogenesis involves transform boundaries. This manifestation of mountain building occurs when two plates "sideswipe" each other rather than collide head-on. The San Andreas fault is one of the most famous of these transform boundaries. In this case, the North American and Pacific plates are grinding alongside each other (the Pacific plate in a northwesterly direction and the North American plate on a southeasterly heading). The Sierra Nevada range in California provides an example of the mountains that form at transform boundaries. The Sierra Nevada was formed (and continues to be formed) by the shearing effects of the plate motions. The mountains are fragments of the plates, pushed upward and outward by the encounter between the two tectonic plates.

COMPUTER MODELS

Because the subterranean movement of tectonic plates cannot be directly observed, the study of these rocks relies heavily on computer models. In a similar vein, scientists studying orogenesis (which for even the youngest mountain ranges took place millions of years ago) rely on computer models to help create

a profile of a region's mountain-building history. For example, scientists estimate that the creation of Wyoming's Laramie range, which spans from Wyoming to the Black Hills of South Dakota, took place about 70 to 80 million years ago through subduction (a process whereby the weight of an ocean or glacier pushes plates downward, allowing for magma to fill the empty space).

In the early twenty-first century, scientists compiled thermal data taken from various points in the Laramie range and entered them into a computer database. Based on the models they generated, researchers created a profile of the degree of magma flow that contributed to the range›s formation.

REMOTE-SENSING SATELLITES

One of the most invaluable tools for geologists and geophysicists studying orogeny is satellite technology. Remote-sensing satellites can compile large volumes of photographs and geodynamic data each time they pass over a target. For example, researchers have used data from satellite flyovers to assess the elevations of certain major mountain ranges. The satellite images showed scientists how natural elements such as water and wind erode mountains and contribute to a reduction in mountain elevation. Satellites have also been called into service to observe the many different features of the Alps. The information collected by the ERTS-1 satellite revealed a complex system of fault lines, folding, and shearing.

GROUND-BASED SENSORS

Geologists and geophysicists commonly use ground-based sensory equipment to better understand the forces involved in building mountains. For example, the rock that is brought to the surface in a divergent boundary scenario maintains a certain magnetic charge. Magnetometers, in one case, detected the magnetic rocks introduced along the Mid-Atlantic Ridge's divergent boundary. (During World War II, scientists used magnetometers to detect German submarines.) Today, magnetometers continue to prove invaluable for surveying the orogeny of mountain ranges.

Scientists also use thermal equipment to observe the different minerals that help form mountains. Such equipment can help differentiate between the elements that foster orogenesis. For example, researchers used thermal imaging equipment to reveal the heat radiating from the minerals found in different locations in the Adirondack Mountains. The use of this thermal equipment helped researchers create a chronology of the orogenesis of mountains in this area. Thermal detection equipment has also helped scientists determine the effects of glaciation on prehistoric mountain ranges. By examining the temperatures of certain components of the bedrock, researchers can determine the rates of glacial development and erosion in a given mountain range.

RELEVANT NETWORKS AND ORGANIZATIONS

Orogeny involves looking at the creation of mountains to understand what is occurring (or has occurred) within the earth's surface. Among the different types of groups involved in the research of orogenesis are governments, universities, and nonprofit organizations and networks.

The U.S. government is typically involved in the study of mountain building in two ways. First, it dedicates resources and personnel directly to such research. For example, the U.S. Geological Survey has geologists, geophysicists, and other professionals analyze plate tectonics, erosion, and volcanic activity, all of which are closely linked to the study of mountain building. Second, the government provides grants and other financial support to independent researchers. The National Science Foundation, for example, provides a wide range of financial support for independent research.

The university is one of the most important institutions involved in orogeny. The scientists working at these institutions have two roles to play: conduct field research on relevant topics and teach undergraduate and graduate students the principles of geodynamics, plate tectonics, and orogenesis. The California Institute of Technology, the University of Cambridge, and the University of Geneva are among the institutions with active programs in the study of mountain building, mostly within geology departments.

Many nonprofit organizations and scientific networks and societies have an interest in the field of orogeny studies. The Geological Society of America, for example, offers publications, conferences, and other venues for experts in this field to share information and research. Other geological societies with similar interests (including orogeny and plate tectonics) are found in Scandinavia, Great Britain, Japan, South Africa, and other areas of the world.

IMPLICATIONS AND FUTURE PROSPECTS

The study of how mountains are formed continues to evolve, especially as available technologies improve. One of the most significant of these technological advances is the use of satellites to observe orogenic changes. The use of the extensive global positioning system makes it possible to carefully monitor the movement of the continents along their boundaries. Furthermore, satellites can track orogenic changes using other shipboard equipment, including thermal imaging and radar systems.

In addition to the ever-evolving use of satellites and state-of-the-art sensory equipment, the study of mountain building is greatly aided by the Internet. Researchers around the world can share, in real time, large volumes of the data they have acquired in the field. As human civilization learns more about plate tectonics using cutting-edge technology, it will also learn more about the mountains and ranges plate tectonics produce.

Michael P. Auerbach

FURTHER READING

Cox, A., and B. R. Hart. *Plate Tectonics: How It Works.* Hoboken, N.J.: Wiley-Blackwell, 1986. This book describes the concepts behind the theory of plate tectonics. The analysis includes a review of the forces that drive the movement of tectonic plates and the deformations of the earth's crust that occur as a result of plate motion.

Egholm, D. L., et al. "Glacial Effects Limiting Mountain Height." *Nature* 460 (2009): 884-887. This article presents the effects of glaciation on mountains in the Himalayan and Andes mountain ranges. The authors describe how glaciers can significantly erode a mountain, reducing its height and reshaping the mountain's appearance dramatically.

Frisch, Wolfgang, Martin Meschede, and Ronald C. Blakey. *Plate Tectonics: Continental Drift and Mountain Building.* New York: Springer, 2010. In this book, the authors provide in-depth reviews of the theories of continental drift and plate tectonics. The authors describe how mountains are borne of plate tectonics and the different manners by which orogenesis takes place.

Johnson, Michael, and Simon Harley. *Orogenesis: The Making of Mountains.* New York: Cambridge University Press, 2012. In this book, the authors provide an introduction to the different processes that form mountains and mountain ranges. The authors also discuss plate tectonics and the applications of orogenesis to petrology, mineralogy, and other scientific fields.

Lamb, Simon, and Anthony Watts. "The Origin of Mountains: Implications for the Behaviour of Earth's Lithosphere." *Current Science* 99, no. 12 (2010): 1699-1718. This article describes the orogenesis that occurs as the result of plate motion. The authors cite examples from the Himalayas, Andes, and other mountain ranges, where the relatively thin lithosphere has been pushed out due to a variety of plate interactions.

Schellart, W. P. "Overriding Plate Shortening and Extension Above Subduction Zones: A Parametric Study to Explain Formation of the Andes Mountains." *Geological Society of America Bulletin* 120, no. 11 (2008): 1441-1454. This article discusses the forces that created the Andes mountain range in South America. Schellart points to the fact that plates that were pushed down by glaciers, a process known as subduction, allowed magma to replace it in an open space, pushing outward to form the range.

See also: Continental Drift; Earthquakes; Earth's Mantle; Geodynamics; Geologic and Topographic Maps; Isostasy; Plate Motions; Plate Tectonics; Relative Dating of Strata; San Andreas Fault; Slow Earthquakes; Stress and Strain; Subduction and Orogeny; Tectonic Plate Margins; Volcanism.

N

NEUTRON ACTIVATION ANALYSIS

Neutron activation analysis uses a flux of neutrons to excite the nuclei of chemical elements in samples, thus causing the excited nuclei to emit characteristic gamma radiation. The technique provides a sensitive method for measuring the amount of chemical elements contained in geological samples, particularly when there is only a small amount of the element in the sample.

PRINCIPAL TERMS

- **cross section:** the effective area that a nucleus presents to an oncoming nuclear particle, which determines the chance that the particle will strike the nucleus, causing a nuclear reaction
- **gamma decay:** the emission of high-energy electromagnetic radiation as a nucleus loses excess energy
- **gamma spectrum:** the unique pattern of discrete gamma energies emitted by each specific type of nucleus; it identifies that nucleus
- **half-life:** the time during which half the atoms in a sample of radioactive material will decay
- **isotope:** atoms of the same chemical element containing equal numbers of protons whose nuclei have different masses because they contain different numbers of neutrons
- **neutron:** an uncharged particle that is one of the two major nuclear constituents having nearly equal masses and different electric charges
- **nuclear reaction:** a change in the structure of an atomic nucleus brought about by a collision of the nucleus with another nuclear particle such as a neutron
- **nucleus:** the tiny central portion of an atom that contains all the positive charge and nearly all the mass of the atom
- **proton:** a particle that carries a single unit of positive charge equal in size to that of the electron; one of the two major nuclear constituents having nearly equal masses and different electric charges

Gamma Radiation

In neutron activation analysis, a sample of interest to the scientist is placed in a beam of neutrons produced by a radioactive source, such as an accelerator, a nuclear reactor, or an alpha-emitter bombarding beryllium. The neutrons interact with nuclei contained in the sample and alter their structures, frequently leaving the nuclei of the sample with excess energy. After a predetermined time, the sample is taken out of the beam of neutrons. The altered nuclei in the sample lose their excess energy by emitting nuclear radiation that is characteristic of each individual type of nucleus. The radiation is detected and allows identification of the nucleus that emitted it. The intensity of a particular characteristic radiation is directly related to the number of nuclei of that species in the sample. Usually, the radiation studied in neutron activation analysis is gamma radiation—that is, high-energy electromagnetic radiation emitted by nuclei without altering their chemical nature. The gamma rays emitted by the sample are directly related to the abundance of a particular chemical element in the sample.

Neutron activation analysis grew out of the systematic study of the interaction of neutrons and nuclei conducted by nuclear physicists to understand the structure of the nucleus, beginning with the work of Enrico Fermi in 1934. Interpretation of the patterns of gamma rays emitted by a sample following irradiation by neutrons requires several types of background information. First, not all types of nuclei react in the same way with a neutron beam. The cross section for the nuclear reaction, which measures the chance that a particular nuclear species will be produced, depends not only on the structure of the nucleus involved but also on the energy of the neutrons used for bombardment of the sample. The reaction cross section must be measured in a separate experiment as the energy of the neutrons is varied. Many reaction cross sections have been measured and are tabulated in the scientific literature.

Second, the gamma radiation from a particular nuclear species has characteristic energies which can be precisely measured using germanium-based detectors. The pattern of energies of gamma rays emitted by a particular type of nucleus identifies that nucleus just as the pattern of visible light emitted by an atom—its spectrum—identifies that atom. Tables of the gamma spectra of nuclei are an important input to neutron activation analysis. Such tables, along with the cross sections for nuclear reactions, are stored in computers and automatically recalled during analysis of the gamma spectra from neutron activation analysis.

HALF-LIVES OF RADIOACTIVE MATERIALS

Emission of nuclear radiation occurs gradually in time at a rate characterized by the half-life of the given decay—that is, the time for half the nuclei in a sample to emit their gamma radiation. Half-lives for nuclear species vary from picoseconds to millions of years. Nuclear species of interest for neutron activation analysis generally have half-lives ranging from seconds to days, as the species must live long enough for the sample to be transported to the detector and must decay quickly enough so that they can be detected in a reasonable amount of time. As the neutrons interact with the sample nuclei, producing new energetic nuclei, the energetic nuclei decay with their characteristic half-lives. Thus, the number of excited nuclei in a sample depends on the half-life of the nucleus as well as on the time it has spent in the neutron beam. Half-lives of nuclear radiation are measured in separate experiments and are tabulated. Analysts keep records of the exposure of the sample to neutrons.

The variations in the half-lives of excited nuclei can be used to identify nuclei present in the sample. Decays with short half-lives happen very rapidly so that the first gamma rays obtained from the sample are mostly those with half-lives less than about five minutes. If the sample then sits for half an hour, the short-lived nuclear species will have decayed, and the gamma rays from the sample will be those from nuclei with longer half-lives. Thus, in neutron activation analysis, gamma rays from the sample are measured at a series of carefully planned time intervals. Computers are used to calculate the effect of half-lives on the gamma spectra that have been recorded.

ISOTOPES

The number of excited nuclei produced during neutron irradiation (and thus the intensity of a particular gamma emission) depends on the concentration of a particular nuclear species in the sample. Because the chemical nature of an atom is not affected by the number of neutrons in its nucleus, most chemical elements are characterized by more than one type of nucleus or isotope—that is, nuclei with the same number of protons in them and thus belonging to the same element but with different numbers of neutrons. Different isotopes have different reaction cross sections and different gamma spectra. Thus, the analyst must know the relative amounts of each isotope of a given element to relate the intensity of a gamma emission to the abundance of a particular element in the sample. These data are well known and readily available.

NEUTRON FLUX

The number of gamma rays at a particular energy level also depends on the number of neutrons that are aimed at the sample. If the neutrons are produced by a radioactive source or an accelerator, they form a beam and are described as a particular number of particles per unit of area and time. If the sample is placed inside a nuclear reactor, the neutrons bombard it from all directions, and the irradiation is described in terms of a neutron flux, or the number of neutrons crossing a square centimeter of the sample each second. The energy distribution of the neutrons that strike the sample must also be recorded because reaction cross sections depend on energy. In an accelerator or a radioactive source, the neutrons produced usually have a well-defined single energy. In a reactor, they will have a distribution of energy levels that must be measured for an individual reactor and for a particular location in the core of that reactor.

COSTS AND BENEFITS OF THE ANALYTICAL METHOD

Neutron activation analysis is expensive because it requires a neutron source and specialized detectors and counting systems, all of which are run by a computer in modern laboratories. The technique is not sensitive to the chemical state of an atom but merely determines the number of atoms of a particular element to be found in the sample; therefore, it is not suitable for determining the chemical state in which elements are present. At the same time, neutron activation analysis is very fast compared to standard quantitative analysis, thus compensating for the expense when fast results are needed. For example, neutron activation analysis using a radioactive source to produce the neutrons has been conducted in the field to determine the copper and manganese content of ores. The technique is advantageous in that samples do not have to be transported to a laboratory for analysis, and results of the analysis can be used to guide drilling operations.

Neutron activation analysis cannot detect every chemical element in the sample, as not all chemical elements produce gamma rays with suitable half-lives or have large cross sections for nuclear reactions. In some cases, strong gamma radiation from abundant elements may mask the weaker signals from less abundant species. The technique for producing neutrons may strongly influence the elements that are detected. For example, the high-energy or fast neutrons produced in accelerators using tritium targets interact strongly with oxygen and silicon, while lower-energy neutrons characteristic of nuclear reactors interact very little with these elements. Thus, fast neutrons are characteristically used for rapid determinations of the silicon and oxygen content of minerals. Frequently, samples are subjected to analysis using more than one sort of neutron source to detect different elements. Rock samples will first be subjected to fast neutron analysis to determine their content of silicon and oxygen and then to analysis using a reactor to detect about twenty-five other elements. Finally, neutron activation analysis may be supplemented by chemical separation of elements for the detection of very rare elements.

Neutron activation analysis is ideally suited for the study of trace elements—that is, relatively rare elements present in samples in small quantities. Because neutrons easily penetrate geological materials, samples for neutron activation analysis require little preparation, and the technique does not destroy the sample, which can thus be saved for display or subjected to further analysis. Therefore, this technique is often applied to samples of archaeological importance, where samples are too precious to destroy in analysis. It offers the researcher the further advantage that many elements in the same data can be identified, and thus elements may be found whose presence in the sample was not initially expected. This scanning for many elements at once is an advantage in problems such as the search for pollutants in river water. For example, extensive studies of

environmental mercury in Sweden have been conducted using neutron activation analysis of minerals, coal, and plant and animal tissues.

The advent of computer-based systems has automated much of the tedious calculation needed to analyze data from neutron activation analysis. The technique is thus accessible to a much wider variety of researchers than was previously the case and promises to find increasing use as a probe of the elemental composition of samples of interest to earth scientists.

TYPES OF EARTH SCIENCE APPLICATIONS

Neutron activation analysis provides a powerful technique for simultaneously determining the amounts of many chemical elements in a geological sample without destroying the sample. Although it requires a source of neutrons and fairly complex instrumentation, it is much faster than conventional chemical analysis and can analyze a sample for several elements at the same time with little sample preparation. This technique is also uniquely sensitive to very small amounts of particular elements and can often detect minute amounts of such elements present in samples that would escape all but extremely detailed and time-consuming chemical analyses using atomic spectroscopic techniques designed to search for that element.

Applications of neutron activation analysis to earth science fall into two broad categories. The first consists of cases in which researchers take advantage of the speed of neutron activation analysis to obtain immediate results on the elemental compositions of their samples. Such work is often done in the field, using a radioactive source to produce the neutrons, and the results of the analysis guide field operations. Similarly, neutron activation analysis may be used to screen a very large number of samples rapidly on a production basis.

The second category of applications of neutron activation analysis utilizes the ability of the technique to determine rapidly very small concentrations of certain chemical elements. One example of this application has been the systematic study of trace elements in rocks of various ages. In the energy industry, neutron activation analysis has been applied to the study of trace elements in coals, thereby providing clues to the origin of particular coal beds. The quality of coal as a heat source varies widely from bed to bed. An understanding of why this variation occurs might lead to new methods of treating coals before burning them in order to reduce pollution. Finally, the ability of neutron activation analysis to scan large numbers of samples for minute quantities of chemical elements has been put to use in the study of sources of pollution, particularly by metals, in the environment. Large numbers of samples of river water or runoff near landfills can be checked for the presence of a wide variety of metals quickly and efficiently using neutron activation analysis.

A particular application of neutron activation analysis is autoradiography. In this case, the sample is irradiated and then placed in contact with a piece of film. The film is developed and records concentrations of radioactivity, showing how particular chemical elements—for example, uranium and thorium—are distributed within the sample.

STUDY OF LUNAR ROCKS

Probably the most famous example of the use of neutron activation analysis in earth science is the study of the lunar rocks brought back to Earth by the Apollo astronauts. Scientists wished to know the chemical composition of these rocks to obtain clues as to their history and to learn whether the moon formed from the same original material as did the earth. At the same time, only relatively small samples were available, as the lunar rocks had to be carried back from the moon in circumstances where the amount of weight was critical; in addition, scientists wanted to save the lunar rocks for future analysis and for public display. Neutron activation analysis does not damage the sample it studies. Even when combined with chemical separation techniques to aid analysis for very scarce elements, the samples needed are very small—on the order of milligrams—as opposed to the gram-sized samples needed for standard chemical analysis. Thus, the lunar rock samples could be subjected to neutron activation analysis to determine their basic elemental composition and still be left intact for display or analysis by other methods. Some of the surprising results from neutron activation analysis of the lunar rock samples include the fact that lunar rocks and soils are very low in oxygen compared to their terrestrial counterparts.

One of the problems in studying the lunar samples was to determine their content of rare-earth elements such as europium, neodymium, or gadolinium. These elements are chemically very similar and thus difficult to separate by quantitative chemical analysis. In contrast, their nuclear structures are very different; therefore, neutron activation analysis is an ideal tool for distinguishing among them. Results of the analysis showed that overall abundances of the rare-earth elements in lunar rocks were fifty to one hundred times greater than is standard for chondritic meteorites, which are meteorites believed to represent the primordial composition of the material from which the solar system formed. At the same time, lunar rocks were depleted in the element europium compared to chondritic meteorites and terrestrial rocks. Explanation of these strange patterns of elemental abundances uncovered by neutron activation analysis supports the theory that the moon formed from a disk of material spun off from the very early Earth by a grazing collision with a very large planetesimal.

Ruth H. Howes

FURTHER READING

Choppin, Gregory R. *Radiochemistry and Nuclear Chemistry*. 3rd ed. Boston: Butterworth-Heinemann, 2001. This widely used college text introduces the reader to the basics of nuclear chemistry and radiochemistry. It explores the theories surrounding those fields and their applications. Well illustrated with clear diagrams and figures, this is a good introduction for someone without a strong background in chemistry.

Fite, L. E., et al. "Nuclear Activation Analysis." In *Modern Methods of Geochemical Analysis*, edited by Richard E. Wainerdi and Ernst A. Uken. New York: Plenum Press, 1971. A comparatively brief summary of the technique of neutron activation analysis, designed for geologists who are not familiar with nuclear physics. While not terribly difficult to read, the chapter stresses the equipment required for use of neutron activation analysis and provides examples from geology.

Keller, C. *Radiochemistry*. New York: John Wiley & Sons, 1988. A general text on the use of radionuclides, both naturally occurring and artificially formed, as in the case of neutron activation analysis. Although the section on neutron activation analysis is brief, the text provides a thorough treatment of the background material needed to understand this relatively complex analytical technique.

Kruger, Paul. *Principles of Activation Analysis.* New York: Wiley-Interscience, 1971. This volume concentrates on the experimental details and on the instrumentation needed to carry out neutron activation analysis. Written as a textbook for the scientist planning to use neutron activation analysis, who is not a specialist in the field. The details of the technique are thoroughly discussed.

Lenihan, J. M. A., S. J. Thomson, and V. P. Guinn. *Advances in Activation Analysis.* Vol. 2. New York: Academic Press, 1972. Although older, this volume provides an excellent series of examples of the use of neutron activation analysis to study a variety of problems not only in the earth sciences but also in the arts and in archaeology. It discusses some of the varied results that can be obtained by using a variety of neutron sources.

Lieser, Karl Heinrich. *Nuclear and Radiochemistry: Fundamentals and Applications.* 2d ed. New York: Wiley-VCH, 2001. Lieser's book gives the reader a basic understanding of the practices and principles involved in the fields of radiochemistry and nuclear chemistry. Although there is quite a bit of chemistry involved in the author's explanations, someone without a chemistry background will still find the book useful. Illustrations, charts, and diagrams help clarify difficult concepts and theories. Bibliography and index.

Medhat, M. E., and M. Fayez-Hassan. "Elemental Analysis of Cement Used for Radiation Shielding by Instrumental Neutron Activation Analysis." *Nuclear Engineering & Design* 241 (2011): 2138-2142. This article provides a recent example of the use of neutron activation analysis. Written in a manner easy to follow with some chemistry background.

Parry, Susan J. *Activation Spectrometry in Chemical Analysis.* New York: Wiley, 1991. Intended for someone with a background in chemistry and related fields, Parry's book contains in-depth descriptions of neutron activation and activation spectrometry. Sections deal with nuclear chemistry and radiochemistry. Bibliographical references and index.

Rakovic, Miloslav. *Activation Analysis.* London: Iliffe Books, 1970. A good introduction to the field, including details of chemical preparations and sample-handling skills. Organized by the chemical element that is being studied rather than by the technique being used. A good source for a person interested in studying a particular chemical element.

Vasilopoulou, T., et al. "Large Sample Neutron Activation Analysis of a Reference Inhomogeneous Sample." *Journal of Radioanalytical & Nuclear Chemistry* 289 (2011): 731-737. An important study in the field of neutron activation analysis. This article discusses the use of large sample sizes in the analysis, and the validity of resulting quantities. Also addresses future needs in the field.

Win, David Tin. "Neutron Activation Analysis." *Assumption University Journal of Technology* 8 (2004): 8-14. Provides a comprehensive overview of neutron activation analysis principles and history, followed by the advantages and limitations of the analysis. Also discusses applications of this analysis to landmine detection and radioisotope power generation.

See also: Electron Microprobes; Electron Microscopy; Elemental Distribution; Environmental Chemistry; Experimental Petrology; Geochemical Cycle; Geologic and Topographic Maps; Infrared Spectra; Mass Spectrometry; Petrographic Microscopes; Radioactive Decay; Radiocarbon Dating; X-ray Fluorescence; X-ray Powder Diffraction.

NOTABLE EARTHQUAKES

When great earthquakes occur, most of the casualties are caused by building collapse, fire, landslides, and tsunamis. Modern concepts of plate tectonics can account for the location of most great earthquakes, and sound planning can do much to minimize damages.

PRINCIPAL TERMS

- **aftershocks:** earthquakes that follow a major earthquake and have nearly the same focus; they are caused by residual stresses not released by the main shock
- **epicenter:** the point on the surface of the earth directly above the focus of an earthquake
- **fault:** a fracture within the earth along which opposing masses of rock slip to produce earthquakes
- **focus:** the area or point within the earth where an earthquake originates
- **intensity:** the strength of shaking that an earthquake causes at a given point; intensity is generally strongest near the epicenter of an earthquake
- **magnitude:** a measure of ground motion and energy release in an earthquake; an increase of one magnitude means roughly a thirtyfold increase in energy release
- **plate tectonics:** the crust of the earth consists of a number of moving plates; most earthquakes occur at plate boundaries where moving plates are in contact
- **seismograph:** an instrument for recording motion of the ground in an earthquake; most seismographs are pendulums that remain static as the ground moves
- **tsunami:** a large sea wave caused by coastal earthquakes, probably generated by submarine landslides; not all coastal earthquakes result in tsunamis

MEASURING STRENGTH OF EARTHQUAKES

Two measures are used for describing the strength of earthquakes: intensity and magnitude. Intensity, generally rated on the twelve-point modified Mercalli scale, is the degree of shaking noted at a given point. Intensity depends on the distance to the focus, the local geology, and the observer. Customarily expressed in Roman numerals, intensity ranges from I (felt by only a few observers) to XII (total destruction; ground motion powerful enough to throw objects into the air).

Magnitude, which is usually expressed in terms of the Richter scale, is a measure of ground motion as measured on seismographs and is related to the total energy of an earthquake. The scale is defined so that an increase of one magnitude corresponds to a tenfold increase in ground oscillation, or approximately a thirtyfold increase in energy release. Earthquakes of magnitude 3 are often unnoticed, those of magnitude 5 produce widespread minor damage, those of magnitude 7 are considered major, and those above magnitude 8 are considered great. The greatest magnitude ever recorded was the Chilean earthquake of 1960, which was measured at 9.5. It is widely believed that there is an upper limit to the Richter scale, but this is a misconception. There is no upper limit to the scale. It appears, however, that the earth's crust cannot store enough elastic energy to generate earthquakes of magnitudes greater than 10.

OCCURRENCE OF STRONG EARTHQUAKES

Clear relationships exist between plate tectonics and the occurrence of great earthquakes. The magnitude of an earthquake generally corresponds to the area of fault surface where slippage occurs. The larger the slippage area, the greater the energy required to overcome friction. Because the ocean basins have a thin crust (about 5 kilometers thick), great earthquakes are rare in the ocean basins. Most great earthquakes are associated with continental crust, which has an average thickness of 40 kilometers.

The greatest earthquakes (magnitude 8.5 and higher) occur where plates converge, such as in Japan or on the west coast of South America. In these regions, one plate dips beneath the other at a shallow angle, resulting in a very large area of fault surface. Rifts, where continental crust is pulled apart, and transcurrent faults, where one block of continental crust slides horizontally past another, also have produced earthquakes above magnitude 8. The most famous transcurrent fault is the San Andreas fault of California. A few great earthquakes have also occurred well within plates. Some are reasonably well

Ships sit in the middle of a street in downtown Banda Aceh on January 8, 2005, nearly two weeks after an earthquake and tsunami devastated the region. The Indonesian death toll from the December 26, 2004, earthquake and tsunami disaster has climbed nearly 6,000 to 107,039, the social affairs ministry's relief coordination center said. (AFP/Getty Images)

understood. Most of the earthquakes of China and central Asia are a response to the collision of India with Asia. Others, such as the Charleston and New Madrid earthquakes in the United States, are poorly understood.

EARLY EARTHQUAKES

Little is known of great earthquakes of the distant past. The casualty figures reported for ancient earthquakes are unreliable. Nevertheless, it can usually be assumed that earthquakes that devastated large areas also inflicted great casualties and destruction. Even for modern earthquakes, damage is often so great that casualty figures can only be estimates. Different sources frequently list casualty figures differing by many thousands.

Perhaps the earliest great earthquake to have a major historical impact struck the Minoan civilization on Crete about 1450 B.C.E. During this earthquake, it appears that all the major complexes on Crete were destroyed. This earthquake possibly was related to the catastrophic eruption of Thera (Santorini), a volcano in the Aegean Sea approximately 120 kilometers north of Crete. The exact order of events is still uncertain.

One of the first earthquakes to be described by historical accounts destroyed the Greek city of Sparta in 464 B.C.E., killing a reported 20,000 people. In A.D. 62, the city of Pompeii in Italy was severely damaged by an earthquake. Pompeii is famous for being buried by an eruption of Vesuvius seventeen years later. An earthquake on July 21, 365, devastated Alexandria, Egypt, killing 50,000 and destroying the Pharos, or lighthouse—one of the Seven Wonders of the World.

The greatest killer earthquake in history struck Shaanxi in north-central China on January 24, 1556. In this region of China, many traditional dwellings were dug into hillsides of loess (wind-deposited silt).

Collapse of these cave homes and landslides triggered by the earthquake reportedly killed 830,000. The area of devastation was so large that the death toll was certainly in the hundreds of thousands.

MODERN EARTHQUAKES

The catalog of well-known modern earthquakes begins with the Lisbon earthquake, also known as the All Saints Day earthquake, of November 1, 1755. The city of Lisbon, Portugal, was demolished by three shocks between 9:30 and 10:00 A.M., with additional major aftershocks at 11:00 A.M. and 1:00 P.M. Approximately 70,000 people were killed by building collapse, fire, and a tsunami. Considerable damage also occurred in nearby Morocco. The Lisbon earthquake is sometimes listed as one of the greatest earthquakes of all time, producing widespread destruction as far away as Algeria and having been felt as far away as the West Indies. In reality, the earthquakes in Algeria and the West Indies were separate events unrelated to the Lisbon earthquake. The earthquake produced effects far beyond the region where the shock was actually felt. Lake oscillations (seiches) were noted all over Western Europe, clocks stopped, and church bells rang. Many of these phenomena were noted and recorded carefully and these observations showed that a wavelike disturbance had traveled outward from Lisbon. The Lisbon earthquake was thus the first earthquake to be studied systematically by modern scientific methods.

One of the strongest earthquakes ever to strike New England occurred off eastern Massachusetts on November 18, 1755, just days after the All Saints Day earthquake. This event occurred just before news of the Lisbon shock reached America. The first earthquakes to be recorded in the United States were those that struck the New Madrid, Missouri, area on December 16, 1811, January 23, 1812, and February 7, 1812. These events are among the few recorded earthquakes of intensity XII, and they took place in a region not generally considered earthquake-prone. Surface effects in the epicentral area were profound. The Mississippi River was churned into turmoil, and large tracts of unstable ground were affected by surface cracks and subsidence. The shocks were felt as far away as New Orleans and caused church bells to ring in Boston. Because of the sparse population in the New Madrid area at that time, only one death was reported. The New Madrid earthquakes, despite

the vast area over which they were felt, were not of extremely large magnitude: They probably had a magnitude between 7.5 and 8. In the central United States, flat-lying and uniform rock layers transmit seismic waves with high efficiency, so that an earthquake at New Madrid is felt over a much larger area than an equally powerful earthquake in a geologically complex region such as California.

On January 9, 1857, a major earthquake (probably magnitude 8) struck Southern California. At least 60 kilometers of the San Andreas fault ruptured near Fort Tejon, north of Los Angeles. A strong earthquake (possibly magnitude 7) struck Charleston, South Carolina, on August 31, 1886. The earthquake was felt over most of the east coast of the United States and killed approximately 110 people. This earthquake was the first in the United States to receive wide scientific attention.

EARLY TWENTIETH CENTURY EARTHQUAKES

For many Americans, the word "earthquake" is synonymous with the San Francisco earthquake of April 18, 1906. The earthquake, with a magnitude of 8.3, was officially reported to have killed about 700, but later estimates have placed the death toll as high as 2,500. In October 1989, an earthquake of magnitude 7.1, known as the Loma Prieta earthquake, would again leave the city with fatalities. The 1906 earthquake triggered fires that could not be fought because of ruptured water mains. As a result, a large area of the city was burned. From a scientific standpoint, the earthquake is important because it revealed the extent of the San Andreas fault. North of San Francisco, fence lines and roads were offset as much as 6 meters by the fault. The fault ruptured for at least 280 kilometers, possibly as much as 400 kilometers.

On September 1, 1923, an earthquake known as the Kwanto earthquake, of magnitude 8.3, destroyed much of Tokyo and Yokohama, Japan. This earthquake is notable for the devastating fire that followed it. The earthquake struck when thousands of open cooking fires were in use all over Tokyo. Traditional Japanese construction, which relies extensively on wood and bamboo, is very resistant to collapse in earthquakes but is also very combustible. The earthquake ignited thousands of fires that coalesced into a firestorm—a self-sustaining whirlwind in which updrafts above the fire draw air in from the outside and keep the fire supplied with oxygen. About 140,000 people died. Forty thousand of those who died had taken refuge in an open square and suffocated from lack of air.

LATER TWENTIETH CENTURY EARTHQUAKES

A little-known earthquake (magnitude 7.9) in southeastern Alaska on July 9, 1958, is remarkable for creating the highest wave ever recorded. The earthquake triggered an avalanche into one arm of Lituya Bay, sending the water 530 meters over a ridge on the other side of the bay. Anchorage, Alaska, was damaged by a magnitude 8.3 earthquake on Good Friday, March 27, 1964. Much of the damage to Anchorage was the result of liquefaction of an unstable layer of clay a few meters below the surface. When the seismic shaking liquefied the clay, the ground above broke up, tilted, or collapsed. A tsunami, reaching up to 30 meters in height, devastated the nearby coast. Of the 131 people killed in Alaska, 122 were killed by the tsunami. The tsunami swept down the coast of North America, causing little damage in most places. At Crescent City, California, however, the bottom topography of the harbor focused the wave, which swept into the center of town, killing twelve people. Surveys of the epicentral region showed that almost 300,000 square kilometers of crust had been measurably deformed. Some points on the coast moved seaward by 20 meters; shorelines were uplifted by 15 meters in places. These motions are among the greatest ever documented for any earthquake.

A magnitude 7.7 earthquake in Peru on May 31, 1970, killed about 70,000 people, including the victims of one of the worst landslide disasters in history. The earthquake triggered a rock and ice avalanche from the summit of 6,768-meter Huascaran, the highest peak in Peru. A portion of the landslide rode over a 250-meter ridge and buried the town of Yungay, killing approximately 20,000 people. This earthquake was the worst earthquake disaster in the Southern Hemisphere.

The greatest earthquake disaster of the twentieth century in terms of loss of life—and the second greatest in history—took place on July 28, 1976, when a magnitude 8.2 earthquake struck Tangshan in northeastern China, an urban area with about 10 million people. According to the most widely accepted estimate, 600,000 people were killed.

The worst earthquake to strike North America

killed 20,000 people in Mexico City on September 19, 1985. The epicenter of the magnitude 8 earthquake was actually on the Pacific coast, some 400 kilometers from Mexico City, yet damage on the coast was light. Buildings on the coast were generally modern, well built, and with foundations on bedrock. Mexico City, in contrast, is built on an ancient lake bed. Unconsolidated sediment shakes badly in earthquakes, accounting for the great damage in Mexico City. Many modern steel-frame buildings were undamaged, while poor-grade masonry suffered badly.

On December 7, 1988, an earthquake measuring magnitude 8 killed an estimated 80,000 people in Soviet Armenia. This event was notable for its political impact, because it happened at a time when the Soviet Union appeared to be moving toward greater political openness. For the first time in many years, the Soviet Union accepted foreign relief efforts after a natural disaster and permitted foreign news coverage at a disaster scene.

On October 17, 1989, a magnitude 7.1 earthquake centered 20 miles from downtown San Francisco at Loma Prieta caused greatest damage in the San Francisco Marina District, which is built upon unstable, water-saturated landfill. The earthquake caused widespread damage to the road system, including collapse of the I-280 Skyway, many landslides along the coastal highway, and at least sixty-three deaths. On January 17, 1994, a magnitude 6.7 earthquake on a previously unknown fault rocked Northridge, California, in the San Fernando Valley for 40 seconds. Damage was estimated at $15-30 billion with 63 dead, thousands injured, nine freeways destroyed, and 250 ruptured gas lines. Power was cut to 3.1 million people, and 40,000 were left without water. A magnitude 7.2 earthquake rocked Kobe, Japan, in January 1995. Although it lasted only 20 seconds, it caused more than 5,000 fatalities, 25,000 injuries, and at least $30 billion in damage.

On August 17, 1999, a magnitude 7.8 earthquake near Izmit, Turkey, 55 miles east of Istanbul lasted for 45 seconds, flattened 60,000 buildings, caused up to $6.5 billion in direct property loss, and killed more than 30,000 people. Nearly 300 aftershocks rocked the region in the next 48 hours.

TWENTY-FIRST CENTURY EARTHQUAKES

India experienced two destructive earth quakes in the early twenty-first century: one in Gujarat on January 26, 2001 (measuring 7.9 and resulting in 20,035 deaths), and one in Bam on December 26, 2003 (measuring 6.6 and resulting in 31,000 deaths). The earthquake that resulted in the Indian Ocean tsunami on December 26, 2004, measured 9.2 on the Richter scale and resulted in 230,000 deaths. Other twenty-first century earthquakes have occurred in Kashmir, Pakistan (October 8, 2005, M7.6, 80,000 deaths); Sichuan, China (May 12, 2009, M7.9, 69,197 deaths); and Port-Au-Prince, Haiti (Jan. 12, 2010, M7.0, 316,000 deaths).

STRONG-MOTION STUDIES

Great earthquakes present special problems and opportunities for geologists. Because of their great energy release, earthquakes are detected clearly by instruments all over the planet; these records frequently reveal details of earth's structure that cannot be detected on the records of smaller earthquakes. The infrequency and unpredictability of great earthquakes, however, mean that instruments and observers are rarely close by when the event occurs, and instruments that are close by are often destroyed.

Ground motion during great earthquakes can be measured by special seismographs called strong-motion seismographs. Strong-motion studies require that instruments be set up in locations that might experience major earthquakes. These instruments are left in place, possibly for years. After remaining dormant for a long time, the instruments must work properly when the earthquake occurs. The need to place and periodically tend instruments that may never record an event makes strong-motion studies expensive.

It is possible to simulate the effects of earthquakes on buildings. During the planning stage, models of the proposed building can be tested on a vibrating table or through computer modeling. Existing buildings can be shaken artificially. The apparatus for testing buildings consists of a set of large, rotating, off-center weights. Sensors at critical points in the building can detect motion without subjecting the building to destructive vibrations. Corrective measures might include reinforcing weak portions of the structure or redesigning connecting wings so that they can vibrate independently.

LONG-TERM SEISMIC STUDIES

Short-term earthquake prediction on the lines of severe weather warning is probably not achievable in the near future. Geologists are pursuing a variety of studies aimed at assessing the long-term likelihood of great earthquakes. One obvious and low-cost approach is simply to compile all historical records of earthquakes. China and the Middle East, areas with the longest and best-written records, show variations in intensity and location of earthquakes on a time scale of centuries. The short historical record of the United States is insufficient for long-term seismic studies.

One way to extend the record of great earthquakes is to look for geological changes created by ancient events. In Japan, uplifted shorelines have been identified with specific historical earthquakes. At Pallett Creek, north of Los Angeles, trenches across the San Andreas fault have revealed evidence of earthquakes over the last 2,000 years. Each earthquake ruptured sediment layers below the then-existing ground surface. Radiocarbon dating (using radioactive carbon in the sediment as a geologic clock) establishes the age of each fault break. The average interval of great earthquakes in this area is approximately 140 years, but actual intervals have ranged from 75 to 300 years.

EARTHQUAKE HAZARDS

Most of the casualties from great earthquakes result from a few basic causes. Building collapse is a major cause of loss of life. Wood-frame buildings, which are flexible, and steel-frame buildings, which are very strong, are the safest kinds of buildings during earthquakes. Non-reinforced masonry and adobe (mud brick) are the most dangerous. Unfortunately, these construction styles are very common in underdeveloped nations. Fire is another major threat in urban areas. Earthquakes overturn stoves and furnaces, rupture gas lines, and create electrical short circuits. At the same time, ruptured water mains and streets blocked with rubble impede fire-fighting efforts. Earthquake-induced landslides are a hazard in mountainous areas and have caused tremendous loss of life.

Tsunamis are a threat in coastal areas. Believed to be generated by submarine landslides, tsunamis are waves of low height and long length that travel at up to 600 kilometers per hour. Because of their breadth and low height, they are entirely unnoticed by ships at sea but can cause great damage when they reach shore, sometimes thousands of kilometers away. Whether a tsunami causes damage depends greatly on its direction of travel, on local tide and weather conditions, and particularly on the bottom topography near shore. Tsunami warnings are routinely issued after large earthquakes.

A tsunami occurred following the Tōhoku earthquake (also known as the Great East Japan earthquake) in the Pacific Ocean on March 11, 2011. The wave caused widespread destruction on the northeast coast of Japan. More than 20,000 people were killed.

EARTHQUAKE MYTHS

There are a few misconceptions about great earthquakes. After a newsworthy earthquake, people often wonder if earthquakes are becoming unusually frequent. In fact, the reverse was true in the twentieth century. There were about two earthquakes per year of magnitude 8 on the average, in contrast to an annual average of eight during the years 1896-1907. One apparent pattern is real, however. Destructive earthquakes are becoming more common. The reason is demographic rather than geologic. Many seismically active regions are in underdeveloped nations where populations, especially in cities, are growing explosively and where construction standards are often poor. The population at risk from earthquakes is steadily increasing.

There are a few geologic misconceptions about earthquakes. Earthquakes frequently cause ground subsidence in areas underlain by poorly consolidated materials, often causing cracks to open on the surface, but stories of fissures opening and engulfing people, buildings, or even entire villages are unfounded. Most of these stories are probably inspired by landslides. Earthquakes and volcanoes tend to occur in the same geologic settings, and there are some recorded cases of major earthquakes associated with the eruption of a nearby volcano. As a general rule, though, earthquakes do not trigger volcanic activity. Also, the earthquakes that accompany volcanic eruptions are generally not very large.

Steven I. Dutch

FURTHER READING

Anderson, D. L. "The San Andreas Fault." *Scientific American* 224 (February 1971): 52. A description of the most famous North American fault, particularly good for its block diagrams showing the complex southern portion of the fault system. *Scientific American* is written for nonspecialists at a college reading level.

Boore, D. M. "The Motion of the Ground in Earthquakes." *Scientific American* 237 (December 1977): 68. A summary of how earthquakes occur, the types of motions they cause, and their effects on structures.

Clarke, Thurston. *California Fault: Searching for the Spirit of State Along the San Andreas.* New York: Ballantine Books, 1996. Clarke traveled the length of the San Andreas fault collecting first-hand accounts from earthquake survivors and predictors. Along with the entertaining stories, Clarke provides historical and scientific information about the fault.

Coffman, Jerry L., Carl A. Von Hake, and C. W. Stover. *Earthquake History of the United States.* U.S. Department of Commerce Publication 41-1. Washington, D.C.: National Oceanic and Atmospheric Administration and U.S. Geological Survey, 1982. The most detailed, general, historical reference on earthquakes in the United States. Contains lists of events by geographic area, descriptions of all widely felt earthquakes, maps of earthquake epicenters, and detailed references. Written at a nontechnical level. A must for any student of earthquakes.

Collier, Michael. *A Land in Motion: California's San Andreas Fault.* San Francisco: Golden Gates National Parks Association, 1999. Filled with beautiful color photographs that accompany text intended for the nonscientist, *A Land in Motion* gives the reader excellent insight into earthquakes and their aftermaths. There are also many diagrams and graphs that explain subduction, faults, and orogeny.

Donlon, Rachael A., ed. *Haiti: Earthquake and Response.* Nova Science Publishers, 2011. This book recounts the events of the 2010 earthquake in Port-au-Prince. Not strongly technical, the editor compiled reports on the response, politics, and reconstruction which provide context for the event. One chapter is devoted to the science behind the earthquake.

Fradkin, Philip L. *Magnitude 8: Earthquakes and Life Along the San Andreas Fault*. Berkeley: University of California Press, 1999. Written for the layperson, this book can sometimes read over-dramatic or unscientific. However, *Magnitude 8* traces the seismic history, mythology, and literature associated with the San Andreas fault.

Molnar, P., and P. Tapponier. "The Collision Between India and Eurasia." *Scientific American* 236 (April 1977): 30. The collision between India and Eurasia causes faulting and great earthquakes over all of China and central Asia. Simple mechanical models duplicate the behavior of the crust remarkably well.

Nash, J. R. *Darkest Hours*. Chicago: Nelson-Hall, 1976. A nontechnical encyclopedia of historical disasters. Descriptions of events are generally accurate, but some errors in geological terminology were noted. Has extensive reference lists for each type of disaster, mostly popular books and periodicals. Individual articles lack references, and specific events can be hard to find. For example, the article on the great 1923 Tokyo earthquake is titled "Japan." Due to its publication date, this book is void of recent major disasters.

Prothero, Donald R. *Catastrophes!: Earthquakes, Tsunamis, Tornadoes, and Other Earth-Shattering Disasters*. Baltimore: Johns Hopkins University Press, 2011. This text provides a detailed and clear explanation of the many natural and anthropogenic disasters facing our planet. Each chapter is devoted to a different catastrophe, including earthquakes, volcanoes, hurricanes, ice ages, and current climate changes.

Reasenberg, Paul A., et al., eds. *The Loma Prieta, California, Earthquake of October 17, 1989: Aftershocks and Postseismic Effects*. Washington, D.C.: Government Publications Office, 1997. A detailed account of the 1989 Loma Prieta earthquake in the San Francisco Bay Area.

Richter, Charles F. *Elementary Seismology*. San Francisco: W. H. Freeman, 1958. The author of this classic 768-page text, who was a seismologist for many years at the California Institute of Technology, developed the Richter scale for measuring the intensity of earthquakes. Judging from his book, Dr. Richter must have been an excellent teacher. Even though this source is outdated, its lucid explanations of basic principles make it a worthwhile reference. Contains excellent and detailed chapters on the complexities of earthquake locating, along with examples, charts, diagrams, and travel-time curves. Some sections using differential equations would be for upper-level college students, but most of the book, including the parts on earthquake locating, would be quite readable to any advanced high school student.

Sutton, Gerard K., and Joseph A. Cassalli, eds. *Catastrophe in Japan: The Earthquake and Tsunami of 2011*. Nova Science Publishers, 2011. Editors compiled a number of reports on the effects of the earthquake and resulting tsunami, including the impact on agriculture and economics. This book focuses on the nuclear crisis following the earthquake and tsunami. One chapter describes the events within the nuclear power plant resulting from the natural disaster.

Wesson, R. L., and R. E. Wallace. "Predicting the Next Great Earthquake in California." *Scientific American* 252 (February 1985): 35. A summary of the major active faults in California, their history, and an assessment of the likelihood of activity in the near future. The most likely location for the next great earthquake is the southern San Andreas fault or one of its branches in the Los Angeles basin.

Woods, Mary C., et al., eds. *The Northridge, California, Earthquake of January 17, 1994*. Sacramento: California Department of Conservation, Division of Mines and Geology, 1995. A look at the 1994 Northridge earthquake and its effects on the San Fernando Valley and Los Angeles.

See also: Deep-Focus Earthquakes; Earthquake Distribution; Earthquake Engineering; Earthquake Hazards; Earthquake Locating; Earthquake Magnitudes and Intensities; Earthquake Prediction; Earthquakes; Elastic Waves; Faults: Normal; Faults: Strike-Slip; Faults: Thrust; Faults: Transform; Plate Motions; Plate Tectonics; San Andreas Fault; Seismometers; Slow Earthquakes; Soil Liquefaction; Subduction and Orogeny; Tsunamis and Earthquakes.

NUCLEOSYNTHESIS

Nucleosynthesis is the process by which the elements are formed in the interiors of stars during the course of their normal evolution. Hydrogen and helium are thought for the most part to have been generated at the origin of the universe itself (nucleogenesis), while all other heavier elements are synthesized via nuclear reactions in stellar cores. The heaviest elements are created during the death throes of massive stars.

PRINCIPAL TERMS

- **big bang theory:** the theory that the universe was created via an initial explosion that resulted in the formation of hydrogen and helium
- **charged-particle reaction:** a nuclear reaction involving the addition of a charged particle—proton or electron—to a nucleus
- **deuterium:** an atom built of one proton and one neutron; an essential stepping-stone in the proton-proton cycle in solar-type stars
- **isotope:** an atom with the same number of protons as another but differing in the number of neutrons and the total weight
- **neutron reaction:** a nuclear reaction in which a neutron is added to increase the atomic mass of the nucleus, forming an isotope
- **nucleons:** positively charged protons and neutral neutrons; large particles that occupy the atomic nucleus
- **supernova:** a massive star that explodes after available energy in the interior is used up and the star collapses

ELEMENTAL SYNTHESIS

Two of the most fundamental questions of modern astrophysics have to do with the origin and composition of the universe's primordial matter: when it came into existence and how it relates to the Einsteinian space-time structure of the present universe. With developments in physics, the problems seem to be divided into two parts: the origin of the simplest elements, hydrogen and helium, during the initial formation of the present universe and the subsequent nucleosynthesis of the other elements in the pressure cookers known as stars.

To understand elemental synthesis, scientists must rely on experimental observations interpreted in the light of current theory. Such data principally have to do with abundances of nuclear species now and in the past. This data set is provided from composition studies of the earth, meteorites, and other planets, and from stellar spectra. The distribution of hydrogen and heavier elements in stars throughout the galaxy, particularly in what are referred to as population I and II stars (younger and older, respectively), indicates how the chemical composition of the Milky Way galaxy has changed over time. From these studies, most theorists conclude that the galaxy has synthesized 99 percent of its own heavy elements and thus that nucleosynthesis occurs during the natural evolution of stars. In the light of variations observed in stars of diverse ages, scientists have formulated theories regarding the formation of elements within stellar structures. A dramatic piece of evidence along that line, for example, was the discovery of technetium in the spectrum of a red giant, all of whose isotopes, being radioactive, are short-lived, indicating that the star in which it is found must be currently producing the element. The study of naturally occurring radioactive isotopes, long- and short-lived, not only allows for measuring the time of galactic and stellar nucleosynthesis but also provides evidence that synthesis of elements heavier than hydrogen must be occurring continuously throughout the universe.

BIG BANG THEORY

Starting with the simplest element, hydrogen, which possesses one proton and one electron and is by far the predominant element in the universe, the study of nucleogenesis has progressed to a consideration of the origin of the universe. Beginning in 1946, Russian physicist George Gamow and others presented the theory that the entire structure started as a gigantic explosion of an extremely dense, hot "singularity," or infinitesimally small object. The explosion would have been so intense as to provide the propellant for all subsequent motion of the outwardly expanding matter and for the creation of the elements. Such a "big bang" concept has come to be accepted almost unanimously, with certain modifications. The discovery of an isotropic microwave background radiation, corresponding to a 3-kelvin temperature residual from the original fireball, lent support to the theory, along with the use of gigantic accelerators in the 1970's, which permits examination of the formation and interactions of the basic constituents and forces of nature.

Such physics has determined that elemental synthesis, via nuclear reactions, combining protons, electrons, and neutrons, could have occurred only when the temperature dropped to below 1 billion kelvins about three minutes after the explosion. Before that point, the energy of motion would have been too great either to form those particles or to let them cling together in electromagnetic interactions. That period of elemental synthesis probably lasted about one hour; eventually the temperature and pressure would have dropped too low to sustain any further reactions. Because of the instability of particles with atomic masses of 5 and 8, only traces of particle combinations beyond a mass of 4, including lithium and beryllium, would have been formed; thus the universe was probably composed of about 75 percent hydrogen and 25 percent helium. The formation of the helium nucleus, with a mass of 4, would have used up all the available neutrons. The reactions would have progressed in a certain order. First, neutrons and protons would combine to produce deuterium; deuterium and protons would then give helium 3; the collision of two helium 3 nuclei would produce a helium 4 nucleus and two protons, releasing energy in the process as gamma rays. This postulated process seems to be in excellent agreement with observational data and theoretical calculations.

STELLAR EVOLUTION

For roughly one million years, radiation was so intense that electrons could not combine with nuclei to form neutral atoms. Only after the radiation pressure became low enough could neutral atoms begin to form galaxies and stars. At that state, the dominating force in the universe became gravity, with the galaxies and stars forming as a result of gravitational contraction. Scientists' understanding of galactic formation remains sketchy, but stellar evolution—from birth in dust-cloud nurseries to death—is well understood through a combination of observational data, laboratory measurements of nuclear reactions and their rates, and copious amounts of theoretical work. As stated best by Indian American astrophysicist Subrahmanyan Chandrasekhar, the working hypothesis generally accepted by astrophysicists is that

the stars are the places where the transmutation of elements occurs, all the elements beyond hydrogen being synthesized there. All the energy available to a star throughout its life span, with minor exceptions, is derived from such transformations.

As the original gas and dust in a nebula collapse and contract, they heat up enormously, until the temperature in the core reaches some 10 million kelvins, at which point thermonuclear proton-proton reactions occur, to form deuterium and give off positrons and radiant energy. Further reactions occur, increasing the helium formed, decreasing hydrogen, and producing energy sufficient to halt the gravitational collapse of the star. For most stars, this stage probably occupies the greater part of their lifetime. The more massive the star is, however, the faster it will exhaust the hydrogen supply at the core and the shorter its time of stability will be. Stars such as the sun are in the range for forming helium. Some interesting side reactions occur also. Some 5 percent of the helium reacts to make beryllium, boron, and lithium. In an even rarer occurrence, proton capture produces the isotope boron 8. The latter is important because it is very sensitive to temperature and therefore acts as a good test of stellar theories; the reaction produces neutrinos, which earthbound astronomers can then study.

CARBON-NITROGEN-OXYGEN CYCLE

In older stars, formed as second, third, or later generations, some heavier elements are present. In these, the so-called carbon-nitrogen-oxygen cycle proposed by Hans Albrecht Bethe works, again turning four protons to helium. Because a higher temperature is necessary to overcome the electrostatic (Coulomb) repulsion barrier, this cycle takes place only in larger stars. In either case, when a significant amount of core hydrogen is used up, with helium ash left, the star will contract. Meanwhile, the hydrogen-containing outer area expands, causing the star to become a red giant; its central temperature rises to 100 million kelvins. At this stage, helium burns to form beryllium, forming one beryllium atom per billion helium atoms. Also produced are carbon, oxygen, and neon, the principal source of energy being the conversion of three heliums to carbon-12 plus gamma radiation. This burn, however, is short-lived, lasting only 10 to 100 million years, as compared to more than 5 billion years for the present sun. Any further synthesis requires much higher thermal energy input than can ever occur.

Beyond this stage, in larger stars, the processes become more complex. When helium is exhausted, contraction starts again. For objects such as the sun, this shrinkage will continue until it is halted by electron degeneracy (a mutual repulsion of tightly squeezed electrons) to form white dwarfs, small and intensely radiative bodies losing their heat into space, with no further nuclear energy available. Many become surrounded by a halo of expanding gases, the so-called planetary nebula; material from the star flows into space as a last gasp of the red giant stage. In larger stars the temperature continues to climb, to 70 million kelvins, eventually causing new sets of elements to form, including magnesium-24, sodium-23, neon-20, silicon-28, and sulfur-32. With further contraction, until the temperature reaches 1 billion kelvins, elements up to and including iron are created. Synthesis stops here, however, because of the energy required to bind more stable nuclei together.

NEUTRON-INDUCED REACTIONS

Additional synthesis does not involve charged-particle reactions but rather neutron-induced reactions, which tie up neutrons and produce energy. Such reactions are called s-processes because they proceed very slowly, taking from 100 to 100,000 years per capture step. This process accounts for the heavier isotopes on even atomic number elements and the distribution of nuclides up to bismuth. This reaction, in conjunction with a p-process involving successful proton reactions, can account for all the stable isotopes up to bismuth. For higher elements, however, a more rapid neutron-capture chain called the r-process is required; it takes place when there is an enormous neutron flux so that many captures can take place in milliseconds. Conditions perfect for such acts occur in supernovas, stars that explode with some of the greatest violence seen in the universe. Type I supernovas are from old, small stars, with masses of 1.2 to 1.5 times that of the sun; in such an explosion the entire star is destroyed, pushing the temperature to 10 billion kelvins. Type II supernovas occur in stars with masses greater than ten times that of the sun. Under contraction, the temperature in the nucleus of the star rises to 5 billion kelvins, and iron and nickel nuclei rapidly absorb neutrons, producing many of the heavier neutron-rich elements. The collapse, which

takes about one second, results in a core mass of neutrons, with explosive ejection of these heavy elements into the interstellar regions. Such explosions, which occur perhaps once in a hundred years in a galaxy, contribute all the material from which other stars, clouds, and planets such as Earth are formed.

DETERMINATION OF ISOTOPIC ABUNDANCES

Since the first theories of the processes of elemental origins were proposed, scientific understanding of nucleosynthesis has progressed greatly, thanks principally to an improved ability to determine abundances of elements particularly in stars and nebulas, and to better understanding of transformation conditions during synthesis. Nuclear physics data on reaction rates, particle formation, and interactions at diverse temperatures and energies, along with clearer notions of strong and weak force interactions, have contributed vital knowledge on both the universe's origin and the generation of elements in stellar bodies during and at the end of a star's life cycle.

Isotopic abundances can be determined from meteorites by the use of mass spectroscopy. In this experimental technique, particles are heated until they break apart into ionic forms; the bodies are then propelled, under the influence of electric and magnetic fields, through a vacuum chamber. The curved path followed depends on the mass and the charge on the elements. Collection at the end of the path allows detailed comparisons to be made, with particular attention to the anomalies that are critical to theories of nucleosynthesis.

ANALYSIS OF EXTRATERRESTRIAL OBJECTS

Spectral analysis has been the principal tool for studying extraterrestrial abundances. In such analysis, light from the observed object is passed through a prism or diffraction grating so that it is broken into all of its component colors; the resulting spectrum ranges from the blue to the red region of the visible section of the spectrum. Invariably, the background will be crossed by dark or bright lines, depending on whether it is an emission or absorption spectrum. These lines, identifiable in the physics laboratory, act as fingerprints, quickly showing such information as which elements are present and their abundances. Observation of material emitted by supernovas, for example, not only shows how heavy elements are enriched in space but also contributes greatly to theories of explosive charged-particle nucleosynthesis.

Similar analyses, using spectroscopes, telescopes, and various light-intensity-enhancing instruments such as charge-coupled devices (CCDs), have been done of other objects, including medium-mass stars with s-process element formation, nova explosions, and mass flows from solar-type stars. The latter can be studied best in the sun, by analyzing the composition of the solar wind with data returned by meteorological and scientific satellites. Detectors placed above the atmosphere can be equipped to detect charged particles such as protons or electrons. Experiments to view the universe in some region of the electromagnetic spectrum besides visible light, such as radio, gamma, infrared, or ultraviolet, also must be placed beyond the disturbing influence of the atmosphere.

ADVANCES IN THEORETICAL STUDY

Much of the information usable for the theoretical study of nucleosynthesis comes from two terrestrial sources. First, experimental studies using nuclear reactors and particle accelerator machines have provided comprehensive measurements of reaction rates and of the actions of the weak force in nature. Increasingly reliable determinations of critical cross sections, representing the space in which reaction occurs between two particles, and of the neutron-capture process, which is responsible for the bulk of nuclei more massive than iron, have become possible with highly refined accelerators and electrical detectors. Theoretical predictions and experimental results are thus more in harmony than ever before.

The second important advance has been in computer technology, which has made possible greatly increased numerical calculations of structures and the evolution of astrophysical objects. The advent of high-speed computers has allowed much greater predictive ability for the standard model of the big bang and regarding the formation of elements at various stages in the stars. Such detailed models, particularly of massive stars, in terms of hydrodynamic phases, have shown, for example, that supernovas are immensely important in the synthesis of heavy elements. Models for actions at extreme temperature and density conditions are very close to what is observed during the expansion, cooling, and mass ejection processes of the dying stars. Computer

technology has helped identify further problems through capture modeling, such as the sites necessary for r-process neutron-capture nucleosynthesis.

IMPORTANCE TO ASTRONOMY

Nucleogenesis and nucleosynthesis are two of the most important topics in astronomy and hence the earth sciences, promising to cast light on not only the evolution of stars but the ultimate origin of the universe as well. The understanding of universal origins has been advanced greatly by the advent of particle accelerators of remarkably high energies. These instruments provide physicists with clearer pictures of the elementary particle structures of the universe and of their interaction under the four forces controlling them. During the creative process of the big bang, these four forces—strong and weak nuclear, electromagnetic, and gravitational—were unified as one, separating only as initial conditions of temperature, pressure, and density changed. Under their actions, radiation and particles ultimately formed, with radiation finally dispersing enough that protons and helium nuclei could combine with electrons to form neutral atoms.

Although the modern understanding of nucleosynthesis is thought to be quite satisfactory, there are still problems unsolved. Certain elemental anomalies have not been explained by either experiments or theory. Predictions of energy fluxes and solar winds from other stars, particularly red giants, represent other unsolved problems. The answer to what causes a dust cloud to begin to contract to form a star is unknown; a widely accepted notion is that the contraction is prompted by the shock wave of a supernova. Problems remain with the big bang theory itself, so that modified theories, such as the "inflationary universe," have been proposed. The investigation of such problems of modern physics and astronomy has led to numerous insights, including the possibility that planets may be by-products of stellar formation; in such a scenario, the galaxy may be filled with planets and, possibly, life-forms. Further fine-tuning of reaction rates, mechanisms, and such experimental topics as element reactions may solve some of the deepest philosophical and scientific mysteries of modern science.

Arthur L. Alt

FURTHER READING

Abell, George O. *Exploration of the Universe.* 4th ed. New York: Holt, Rinehart and Winston, 1982. One of the best standard textbooks on astronomy available. Covers in detail the life history of stars, particularly those in which heavy elements are formed. Separate sections on white dwarfs, neutron stars, black holes, and supernovas. Glossary and references. Excellent diagrams and pictures. The late Abell's text has since been revised and is available as the seventh edition by David Morrison, Sidney Wolff, and Andrew Fraknoi, published in 1995 by Saunders College Publishing.

Arnett, David W. *Supernova and Nucleosynthesis: An Investigation of the History of Matter, from the Big Bang to the Present.* Princeton: Princeton University Press, 1996. This book sheds light on discoveries concerning the origin of hydrogen and other elements through the actions of stars. Deals extensively with reaction rates, primordial hydrogen, galactic chemical composition, massive stars, and supernovas. Extensive bibliography.

Bowers, Richard, and Terry Deeming. *Astrophysics.* Vol. 1, *Stars.* Boston: Jones and Bartlett, 1984. A detailed exposition on the characteristics of stars. The writer extends basic data into an understanding of how stars evolve differently depending on their original mass and brightness. Extensive sections on formation of elements during diverse stages of life cycles. Difficult reading unless one ignores the mathematics.

Brush, Stephen G. *Transmuted Past: The Age of the Earth and the Evolution of the Elements from Lyell to Patterson.* New York: Cambridge University Press, 1996. A look into modern planetary physics, this book traces the evolution of the elements and the solar system. Intended for the reader with some background knowledge in astronomy, this book is well illustrated and includes a bibliography and index.

Clayton, Donald D. *Principles of Stellar Evolution and Nucleosynthesis.* Chicago: University of Chicago Press, 1984. Arguably one of the best, most complete works on the evolution of stars, tracing the life histories of different-sized objects. Although outdated, it discusses element formation and unsolved problems in the field. It is a good starting point for the subject. Mathematics spreads throughout the work, but the advanced layperson should find it understandable. Additional references.

Hartmann, William, Pamela Lee, and Tom Miller. *Cycles of Fire: Stars, Galaxies, and the Wonder of Deep Space.* New York: Workman Publishing, 1988. A delightful work encompassing the history of stars from birth to death. Touches on a multitude of topics, including black holes, white dwarfs, binary stars, and the origin of the universe and planets. The text, though well written, is outdated. However, the attractive illustrations are by far the book's greatest asset. Glossary and some references.

Henbest, Nigel. *The Exploding Universe.* New York: Macmillan, 1979. An overview of the violent nature of the universe, this book deals with how atoms, elements, stars, planets, and other objects are formed and die. Detailed sections on the fundamental forces and particles of nature and their relationships to the formation of the elements. Glossary. Fairly easy reading.

McWilliam, Andrew, and Michael Rauch, eds. *Origin and Evolution of the Elements.* New York: Cambridge University Press, 2010. Covers stellar nucleosynthesis through chemical evolution of the cosmos. Written for the graduate student or professional astronomer.

Ozima, Minoru. *Geohistory: Global Evolution of the Earth.* New York: Springer-Verlag, 1987. A well-written book that deals with the origin of the earth, its waters, atmosphere, and rocks. Addresses the issue of the internal structure of the earth as it has changed over time. Abundant diagrams. Bibliography. For the more advanced layperson.

Pagel, B. E. J. *Nucleosynthesis and Chemical Evolution of Galaxies.* 2d ed. Cambridge: Cambridge University Press, 2009. An overview of the chemical composition of the universe, this book examines the origin and evolution of atoms, elements, planets, and stars, and discusses how all of these pieces work together to form galaxies. Illustrations and diagrams help explain difficult processes and concepts. Intended for the college reader.

Prantzos, N., E. Vangionu-Flam, and M. Cassae. *Origin and Evolution of the Elements.* Cambridge: Cambridge University Press, 1993. A look at how elements are formed and their processes of evolution, this book follows the life of the elements and documents each phase as it occurs. A good introduction to nucleosynthesis, it is well illustrated and includes a bibliography and index.

Rolfs, Claus E., and William S. Rodney. *Cauldrons in the Cosmos: Nuclear Astrophysics*. Chicago: University of Chicago Press, 1988. Reprinted in 2005. This book represents the then-current state of knowledge on the stars, pertaining to the synthesis of the elements. Starting with basic data, it tracks the lives of stars of varying masses, detailing the elements created at each step. Good pictures and diagrams; extensive bibliography. Although there is some mathematics, the text is quite clear.

Schramm, David, ed. *Supernovae*. Dordrecht, Netherlands: Reidel Press, 1977. A detailed work on supernovas, particularly the Crab Nebula. Traces how stars explode and how elements are formed in the final death throes. Presents a good overview of theories with a minimum of mathematics. Contains helpful illustrations and bibliography.

Shklovskii, Iosif S. *Stars: Their Birth, Life, and Death*. San Francisco: W. H. Freeman, 1978. A basic review of stars—how they work and why they exist and die. Extensive sections on origins of elements at various stages in stellar life, focusing on the production of heavy elements in star deaths. Well written. Numerous line drawings provide clarity. Bibliography.

Spitaleri, C., C. Rolfs, and R. C. Pizzone, eds *Fifth European Summer School on Experimental Nuclear Astrophysics*. American Institute of Physics, 2010. Contains multiple articles covering topics in nuclear physics, stellar nucleosynthesis, and big bang nucleosynthesis. Highly technical and dense; written for the advanced graduate student.

Taylor, R. J., ed. *Stellar Astrophysics*. Philadelphia: Institute of Physics, 1992. This multiauthored work examines advances in the study of the stars. Essays track the lives of stars of varying masses, detailing the elements created at each step and the processes that occur throughout the phases. Useful illustrations and diagrams.

See also: Elemental Distribution; Environmental Chemistry; Fluid Inclusions; Freshwater Chemistry; Geochemical Cycle; Geothermometry and Geobarometry; Oxygen, Hydrogen, and Carbon Ratios; Phase Changes; Phase Equilibria; Radiocarbon Dating; Water-Rock Interactions.

O

OCEAN DRILLING PROGRAM

The series of ocean-drilling efforts called the Ocean Drilling Program (ODP) has revolutionized understanding of the earth's structure, climate, and available minerals. It has also allowed researchers to collect data about the earth's cosmological history.

PRINCIPAL TERMS

- **abyssal plains:** flat areas that make up large areas of the ocean floor
- **basalt:** rock formed from recrystallization of molten rock; most of the rock in the mid-ocean ridges and underlying the abyssal plains is basaltic
- **hydrothermal vents:** areas on the ocean floor, typically along fault lines or in the vicinity of undersea volcanoes, where water that has percolated into the rock reemerges much hotter than the surrounding water; such heated water carries various dissolved minerals, including metals and sulfides
- **mantle:** the thick rock layer between the earth's crust and the core below
- **marker fossil:** a species that existed in a wide area but died out in a short time; finding such a fossil fixes the date of the strata in which it is found
- **methane hydrate:** mineral formed when methane (natural gas) is trapped within the structure of water ice crystals; extensive ocean-floor deposits of methane hydrate might influence climate and could become a major resource
- **plate tectonics:** mechanism that allows continental drift; the earth's crust consists of individual shifting plates that form at mid-ocean ridges and other locations and are destroyed where plates collide and send material back into the mantle

MOHOLE PROJECT

Scientific drilling on the ocean floors began almost as a stunt, but it evolved into a long-term research program that revolutionized geology, contributed vital clues about climate change and about the extinction of the dinosaurs, discovered a major new type of subsea hydrocarbon deposit that may fuel the world, and aided in the study of rich metal ores that may also be tapped. Along the way, scientific drilling pioneered many techniques that have

been used in offshore drilling for petroleum and natural gas.

Drilling samples have been a major part of geology since the second half of the nineteenth century. Drillers seeking water or oil could sample pieces of rock drilled from varying depths and log them into their drill records. For scientific purposes, drilling with circular drills allowed the cutting of a long cylinder or "core" that could be pulled up and carefully measured for position and composition. By the middle of the twentieth century, hard-rock miners were using cores to sample for minable ores. Data from these drillings were correlated into three-dimensional maps of distinctive strata showing dips and faults.

In the late 1950's, a number of geologists envied the space program with its romantic goal of flight to the moon. In 1959, they proposed a similarly dramatic program called the Mohole project. In 1909, Andrija Mohorovičić had analyzed seismic waves from earthquakes and concluded that the rock of the crust changed significantly about 16 to 40 kilometers below the surface as it changed to partially melted mantle. There was speculation that rocks might be different at this so-called Mohorovičić discontinuity, or Moho, and researchers proposed drilling all the way to the Moho for samples. Drilling such a tremendously deep hole would be expensive and maybe impossible. Several marine geologists suggested that oceanic drilling to the Moho would be cheaper because the crust is thinner under the ocean floor. Also, a drill core through the sea floor might yield a complete fossil record of tiny marine shells.

Drilling into the ocean floor required several major innovations. The drilling locations were in deep water, so the drilling platform had to be an oceangoing vessel rather than a tower resting on the bottom. Anchoring in those depths would be difficult, so the vessel had to actively maintain position.

This "dynamic positioning" had to keep the drill ship within two ship lengths of straight, or the drill string would break. While out of sight of land, the drilling vessel crew had to navigate within this small area with no landmarks whatsoever. This was managed originally by taut, moored buoys and later by satellite position-finding and acoustic beacons on the sea floor. Since the drill bit would hang as much as several thousand meters below the drill ship, bottom-hole assemblies were required for getting the drill started and for returning to the same hole. Finally, because the drill platform rose and fell with the waves, the rig needed a heave compensator.

Project Mohole started with engineering tests in 1961 in waters off California by *Cuss I*, a drilling barge developed for offshore oil drilling and built by Global Marine. (A number of famous drill ships from this company begin with the abbreviation "Glomar.")

Initial tests off Baja California, Mexico, were promising, but the project eventually ended when technical difficulties and political management problems caused projected drilling costs to increase significantly. It was later discovered that the Moho outcrops at the Earth's surface in certain areas and that most areas of the ocean are comparatively young geologically, so continuation of the Mohole project would have been of less use than its backers had hoped.

However, the possibility of deep-ocean drilling had been demonstrated. Oil drillers, who had previously worked in depths of fewer than 100 meters, saw new possibilities. Project Mohole stirred interest in dynamic positioning for oil and gas exploration; the offshore drilling industry subsequently developed many technologies that passed back to the scientific drillers. Oceangoing rigs allowed exploratory drilling in deep areas of the continental shelf before making

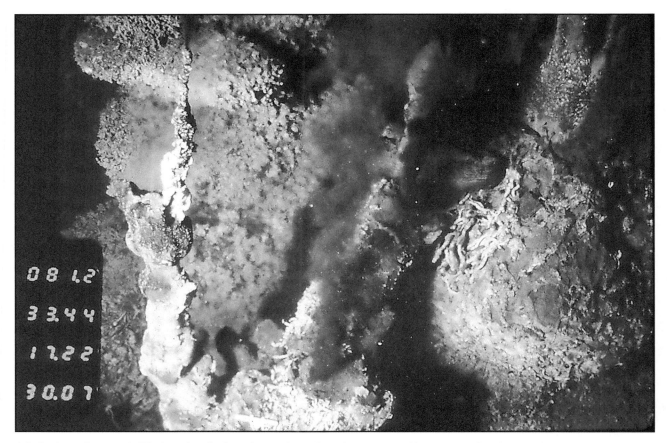

A hydrothermal vent with "black smokers," where plumes of superheated water meet colder waters and dark minerals begin to precipitate out and form "chimneys." (National Oceanic and Atmospheric Administration)

multibillion-dollar investments in production platforms. Thus, the Mohole project was a major factor in opening many offshore oil fields.

Furthermore, the Mohole demonstration had obtained one major data point. A mysterious lower layer visible on sonar graphs was found to be not another layer of sedimentary rock but rather basalt, an igneous rock formed by recrystallization of molten rock and often associated with volcanoes. This discovery suggested that the oceans were younger than expected and that the Mohole project, as originally planned, would have been able to gather minimally useful data about the geological history of the Earth. Conversely, it gave support to the then-radical theory of continental drift.

Testing continental drift required not one deep hole to the Mohorovičić discontinuity but many shorter holes surveying many areas. These shorter holes were still a tremendous advance. Before oceanic drilling, the only data about the sea floor came from dredge hauls and piston cores. Dredges only pull a jumbled mass of material from the first few centimeters of the sea floor. A piston core is essentially a weighted pipe that is allowed to fall as fast as possible to the sea floor; its weight and momentum drive it into soft sediment. Its limitations are that it penetrates only a few meters and that it cannot penetrate hard surfaces.

JOIDES

In 1964, the Joint Oceanographic Institutions for Deep Earth Sampling (JOIDES) was formed; it has become an international organization that includes universities and government research organizations. In April and May of 1965, JOIDES used the drill ship *Caldrill I* to test upgraded methods by drilling six holes on the Blake Plateau off the coast of Florida to sub-bottom depths of more than 1,000 meters. Based on that success, JOIDES proposed an eighteen-month program of scientific drilling in the Atlantic and Pacific Oceans to the U.S. National Science Foundation. The resultant Deep Sea Drilling Project was operated by the Scripps Institution of Oceanography. The DSDP drill ship *Glomar Explorer* was capable of drilling 760 meters in 6,100 meters of water. It began operations in July 1968, and ultimately made ninety-six voyages (or drilling legs) for JOIDES that focused on sites on or near the mid-ocean ridges. The cores retrieved revolutionized geology by proving the theory of continental drift.

In 1978, the *Glomar Explorer* was replaced by the *JOIDES Resolution* run by Texas A&M University as part of the one-third internationally funded Ocean Drilling Program (ODP). The ship was designed to drill in 8 kilometers of water for a total drill-string length of 9 kilometers, and it can drill 2 kilometers into the sea floor in shallower waters. Both the *Glomar Explorer* and the *JOIDES Resolution* have worked in concert with remote instruments and piloted instruments in a number of revolutionary developments.

PLATE TECTONICS

Drilling in marine strata has yielded tremendous advances in geological knowledge for several reasons. First, oceans cover more than 70 percent of the earth, so a proportionate number of discoveries should be expected from oceanic data. Second, tens of thousands of drill cores have been logged on land, so many of the land discoveries have already been made. Third, land is often subject to erosion, so there are more gaps than in marine strata. Fourth, land strata have been subjected to more compression, heat, and chemical attack—all of which could confuse possible geological data—than seafloor strata. Fifth, some of the mechanisms that happen under several kilometers of water in the oceans happen under several kilometers of rock on land, so these mechanisms are easier to study with marine drilling. Finally, drilling on land requires disassembly and reassembly of equipment, often in difficult terrain; a drill ship simply sails to the next drill site.

In 1620, Francis Bacon noted similarities between the coastlines of South America and Africa and suggested that they had once been interconnected. In the early twentieth century, Alfred Wegener proposed the former supercontinent of Pangaea to explain similar fossils from areas now widely separated. However, these theories lacked a provable mechanism for moving entire continents through the rock underlying the ocean floor.

It was then noted that the Mid-Atlantic Ridge had bands of different magnetic orientation on either side. These bands corresponded to reversals of the earth's magnetic field that occur every several thousand years. It appeared that new melted rock was crystallizing (and thus freezing a weak magnetic field) at the ridge and was then shoved away from the ridge by new rock. The crystallizing rock that cooled and hardened out of magma would quite likely be basalt. The Mohole finding of basalt in its deeper cores was encouraging, but conclusive proof required correlating the age from a large number of drill samples. The initial *Glomar Explorer* drill cores allowed geologists to concentrate on that dating using marker fossils and radioisotopic dating. Data from the cores showed that the ocean floor approaching the Mid-Atlantic Ridge was progressively younger. Thus, it appeared that Europe, North America, Africa, and South America were joined until sometime between 200 and 170 million years ago, when the Atlantic Ocean began opening.

This revolutionary discovery allowed development of plate tectonics theory, which proposes that large plates of connected rock exist on the earth's surface. Plate tectonics has many implications. Because the earth cannot expand indefinitely, the idea that new rock accretes along mid-ocean ridges means that plate material must be disappearing elsewhere. That explains the existence of deep ocean trenches (such as the Mariana Trench) where oceanic strata are diving steeply down toward the mantle and some of the heated rock spurts up, forming island arcs. One plate riding over another provides the mechanism for raising up mountains, such as the Andes Mountains, which are riding up over the edge of the Pacific plate. A more extreme example is the Indian subcontinent, which is burrowing under the Tibetan Plateau and lifting the Himalaya.

HYDROTHERMAL VENTS

Continued study of plate tectonics along the mid-ocean ridges and other plate boundaries led to the unexpected discovery of hydrothermal vents, which are somewhat like geysers and hot springs but which are found on the ocean floor; however, the scale and results of the activity surprised researchers. In 1965, a sample of rocks dredged from the sea floor yielded rock samples of two types. One was depleted in certain chemicals that were present in the other type. Also in the 1960's, it was noted that thermal readings in sediments near plate boundaries did not have as much heat as expected. Hydrothermal vents were suggested as the mechanism for moving the large amounts of heat. Seawater percolating down to the molten rock beneath the sea floor would be heated and return to the ocean as hot springs. The heated water would be much less dense than the near-freezing seawater near the ocean floor, so convection would drive the process.

In 1977, researchers in a piloted submersible discovered an active field of hydrothermal vents with "black smokers," where plumes of superheated water meet colder waters and dark minerals begin to precipitate out and form "chimneys." This discovery revealed the importance of such vents in circulating heat from the Earth's core, circulating minerals into and out of the ocean waters, and sustaining life driven by chemosynthesis (chemical energy) rather than photosynthesis (light energy).

However, only drilling could supply data on subsurface events. For example, bacteria living by chemosynthesis have been found several hundred meters below the ocean floor, suggesting that such life-forms may be more biomass-based than photosynthesis-based life. Also, drilling in mounds built by successive chimneys revealed large amounts of carbonate salt that eventually dissolves when the area cools. Thus, old, inactive hydrothermal fields might have rich mining potential. Finally, the sulfide-enriched upper layer may be underlain by a copper-enriched layer, also with good mining potential. More important, comparable areas of former seabed have been lifted up and become land. These areas, such as the Klamath Mountains of northwestern North America and areas of Cyprus, have already served as mining areas, and better knowledge of the mineral-formation processes will probably improve land mining long before any ocean mining.

MARKER FOSSILS

Probably the most important dividends from the ODP are carefully logged and partially analyzed cores from every ocean except the Arctic. By the end of the twentieth century, there were more than 240 kilometers of cores available for new studies and new instruments. These cores have helped provide increasingly fine chronological correlation among strata from various locations. Oceanic core data allow the identification of "marker fossils" that (ideally) lived for a short time but over a wide area. Those marker fossils can then be used to date match with other oceanic cores and land cores. This correlation is crucial for tracking individual layers of rock through folds and faults. The dated and mapped layers allow one to calculate what might have happened in the past, what might happen in the future, and where ores may exist at present.

The types of fossils in the cores are generally tiny microfossil remains of plankton that lived in the surface waters. The types of fossils indicate the climate in those waters when they were deposited. Likewise, percentages of carbon and oxygen isotopes vary with sea surface temperature. Data from isotopes and microfossils can be checked against one another and against climate data from land to calculate past climates (paleoclimatology). With better estimates of past climate, climatologists may better project possible climate change in the future.

Ocean drilling provides another climate indicator: sediments from land. Drilling confirmed massive catastrophic flows when ice dams broke at the end of the last ice age, releasing torrents of water down the Columbia and St. Lawrence River Basins. Drilling also shows the timing, direction, and quantity of windborne sand and dust from places such as North Africa and Central Asia. Cores around Antarctica have corroborated that Antarctica's continental ice sheet has existed for millions of years; however, continued studies are needed to detail fluctuations of ice level.

Drill cores also confirmed that turbidity currents (dense masses of sediment-laden water) are a major mode of deposition in the ocean. (The low relief of the abyssal plains in the Atlantic Ocean, which has a high amount of sediment compared to total area, is caused by turbidity flows filling in low areas.) Turbidities indicate past ocean currents, another paleoclimate factor.

Petroleum and natural gas traces in deep waters are another major discovery of ocean drilling. However, the *Glomar Challenger* and *JOIDES Resolution* had to stop drilling whenever they encountered such traces. These craft have room for laboratories because they were not fitted with a "riser," which collects anything coming out of the drill hole and pumps down drilling mud. The riser can be used to stop a flow of fluid from a borehole. Therefore, the riserless drill ships risked causing an uncontrolled leak into the ocean if they continued drilling in areas with any sign of petroleum or natural gas.

METHANE HYDRATES

Still another hydrocarbon deposit researched by ocean drillers is methane hydrate. Scattered reports of flammable ice and fizzing material brought up by piston cores led to a theory that methane (natural gas) might be frozen into the structure of ice in deposits beneath the deep ocean floors. These deposits could form because the waters are only slightly above the freezing

point of fresh water and they have high pressure—the two requirements for methane hydrate. The methane source would probably consist mostly of the decay of biological material in sediments. Sonar readings even showed layers in seafloor sediments that might be such hydrates.

Drill cores from several sites confirmed extensive methane hydrate deposits, and their analysis led to several conclusions. Hydrates may contain several times as much energy as all other fossil fuel deposits combined. Hydrates might act as a cap rock to contain conventional gas and petroleum. It might be some time before hydrates are exploited because they are often spread thinly in sub-seafloor ocean strata. Hydrates might be a major factor in climate change because disturbances on the ocean floor or lowered sea levels might release methane, a greenhouse gas; conversely, raised sea levels would increase the area of high pressure, allowing more hydrates to form, thus decreasing the flow of greenhouse gases to the atmosphere. Finally, hydrates are stronger than unfrozen seafloor muds and oozes. Consequently, a hydrate-melting event could cause a major collapse of sediments on the continental slope and result in a major turbidite flow.

Drill cores also corroborated the asteroid-impact explanation for extinction of the dinosaurs. A thin dead zone marked the end of the Cretaceous period, the last great age of dinosaurs, and the beginning of the Tertiary. This Cretaceous-Tertiary (K-T) boundary zone was noted in a few rock outcrops in Europe, and rock from this layer had elevated levels of platinum-group metals (similar to elevated levels in asteroids) in that strata. However, the theory was doubted until multiple drill cores closed in on the probable impact site on the Yucatán Peninsula of Mexico with thicker corresponding deposits off Florida and in the Caribbean Sea. Besides the platinum-group-enriched dead layer, those deposits had large amounts of shocked quartz, suggesting cosmic impact.

Sea levels over time have also been deduced from oceanic cores. One major sea-level surprise was massive evaporite deposits in the Mediterranean Sea from when sea levels were low enough to make the Strait of Gibraltar dry land and the Mediterranean a salty inland sea. Finally, remote stations have been implanted in several oceanic boreholes. These stations take seismic data to augment land seismic stations, and they also have temperature sensors.

Roger V. Carlson

FURTHER READING

Bascom, Willard. *The Crest of the Wave: Adventures in Oceanography.* New York: Doubleday, 1988. Chapter 8, "The First Deep Ocean Drilling," describes the technical issues faced by the Mohole project, as well as the changes in attitudes that the project prompted.

Broad, William J. *The Universe Below: Discovering the Secrets of the Deep Sea.* New York: Simon & Schuster, 1997. Provides an entertaining general overview of oceanic exploration, including descriptions of oceanic drilling, associated submersible explorations near those sites, and implications for possible future mining.

Cone, Joseph. *Fire Under the Sea: The Discovery of the Most Extraordinary Environment on Earth, Volcanic Hot Springs on the Ocean Floor.* New York: William Morrow, 1991. A detailed history of the exploration and description for the lay reader of hydrothermal vents on the ocean floor and the plate tectonics that cause them. The book is an excellent introduction to the technical articles listed in its extensive reference list.

Davies, Thomas A. "Scientific Ocean Drilling." *Marine Technology Society Journal* 32 (Fall 1998): 5-16. Davies summarizes the history of the Ocean Drilling Program, major accomplishments, the technologies involved, and proposed work for the future.

Exon, Neville. "Scientific Drilling Beneath the Oceans Solves Earthly Problems." *Australian Journal of Maritime and Ocean Affairs* 2 (2010): 37-47. Presents an overview of the purpose and future expeditions of the Integrated Ocean Drilling Program. Discusses the importance of deep ocean drilling to understanding past and present earth processes.

Graham, Bob, et al. *Deep Water: The Gulf Oil Disaster and the Future of Offshore Drilling.* National Commission on the BP Deepwater Horizon Oil Spill and Offshore Drilling, 2011. A government report addressing the disaster in the Gulf of Mexico, the events leading up to it, and the repercussions that followed. The committee focused its efforts on future prevention of such disasters in the third part of this report. Provides great detail from key players, scientists, and local residents.

Huey, David P., and Michael A. Storms. "Novel Drilling Equipment Allows Downhole Flexibility." *Oil and Gas Journal* 93 (January 16, 1995): 63-68. Details the ocean-drilling innovations developed by the DSDP and ODP.

Moore, J. Robert. *Oceanography: Readings from Scientific American.* San Francisco: W. H. Freeman, 1991. Contains several classic articles written about the sea floor, plate tectonics, and studies of microfossils (micropaleontology).

Normile, Dennis, and Richard A. Kerr. "A Sea Change in Ocean Drilling." *Science* 300 (2003): 410. This article provides a short history of the purpose and accomplishments of the Ocean Drilling Program and discusses the future efforts of the Integrated Ocean Drilling Program.

Zierenberg, R. A., et al. "The Deep Structure of a Sea-Floor Hydrothermal Deposit." *Nature* 392 (April 2, 1998): 485-488. The authors detail an important set of results about ocean drilling used to study metallic minerals associated with the Juan de Fuca spreading center off the coast of Seattle.

See also: Continental Drift; Deep-Earth Drilling Projects; Earth's Interior Structure; Engineering Geophysics; Environmental Chemistry; Geochemical Cycle; The Geoid; Ocean-Floor Drilling Programs; Plate Tectonics; Seismic Reflection Profiling; Volcanism.

OCEAN-FLOOR DRILLING PROGRAMS

Ocean-floor drilling programs have allowed geologists and oceanographers to extend their knowledge of the earth's history by analyzing long marine sediment cores and basement rock cores recovered from the sea floor. Data from ocean-floor drilling have provided evidence supporting the theories of seafloor spreading and plate tectonics and have permitted the investigation of the paleoclimatic and paleoceanographic history of the earth.

PRINCIPAL TERMS

- **abyssal plains:** flat-lying areas of the sea floor, located in the ocean areas far from continents; they cover more than half the total surface area of the earth
- **basalt:** a dark-colored, fine-grained rock erupted by volcanoes, which tends to be the basement rock underneath sediments in the abyssal plains
- **chert:** a hard, well-cemented sedimentary rock that is produced by recrystallization of siliceous marine sediments buried in the sea floor
- **correlation:** the demonstration that two rocks in different areas were deposited at the same time in the geologic past
- **deposition:** the process by which loose sediment grains fall out of seawater to accumulate as layers of sediment on the sea floor
- **mid-ocean ridge:** a continuous mountain range of underwater volcanoes located along the center of most ocean basins; volcanic eruptions along these ridges drive seafloor spreading
- **paleoceanography:** the study of the history of the oceans of the earth, ancient sediment deposition patterns, and ocean current positions compared to ancient climates
- **plate tectonics:** a theory that the earth's crust consists of individual, shifting plates that are formed at oceanic ridges and destroyed along ocean trenches
- **seafloor spreading:** a theory that the continents of the earth move apart from each other by splitting of continental blocks, driven by the eruption of new ocean floor in the rift

ORIGINS OF TECHNOLOGY

Most of our knowledge of the history of the earth comes from the study of sedimentary rocks, as sediments contain the preserved fossil remains of ancient plants and animals, while sedimentary structures record the processes of deposition. Sedimentary rocks exposed on land tend to have an incomplete record because they may be deformed by folding and faulting, which may destroy both the fossils and the sedimentary structures, and because they may be eroded by wind, water, and ice moving across the surface. In contrast, marine sedimentary sequences contain a more complete record of accumulation because they accumulate in a lower-energy environment, which is not as affected by folding, faulting, erosion, and post-depositional alteration as are terrestrial sediments. As a result, the deep-sea sediments tend to preserve a continuous record of sediment deposition in the ocean basins.

Ocean drilling techniques were originally developed in the 1950's by petroleum exploration companies searching for shallow-water hydrocarbon and petroleum deposits located on the continental shelves. These industrial exploration methods were adapted in the 1960's to obtain long sediment cores from the sea floor in deep-water areas on the continental slopes and abyssal plains. By drilling through the entire sediment record into the harder basement rocks below marine sediments, geologists hoped to acquire the complete history of sediment deposition within an ocean basin from the time that sediments were first deposited atop volcanic basalts.

GLOMAR CHALLENGER

Preliminary attempts to drill the ocean floor included engineering tests for Project Mohole by the drilling barge *Cuss I*, which in 1961 drilled marine sediments off La Jolla, California, and at the deep-water Experimental Mohole Site east of Guadalupe Island, off Baja California, Mexico, in a water depth of 3,566 meters. Although further Project Mohole development was not undertaken because of a combination of political conflicts and increasing cost estimates for the project, in 1964 four American universities formed a consortium to initiate a program of scientific deep-sea drilling. JOIDES, the Joint Oceanographic Institutions for Deep Earth Sampling, successfully operated a drilling program in April-May, 1965, using the vessel *Caldrill I* to drill six holes on the Blake Plateau off Florida to sub-bottom depths of more than 1,000 meters, with continuous core recovery.

Following these successful trials, JOIDES proposed an eighteen-month program of scientific drilling in the Atlantic and Pacific oceans, to be called the Deep Sea Drilling Project (DSDP), operated by the Scripps Institution of Oceanography of La Jolla, California, using the drilling vessel *Glomar Challenger*. *Glomar Challenger* left Orange, Texas, on July 20, 1968, on Leg 1 of the Deep Sea Drilling Project. The results of DSDP drilling on the first nine cruises in the Atlantic and Pacific oceans caused the National Science Foundation to extend the drilling program beyond the initial eighteen-month period, with further drilling in the Indian Ocean and in the seas surrounding Antarctica.

When DSDP began operations, many other American oceanographic institutions joined JOIDES in support of the drilling program, and the success of DSDP also attracted scientific participation and financial support from foreign countries. The International Program of Ocean Drilling (IPOD) started in 1975 when the Soviet Union, the Federal Republic of Germany, France, the United Kingdom, and Japan joined JOIDES, with each country providing $1 million yearly to support drilling programs. DSDP/IPOD drilling activities continued through the early 1980's, leading to international scientific exchange of information between oceanographers.

SEDCO/BP 471

Because the initial JOIDES proposal was only for eighteen months, it was never expected that ocean drilling would continue for fifteen years. Because of demands for ocean drilling in deeper waters and in high-latitude polar areas, JOIDES proposed in the early 1980's that a larger drilling vessel be acquired for continued drilling. The last cruise of the *Glomar Challenger*, DSDP Leg 96, ended in Mobile, Alabama, on November 8, 1983, with the retirement of the drilling vessel from service. In 1983, responsibility for scientific supervision of the international project, now called the Ocean Drilling Program (ODP), passed from the Scripps Institution to Texas

A&M University, and the drilling vessel *Sedco/BP 471* replaced *Glomar Challenger*. The first cruise of a ten-year ODP drilling program began on March 20, 1985, when the *Sedco/BP 471*, informally called the *JOIDES Resolution*, left port to begin drilling on ODP Leg 101. Leg 186 was scheduled for completion in the year 2000, and planning continued for cruises up to Leg 201 in 2002.

The results of each cruise, or leg, have been published in a series of books, entitled *Initial Reports of the Deep Sea Drilling Project*, which are published by the U.S. Government Printing Office. The cores recovered from the DSDP and ODP holes represent an invaluable record of the history of ocean sediment deposition around the globe. These recovered sediment cores are studied by a variety of scientists, who are interested in the sediment type, fossil content, geochemistry, magnetic orientation and strength, shear strength, and other sedimentary properties of the samples.

A dockworker walks past the Global SanteFe Development Driller II platform at Singapore's Jurong Shipyard on December 6, 2004. The $285 million platform is the second of two Friede and Goldman-designed fifth-generation, dynamic positioning, ultra-deepwater, semi-submersible rigs built by Jurong Shipyard for GlobalSanteFe Corporation. (Bloomberg via Getty Images)

DRILLING PROCEDURES

In shallow water, drilling is accomplished either by building a drilling platform directly atop the sea floor or by firmly anchoring a drilling vessel to the bottom. In deep-water ocean drilling, however, it is not possible to anchor the drilling vessel to the bottom, so the technique of dynamic positioning is used to maintain the position of the vessel above the hole being drilled. In dynamic positioning, an acoustic beacon emitting sounds at either 12.5 kilohertz or 16 kilohertz is dropped to the sea floor. Four hydrophones, located at different points on the hull of the drillship, receive the signal from the acoustic beacon at slightly different times, depending on the position of the hull relative to the beacon. The position of the ship is maintained by a shipboard computer, which interprets the information from the hull hydrophones and controls the position of the ship by driving both the main propellers and two laterally oriented propellers, or hull thrusters. If the vessel is pushed off location by waves or surface currents, the shipboard computer attempts to compensate by using the propellers and hull thrusters to maintain the ship's location relative to the seafloor beacon.

To drill sediment and rock samples from the sea floor, a drill bit is attached to the bottom of a 9.5-meter-long piece of hollow cylindrical stainless steel drill pipe. More individual lengths of pipe are connected on the rig floor of the drillship to make a "drill string," which extends from the vessel through the water down to the sea floor, where coring may begin. Usually, about 450-510 lengths of drill pipe are required simply to reach the bottom, and the assembly of this drill string may take twelve hours before bottom drilling may be started. Once the string reaches the sea floor, the drill pipe is rotated by hydraulic motors on the rig floor, and the rotary action combined with the weight of the drill string causes the drill bit to spiral down into the sea floor. Sharp iron carbide or diamond cutting teeth on the drill bit assist the penetration of the drill string into the sediment and the rock on the ocean bottom.

Samples of sediment and rock are retrieved from the sea floor by drilling a hole about 25 centimeters in diameter, using a drill bit with a 7.5-centimeter hole in its center. In effect, sediment is cored by "drilling the doughnut and saving the doughnut hole": Rotating the drill string grinds the outer ring of sediment to small pieces against the diamond teeth of the drill

bit, while the material in the center of the drill hole is saved as a core of drilled sediment 6.6 centimeters in diameter. As the drill string is lowered deeper into the drilled hole, the core is pushed up into a plastic core liner in a steel "core barrel" within the lowest stand of drill pipe. After 9.5 meters of the sea floor has been drilled, the core barrel is pulled up to the rig floor by a cable lowered through the drill string. By the time ODP ended in 2004, Joides Resolution had completed 110 expeditions, collecting about 2,000 deep sea cores from major geological features in the oceans of the world.

ATLANTIC OCEAN DRILLING

The first three cruises of the *Glomar Challenger* provided information proving that seafloor spreading had occurred in the Atlantic Ocean. A series of DSDP holes across the Mid-Atlantic Ridge showed that the age of bottom sediments increased with distance from the ridge crest and indicated that the ages of sediments with depth correlate from one hole to the next. The total thickness of sediment atop basaltic basement also increased with greater distance from the ridge crest, on both the east and west sides of the Mid-Atlantic Ridge. Further DSDP and ODP drilling has provided evidence that seafloor spreading has occurred in all the earth's ocean basins. In addition, ocean drilling has confirmed the relative youth of the ocean basins, as predicted by plate tectonics; the oldest sea floor yet discovered is Early Jurassic in age (160 million years old), compared to continental rocks, which may be as old as 4.5 billion years.

PACIFIC OCEAN DRILLING

Glomar Challenger and *JOIDES Resolution* have operated from the Norwegian Sea to the Ross Sea off Antarctica and have drilled holes in water depths from 193 meters on the Oregon continental shelf to 7,050 meters in the Mariana Trenchoff Guam, in the western Pacific Ocean. The deepest hole through seafloor sediment deposits is more than 1,750 meters below the sea floor, and one site in the equatorial Pacific Ocean west of South America (DSDP Hole 504B) has been drilled through 300 meters of sediment and 1,500 meters of volcanic basement rock.

GLOBAL STRATIGRAPHIC CORRELATION

Seafloor drilling has indicated that deep-sea sediments contain long sequences of well-preserved microfossils, which may be used for global strati-graphic correlation, in contrast to the fragmentary record preserved on land, where structural defor-mation of sediment deposits may complicate the problem of correlating different sedimentary se-quences. Analysis of these sediments has revealed the history of deposition in the different ocean basins and has provided information on ancient climates and oceanographic conditions (such as the position, strength, and temperature of past ocean currents).

Sediment cores have indicated the presence of great shifts in oceanic climate conditions during the geologic past and have demonstrated that the Antarctic continent has been covered by glacial ice caps for at least 40-50 million years, rather than the 5 million years accepted prior to DSDP drilling near Antarctica. Another startling result of ocean drilling has been the discovery that the Mediterranean Sea dried up between 12 and 5 million years ago. Massive salt and evaporite mineral deposits below the Mediterranean basin indicate that the Strait of Gibraltar connection to the Atlantic Ocean was blocked during this time. Blockage of the Gibraltar connection allowed the water in the basin to evapo-rate, causing the deposition of vast salt and evaporite mineral deposits as the Mediterranean dried up.

STUDY OF SEAFLOOR BASEMENT ROCKS

Not all the information provided by ocean drilling has been concerned with the sediment column. Drilling into basement rocks has allowed geophysi-cists to compare the structure of seafloor basement to that of layered igneous-rock deposits that have been uplifted above sea level on the edges of certain continents. Similarly, direct drilling through these basalt and gabbro layers has allowed a comparison of the rock type to sound velocities measured by ma-rine geophysicists. Some other results of hard-rock seafloor drilling have been the investigation of sedi-ment and mineral deposition by hydrothermal pro-cesses at rapidly opening mid-ocean ridge segments, such as the sulfides deposited by high-temperature fluids emitted by "black smoker" and "white smoker" structures near the Galápagos Islands west of South America. Drilling of bare basement rocks along mid-ocean ridges in the Atlantic, Pacific, and Indian oceans has enabled geochemists and igneous petrolo-gists to study the frequency at which seafloor volcanic rocks are produced at individual ridge segments and

to determine whether temporal changes occur in the chemistry of basalts erupted from one location on the ridge. These studies of seafloor basement rocks may be applied to mineral exploration of marine rocks that have been uplifted above sea level and ex-posed on continents.

IMPROVEMENTS IN DRILLING TECHNOLOGY

In addition to their scientific results, DSDP and ODP operations have resulted in improvements in drilling technology by developing the ability to re-enter sea floor boreholes, by devising techniques for "bare-rock" drilling on the sea floor, and through the development of new coring bits. During DSDP Leg 1, it was discovered that existing drill bits could not pene-trate hard chert beds; thus, they also would not be able to penetrate through deeper igneous rocks below sea-floor sediments. The drag bits were solid, consisting of a central opening and radial curved ridges of steel or tungsten carbide, capped with industrial diamonds and designed to churn through soft sediments. As a result, DSDP began a drill-bit design program, which led to the development of roller bits capable of pen-etrating both chert layers and seafloor basalts. These bits consist of four conical cutting heads studded with tungsten carbide or diamond cutting teeth, situated around the central core opening in the bit.

Another important technical development of DSDP, first successfully accomplished on Leg 15, was the ability to reenter a drilled borehole on the sea floor. Even with roller bit designs, drill bits wear out from the stresses of rotary coring through seafloor sediments and rocks. When a bit fails, the entire drill string has to be "tripped," or pulled up to the vessel to replace the bit at the lower end of the string, which in most deep-ocean drilling sites requires pulling the string up not only several hundred meters from below the sea floor but also through 2,000-5,000 meters of seawater. During early DSDP legs, bit failure forced the abandonment of a hole because after the fatigued bit was replaced, it was impossible to reenter the original borehole. Successful reentry techniques were facilitated by the development of a steel reentry cone 6 meters in diameter, topped with three sonar reflectors and a rotating sonar scanner that can be lowered through the drill string. In areas where hardened sediment layers are anticipated, requiring bit replacement to complete drilling, the reentry cone is placed on the sea floor prior to drilling the initial bore-hole. As bits become worn, they may be replaced and the hole reentered by using the sonar scanner to locate the reentry cone (and thus the original hole).

DSDP and ODP drilling specialists have also devised methods to enable drilling in hard seafloor areas, such as mid-ocean ridges, which were not previously drillable by existing techniques. Development of a seafloor "guide base" for drilling has allowed successful drilling and reentry of boreholes in these areas and has permitted the implantation of seafloor sensing devices, such as earthquake-measuring seismometers, in these holes.

Role in Development of Paleoceanography

Ocean-floor drilling programs have enabled scientists to correlate apparently unconnected phenomena through the theory of plate tectonics, a global synthesis of geology and oceanography. Ocean drilling has provided verification of the seafloor spreading hypothesis as it applies to plate tectonics and has indicated that seafloor spreading has occurred in all the earth's ocean basins.

Before long sediment cores could be acquired from the sea floor, scientists' knowledge of seafloor geology was sparse, based on limited samples available from dredging and shallow coring of the sea floor by oceanographic vessels. Prior to DSDP, global stratigraphic correlation was based on a fragmentary record preserved on land, where structural deformation of sediment deposits may complicate the problem of correlating different sedimentary sequences; DSDP drilling, however, has revealed that deep-sea sediments contain long sequences of well-preserved microfossils. Furthermore, seafloor sediment cores have revealed the history of the ocean basins and have provided information on ancient climates and oceanographic conditions (such as the position, strength, and temperature of past ocean currents). A new science, paleoceanography, has been developed based on this information from DSDP and ODP drilling. Analysis of the earth's ancient climates may provide information to predict future shifts in the biosphere.

Industrial Applications

Ocean drilling has also provided evidence for the existence of deep-water hydrocarbon accumulations, which has enabled petroleum exploration companies to drill petroleum deposits on the continental slopes and may eventually lead to the discovery of significant hydrocarbon deposits in ocean basins. If future technology is developed, humans may be able to exploit these deep-water petroleum resources. Furthermore, scientific ocean drilling has enabled geologists to understand the processes controlling the deposition of "black shale" deposits and other high-productivity seafloor sediments, which may be altered by burial into source beds for the generation of petroleum hydrocarbons. Understanding of the processes affecting the formation and distribution of these sediments may assist in future exploration for fossil fuel resources. In addition, studies of seafloor basement rocks may lead to a more complete understanding of the nature of mineral deposition at mid-ocean ridges, which may be applied to mineral exploration of similar marine rocks that have been uplifted above sea level and exposed on continents.

Finally, deep-ocean drilling has led to technological innovations in the tools and techniques used to sample the sea floor. These methods have been adapted by industrial companies exploring for hydrocarbons buried beneath marine sediments and for mineral deposits on the sea floor. In 2004 the Ocean Drilling Program was transformed into the Integrated Ocean Drilling Program (IODP). The final report of ODP was published as the "Ocean Drilling Program Final Technical Report 1983-2007." The IODP is an international program utilizing a variety of drilling platforms to investigate the earth deep below the sea floor.

Dean A. Dunn

Further Reading

Bascom, Willard. *A Hole in the Bottom of the Sea.* Garden City, N.Y.: Doubleday, 1961. A history of the Mohole Project, which planned to drill through oceanic rocks down to the crust-mantle boundary.

Condie, Kent C. *Plate Tectonics and Crustal Evolution.* 4th ed. Oxford: Butterworth Heinemann, 1997. An excellent overview of modern plate tectonics theory that synthesizes data from geology, geochemistry, geophysics, and oceanography. A very helpful tectonic map of the world is enclosed. The book is nontechnical and suitable for a college-level reader. Useful "suggestions for further reading" follow each chapter.

Cramp, A., et al., eds. *Geological Evolution of Ocean Basins: Results from the Ocean Drilling Program.* London: The Geological Society, 1998. Intended for the reader with some scientific background, this book offers many articles describing the processes and results of ocean drilling.

Davidson, Jon P., Walter E. Reed, and Paul M. Davis. *Exploring Earth: An Introduction to Physical Geology.* 2d ed. Upper Saddle River, N.J.: Prentice Hall, 2001. An excellent introduction to physical geology, this book explains the composition of the earth, its history, and its state of constant change. Intended for high-school-level readers, it is filled with colorful illustrations and maps.

Exon, Neville. "Scientific Drilling Beneath the Oceans Solves Earthly Problems." *Australian Journal of Maritime and Ocean Affairs* 2 (2010): 37-47. Presents overview of the purpose and future expeditions of the Integrated Ocean Drilling Program. Discusses the importance of deep ocean drilling to understanding past and present earth processes.

Hamblin, William K., and Eric H. Christiansen. *Earth's Dynamic Systems.* 10th ed. Upper Saddle River, N.J.: Prentice Hall, 2003. This geology textbook offers an integrated view of the earth's interior not common in books of this type. The text is well organized into four easily accessible parts. The illustrations, diagrams, and charts are superb. Includes a glossary and laboratory guide. Suitable for high school readers.

Hsu, Kenneth J. *The Mediterranean Was a Desert.* Princeton, N.J.: Princeton University Press, 1983. A personal account of DSDP Leg 13 drilling in the Mediterranean Sea basin during 1970, as written by one of the two chief scientists on the drilling vessel *Glomar Challenger.*

Nierenberg, William A. "The Deep Sea Drilling Project After Ten Years." *American Scientist* 66 (January/February 1978): 20-29. A review of the significant technical and scientific developments of the DSDP, written by the director of Scripps Institution of Oceanography.

Normile, Dennis, and Richard A. Kerr. "A Sea Change in Ocean Drilling." *Science* 300 (2003): 410. This article provides a short history of the purpose and accomplishments of the Ocean Drilling Program and discusses the future efforts of the Integrated Ocean Drilling Program.

Peterson, M. N. A., and F. C. MacTernan. "A Ship for Scientific Drilling." *Oceanus* 25 (Spring 1982): 72-79. Summary of the technical aspects of ocean drilling on *Glomar Challenger,* written for a nonscientific audience by the director of the DSDP.

Segar, Douglas. *An Introduction to Ocean Sciences.* 2d ed. New York: Wadsworth, 2007. Comprehensive coverage of all aspects of the oceans and the oceanic crust. Readable and well illustrated. Suitable for high school students and above.

Shor, Elizabeth Noble. *Scripps Institution of Oceanography: Probing the Oceans, 1936 to 1976.* San Diego, Calif.: Tofua Press, 1978. The book recounts the formation of the Scripps Institution of Oceanography and provides a history of the oceanographic research performed by its scientists. Chapter 12 covers the history of the Deep Sea Drilling Project.

Van Andel, Tjeerd H. "Deep-Sea Drilling for Scientific Purposes: A Decade of Dreams." *Science* 160 (June 28, 1968): 1419-1424. A classic article summarizing the results of scientific ocean drilling up to the start of the Deep Sea Drilling Project.

Warme, John E., Robert G. Douglas, and Edward L. Winterer, eds. *The Deep Sea Drilling Project: A Decade of Progress.* Tulsa, Okla.: Society of Economic Paleontologists and Mineralogists, 1981. A volume of scientific papers discussing the results of oceanographic research based on sediments and rocks recovered by DSDP. Best suited to those with some scientific background.

West, Susan. "Diary of a Drilling Ship." *Science News* 119 (January 24, 1981): 60-63.

_____. "DSDP: Ten Years After." *Science News* 113 (June 24, 1978). Summarizes the choices facing the DSDP in 1978: whether to continue ocean drilling with *Glomar Challenger* or to seek a larger and more sophisticated drilling vessel.

_____. "Log of Leg 76." *Science News* 119 (February 21, 1981): 124-127. These articles tell a reporter's story of seven days aboard *Glomar Challenger* during Leg 76 drilling operations in the Atlantic Ocean off Florida.

Wilhelm, Helmut, et al., eds. *Tidal Phenomena.* Berlin: Springer, 1997. A collection of lectures from leaders in the fields of earth sciences and oceanography, *Tidal Phenomena* examines earth's tides and atmospheric circulation. Complete with illustrations and bibliographical references, this book can be understood by someone without a strong knowledge of the earth sciences.

See also: Deep-Earth Drilling Projects; Engineering Geophysics; Ocean Drilling Program; Plate Tectonics; Relative Dating of Strata; Seismic Reflection Profiling; Volcanism.

OXYGEN, HYDROGEN, AND CARBON RATIOS

Oxygen-18/oxygen-16, deuterium/hydrogen, and carbon-13/carbon-12 ratios in rocks are heavy- to light-isotope ratios of oxygen, hydrogen, and carbon, respectively. These ratios can give clues to the geologic conditions under which rocks were formed.

PRINCIPAL TERMS

- **carbonate:** a mineral containing the carbonate ion, which is composed of one carbon atom and three oxygen atoms
- **geothermometers:** minerals whose components can be used to determine temperatures of mineral formation
- **ion:** an atom that has either lost or gained electrons
- **isotopes:** atoms with an identical number of protons but a different number of neutrons in their nuclei
- **isotopic fractionation:** the enrichment of one isotope relative to another in a chemical or physical process; also known as isotopic separation
- **limestone:** a sedimentary rock composed predominantly of calcite
- **mineral:** a natural substance with a definite chemical composition and an ordered internal arrangement of atoms
- **radioactive isotope:** an isotope of an element that naturally decays into another isotope
- **stable isotope:** an isotope of an element that does not change into another isotope

STABLE ISOTOPES

Rocks are divisible into three major types: igneous, sedimentary, and metamorphic. Igneous rocks are formed from hot, molten rock material called magma. Sedimentary rocks are formed by the compaction and cementation of mineral grains and rock fragments, which collectively are called sediments. Metamorphic rocks are formed by the alteration of rocks caused by increased heat and pressure and interaction with water in pore spaces and fractures. The temperature of formation and the types of water that interacted with the rocks are among the geologic conditions that scientists can determine from stable isotope ratios in rocks.

Rocks are aggregates of minerals. Minerals are natural compounds made of atoms that are arranged in an ordered fashion. The atoms in minerals exist in different isotopes. Some isotopes are unstable, or radioactive; others are stable. Radioactive isotopes, or parent isotopes, change into isotopes of other elements, or daughter isotopes. This change, which is called radioactive decay, occurs at a constant rate and can be determined via experimental work. The amounts of parent and daughter isotopes in rocks and the decay constant of the isotope are used by scientists to determine rocks' ages. Unlike radioactive isotopes, stable isotopes do not decay into isotopes of other elements. Thus, stable isotopes are not used for age-dating purposes. Rather, the ratios of stable isotopes—particularly of low atomic number elements, such as hydrogen, carbon, and oxygen—are used to determine the geologic conditions under which rocks are formed.

IONIC AND COVALENT BONDS

The three elements oxygen, hydrogen, and carbon are key components of the minerals which make up rocks. Oxygen, the most common element in the crust, the topmost layer of the earth, is found in all kinds of rocks. Carbon is abundant in coal. Carbon and oxygen are important components of a fairly common group of sedimentary rocks called limestones and their metamorphosed product, marble. Hydrogen and oxygen are constituents of water and can be used to characterize the types of watery solution that interact with rocks.

Oxygen, hydrogen, and carbon form bonds that range from ionic to covalent. Ionic bonds are formed by the electrostatic attraction of adjacent atoms; covalent bonds are formed by the sharing of electrons and are stronger bonds. Scientists have found that bonding characteristics affect the behavior of isotopes in natural conditions; elements which form only one type of bond in all conditions are not useful for isotopic work.

ISOTOPIC SEPARATION

Oxygen, hydrogen, and carbon all have low atomic weights. The relative mass difference between the heavy and light isotopes is large for such elements, unlike the elements of high atomic weight. For example, deuterium, the heavy stable isotope of hydrogen, is heavier than the light isotope of hydrogen by about 100 percent. In contrast, the stable strontium isotopes differ from each other by only 1.2 percent. Some conditions and processes favor the incorporation of heavy isotopes into a material; others favor light isotopes. Thus, depending on the geologic conditions, there will be a difference between the heavy- to light-isotope ratio in one mineral and the ratio of the same isotopes in another mineral or in the source material. This syndrome is called isotopic separation or fractionation. Isotopic separation is significant and detectable only for elements with low atomic numbers, such as oxygen, carbon, and hydrogen. The conditions and processes that cause separation include temperature, evaporation and diffusion, and oxidation reduction reactions.

A mineral is a compound formed by the chemical bonding of atoms. The internal energy of a mineral is controlled by factors including the vibration of atoms. The atoms of light isotopes vibrate with higher frequencies than do atoms of heavy isotopes. Since the bond strength between atoms depends on the vibration frequencies of atoms, light isotopes are weakly bonded. Such bonds can be broken comparatively easily upon dissolution or by bacterial action. Significantly, with increased temperature, the vibrational frequencies of all isotopes of the same element become nearly equal. Therefore, all other factors being equal, a mineral which is formed at low temperatures will contain a higher heavy- to low-isotope ratio than a similar mineral which is formed at high temperatures. A change in temperature during mineral formation can cause isotopic separation. Consequently, scientists can use isotopic separation to determine temperatures of mineral formation; in other words, isotope ratios can be used as geothermometers. Commonly, oxygen-isotope ratios in two different oxygen-bearing minerals which were formed at the same time and from the same source are used to determine the temperature at which a rock formed.

WATER-ROCK INTERACTIONS

Applications of oxygen isotope work have provided scientists with additional insights into the formation of rocks. It is now known that groundwater interacts with magmas. Part of this water is incorporated into the magma, and part of it circulates through rocks, changing their nature (metamorphosing them) in

the process. Furthermore, it is now possible to determine whether a magma has been contaminated by the incorporation and subsequent melting of roof rocks.

Evaporation and diffusion also cause isotopic separation. Water is a compound of oxygen and hydrogen. During evaporation, the light isotopes of oxygen (oxygen-16) and of hydrogen break through the water surface and escape to the atmosphere in the form of water vapor, while the heavy isotopes, oxygen-18 and deuterium, concentrate in reservoirs. Similarly, molecules with lighter isotopes move across a boundary (diffuse) faster than the same molecules with heavier isotopes. Evaporation and diffusion lead to different waters' having different isotopic ratios. Scientists have determined oxygen-18/oxygen-16 and deuterium/hydrogen ratios in many kinds of water: lakes, rivers, rain, snow, oceans, and water that comes from molten rocks. When these waters enter into interconnected pore spaces and fractures of rocks and circulate through the rocks at depth, they become hot solutions. These solutions interact with the rocks, leading to the diffusion (movement) of atoms from rocks to the solutions and from the solutions to the rocks. Such water-rock interaction changes the nature of both the rocks and the solutions. From stable isotope ratios, scientists can determine the nature of the rock, the nature of solution, and the type of water-rock interaction.

OXIDATION AND REDUCTION

Oxidation reactions are another cause of isotopic separation. Atoms are said to be in an oxidized state if they have a lower number of electrons and in a reduced state if they have a higher number of electrons when compared with other atoms of the same element. For example, the element carbon can occur in the form of C^{+4}, C, and C^{-4}. C^{+4} is a highly oxidized state of carbon, a positively charged ion formed by the loss of four electrons. Such ions combine with negative ions of other elements, such as oxygen ions, to form compounds such as carbon dioxide. In a highly reduced state, carbon atoms gain four electrons, forming the negatively charged carbon ion, C^{-4}. These ions combine with positively charged ions to form compounds such as methane.

Oxidized carbon is enriched in the heavy carbon isotope carbon-13, and reduced carbon is enriched in carbon-12. Thus, if the same source material were to permit the formation of two compounds, one with oxidized carbon and the other with reduced carbon, the one with the oxidized carbon will be enriched in carbon-13. The application of this simple principle is quite involved, however, because carbon can cycle through living organisms and other environments.

Green leaves of plants photosynthesize carbon dioxide, reducing the carbon and making it part of organic compounds. This reduction is done in stages, which are different in different plants. Thus, some plant types can be distinguished by their carbon-13 values. Generally, land plants have lower values of carbon-13 than marine plants; however, marine algae have values within the range of land plants. Evaporation leads to the enrichment of seawater in carbon-13. Condensation in clouds leads to the enrichment of carbon-13 in raindrops as compared with water vapor. Consequently, repeated evaporation and rain cause seawater to be richer in carbon-13; the atmosphere is lower. Carbon dioxide in soils and groundwater has an even lower carbon-13 value, because the carbon there has cycled through decaying plants.

DIAGENESIS

It appears that freshwater limestone should have lower carbon-13 values than marine limestone. Also, marine limestone should have variable carbon-13 values, depending on the amount of carbon inherited from algae (lower values indicating higher algal content). Most limestones, however, undergo a change called diagenesis, which involves the reconstitution of carbonate minerals under conditions different from the original ones. That makes it difficult to determine precisely the original conditions under which the rock formed or the subsequent conditions that resulted in the diagenetic change. Commonly, low values of carbon-13 are obtained from the diagenesis of marine limestones.

ANALYTICAL TECHNIQUES

Scientists use many analytical techniques, including mass spectrometers, mass spectrographs, and ion microprobes, to determine isotopic ratios. All these techniques utilize the fact that different isotopes separate from each other and arrive at a detector at different speeds when they travel through a magnetic field from an ionization chamber, in which a sample containing the elements is bombarded by

electrons or by (in the case of the microprobe) a beam of negatively charged oxygen atoms.

The separated isotopes are detected electronically in mass spectrometers. In mass spectrographs, they are detected by nonelectronic methods, such as photographic devices. For these methods of analysis, the elements or molecules of interest are chemically separated and introduced into an ionization chamber in a gaseous form, or they are deposited as solids on filaments that are then vaporized in the ionization chamber. In the ion microprobe method, chemical separation of the sample is not necessary, and the original sample does not have to be destroyed. A small sample of a rock is polished and then coated with gold or carbon. A beam of negatively charged oxygen is focused on the sample, over an area of less than 0.01 millimeter. That causes ionization of the sample; the ions are accelerated through a magnetic field to a detector, and the isotopes are measured by a mass spectrometer or spectrograph.

MASS SPECTROMETERS AND SPECTROGRAPHS

There are many different spectrometers and spectrographs, but the principles involved can be understood by considering one type of spectrometer. In this device, an appropriate voltage applied across a filament, possibly of tungsten, produces a stream of electrons. These electrons bombard a sample and cause the removal of electrons from the atoms of the sample. The resulting positive ions of different isotopes are accelerated by a high-voltage field and are collimated into a beam. Since the kinetic energies of isotopes of the same element are identical, the lighter isotopes travel faster than the heavier isotopes. The ion beam passes through a magnetic field, which is constructed in such a way that different isotopes are separated from each other as they exit the field and enter a collector cup. The accelerating voltage and the magnetic field can be adjusted so that an ion beam of one isotope can be focused through a collector slit to enter a detector cup. The focused ions are neutralized by electrons which flow through a resistor, and the voltage difference across the resistor can be measured with a voltmeter. The ensuing electrical signals can be digitized or, more commonly, displayed on a strip-chart record. A series of peaks and valleys—each peak representing an isotope, with the peak height being proportional to the abundance of the isotope—can be recorded by adjusting the accelerating voltage or the magnetic field, which would vary the ions being focused through the collector slits. In this way, the various isotopes and their relative abundances can be determined. In turn, the ratios of heavy to light isotopes can be calculated.

Modern commercial mass spectrometers, equipped with multiple collectors for the simultaneous detection of different isotopes and with digital computers, have improved both the speed of acquisition and the reliability of isotopic data.

APPLICATIONS OF STABLE ISOTOPIC GEOCHEMISTRY

Many kinds of chemical analysis can be used to determine the concentration of ions in groundwater. If different ions originate from different source areas, then these areas can be distinguished by such analysis. When the source areas produce the same ions, the isotopic ratios of the ions may be different because of different environmental conditions. In such instances, stable isotope geochemistry can be used to specify the source area, such as a groundwater-polluting factory. Stable isotope geochemistry, then, is useful in groundwater pollution studies. One of the uses of mass spectrometers is to identify elements by their isotopes, and by using a procedure called isotopic dilution, scientists can identify elements whose amounts in a sample are extremely small. The isotope dilution method provides the added advantage of identifying the groundwater pollutants.

Scientists have used stable isotopic geochemistry to study the diet of prehistoric humans. The remains of wood in ancient campfires have been analyzed for their stable isotopic ratios. Such ratios can help the researcher to distinguish between land and marine plants and between groups of land plants. The ion microprobe, with its ability to analyze particles smaller than 0.01 millimeter, has a potential application in forensic science, where trace amounts of hair and clothes are used to identify criminals. The isotopic dilution method is another procedure that can be applied to such an effort.

Habte Giorgis Churnet

FURTHER READING

Albarede, Francis. *Geochemistry: An Introduction.* 2d ed. Boston: Cambridge University Press, 2009. A good introduction for students looking to gain some knowledge in geochemistry. Covers basic topics in physics and chemistry; isotopes, fractionation, geochemical cycles, and the geochemistry of select elements.

Berner, Elizabeth K., and Robert A. Berner. *Global Environment: Water, Air, and Geochemical Cycles.* Upper Saddle River, N.J.: Prentice Hall, 1996. This book discusses the processes that sustain life and effect change on the earth. Topics include the hydrologic cycle; the atmosphere; atmospheric carbon dioxide and the greenhouse effect; acid rain; the carbon, sulfur, nitrogen, and phosphorus cycles; weathering; lakes; rivers; and the oceans. It is understandable to the college-level reader.

Cazes, Jack. *Ewing's Analytical Instrumentation Handbook.* 3rd ed. New York: Marcel Dekker, 2005. Updated from Ewing's original publication to include more current methods, techniques and analyses.

Ewing, G. W., ed. *Chemical Instrumentation.* Easton, Pa.: Chemical Education Publishing, 1971. An excellent publication on instrumentation written for a college-level audience.

Faure, Gunter. *Isotopes: Principles and Applications.* 3rd ed. New York: John Wiley & Sons, 2004. Originally titled *Principles of Isotope Geology.* An excellent book that integrates theory with practice and is easy to read.

Greenwood, Norman Neill, and A. Earnshaw. *Chemistry of Elements.* 2d ed. Oxford: Butterworth-Heinemann, 1997. An excellent resource for a complete description of the elements and their properties. The book is filled with charts and diagrams to illustrate chemical processes and concepts. Bibliography and index.

Hoefs, J. *Stable Isotope Geochemistry.* 6th ed. New York: Springer-Verlag, 2009. This useful source provides ample field examples.

Krauskopf, Konrad B. *Introduction to Geochemistry.* 3rd ed. New York: McGraw-Hill, 2003. A section in this book provides a distillation of the nature and uses of stable isotope work.

Krebs, Robert E. *The History and Use of Our Earth's Chemical Elements: A Reference Guide.* 2d ed. Westport, Conn.: Greenwood Press, 2006. This book defines geochemistry and examines its principles and applications. A good resource for the layperson interested in the field of geochemistry and in the earth's elements. Accessible to high school readers. Illustrations, charts, and bibliography.

Richardson, S. M., H. Y. McSween, Jr., and Maria Uhle. *Geochemistry Pathways and Processes.* 2d ed.

Englewood Cliffs, N.J.: Prentice-Hall, 2003. A chapter in this book offers an accurate summary of stable isotope geochemistry.

Santamaria-Fernandez, Rebeca. "Precise and Traceable Carbon Isotope Ratio Measurements by Multicollector ICP-Ms: What Next??" *Analytical & Bioanalytical Chemistry* 397 (2010): 973-978. This article reviews a new method of mass spectrometry used to measure carbon ratios. Provides a good description of mass spectrometry methodology for the intermediate or beginner MS analyzer.

Szczepanek, MalGorzata, et al. "Hydrogen, Carbon, and Oxygen Isotopes in Pine and Oak Tree Rings from Southern Poland as Climatic Indicators in Years 1900-2003." *Geochronometria: Journal on Methods & Applications of Absolute Chronology* 25 (2006): 67-76. This article discusses the use of carbon, hydrogen, and oxygen ratios in isotope-dating methods. Written technically for a reader with a scientific background, but still accessible to the advanced layperson.

Valley, J. W., H. P. Taylor, Jr., and J. R. O'Neil, eds. *Reviews in Mineralogy*. Vol. 1b, *Stable Isotopes in Higher Temperature Geologic Processes*. Washington, D.C.: Mineralogical Society of America, 1986. A collection of work from authorities on various aspects of the topic, with a primary emphasis on igneous and metamorphic rocks and ore deposits.

See also: Carbon Sequestration; Elemental Distribution; Environmental Chemistry; Fluid Inclusions; Freshwater Chemistry; Geochemical Cycle; Geothermometry and Geobarometry; Nucleosynthesis; Phase Changes; Phase Equilibria; Rock Magnetism; Water-Rock Interactions.

P

PETROGRAPHIC MICROSCOPES

The petrographic microscope is an essential tool for studying the mineral content and texture of fine-grained rocks. It also provides a rapid and accurate means for identifying minerals through their optical properties.

PRINCIPAL TERMS

- **anisotropic crystal:** a crystal with an index of refraction that varies according to direction with respect to crystal axes
- **birefringence:** the difference between the maximum and minimum indices of refraction of a crystal
- **crystal axes:** directions in a crystal structure with respect to which its molecular units are organized
- **index of refraction:** the ratio of the speed of light in a vacuum to its speed in a particular transparent medium
- **interference:** the combining of waves or vibrations from different sources so that they either are in step and reinforce each other or are out of step and oppose each other
- **interference color:** a color in a crystal image viewed under crossed polars, caused by subtraction (cancellation) of other colors from white light by interference
- **interference figure:** a shadow shape caused by the blocking of polarized light from certain areas of a crystal image
- **polarization:** a method of filtering light so that only rays vibrating in a specific plane are passed
- **principal vibration directions:** directions in a crystal structure in which light vibrates with maximum or minimum indices of refraction
- **retardation:** the progressive falling behind of part of a ray vibrating in a slower direction compared to a part vibrating in a faster direction

POLARIZED LIGHT

The crystals of many rocks and deposits are too small to be distinguished—much less identified—by the naked eye. Individual crystals can be distinguished under an ordinary microscope, but identification in this manner is still difficult. A powerful improvement was discovered in 1828 by Scottish geologist William Nicol when he applied to the microscope his newly invented polarizing prisms cut from calcite crystals and found that different minerals have very distinctive appearances in polarized light. The modern petrographic microscope is a refinement of Nicol's discovery.

The petrographic microscope is similar to a standard biological microscope but has adaptations for use with polarized light. A polarizer (or "polar") beneath the condenser lenses polarizes the light before it passes through the specimen. A circular, rotatable stage allows the slide and specimen to be turned with respect to the polarized light. A second, removable polarizer called the analyzer, oriented crosswise to the lower polar and above the specimen, can be inserted into the light path in the tube, providing "crossed polars." There are other accessories used for special purposes.

A polarizing microscope adapted to view reflected rather than transmitted light is commonly used by metallurgists to study the identity, size, and texture of metal crystals produced in industrial processes. Economic geologists use the reflection microscope to identify opaque, metal-bearing minerals and to determine their abundance in samples from an ore body.

INTERFERENCE COLORS

Under the polarizing microscope, crystals show bright, distinctive colors called interference colors. As the stage is turned, these colors move and change or become dark ("extinct") in ways that can be used to identify the minerals. It is the interaction of light with the crystal structure that causes the distinctive behavior.

Light is an electromagnetic wave. As a ray of light travels along, its electric field strength oscillates back and forth transverse to the path of the ray, somewhat like the vibration that travels along a horizontal rope when it is shaken up and down at one end. The vibration direction and the travel path are perpendicular. The distance traveled by the ray between one maximum of the transverse field and the next is the wavelength of the light. Each color of visible light has its own particular wavelength, ranging from roughly 700 nanometers for red to 400 nanometers for violet (a nanometer is one-billionth of a meter). White light is a mixture of all colors.

INDEX OF REFRACTION

All light travels at the same speed in a vacuum, but in a transparent medium, it is slowed by interaction with matter. The speed is characteristic of a given medium and is indicated by its index of refraction; the greater its index, the more slowly light moves through it. The speed in air (index 1.0003) is essentially the same as in a vacuum (1.0000); however, in quartz (index 1.54), it is only 65 percent of that value, and in diamond (index 2.42), it is only 40 percent.

The atoms or molecules in crystals are arranged in a strict order, repeated over and over to make the crystal structure (or "lattice"). The structure is responsible for many characteristic features of minerals, such as the natural shapes and faces of crystals and the likelihood of breaking along flat surfaces called cleavages. The pattern of any given mineral is very distinctive, so the ways in which light interacts with the different structures can be used to identify the various minerals.

Minerals are classified as isotropic or anisotropic. In isotropic crystals, light can travel in any direction at the same speed, and the index of refraction has the same value for all orientations. Isotropic minerals all have a molecular structure and spacing that are identical along each of three perpendicular crystal axes. Some common examples are halite (rocksalt) and garnet. Glass substances, which are equally disordered in all directions, are also isotropic.

In anisotropic minerals, the molecular structure and spacing are different along one or more crystal axes. In such crystals, the interaction of light vibrations with matter (and therefore, the speed of light and the index of refraction) depends on the direction of travel. The analyst, to identify the minerals, must know the detailed differences among the crystal structure systems and how light interacts

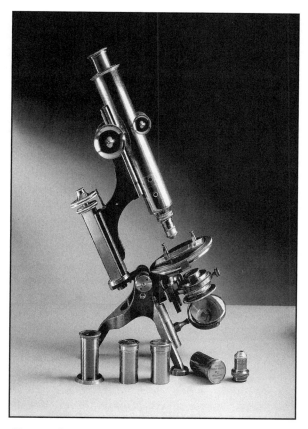

Close-up of a petrographic microscope. (De Agostini/Getty Images)

with them. For the present purposes, however, a single fact is important: The speed of light traveling along most paths through anisotropic crystals depends on the direction of vibration. Even along the same path, light vibrating in one direction may be faster or slower than light vibrating in another, with both vibration directions being perpendicular to the travel path.

The polarizer under the stage allows the analyst to select the vibration direction. Without it, the rays of light rising vertically through the specimen vibrate parallel to the stage but with random orientation. The polarizer absorbs all these rays except those vibrating in one specific direction, which is usually fixed back-to-front or sideways in the field of view. The analyst turns the specimen on the stage to change the orientation of the crystals to the polarized light.

RETARDATION

An anisotropic crystal viewed down the microscope tube, in general, has one direction with maximum index of refraction and another perpendicular to it with minimum index. A ray vibrating parallel to the first would be the slowest, while the other would be the fastest. A ray oriented in any other direction actually separates into two parts, each part vibrating in one of the two directions but following the same vertical path. The part vibrating in the higher index direction is slower than is the other and falls progressively farther behind; it is said to be retarded. The amount of retardation depends on how far the parts travel (the thickness of the crystal) and on the difference in their speeds (and, therefore, on the difference of their indices of refraction, called the birefringence). Because each color has its own particular wavelength, the amount of retardation affects whether the vibrations of the two parts of a given color are in step with each other (in phase) as they exit the crystal or determines how much they are out of step.

The analyzer (the upper polarizer) blocks all light vibrating parallel to the lower polarizer because it is oriented at right angles. Thus, the glass of the slide and any isotropic crystals appear black, as they do not alter the polarization. Similarly, the two separated parts of a ray from an anisotropic crystal, if they happen to emerge in step, recombine in the original polarization, and this light is blocked as well. Although if one part is retarded out of step with the other (so that, recombined, they have a rotating "elliptical polarization"), the analyzer in effect deals with each part individually. It resolves each part once more and allows only those portions parallel to the analyzer vibration direction to pass. The passed portions of each part now vibrate in the same plane but are out of step with each other. Depending on how much they are out of step, the vibrations may reinforce each other and strengthen the color or oppose each other and weaken the color. Colors that are weakened or canceled are subtracted from the original white light, and what remains to be viewed is the complementary color. A sheet of mica placed between crossed Polaroids shows this effect well even without a microscope.

MICHEL-LEVY CHART

The interference colors that result from this process are one of the most striking features of crystals viewed under crossed polars. Because they result from the subtraction of specific colors from white light—some more and some less—they fall in a sequence that is distinctly different from an ordinary spectrum. Beginning with black when there is no retardation (the passed rays are in step), as retardation increases, the colors go through gray and white to orange and red for the first "order," then through several cycles from red through blue for higher orders, eventually merging to pinks and greens, and finally to more or less white for very high orders. The sequence is displayed on a Michel-Levy chart (which shows the sequence of colors as a function of birefringence). The chart is named after French geologist Auguste Michel-Levy. The colors that actually appear in a given crystal give important information about the mineral.

SAMPLE PREPARATION

Samples for microscopic examination are usually prepared either as a powder or as a thin section. The powder is made by crushing a mineral grain and screening it very finely; a small amount of the powder is then placed on a microscope slide with a drop of oil. The thin section is made by sawing a slice off of the sample, gluing it to a slide, then further sawing and grinding the slice until it is only 0.03 millimeter thick. In such thin samples, most minerals are transparent or translucent, although metals and many sulfide minerals are still opaque. Special reflection techniques can be used to examine opaque minerals.

Microscopic analysis of crushed mineral grains (powder) is the most efficient way to identify any mineral (and some nonmineral substances) whose crystals are large enough to be distinguished with a microscope. Thin sections are less efficient, but they have other advantages because they preserve the structure of the original sample; they are essential for the study of fine-grained rocks. By calculating the relative abundance of each kind of mineral, examining the shapes of grains and the ways they contact each other, and studying the distribution of grains and larger structures like bedding, the analyst can identify the rock type, estimate its properties, and interpret clues to its history. For example, a thin section of sandstone under the microscope would show the shape of the sand grains, fine details of its bedding, the amount of cement between grains, the amount of empty space, or porosity, and the presence and distribution of any mineral grains besides the quartz sand. This information could be used to estimate its mechanical strength for engineering purposes, its ability to hold water or oil, or its potential as a quarry stone, raw material for glassmaking, or ore of uranium.

MINERAL IDENTIFICATION

With either preparation, the first goal is to identify the minerals present by observing their visible properties. Features of shape, such as a characteristic crystal form, habit, cleavage, or fracture (keeping in mind that only a cross section is visible), give the first clues to identity. For example, garnets often exhibit a polygon-like cross section of their characteristic crystal form, and mica usually shows its perfect one-directional cleavage. Typical colors may be present (with polars uncrossed), although they are much fainter than in a hand sample. Some minerals, like tourmaline or biotite mica, change color as they are turned in the polarized light; these minerals are called pleochroic.

The relief of a crystal indicates the contrast between its index of refraction and that of its surroundings. A mineral with high relief appears to stand out from its background and have very distinct boundaries, while one with low relief is hard to distinguish from its background. If neighboring minerals or a medium (mounting or immersion) of known index are present, the analyst can estimate the index of refraction of an unknown mineral from its relief. The analyst can measure the index of minerals in powdered form exactly by comparison with standard index oils (called the immersion method). If the index of the mineral matches the oil closely, the grain boundary almost disappears. Anisotropic crystals require a different oil for each vibration direction. Having measured the indices, the analyst can then consult a table to identify the mineral.

If the indices cannot be measured directly, as in a thin section, the birefringence (the difference between maximum and minimum indices) gives useful information for identification. The interference colors in a crystal depend on its birefringence and its thickness (usually approximately 0.03 millimeter in a thin section). The analyst compares the highest interference colors found in a crystal to a Michel-Levy chart, determines the corresponding birefringence, and consults a table to identify the mineral.

The relationships between the vibration directions and visible features such as crystal faces and lines of cleavage give another clue to identity. At every quarter turn as an anisotropic crystal is turned on the stage, there is a point at which the crystal becomes completely dark, or extinct. Extinction occurs whenever the crystal's vibration directions are parallel to the polarizer or analyzer. The angle between an extinction direction and a crystal face or cleavage can distinguish between many otherwise similar minerals, such as the pyroxenes and amphiboles. Isotropic minerals like garnet are extinct at all positions of the stage.

INTERFERENCE FIGURES

Interference figures provide another powerful means of identifying crystals. They are shadows with distinctive shapes that appear with crossed polars and diverging light because polarized light is blocked from certain areas of the crystal image. Special lenses are used to cause the light to diverge and to change the focus of the eyepiece.

The shadow figures, which depend on the nature and orientation of the crystal, take the shapes of Maltese crosses or sweeping curves that move in distinctive ways as the stage is turned. The analyst can use them to determine many details about the crystal structure, the relationships of the vibration directions, and other features useful for identification.

COMMERCIAL AND PUBLIC-SAFETY APPLICATIONS

The petrographic microscope is an important tool for identifying many kinds of minerals and other substances that cannot easily be distinguished by ordinary physical and chemical tests. It has been used, for example, to determine the nature of corrosion products on metal surfaces. The corrosion products indicate which chemical reactions might be responsible for the damage and, therefore, how the surfaces might be protected. In another application, the microscope has been used to study the different materials traded commercially or displayed in museums as "jade." Officially, the name "jade" is applied to rocks composed of either an amphibole called nephrite or a pyroxene called jadeite, but the microscope revealed that much of what has been called jade is really composed of other minerals similar in appearance. The study showed historically significant patterns in the use of different kinds of jade in various cultures.

The petrographic microscope has many applications to areas in which geology touches on the economy or on public safety. Rock that has been sheared and fractured, as by faulting, shows distinctive texture and structure in thin section. Knowledge of these features in the rock of a given region can be important in the prediction of earthquake potential or in the evaluation of stability for engineering projects. Thin sections also show the amount of empty space, or porosity, between the grains in a rock, which is essential for estimating the potential of the rock for bearing oil, for carrying groundwater, or for allowing the passage of pollutants and radioactive waste. In the mining industry, thin sections are used to identify and evaluate the abundance of ore minerals and also to determine their grain size and how they are locked into the rock structure; all these factors determine whether the minerals can be recovered at a profit.

There are many anisotropic substances besides minerals. Whenever there is a distinct alignment of long molecules in a substance, polarized light may interact with it and reveal interference colors. Some biological tissues, structures in cells, plastics, and glasses have such anisotropic structures, and polarized light is useful for studying them. In one application, polarized light is used to study the distribution of stress in engineering structures such as machine parts and architectural members. The structure is modeled with a plastic such as Lucite and viewed through crossed polars. When the model is placed under load, the plastic develops interference colors that are concentrated at points of maximum stress.

Fiber-optic systems for transmission lines and optical switching devices developed for telephone and computer communications depend on the differences of the indices of refraction of their various parts. The polarizing microscope, which shows the differences by interference colors, is a key instrument for designing and testing such systems. Although microscope systems using other kinds of radiation are becoming widely employed, the polarizing light microscope will continue to hold a central importance both in the field of geology and outside it.

James A. Burbank, Jr.

FURTHER READING

Barker, James. *Mass Spectrometry.* 2d ed. New York: Wiley, 1999. A college text concerning the field of mass spectrometry and its protocol and applications. There is a fair amount of analytical chemistry involved, so the reader without a scientific background may have a difficult time. Bibliographical references and index included.

Craig, James R., and D. J. Vaughan. *Ore Microscopy and Ore Petrography.* 2d ed. New York: John Wiley & Sons, 1995. The first three chapters (on the reflection microscope, preparation of polished specimens, and qualitative properties of minerals) give a compact overview of how the technique of reflection microscopy is used. Knowledge of basic mineralogy is assumed. Most of this college-level text is beyond the interests of the casual reader. The index is thorough, but the tables and references are technical.

Gribble, C. B., and A. J. Hall. *Optical Mineralogy: Principles and Practice.* New York: Chapman and Hall, 1993. A popular college textbook that provides a basic and comprehensible description of the theories, protocols, and applications involved in the field of optical mineralogy. Bibliography and index. This text covers light microscopy in depth, but lacks coverage of electron microscopy.

Hecht, E. *Optics.* 4th ed. Reading, Mass.: Addison-Wesley, 2001. A college-level text with considerable advanced mathematics, so well written and illustrated that a courageous nonmathematical reader can ignore the equations and still gain insight on many topics, especially polarization. Interesting examples and home experiments. Technical bibliography. An excellent index including many historical references.

Jambor, J., et al., eds. *Advanced Microscopic Studies of Ore Minerals.* Nepean, Ontario: Mineralogical Association of Canada, 1990. A collection of essays and lectures written by scientific experts in their respective fields, this volume is at times somewhat technical. It deals with topics such as mineral microscopy, the optical properties of ores, and optical microscopy. Anyone interested in petrographic microscopy is sure to find this a useful reference tool.

Kerr, Paul E. *Optical Mineralogy.* 4th ed. New York: McGraw-Hill, 1977. A college-level textbook emphasizing the identification of minerals in and interpretation of thin sections. The chapters on theory presume a knowledge of basic optics and mineralogy but provide a good summary of applications to the microscope. An ample bibliography, a selective index, excellent tables, and individual mineral descriptions make this a solid reference work for identifying common minerals.

Kile, Daniel E. *The Petrographic Microscope: Evolution of a Mineralogical Research Instrument*. The Mineralogical Record, 2003. Describes the techniques and applications of the petrographic microscope.

Klein, Cornelis, and Barbara Dutrow. *Manual of Mineral Science*. 23rd ed. New York: John Wiley & Sons, 2008. A classic college-level introduction to mineralogy, updated numerous times since its original publication in 1912. Contains a thorough discussion of crystal systems and concise descriptions of all common minerals, including essential optical data. Chapters 13 and 14 contain a summary of optical microscopy, X-ray and electron imaging methods, and mass spectrometry. Well illustrated and indexed, with key references after each chapter.

MacKenzie, W. S., and C. Guilford. *Atlas of Rock-forming Minerals in Thin Section*. New York: Halsted Press, 1980. The bulk of this short atlas consists of excellent color photographs of thin sections designed to show how common minerals appear with crossed and uncrossed polars. The minimal text identifies the minerals and key points of interpretation. Illustrates typical features but also shows the beauty of rocks in thin section.

Nesse, William D. *Introduction to Optical Mineralogy*. 3rd ed. New York: Oxford University Press, 2004. A good introduction to mineralogy and optical mineralogy. Nesse clearly discusses the procedures and protocols of optical mineralogy and petrographic microscopy. Illustrations, bibliographic references, and an index.

Rochow, T. G., and E. G. Rochow. *An Introduction to Microscopy by Means of Light, Electrons, X Rays, or Ultrasound*. New York: Plenum Press, 1979. Thorough coverage of basic principles and the construction of various types of microscope. Very readable (high school level), with little scientific background presumed. The reader willing to cross-reference in this well-indexed volume will find technical terms carefully defined. Abundant useful illustrations, including views of thin sections. Michel-Levy chart included. Broad bibliography.

Sinkankas, John. *Mineralogy*. New York: Van Nostrand Reinhold, 1975. A wide-ranging and effective introduction to the nature and properties of crystals, written for the amateur. Very helpful illustrations. Chapter 8 contains an easy approach to many concepts needed to understand the petrographic microscope but applied instead to the simpler polariscope. Roughly one-half of the 585 pages are devoted to individual minerals. Useful tables and selected bibliography, with index.

Stoiber, Richard E., and S. A. Morse. *Microscopic Identification of Crystals*. Reprint. Malabar, Fla.: Robert E. Krieger, 1981. A compact, college-level text emphasizing the use of immersion oils for identification. The explanation of polarization and interference is clear and detailed, with many helpful illustrations, using a mostly geometrical approach.

Walker, Hollis N., D. Stephen Lane, and Paul E. Stutzman. *Petrographic Methods of Examining Hardened Concrete: A Petrographic Manual*. Federal Highway Administration, U.S. Department of Transportation, 2006. Provides a succinct overview and general procedures of petrography followed by more specific details on studying concrete. Chapters 12-14 cover petrographic microscopy and sampling procedures.

Wenk, Hans-Rudolf, and Andrei Bulakh. *Minerals; Their Constitution and Origin*. New York: Cambridge University Press, 2004. This text provides an overview of mineralogy and petrology in a clear, concise manner. Contains an excellent chapter covering petrographic microscopy and mineral classification.

Winter, J. D. *Principles of Igneous and Metamorphic Petrology*. 2d ed. Pearson Education, 2010. A good undergraduate text, which covers techniques of modern petrology. It is recommended to have a geological dictionary on hand, as this book has some parts with very technical writing.

See also: Earthquake Engineering; Electron Microprobes; Electron Microscopy; Environmental Chemistry; Experimental Petrology; Fission Track Dating; Fluid Inclusions; Geologic and Topographic Maps; Neutron Activation Analysis; Rock Magnetism; X-ray Fluorescence; X-ray Powder Diffraction.

PHASE CHANGES

Phase changes among liquids, solids, or gases are important in many geologic processes. The formation of ice from water, of minerals from magma, of gases bubbling out of magma, and of halite (sodium chloride) precipitating out of a lake are examples of phase changes in nature. Many of these phase changes aid in the understanding of deposits that are of economic importance.

PRINCIPAL TERMS

- **gas:** a substance that can spontaneously fill its own container
- **igneous rock:** a rock formed from molten rock material (magma or lava)
- **liquid:** a substance that flows
- **magma:** a liquid, usually composed of silicate material and suspended mineral crystals, that occurs below the earth's surface
- **metamorphic rock:** a rock in which the minerals have formed in the solid state as a result of changing temperature or pressure
- **phase:** that part of nature that has a more or less definite composition and thus homogeneous physical properties, with a definite boundary that separates it from other phases
- **precipitate:** the process in which minerals form from water or magma and settle out of the liquid
- **sedimentary rock:** a rock that has formed from the accumulation of sediment from water or air; the sediment might be fragments of rocks, minerals, organisms, or products of chemical reactions
- **solid:** a substance that does not flow and has a definite shape

POLYMORPHS

A phase is a physically distinct and mechanically separable portion of a mixture. Different phases in a mixture may be of differing chemical composition, or they may be identical in composition. Thus, the rock, granite, containing the minerals quartz, feldspar, and mica, is made up of three physically distinct components or three phases. Also, water, ice, and vapor are three distinct phases even though they share the same chemical composition.

Three different minerals in a rock—for example, quartz, plagioclase, and alkali feldspar—constitute three separate phases. The number of separate mineral grains is not the same, however, as the number of phases. There might be 231 grains of quartz, 257 grains of alkali feldspar, and 199 grains of plagioclase in a given rock, but the rock does not contain this many different phases. Instead, there are only three phases in the rock, corresponding to the three different minerals with the same composition and physical properties. Minerals of the same composition with different crystal structures are called polymorphs. Calcite and aragonite are examples of polymorphs of calcium carbonate. Water, ice, and water vapor are polymorphs of dihydrogen oxide, and the spectacularly diverse minerals graphite and diamond are polymorphs of carbon. Ice may exist as several different phases depending upon the temperature and pressure.

LIQUID PHASES

There are also many liquid phases. Water is a liquid with the same composition as ice. Ice cubes and the water in which they float may be considered as two separate phases, since the ice and water have different physical properties (ice is lighter or less dense than water, for example) and are separated by boundaries. Melted rocks form liquid rock material called magma. The magma may be considered one phase, while any minerals suspended in it are considered separate phases.

Two or more liquids may coexist as separate phases if they do not mix. Oil and water form separate layers with a boundary between them, so they are separate phases. Being less dense than the water, the oil floats on top of the water. Similarly, carbonate-rich magmas may not mix with many silicate-rich magmas, and they may form separate liquid phases. Water and ethyl alcohol, in contrast, mix in all proportions and thus form only one homogeneous phase with no boundary surfaces. In a similar fashion, two silicate magmas of somewhat different composition may mix and form a homogeneous magma of an intermediate composition.

Ice changes to water at 0 degrees Celsius; at one atmosphere pressure and 100 degrees Celsius, water changes to steam. Such phase changes may differ with changes in atmospheric pressure. At about 200 times atmospheric pressure, the boiling point of water is more than 300 degrees Celsius, and the freezing point is less than 0. At a pressure of less than 0.006 atmosphere, liquid water is not stable; rather, ice changes directly to water vapor at less than 0 degrees Celsius without any intervening water phase. There is even one temperature (0.1 degree Celsius) and pressure (0.006 atmosphere), called the triple point, at which ice, water, and steam coexist.

The phase relations of water have direct application to understanding the formation of certain features on Mars. Some features appear to have been formed by a running fluid such as water. The atmospheric pressure of Mars is currently too low for the planet to have any running water. Billions of years ago, however, Mars's atmospheric pressure might have been high enough to permit stabilized water to exist there. Thus, water could have been an erosional agent on Mars early in that planet's history.

SOLID AND GAS MIXING

Though it may seem difficult to visualize, some solids of different composition may be able to mix partially or completely in all proportions. The silicate mineral olivine, for example, can accommodate any ratio of magnesium to iron into its composition; the magnesium end member is said to have a complete solid solution with the iron end member. Gases, in contrast to solids and liquids, mix in all proportions. The earth's atmosphere, for example, is a fairly homogeneous mixture of nitrogen (the predominant gas) and oxygen. There are also small amounts of other gases, such as water vapor and carbon dioxide.

PHASE CHANGES IN METAMORPHIC ROCKS

Important phase changes occur among solids in metamorphic rocks. Metamorphic rocks were formed from other rocks by chemical reactions in the solid state because of differing temperatures and pressures. The minerals kyanite, andalusite, and sillimanite are different aluminum silicate minerals with the same composition occurring in metamorphic rocks. Phase changes among these three solids depend on temperature and pressure, as in the ice-water-steam system. No liquid or gas, however, is involved in the aluminum silicate minerals. The triple

Kyanite (aluminum silicate) is a polymorph with andalusite and sillimanite, meaning that it shares the same chemistry but has a different crystal structure. (Charles D. Winters/Photo Researchers, Inc.)

point of the aluminum silicate minerals is at about 600 degrees Celsius and nearly 6,000 times atmospheric pressure (6 kilobars), so changes among these minerals take place only deep within the earth. Sillimanite is stable from about 600 degrees Celsius and 5-6 kilobars up to more than 800 degrees and 1-11 kilobars. In contrast, andalusite is stable at pressures up to only 6 kilobars over a wide range of temperatures (200 to 800 degrees). Kyanite is also stable over a wide temperature range but at a higher pressure for a given temperature than is the case for either andalusite or sillimanite. For a geologist, then, knowing which of these aluminum silicate minerals is present in a rock helps to show the range of temperature and pressure at which the rock formed. There are also solid-to-liquid phase changes in sedimentary systems. A variety of minerals may crystallize or precipitate from water to form sediments.

PHASE CHANGES IN IGNEOUS ROCKS

In igneous rocks, too, there are many phase changes between solids and liquids. Igneous rocks form from the crystallization of minerals from magma or melted rock, usually of silicate composition. As magma slowly cools within the earth, it forms minerals that gradually either sink or float in the magma, depending on their density (weight in relation to volume). The minerals that are heavier or denser than the magma gradually sink, and the lighter or less-dense minerals gradually float upward. The magma composition gradually changes as the minerals are extracted, because the minerals' compositions are different from that of the magma. Magma forms by the melting of solid rock in the lower crust or upper mantle of the earth. The magma's composition will depend on the composition of the rock melted, the pressure, and the degree of melting. Also, the magma composition will be different from that of the solid. The melting of a typical rock in the upper mantle, for example, will produce a basaltic magma. A basaltic magma will produce a dark, fine-grained rock of low silica content, called basalt, when extruded at the surface of the earth. The melting of a silica-rich rock in the continental crust is more likely to produce magma with a high silica content. These high-silica magmas will crystallize to light-colored rocks called dacite or rhyolite when extruded at the surface.

SATURATION POINTS

The maximum amount of a mineral that may be dissolved in water is called its saturation point. Different minerals have different saturation points in water. Considerably less calcite (calcium carbonate mineral) may dissolve in water than gypsum (calcium sulfate mineral). Even more halite (sodium chloride) may dissolve in water than gypsum. If the saturation points for these minerals are exceeded, the minerals will begin to crystallize or precipitate and sink to the bottom of the water. Saturation points of minerals may be exceeded when water evaporates or when the temperature changes. If seawater is present in a bay in which evaporation exceeds the influx of new seawater, calcite, gypsum, and halite may precipitate, in that order, as the water gradually evaporates. Vast amounts of salt deposits of halite and gypsum are believed to have formed in this fashion during the geologic period called the Permian (about 250 million years ago) in Kansas and Oklahoma. Such salt deposits are not nearly as common as are limestones. Limestones are sedimentary rocks composed of mostly calcite. The calcium carbonate is believed to have precipitated in warm, shallow seas either by inorganic precipitation or by organisms forming calcite or aragonite. The precipitation of calcite or aragonite is aided by the evaporation of seawater in shallow seas and by warming of the water.

EXPERIMENTS AT ATMOSPHERIC PRESSURE AND TEMPERATURE

A variety of techniques are used to study phase changes. The technique selected to study phase changes depends on the pressure, temperature, and types of phases. The easiest phase changes to study are those involving precipitation of minerals from water solutions at atmospheric pressure and temperature. One of the intriguing problems in the study of sedimentary rocks, for example, is why among ancient rocks so much limestone that is made up of calcite and dolostone is composed of dolomite (a calcium/magnesium carbonate mineral), as modern sediments seem to be forming mostly aragonite and calcite. Little dolomite is apparently forming today. Experiments in the laboratory have helped geologists to explain such observations. The precipitation of calcite and aragonite in the laboratory is temperature-dependent. A temperature of about 35 degrees Celsius, for example, favors precipitation of needlelike crystals of aragonite. In contrast, a lower temperature of 20 degrees favors precipitation of mostly stubbier crystals of calcite.

To identify the minerals, scientists observe them under a microscope or by X-ray diffraction. X-ray wavelengths and the distances between atoms in the minerals are about the same, so the X-rays will be reflected off planes of atoms in the mineral. The angle of reflection depends on the distance between the atoms and the wavelength of the X-rays. Since every mineral has different spacings between atoms, the reflections of different minerals have different angles and serve as "fingerprints" for the minerals. Thus, calcite and aragonite can easily be distinguished. However, it is difficult to produce dolomite under any conditions in the laboratory. Such laboratory observations are consistent with the observed abundance of aragonite needles in warm, shallow seas and with the greater abundance of calcite forming from cooler waters. They are also consistent with the lack of observed dolomite formation.

The only dolomite that can form in the laboratory is produced by the conversion of calcite to dolomite in contact with concentrated waters with a high magnesium-to-calcium ratio. Thus, only under special geologic conditions below the land surface will calcite convert—slowly—into dolomite. Waters high in magnesium moving through calcite-rich rocks below the surface will convert the calcite into dolomite over long periods of time. This process may be occurring in certain places presently, though it simply cannot be observed.

EXPERIMENTS AT HIGHER TEMPERATURES AND PRESSURES

Other experiments in furnaces at high pressure tell geologists that aragonite is in reality stable only at a pressure much higher than atmospheric pressure. Aragonite is unstable at atmospheric pressure, so any aragonite forming presently should slowly revert to the more stable calcite with time. This fact explains why there is no aragonite in ancient rocks.

Experiments involving phase changes at higher pressures and temperatures are more difficult to carry out because of the problem of controlling and measuring the temperature and pressure. In some experiments, a cylindrical container or hydrothermal vessel composed of a special steel alloy is hollowed out in the center so that a sample container

may be placed inside of it. The sample container is composed of pure gold or platinum and is sealed at one end. The sample and some water are placed in the container, and the other end is sealed. The container is placed inside the hydrothermal vessel, and water is pumped into the container and heated to the desired temperature. As the temperature gradually rises, the water vapor must periodically be released so that the pressure does not rise too high and rupture the hydrothermal vessel. The gold or platinum container distorts easily and transmits the pressure to the sample inside the container. After the experiment has continued for the desired length of time, the container is suddenly cooled so that the sample is frozen in the state it had reached at the higher temperature and pressure.

Suppose the experimenter is studying the melting of rocks. After the sudden cooling, he may find that some of the sample is glass with embedded crystals of one or more minerals. Presumably the glass represents liquid that was quickly "frozen." The minerals may be identified by observing them under a microscope or by using X-ray diffraction. The exact mineral and glass composition may be determined through the use of an electron microprobe. In this technique, a narrow electron beam is focused on a part of the material to be analyzed, causing electrons of various elements to be removed from the atoms. Other electrons take the places of the removed electrons. X-rays of certain specific energies are then emitted; because this radiation is characteristic of a given element, that element may now be identified. The number of gamma rays or their intensity depends on the amount of the element in the sample; thus, the concentration of the element may be determined.

The mineral and glass composition at a series of temperatures and pressures may be determined during the gradual solidification of a silicate liquid—for example, to understand how the crystallization of the minerals may change the composition of the liquid. These liquid changes in the experiment may then be related to the changing composition of a series of natural lavas to see whether they might have formed by a similar process.

OCCURRENCE IN FAMILIAR PROCESSES

Phase changes occur in familiar processes every day. The phase changes from ice to water to water vapor are familiar to most people. Ice is less dense than water, so the ice takes up more space than does the original water. A drink placed in a freezer may explode as a result of this phase change. This effect is avoided in automobile cooling systems when ethylene glycol is mixed with the water; the freezing point of this mixture is much lower than that of water. Ice floats in water because of its lower density. What if ice were denser than water? Then ice would surely sink to the bottom of lakes and oceans, and it might remain there the year round. Profoundly different oceanic, lake, and atmospheric circulation and very different climates and ecosystems would be the result. In fact, marine life would not exist in colder climates as lakes would freeze completely with no layer of ice at the top for insulation.

People cook with boiling water. It is generally known that the boiling point of water decreases with increased elevation as the pressure is reduced. Cooking time must thus be increased to compensate for this lowered boiling temperature at higher elevations. Alternately, salt could be added to the water to raise the boiling point.

Water vapor in the atmosphere can increase only up to a certain maximum point, called the saturation point. This saturation point varies with temperature. More water vapor may be contained in warmer air. Rainfall results when warm, saturated air rises and cools. The cooler air cannot hold as much moisture as can warmer air, so rain falls.

ROLE IN UNDERSTANDING GEOLOGICAL PROCESSES

An understanding of phase changes is essential for an understanding of geological processes. The concentration of elements in geologic systems involves one or more phase changes. The so-called fractional crystallization process, for example, involves the precipitation of minerals from a slowly cooling magma. The minerals either sink or rise in the magma, depending on whether they are heavier or lighter than the same volume of the magma. Some elements are more concentrated in the minerals than in the magma; others are more concentrated in the magma than in the minerals. Some elements may boil out of magma with water vapor and become concentrated in hydrothermal deposits. Common table salt (sodium chloride) forms vast deposits where large, saline bodies of water evaporated slowly over long periods of geologic time, much as the Great Salt Lake in Utah is doing today. Animal matter may slowly change

to petroleum or natural gas when buried gradually below the surface of the earth. As large swamps are gradually buried, they may be transformed into coal. Buried deep, some of this material may change to graphite (all carbon). The mineral diamond (also all carbon) may have existed as graphite before it was transformed to diamond at even greater pressure within the earth.

Quartz (silica or SiO_2) is a common mineral at the earth's surface. The silica polymorphs trydimite and cristobalite are found in volcanic rocks formed at high temperatures. Under extreme pressure, quartz is converted to high-density polymorphs known as stishovite and coesite, also called shocked quartz. The only known occurrence of pressure high enough to cause this change is the impact of meteors on the planetary surfaces. The presence of coesite is used as evidence that various suspect structures were caused by meteor impact. Thus, the different polymorphs of quartz provide evidence concerning the origin of the rock in which it is found.

Robert L. Cullers

FURTHER READING

Bowen, Robert. *Isotopes and Climates.* London: Elsevier, 1991. Bowen examines the role of isotopes in geochemical phases and processes. This text does require some background in chemistry or the earth sciences but will provide some useful information about isotopes and geochemistry for someone without prior knowledge in those fields. Charts and diagrams help clarify difficult concepts.

Brownlow, Arthur H. *Geochemistry.* Englewood Cliffs, N.J.: Prentice-Hall, 1995. A variety of phase changes are discussed in this introductory text in geochemistry. Suitable for a college student who has taken introductory courses in geology and chemistry. Many illustrations.

Carey, Van P. *Liquid Vapor Phase Change Phenomena.* 2d ed. Flourence, Ky.: Taylor & Francis, 2007. A well-written text that thoroughly covers two-phase flow. Easy to understand with some science background.

Ciccioli, Andrea, and Leslie Glasser. "Complexities of One-Component Phase Diagrams." *Journal of Chemical Education* 88 (2011): 586-591. An easily accessible article for a reader with little chemistry background. Many figures and good descriptions complement the text. A vital read for anyone striving to learn more about phase change than what is provided in the average textbook.

Ehlers, Ernest G. *The Interpretation of Geological Phase Diagrams.* San Francisco: W. H. Freeman, 1972. An excellent and very detailed discussion of the principles used to interpret geological phase diagrams in geology. Suitable for college-level students with basic knowledge of mineralogy and chemistry.

Ernst, W. G. *Earth Materials.* Englewood Cliffs, N.J.: Prentice-Hall, 1969. This book is part of a series which supplements introductory textbooks in geology. It discusses mineralogy, igneous rocks, sedimentary rocks, and metamorphic rocks in more detail than do most introductory textbooks. Features good treatments of phase changes and phase diagrams in all the rock types. Accessible to the college-level student who has studied general geology.

_____. *Petrologic Phase Equilibria.* San Francisco: W. H. Freeman, 1976. A detailed treatment of phase diagrams in geologic processes. Appropriate for college students with background in chemistry and mineralogy. Illustrated.

Hamblin, William K., and Eric H. Christiansen. *Earth's Dynamic Systems.* 10th ed. Upper Saddle River, N.J.: Prentice Hall, 2003. This geology textbook offers an integrated view of the earth's interior not common in books of this type. The text is well organized into four easily accessible parts. The illustrations, diagrams, and charts are superb. Includes a glossary and laboratory guide. Suitable for high school readers.

Hillert, Mats. *Phase Equilibria, Phase Diagrams and Phase Transformations.* 2d ed. New York: Cambridge University Press, 2008. Extremely theoretical and technical. Suited for a graduate-level student. The writing is dense and exhaustive.

Krauskopf, Konrad B. *Introduction to Geochemistry.* 3rd ed. New York: McGraw-Hill, 2003. In this well-written introductory geochemistry text, a variety of phase changes are discussed in detail. College students who have taken geology and chemistry courses will find it helpful. Contains many figures.

Mason, Brian, and Carleton B. Moore. *Principles of Geochemistry.* 2d ed. New York: John Wiley & Sons, 1982. An introductory college-level text in geochemistry. Includes some discussion of phase changes. Illustrated with many figures.

Thompson, Graham R. *An Introduction to Physical Geology.* Fort Worth: Saunders College Publishing, 1998. This college text provides an easy-to-follow look at physical geology. Thompson walks the reader through each phase of the earth's geochemical processes. Illustrations, diagrams, and bibliography included.

See also: Elemental Distribution; Environmental Chemistry; Fluid Inclusions; Freshwater Chemistry; Geochemical Cycle; Geothermometry and Geobarometry; Glaciation and Azolla Event; Mass Spectrometry; Neutron Activation Analysis; Nucleosynthesis; Oxygen, Hydrogen, and Carbon Ratios; Phase Equilibria; Radioactive Decay; Radiocarbon Dating; Volcanism; Water-Rock Interactions.

PHASE EQUILIBRIA

The mineral assemblages in most igneous and metamorphic rocks preserve a record of the chemical equilibrium related to the initial rock-forming process. Phase equilibria studies attempt to determine quantitatively the pressure and temperature conditions of rock formations from these mineral assemblages.

PRINCIPAL TERMS

- **degree of freedom:** the variance of a system; the least number of variables that must be fixed to define the state of a system in equilibrium, generally symbolized by F in the phase rule ($P + F = C + 2$), where P is the number of phases and C is the number of chemical components
- **equilibrium:** the condition of a system at its lowest energy state compatible with the composition (X), temperature (T), and pressure (P) of the system; the smallest change in T, P, or X induces a state of disequilibrium that the system attempts to rectify
- **isochemical processes:** processes that leave rock compositions unchanged; in thermodynamic terms, a system in which X remains constant even if T and P change
- **mole:** the amount of pure substance that contains as many elementary units as there are atoms in 12 grams of the isotope carbon-12
- **phase:** any part of a system—solid, liquid, or gaseous—that is physically distinct and mechanically separable from other parts of the system; a boundary surface separates adjacent phases
- **phase diagrams:** graphical devices that show the stability limits of rocks or minerals in terms of the variables T, P, and X; the simplest and most widely used are P-T diagrams (X = constant) and T-X diagrams (P = constant)
- **system:** any part of the universe (for example, a crystal, a given volume of rock, or an entire lithospheric plate) that is set aside for thermodynamic analysis; open systems permit energy and mass to enter and leave, while closed systems do not
- **thermodynamics:** the science that treats transformations of heat into mechanical work and the flow of energy and mass from one system to another, based on the assumption that energy can neither be created nor destroyed (the first law of thermodynamics)

CHEMICAL EQUILIBRIUM

The traditional methods of studying rock bodies are descriptive in nature and involve mapping large-scale outcrop features in the field and detailed microscopic observations of rock textures and mineralogy in the laboratory. These methods, which are often successful by themselves, are supplemented by a second, more theoretical approach wherein rocks are treated as chemical systems and the principles of phase equilibria are applied to determine the conditions of their origin. A full and complete description of a rock body is still required, but that is no longer the main goal of petrologic study. The principles of phase equilibria are simply the laws that govern attainment of equilibrium of chemical reactions such as A + B = C + D, where A and B are known as reactants, and C and D are known as products.

Before exploring how these principles cast light on rock-forming processes, the concept of chemical equilibrium must first be developed. By analogy with gravitational potential, there must exist a similar tendency in chemical systems to lower their energy state through chemical reactions. Reasoning along these lines, J. Willard Gibbs introduced the term "chemical potential" to describe the flow of chemical components from one site (of high potential) to another (of lower potential) during reactions that lead a chemical system toward its lowest energy state. The total energy available to drive a chemical reaction must therefore be the sum of the chemical potentials of each component in the system multiplied by the number of moles of each component. The usual definition of chemical potential of a phase (or pure substance) is "the molar free energy" (or free energy per mole). This simple statement leads to a workable, three-part definition of chemical equilibrium, which is central to the understanding of phase equilibria.

First, if the chemical potential of a component is the same on either side of a reaction equation, the component can have no tendency to participate in the reaction. Second, in a multicomponent system consisting of several phases under uniform temperature (T) and pressure (P), equilibrium must prevail when the chemical potential of each component is the same in all phases in which the component is present. Third, the condition of equilibrium is one of maximum chemical stability. The second part of the definition is equivalent to saying that, for a given chemical reaction, the free energy of the reactants must equal the free energy of the products if a condition of chemical equilibrium prevails under fixed conditions of T and P. If either T or P changes, the system is no longer in equilibrium, but it will immediately adjust itself in such a way as to "moderate" the effect of the disturbing factors. The last statement is known as the moderation theorem, or Le Châtelier's principle. If the free energy on the product side of an equation is less than the free energy on the reactant side, the reaction will be spontaneous. If the opposite is true, no reaction is possible. Scientists are thus able to predict the result of any chemical reaction if the free energy of the reactants and products under the reaction conditions (T, P) is known.

EQUILIBRIUM STATE OF A SYSTEM

The most significant processes of rock formation—magmatism, metamorphism, and sedimentation—all involve large-scale flow of energy and movement of matter that produce an uneven distribution of chemical potential. Inevitably, the result must be chemical reactions tending to restore these natural systems to a state of equilibrium. The equilibrium state of a system is governed by its bulk composition (X), temperature (T), and pressure (P). For most geological processes, T and P change slowly relative to the rates of most chemical reactions, which means that most rock-forming reactions may be considered to take place under constant T and P and, if X also remains constant, most such reactions should easily attain chemical equilibrium.

In geology, the major concern is not so much with achievement of equilibrium but rather with preservation of equilibrium mineral assemblages through hundreds of millions of years, which must follow before deep-seated rocks are finally exposed at the surface. Rocks formed at depth must clearly undergo significant reductions in T and P prior to exposure at the surface, and there are several mechanisms that may induce changes in X during this lengthy period. Geologists are acutely aware of the implications of the moderation theorem: Retrograde metamorphism, mineralogical inversions and exsolutions,

hydrothermal alteration, and weathering are but a few of the processes that could trigger reequilibration in rock bodies before they are exposed for study. Fortunately, microscopic studies coupled with Gibbs's pioneering work in phase equilibria provide the means to discern whether a given mineralogical assemblage preserves a former equilibrium.

GIBBS PHASE RULE

The "phase rule," also called the Gibbs phase rule, was initially derived by Gibbs in 1878 from the mathematical formulas of thermodynamics. The phase rule, which is fully applicable to all chemical systems, expresses the relationship between the governing variables T, P, and X, and the number of phases that may coexist in a state of equilibrium. Usually the phase rule is expressed in equation form as $P + F = C + 2$, where P = number of phases, C = number of

Josiah Willard Gibbs (1839-1903) formulated the Gibbs phase rule, which describes the equilibrium of heterogeneous systems. (Photo Researchers, Inc.)

chemical components, and F = degrees of freedom possessed by the system (normally T, P, X). Phases are chemically pure, physically separable subparts of the system and may be gases, liquids, or solids. In the formal sense, C is the minimum number of chemical entities needed to define completely the composition of each reactant and product phase in a given reaction.

Although the objective of phase rule applications is to determine F for major rock-forming reactions, a far simpler situation that could be experimentally verified in any high school laboratory or even in an ordinary kitchen may be considered. Pure water (H_2O) boils at T = 100 degrees Celsius at sea level (P = 1 bar; atmospheric pressure). The effect of dissolving common salt (NaCl) in water is to raise the boiling temperature approximately 0.8 degree Celsius for each mole percent of NaCl in the liquid phase. The steam given off by boiling is pure H_2O and, therefore, the salt concentration in the remaining liquid must progressively increase with temperature during boiling.

To apply the phase rule to this simple system, one first must tally up the participating phases: there is steam and there is liquid salt solution, and one must conclude that P = 2. Both pure water and salt are required to form these coexisting phases and, therefore, C = 2. The phase rule for this process (boiling), under conditions of fixed pressure (P = 1 bar), tells one that

$$F = C - P + 2$$
$$= 2 - 2 + 2$$
$$= 2$$

Therefore, the system has two degrees of freedom and is said to be "divariant," which means that, because one degree of freedom is utilized by fixing P = 1 bar, only one additional variable need be known to specify completely the state of the system. That may be either T (boiling temperature) or X (composition of the boiling solution); in other words, T and X are dependent variables at constant P. This T-X compositional dependence is easily determined for P = 1 bar by direct experiment. If the resulting T-X data were graphically plotted, the diagram would indicate, for example, that the boiling temperature for an 8 mole percent solution is close to 106 degrees Celsius. Conversely, if it were known only that the boiling temperature of a salt solution were 106 degrees Celsius, that would necessitate a solution concentration of 8 mole percent at P = 1 bar.

APPLICATION OF THE PHASE RULE

To apply the phase rule to a reaction that has some geological significance, consider the appearance of diopside in siliceous dolomite during contact metamorphism by the reaction

$$CaMg(CO_3)_2 + 2\ SiO_2 = CaMgSi_2O_6 + 2\ CO_2$$
$$(dolomite)\quad (quartz)\quad (diopside)\quad (gas)$$

The reaction involves three mineral phases and a fugitive gas phase, which is necessarily lost from the rock if diopside appears; P = 4. Note that the Ca:Mg (calcium-to-magnesium) ratio is the same in the reactant phase (dolomite) and the product phase (diopside); consequently, the minimum number of components needed to define the compositions of the four phases in the reaction is three. Therefore, C = 3. Substituting these values into the phrase rule, one obtains

$$F = C - P + 2$$
$$= 3 - 4 + 2$$
$$= 1$$

and the reaction has one degree of freedom and is said to be univariant. For any given reaction pressure, the phases appearing in that reaction can coexist in equilibrium at one—and only one—temperature.

Univariant reactions are of great interest to petrologists because it is frequently possible to estimate the depth of a rock-forming reaction and, therefore, P, from field relationships. It is then a simple matter to estimate the reaction temperature from the P-T diagram for the univariant reaction involved. For example, suppose that field relationships lead to the conclusion that diopside-bearing contact metamorphic rocks formed by the reaction above at an estimated depth of 4 kilometers. The pressure equivalent of this depth is about 1,000 bars. The experimentally derived P-T diagram indicates a reaction temperature of about 450 degrees Celsius.

MINERALOGICAL PHASE RULE

The principles of phase equilibria were first applied to rocks in 1911 by Swiss geochemist V. M. Goldschmidt in his classic account of contact metamorphism in the Oslo area in Norway. Countless similar attempts have followed, with most authors concluding, as did Goldschmidt, that rocks, in general, record a state of chemical equilibrium governed by the temperature and pressure prevailing at their time of origin. The major generalization emerging from eighty years of such studies is that rocks with a large number of mineral phases tend to have a low number of degrees of freedom. Goldschmidt recognized that, at Oslo, metamorphism must be controlled mainly by T and P and that divariant equilibrium (F = 2) is the general case for isochemical rock-forming reactions. For F = 2, the phase rule reduces to P = C. Dubbed "the mineralogical phase rule," this equation cannot be mathematically derived from thermodynamic principles. It simply reflects the common case in nature where rock-forming processes are approximately isochemical and their phase equilibria are controlled by both T and P operating as independent variables. It follows that rocks recording univariant equilibria are more restricted in occurrence than those recording divariant equilibria. Similarly, the rarity of rocks recording invariant equilibria (F = 0) can be understood, as the phase rule requires that both a unique T and a unique P be maintained during their formation. Rocks of complex mineralogy are occasionally encountered in which the number of phases, P, exceeds C + 2 and, consequently, the value of F is negative. Phase rule departures, indicated by negative F values, are a sure sign of disequilibrium and serve as a reminder that not all mineral assemblages can be treated by the methods of phase equilibria.

PETROGENETIC GRIDS

The temperature and pressure range in which any mineral may exist is limited, and it is the task of experimental petrology to ascertain these limits for rock-forming minerals. The data resulting from this experimental work are utilized to construct "phase diagrams." Such diagrams show the effects of changing P, T, X values on mineral stability fields and are, therefore, simply graphical expressions of the phase rule. In metamorphic petrology, the major use of phase equilibria data has been for construction of "petrogenetic grids." A petrogenetic grid, initially conceived by Canadian petrologist N. L. Bowen in 1940, is a P-T diagram on which experimentally derived univariant reaction curves are plotted for a particular metamorphic rock type (for example, blueschists, marbles, calcsilicates, or pelites). The value of such "grids" lies in the fact that each natural equilibrium assemblage recognized in the field will fall within a definite P-T pigeonhole and thereby inform the field geologist immediately of the P-T conditions of the metamorphic terrane under study. This goal, so simple in concept, has proved elusive even

after half a century of vigorous experimental, theoretical, and field effort. The problem lies in the lack of truly univariant reactions. For nearly seventy years, field geologists mapped isograds recording the "first appearance" of notable zone minerals such as biotite, garnet, staurolite, kyanite, and sillimanite under the impression that they represented the intersection of the ground surface with a plane of univariant equilibrium. Virtually all such "isograds" have proved to be the result of divariant equilibria and thus plot as a "band"—which may be rather wide—on a P-T diagram. This undesirable result has the effect of "smearing" grid boundaries and rendering them less useful as metamorphic indicators.

The general absence of univariant reactions in metamorphic rocks was eventually recognized because of theoretical and experimental advances in phase equilibria studies. The problem stems from the fact that most mineral phases participating in metamorphic reactions are solid solutions of variable composition and the most common reactions lead to the release of a fluid, the composition of which may vary with time. Each of these effects introduces an additional degree of freedom in phase rule terms, and, as a result, virtually all important reactions are divariant.

In spite of these difficulties, petrogenetic grids, based on divariant and quasi-univariant equilibria, have gradually evolved for all major metamorphic rock types. These are not the simple, quantitative grids envisioned by Bowen, but they do provide quick, reliable, and fairly narrow estimates of P-T conditions for common metamorphic mineral assemblages. Modern grids, continually subject to refinement, are phase equilibria's greatest contribution to metamorphic petrology.

APPLICATION TO IGNEOUS PETROLOGY

In the area of igneous petrology, phase equilibria methods and data have become indispensable. Natural rocks, spanning the compositional spectrum, are melted under strictly controlled laboratory conditions to determine solidus and liquidus temperatures at pressures ranging from 1 to 35,000 bars. The results of such experiments place tight constraints on depths and temperatures of magma generation. They also permit the experimentalist to explore P-T effects on partial melting (anatexis) in terms of melt composition and refractory solid phases. The resulting phase diagrams, like metamorphic grids, permit petrologists to "see" deep into the crust and upper mantle and to test hypotheses dealing with the origin of magma.

For nearly a century, igneous petrologists have studied crystal-melt equilibria of simplified, synthetic melts as models for complex, natural magmas. The objective is to reduce the number of equilibrium phases by elimination of minor components of real magmas. Studies of this type were introduced by Bowen at the Geophysical Laboratory of the Carnegie Institution in Washington, D.C. Through its many subsequent researchers, this laboratory published hundreds of phase diagrams and earned a reputation for meticulous and exhaustive experimental work.

Phase equilibrium studies have provided a rather complete understanding of two fundamentally different modes of magma crystallization. Equilibrium crystallization occurs when P-T-X conditions change so slowly that chemical reactions within the melt are able to maintain the state of chemical equilibrium. Conversely, fractional crystallization results when changes in P-T-X conditions outpace the compensating reactions. This disequilibrium process greatly influences the behavior of natural magmas and extends the range of melt compositions that can be derived from a given parent magma. This latter type of behavior, recognized through the early phase equilibria studies of the Geophysical Laboratory, is the major factor in explaining the compositional diversity of igneous rocks.

The relatively simple phase diagrams of synthetic systems unraveled the complexities of sequential crystallization, cast light on the mechanics of crystal nucleation, and exposed the crucial role that water plays in magmatic processes. Collectively, these diagrams are the foundation of modern igneous petrology.

GOAL OF PHASE EQUILIBRIA STUDIES

The refined symbolic notation and elegant mathematical derivations of thermodynamics are likely to remain unappreciated by the majority of laypersons and geologists alike. It is precisely these formalisms, however, that place phase equilibria on a quantitative footing and permit calculation of mineral stability fields from compositional data. Future development in the area of phase equilibria will follow this theoretical line.

The qualitative form of phase equilibria is expressed in phase diagrams rather than equations. Such diagrams have been a major part of petrology

since the 1950's. Historically, emphasis in phase equilibria studies has been on high-temperature igneous and metamorphic rocks, which are most likely to preserve former equilibrium mineral assemblages. This preservation is the fundamental prerequisite for any application of phase equilibria methods. For this reason, the phase equilibria approach has generally not been applied to sedimentary rocks, except for saline deposits formed by intense evaporation of seawater and record chemical equilibrium.

The phase diagrams and sophisticated calculations utilized in phase equilibria studies are often imposing, but that merely reflects the compositional complexity of natural rocks and minerals. What must be appreciated is that the goal of such studies is both simple and practical: to determine how rocks form. All processes taking place on or within the earth (as well as all other solar system bodies) involve the flow of energy and mass. If scientists wish to advance beyond simply describing these processes—that is, to understand the chemical nature of the world—the phase equilibria approach must be employed.

Gary R. Lowell

FURTHER READING

Aharonov, Einat. *Solid-Fluid Interactions in Porous Media: Processes That Form Rocks.* Woods Hole: Massachusetts Institute of Technology, 1996. Aharonov examines the processes involved in rock formation. This is a technical book at times, it but can be understood by the careful reader.

Angrist, Stanley W., and Loren G. Hepler. *Order and Chaos: Laws of Energy and Entropy.* New York: Basic Books, 1967. An elementary treatment of the basic thermodynamic concepts that lie behind phase equilibria studies. One of very few books that present this topic in nonmathematical terms. Aimed at high school readers.

Best, Myron G. *Igneous and Metamorphic Petrology.* 2d ed. Malden, MA: Blackwell Science, 2003. A widely used text for undergraduate geology majors. The treatment skillfully balances the traditional and phase equilibria approaches to petrology. Chapter 1 introduces basic thermodynamic concepts from the geological perspective. Chapters 8 and 14 summarize phase equilibria applications to igneous and metamorphic processes, respectively. For college-level readers with some background in geology and chemistry.

Blatt, Harvey, and Robert J. Tracy. *Petrology: Igneous, Sedimentary, and Metamorphic.* 3rd ed. New York: W. H. Freeman, 2005. Undergraduate text in elementary petrology for readers with some familiarity with minerals and chemistry. Thorough, readable discussion of most aspects of phase equilibria. Abundant illustrations and diagrams, good bibliography, and thorough indices.

Bowen, Norman L. *The Evolution of the Igneous Rocks.* Mineola, N. Y.: Dover, 1956. A reprint of the 1928 classic that first brought the phase equilibria approach to American geologists. Still valuable as an introduction to igneous processes, though some parts are dated. Suitable for college-level readers who have had a course in physical geology.

Bowen, Robert. *Isotopes and Climates.* London: Elsevier, 1991. Bowen examines the role of isotopes in geochemical phases and processes. This text does require some background in chemistry or the earth sciences but will provide some useful information about isotopes and geochemistry for someone without prior knowledge in those fields. Charts and diagrams help clarify difficult concepts.

Carey, Van P. *Liquid Vapor Phase Change Phenomena.* 2d ed. Flourence, Ky.: Taylor & Francis, 2007. A well-written text that thoroughly covers two phase flow. Easy to understand with some science background.

Ciccioli, Andrea, and Leslie Glasser. "Complexities of One-Component Phase Diagrams." *Journal of Chemical Education* 88 (2011): 586-591. An easily accessible article for a reader with little chemistry background. Many figures and good descriptions complement the text. A vital read for anyone striving to learn more about phase change than what is provided in the average textbook.

Ehlers, Ernest G. *The Interpretation of Geological Phase Diagrams.* San Francisco: W. H. Freeman, 1972. Aptly titled, this book is a "must" for those interested in the practical side of phase equilibria. A well-indexed guide to most chemical systems relevant to igneous petrology. Theoretical aspects of the subject are omitted, except for a terse presentation of the phase rule in Chapter 1. For college-level readers with a knowledge of mineralogy.

Emsley, John. *The Elements.* 3rd ed. Oxford: Oxford University Press, 1998. Emsley discusses the properties of elements and minerals, as well as their distribution in the earth. Although some background

in chemistry would be helpful, the book is easily understood by the high school student.

Fermi, Enrico. *Thermodynamics.* Mineola, N.Y.: Dover, 1956. A reprint of the 1937 work by one of the most prominent physicists of the twentieth century. Concepts are presented simply and without regard to applications. For college-level readers with some background in calculus; others may still find the text valuable as the emphasis on mathematical derivation is minimal compared with modern books on the subject.

Fyfe, W. S., F. J. Turner, and J. Verhoogan. *Metamorphic Reactions and Metamorphic Facies.* New York: Geological Society of America, 1958. This monumental book established the thermodynamic approach in metamorphic petrology, and its influence is seen in all subsequent texts on the subject. A major reference for any serious student of metamorphism. Will be found in all university libraries.

Greenwood, Norman Neill, and Allen Earnshaw. *Chemistry of Elements.* 2d ed. Oxford: Butterworth-Heinemann, 1997. An excellent resource for a complete description of the elements and their properties. The book is filled with charts and diagrams to illustrate chemical processes and concepts. Bibliography and index.

Grotzinger, John, et al. *Understanding Earth.* 5th ed. New York: W. H. Freeman, 2006. An excellent general text on all aspects of geology, including the formation of igneous and metamorphic rocks. Contains some discussion of the structure and composition of the common rock-forming minerals. The relationship of igneous and metamorphic petrology to the general principles that form the basis of modern plate tectonics theory is discussed. Suitable for advanced high school and college students.

Hillert, Mats. *Phase Equilibria, Phase Diagrams and Phase Transformations.* 2d ed. New York: Cambridge University Press, 2008. Extremely theoretical and technical. Suited for a graduate-level student. The writing is dense and exhaustive.

Powell, Roger. *Equilibrium Thermodynamics in Petrology: An Introduction.* New York: Harper & Row, 1978. This text, as well as many others of recent vintage, treats phase equilibria as a problem-solving methodology rather than as a body of knowledge derived through field and laboratory work. Written for advanced students of geology, but Chapters 1 and 2 are suitable for general readers seeking an introduction to equilibrium concepts and phase diagrams.

Van Ness, H. C. *Understanding Thermodynamics.* Dover Publications, 1983. Discusses the abstract properties of thermodynamics, making them more understandable. A short book of only 100 pages that covers topics quickly.

See also: Elemental Distribution; Environmental Chemistry; Fission Track Dating; Fluid Inclusions; Freshwater Chemistry; Geochemical Cycle; Geothermometry and Geobarometry; Nucleosynthesis; Oxygen, Hydrogen, and Carbon Ratios; Phase Changes; Plate Tectonics; Rock Magnetism; Water-Rock Interactions.

PLATE MOTIONS

To trace the geological history of the earth, it is necessary to know how the tectonic plates have moved around upon its surface. Using geological evidence, scientists can determine their relative locations at various times in the past. Such information can help in understanding the distribution of geological provinces and also in locating economically important formations.

PRINCIPAL TERMS

- **declination:** the angle in the horizontal plane between true north and the direction that the magnetization of a rock points
- **Euler pole:** the point on the surface of the earth where an axis, about which a rotation occurs, penetrates that surface
- **frame of reference:** a part of the planet, with respect to which all velocities are quoted
- **hot spot:** a point on the earth's surface, unrelated to plate boundaries, where volcanic activity occurs
- **inclination:** the angle in the vertical plane between horizontal and the direction of magnetization of a rock
- **Pangaea:** a supercontinent consisting of all the present continental fragments; it existed approximately 200 million years ago
- **relative velocity:** the velocity of one object measured relative to another
- **triple junction:** a point where three plate boundaries meet
- **vector:** a quantity that is defined by both magnitude and direction
- **velocity:** speed and direction of motion

RELATIVE VELOCITY

A central tenet of the plate tectonics theory is that plates are moving across the surface of the earth. Though all motion is relative in the context of plate tectonics, motion must be defined with respect to a given frame of reference. There is also the difficulty of treating an enormous period of time over which the plate motion has taken place. Some of the geological methods available to earth scientists can be utilized to find the position of a plate millions of years ago, while other methods can yield its present velocity. It is difficult, however, to resolve these two pieces of information into a consistent pattern describing the history of the plate's motion.

To measure a plate's motion, the first step is to find its velocity with respect to an adjoining plate. This is known as relative velocity. Such a relative velocity is actually a linear velocity and, like any velocity, is a vector, which means that it is described not only by the speed of the plate (the magnitude of the vector) but also by the direction in which the plate moves. Some of the methods that geologists employ to find plate velocities give both magnitude and direction; others provide only one of these quantities.

EULER POLES

Plates do not move in straight lines, as they are constrained to be on the surface of a globe. In fact, the plates are moving along curved paths, so their velocities should be described as angular velocities, strictly speaking. If one is considering only a very small area on the earth's surface, then linear velocities are an acceptable approximation. While linear velocities are quoted in units of millimeters per year, angular velocities are quoted as degrees per year, or radians per year. Furthermore, angular velocities are described as a change in angle per unit time around a pivot point (or axis). An everyday analogy might be a door: When a door opens, it pivots at the hinge, and the entire door moves at a particular angular velocity around this pivot. Note that the linear velocity of various parts of the door varies. Near the hinge, the distance moved in the time taken to open the door is small, so the linear velocity here is small, too. The door handle moves a much greater distance in the same time, so its linear velocity is greater. Note also that as the door opens the linear velocity of any point on the door changes continually, as any point on the door is constantly changing direction. Considering the door handle, at every instant during the opening its direction of motion is changing (even though the speed may be constant); hence, the velocity is also changing.

Now consider a plate on the surface of the earth. The linear velocity of the plate has a small magnitude near the pivot point around which it moves. This pivot point is called an Euler pole (named after the Swiss mathematician, Leonhard Euler, who developed this concept). Farther away from the Euler pole, the magnitude of the linear velocity increases.

Suppose that two plates are spreading apart and that the pivot point is the north geographic pole. The mid-ocean ridge between the plates would lie on a line of longitude. The linear velocity of one plate with respect to the other (at any instant) would be zero at the Euler pole and increase to a maximum at the equator. On the other side of the equator, the linear velocity would decrease until it reached zero again at the South Pole, where another Euler pole would be located. In fact, the two Euler poles are just the points where the axis around which the rotation is taking place penetrates the earth's surface.

RECONSTRUCTING ANCIENT LANDMASSES

Knowledge of relative velocities and Euler poles enabled the reconstruction of the position of the continental landmasses in the past. Approximately 200 million years ago, the continents were grouped in a single supercontinent called Pangaea. Pangaea then split into a northern fragment (Laurasia) and a southern part (Gondwanaland) separated by the Tethys Sea. Since then, the fragmentation has continued, and the plates have shifted such that the continents have drifted to the positions they occupy in the modern age. Some of the continental fragments have drifted quite rapidly, such as the Indian subcontinent, which broke from Africa and Antarctica and drifted north until colliding with Asia to form the Himalayan Mountains.

Deposits of oil or coal in the earth may have originally formed before the breakup of Pangaea. Therefore, if the location of one such deposit is known it may be possible to determine the full paleogeographic extent of the environment that gave rise to the deposit in the past by reconstructing the ancient landmasses. By using this method, geologists can predict where further economically valuable sites of oil and coal may lie, even if these sites are currently thousands of miles from the known deposit on a separate continent.

PREDICTING FUTURE PLATE MOVEMENT

Future plate movement can be predicted using their known present relative velocities. For example, the Atlantic Ocean will continue to open, mostly at the expense of the shrinking Pacific Ocean. Australia and Africa will continue to move north, as will Baja California and parts of southern California as the San Andreas fault lengthens. In roughly 10 million years, Los Angeles and San Francisco will become neighbors.

Although plate motions are very slow, the consequences of those motions can often be very abrupt and dramatic. The study of plate tectonics has led to an understanding of why certain regions of the earth are prone to earthquakes and volcanic eruptions. It would beneficial to everyone to be able to predict when these events will take place, and some of the methods that are used to determine plate motions can give direct information concerning these phenomena. In a region such as Southern California, measurements of relative velocities along the San Andreas fault can be applied to the forecasting of earthquakes, and are, therefore, of direct interest to the local population.

ABSOLUTE PLATE MOTIONS

Although relative velocities are useful, the absolute motions of plates can be defined from a frame of reference that is geologically determinable. What is needed is a frame of reference fixed with respect to the interior of the planet, beneath the lithosphere. This region of the interior is termed the mesosphere. It appears that there are locations on the earth's surface that are in some way tied to the mesosphere: hot spots. Hot spots are places where there is volcanic activity that is unrelated to plate boundary activity. These regions are often typified by lavas that are geochemically dissimilar to those formed at either mid-ocean ridges or island arcs, and it has been theorized that the dissimilarity results from the fact that their magma source is much deeper.

Perhaps the best example of a hot spot trace is the Hawaii-Emperor chain of seamounts, which is basically a chain of extinct volcanoes except for the island of Hawaii itself. As one moves away from Hawaii along this seamount and island chain, the lava flows become progressively older. In the hot spot hypothesis, this is explained by the postulation that each island (or seamount) formed over the hot spot, but that the motion of the Pacific plate over the hot spot continually moved the islands away from the magma source—rather like a conveyer belt moving over a static Bunsen burner, leaving a progressively lengthening scorch mark. The Hawaii-Emperor chain has a bend in it, at about the location of Midway Island, which is interpreted as meaning that the Pacific plate motion changed direction at the time that that island formed (some 37 million years ago). Because it is possible to date the lava flows on these islands, the velocity of the Pacific plate relative to this particular hot spot can be calculated (its direction being obtained from the bearing of the seamount chain). The same can be done for other hot spots, so that geologists can ultimately find the velocities of the hot spots relative to one another. The result of this procedure is the discovery that the hot spots move with respect to one another but at rates much slower than do the plates. Assuming that these relative motions are insignificant and that the hot spots are in reality fixed with respect to their proposed source, the mesosphere, then the mean hot spot frame of reference can be defined and all plate motions calculated with respect to that. In fact, this method essentially determines the velocities of the plates with respect to a mantle velocity that best simulates all the known hot spot traces. Absolute plate velocities determined by this method are commonly given in contemporary global plate motion analyses. Studies in 2006 reviewed evidence for a westward motion of the continents due to the tidal action of the moon.

THEORETICAL IMPLICATIONS

The analysis of absolute plate motions has a bearing on theories concerning why the plates move. One group of plates is apparently moving quite slowly, with velocities of between 5 and 25 millimeters per year. This group includes the Eurasian plate, the North and South American plates, and the African and Antarctic plates. In contrast, the Indian, Philippine, Nazca, and Pacific plates move much more rapidly, and the Cocos plate has a velocity of roughly 85 millimeters per year. This observation has led to the realization that it is the plates with actively subducting margins that move the fastest. None of the slower group has a significant percentage of its margin being subducted, whereas all in the faster group do. The implication may be that the subduction process itself plays an important role in driving plate motions. This idea is in contradiction to the earlier hypothesis that the lithospheric plates rode on the back of giant convection cells within the mantle. If this latter view were correct, one might expect the larger plates to move faster (although that is debatable, depending on the geometry of the convecting cells). At the least, a passive plate theory such as that would not produce the correlation noted earlier.

It seems that the plates are not passive players in the plate tectonic cycle but rather an active part of convection.

DETERMINING EULER POLE LOCATIONS

Geologists have a variety of ways to determine how plates have moved in the past and how they may move in the future. Crucial to this endeavor is determining the location of Euler poles, but finding the location of an Euler pole for the relative motion between two plates can be difficult. As indicated in the previous section, however, if one follows a line of longitude along which a ridge lays, one must eventually arrive at the Euler pole. Unfortunately, ridges do not always lay on the geographic longitude lines of the earth. The ridge system separating two plates describes its own set of longitude lines, which may not correspond with geographic longitude lines. To distinguish them, these longitude lines can be referred to as great circles. In fact, any circle that is drawn around the earth is a great circle. All great circles would be identical in length on a perfectly spherical earth. Latitude lines, in contrast, are not great circles, with the exception of the equator, and vary considerably in length. They are referred to as small circles.

Mid-ocean ridge segments are offset by transform faults. Therefore, a set of great circles can be drawn through the various segments of the ridge system, revealing the Euler pole. The transform faults can be used. Small circles drawn through these define the position of the Euler pole as well. This latter case has the added advantage that the fracture zones on either side of the transform fault effectively extend their length and make the geometric construction easier, as it is advantageous to have as long a feature as is possible to which to fit the circle in order to cut down on the errors inevitably involved with any line-fitting method.

VECTOR ADDITION

When a plate does not have a mid-oceanic ridge system separating it from a neighboring plate, other methods must be employed. Such is the case with the Philippine plate, which is surrounded entirely with subduction zones. In this instance, finding its velocity with respect to its neighboring plates and the Euler pole around which the rotation occurs is much more difficult. The motion of the Philippine plate is usually found by adding the velocities of all the other plates on the earth's surface and finding the resultant. The velocity that exactly cancels this resultant is taken to be the velocity of the Philippine plate.

Locations where three plates meet at a single point (triple junctions) can be analyzed by vector addition as well; if the velocities of two of the plates are known, then the velocity of the third can be determined. Significantly, the relative velocity of the triple junction itself can be determined; from that number it can be determined whether any of the plate boundaries is lengthening or shortening. In this fashion geologists were able to determine that the San Andreas fault is lengthening. At its southern end, the triple junction (a convergence of all three types of plate boundaries) migrates south, and at its northern end, the triple junction (two transform faults meet a subduction zone) moves north. This deduction led to the realization that part of the ancient Pacific sea floor, the Farallon plate and part of the East Pacific Rise, had been subducted down a trench that used to lie offshore of western North America. The two remnants of this older plate are the Cocos plate to the south and the Gorda (or Juan de Fuca) plate to the north.

INSTANTANEOUS VELOCITIES

The relative motions of two plates can also be measured by more direct approaches. One technique is to try to measure directly the changes in positions occurring over a few years, which, from a geological point of view, is instantaneous. These are referred to as instantaneous velocities. One example of how that may be done is using geodetic measurements—essentially surveying the region across a plate boundary at regular intervals and, therefore, observing the motion. This technique does not lend itself to the examination of mid-ocean ridges but has been extensively used in studying the San Andreas fault in California. The results of these measurements give the magnitude of the relative linear velocity between the Pacific plate and the North American plate to be between 50 and 75 millimeters per year, the direction of this relative velocity being known from the bearing of the fault line. These numbers agree quite well with other estimates based upon geological evidence, such as the separation of once-continuous geological features that has taken place over much longer time periods.

Another example is the use of satellite laser ranging (SLR). This technique employs a laser beam bounced off a satellite, which affords a method to calculate the distance between two points on the surface of the earth with great accuracy. The distance between two points on separate plates is regularly found, and hence the velocity between the points is calculated. By this method, the relative velocity between North America and Europe has been found to have a magnitude of approximately 15 ± 5 millimeters per year. Once again, that is in agreement with geological data for much longer time periods. In some cases, however, the agreement between the results of SLR and geological evidence is not as good. In the Zagros Mountains of Iran, the two methods do not agree, implying that the instantaneous velocity indicates a change in the relative motion of the two plates on either side of this plate boundary.

FINITE VELOCITIES

The velocities calculated by geological means over much longer time spans are referred to as finite velocities. One major technique used to determine finite velocities depends upon the Vine and Matthews theory of seafloor spreading. The geomagnetic polarity record is now well established and the dates of the geomagnetic reversals known (although this information is still undergoing refinement and short polarity episodes are sometimes added to the known record). This time scale can be used to identify marine magnetic anomalies caused by the magnetization of the sea floor and affords a method by which to date a point on the sea floor. If one measures the distance between two locations of the same age, located on either side of a mid-ocean ridge, then it is quite simple to calculate the relative velocity between the two plates (the direction of the motion being, in most cases, perpendicular to the ridge or parallel to the transform faults). If the separation of two points on the ocean floor, on the same plate but of different ages, is measured, then geologists can still find the "half spreading rate" of the ridge (the amount of new crustal material added per year at the ridge), but this is not a relative velocity.

REMANENT MAGNETIZATION

Because the oceanic crust is relatively young, with the oldest crust being approximately 160 million years old (as compared to 4.6 billion years of earth history), the techniques described above are not applicable to the majority of the history of the earth.

To work out plate motions for older periods, other methods must be employed. The most prevalent of these methods is the use of the remanent magnetization of rocks. Remanent magnetization is acquired when rocks form, and it is oriented parallel to the geomagnetic field at the time and place at which they are forming. This magnetization is retained in much the same way that a bar magnet retains its magnetization. If the rock is subsequently moved, by being carried along with a moving plate, the rock may end up at a location where the direction of the geomagnetic field is substantially different from that of its own magnetization. It is this difference between the field and magnetization directions that was critical in proving that continents do indeed drift across the earth's surface and that was a contributing factor in the acceptance of this theory by geologists.

The angle that the geomagnetic field makes with the horizontal varies considerably, from vertically up at the south magnetic pole to horizontal at the equator to vertically down at the north magnetic pole. This angle is referred to as the inclination. By calculating the inclination of the magnetization of a rock, the latitude at which the rock formed, called the paleolatitude, can be ascertained. If a series shows paleolatitudes for successively older rocks, the latitudinal motion of the plate over time can be traced. Unfortunately, the same cannot be done for longitude, for the simple reason that while latitude is an inherent property of a spinning planet, longitude is not.

Plates not only shift in latitude but also rotate as they move with respect to one another. The angle between true north and a rock's magnetization is referred to as the declination, and it is this angle that allows such rotations to be determined. It is interesting that geologists have been able to delineate rotations of small blocks near the edges of the major plates, which means that a considerably more complex story unravels concerning the interactions at plate boundaries. In both Southern California and Southeast Asia, there are numerous microplates that may have rotated between larger plates.

Ian Williams

FURTHER READING

Condie, Kent C. *Plate Tectonics and Crustal Evolution.* 4th ed. Oxford: Butterworth Heinemann, 1997. An excellent overview of modern plate tectonics theory that synthesizes data from geology, geochemistry, geophysics, and oceanography. A very helpful tectonic map of the world is enclosed. The book is nontechnical and suitable for a college-level reader. Useful "suggestions for further reading" follow each chapter.

Cox, Allan, and R. B. Hart. *Plate Tectonics: How It Works.* Palo Alto, Calif.: Blackwell Scientific, 1986. A well-illustrated and detailed account of the methodology of plate tectonics. Includes information on many different aspects of plate tectonics theory and supplies explanations of the mathematical techniques utilized in solving plate tectonic problems. Suitable for those with a good mathematical background.

Dewey, J. F. "Plate Tectonics." In *Continents Adrift and Continents Aground.* San Francisco: W. H. Freeman, 1976. This article appears in a book of articles reprinted from *Scientific American.* The first part of the article gives a succinct explanation of plate rotations and includes several excellent diagrams. While not giving a full mathematical treatment, the article does approach some complex ideas in an understandable fashion. Suitable for high school readers who have some prior knowledge of the subject.

Frisch, Wolfgang, Martin Meschede, and Ronald C. Blakey. *Plate Tectonics: Continental Drift and Mountain Building.* New York: Springer, 2010. This textbook provides a basic overview of continental drift and plate tectonics focusing on the resulting changes in the earth's surface.

Grotzinger, John, et al. *Understanding Earth.* 5th ed. New York: W. H. Freeman, 2006. This comprehensive physical geology text covers the formation and development of the earth. Readable by high school students, as well as by general readers. Includes an index and a glossary of terms.

Kearey, Philip, Keith A. Klepeis, and Frederick J. Vine. *Global Tectonics.* 3rd ed. Cambridge, Mass.: Wiley-Blackwell, 2009. This college text gives the reader a solid understanding of the history of global tectonics, along with current processes and activities. The book is filled with colorful illustrations and maps.

Press, Frank, and Raymond Siever. *Earth.* 4th ed. New York: W. H. Freeman, 1986. A general geology text. The chapter on global plate tectonics is quite thorough and contains "boxes" that explain the motions of the plates. Hot spots are explained elsewhere in the text and not related to plate motions. The diagrams, although only two-tone, are quite detailed. The text is appropriate for advanced high school readers.

Prichard, H. M. *Magmatic Processes and Plate Tectonics*. London: Geological Society, 1993. Although fairly technical, this special publication has relevant information about plate motions and plate tectonics. The maps and graphics help to illustrate the ideas presented.

Uyeda, Seiya. *The New View of the Earth*. San Francisco: W. H. Freeman, 1978. A very readable account of the development of plate tectonics up to the early 1970's. Does not go into mathematical detail concerning plate motions but does give many examples. Suitable for high school readers.

Wyllie, Peter J. *The Way the Earth Works*. New York: John Wiley & Sons, 1976. A good introductory geology text written from the point of view of plate tectonics. The author does not go into detail concerning the mathematics involved with determining plate motions. Well illustrated and easy to read. Information concerning plate motions is disseminated throughout the text. Suitable for high school readers.

See also: Continental Drift; Creep; Earthquakes; Earth's Core; Earth's Differentiation; Earth's Mantle; Heat Sources and Heat Flow; Mantle Dynamics and Convection; Plate Tectonics; Plumes and Megaplumes; Subduction and Orogeny; Tectonic Plate Margins.

PLATE TECTONICS

Plate tectonics theory holds that the earth's surface is composed of major and minor plates that are being created at one edge by the formation of new igneous rocks and consumed at another edge as one plate is thrust, or subducted, below another. This theory accounts for the cause of earthquakes, the formation of volcanoes and mountain belts, the growth and fracturing of continents, and many types of ore deposits.

PRINCIPAL TERMS

- **andesite:** a volcanic rock that occurs in abundance only along subduction zones
- **basalt:** a dark-colored, fine-grained igneous rock
- **continental rift:** a divergent plate boundary at which continental masses are being pulled apart
- **convergent plate margin:** a compressional plate boundary at which an oceanic plate is subducted or two continental plates collide
- **divergent plate margin:** a tensional plate boundary where volcanic rocks are being formed
- **earthquake focus:** the area below the surface of the earth where active movement occurs to produce an earthquake
- **oceanic rise:** a type of divergent plate boundary that forms long, sinuous mountain chains in the oceans
- **subduction zone:** a convergent plate boundary where an oceanic plate is being thrust below another plate
- **transform fault:** a large fracture transverse to a plate boundary that results in displacement of oceanic rises or subduction zones

PLATE BOUNDARIES

According to plate tectonics theory, the earth's crust is composed of seven major rigid plates and numerous minor plates with three types of boundaries. The divergent plate margin is a tensional boundary in which basaltic magma (molten rock material that will crystallize to become calcium-rich plagioclase, pyroxene, and olivine-rich rock) is formed so that the plate grows larger along this boundary. The rigid plate, or lithosphere, moves in conveyer-belt fashion in both directions away from a divergent boundary across the ocean floor at rates of 0-18 centimeters per year. The lithosphere consists of the crust and part of the upper mantle and averages about 100 kilometers thick. It is thicker under continental crust than over oceanic crust. The lithosphere seems to slide over an underlying plastic layer of rock and

magma called the asthenosphere. Eventually, the lithosphere meets a second type of plate boundary, called a convergent plate margin. If a lithospheric plate containing oceanic crust collides with another lithospheric plate containing either oceanic or continental crust, then the oceanic lithospheric plate is thrust or subducted below the second plate. If both intersecting lithospheric plates contain continental crust, they crumple and form large mountain ranges, such as the Himalaya or the Alps. Much magma is also produced along convergent boundaries. A third type of boundary, called a transform fault, may develop along divergent or compressional plate margins. Transform faults develop as fractures transverse to the sinuous margins of plates, in which they move horizontally so that the plate margins may be displaced many tens or even hundreds of kilometers.

OCEANIC RISES

Divergent plate margins in ocean basins occur as long, sinuous mountain chains called oceanic rises that are many thousands of kilometers long. The rises are often discontinuous, as they are displaced over long distances by transform faults. The two longest oceanic rises are the East Pacific Rise, which runs from the Gulf of California south and west into the Antarctic, and the Mid-Atlantic Ridge, which runs more or less north-south across the middle of the Atlantic Ocean. The oceanic rises are deep-sea mountain ranges, and there is a rift valley that runs down the middle of the highest part of the mountain chain. The rift valley apparently forms along the ocean rises as the plates move outward from the rises in both directions and pull apart the lithosphere. The oceanic floor descends from a maximum elevation at the oceanic rises to a minimum in the deepest trenches along subduction zones. Thus, the lithosphere moves downhill from the oceanic rises to the convergent plate margins. It is thought that the lithosphere gradually cools and contracts as it moves from the oceanic rises to the convergent margins.

The oceanic rises are composed of piles of basalts forming gentle extrusions. There is high heat flow out of oceanic rises because of the large volume of magma carried up toward the surface. The magnetic minerals in the lavas are frozen into alignment with the earth's magnetic field. Half the magnetized lavas move out from the oceanic rises in one direction, and the other half move out in the opposite direction. The magnetic field of the earth appears to reverse itself periodically over geologic time. The last magnetic reversal occurred about 730,000 years ago. This last reversal can now be observed at the same distance in both directions away from the oceanic rises. A series of such magnetic reversals can be traced back across the Pacific Ocean floor for a period of about 165 million years. Many shallow-focus earthquakes occur at depths of up to 100 kilometers below the surface, along the rises and transform faults. Presumably they result from periodic movement that releases tension in the lithosphere.

CONTINENTAL RIFTS

A second type of divergent plate margin, called a continental rift zone, occurs in continents. Examples are the Rio Grande Rift, occurring as a sinuous north-south belt in central New Mexico and southern Colorado, and the East African Rift, occurring as a sinuous north-south belt across eastern Africa. These rift zones occur as down-dropped blocks forming narrow, elongate valleys that fill with sediment. The rift valleys often contain rivers or elongated lakes. They are characterized by abundant basalts with high potassium content and, often, smaller amounts of more silica-rich rocks called rhyolites. Rhyolites are light-colored volcanic rocks containing the minerals alkali feldspar (potassium, sodium, and aluminum silicate), quartz (silica), sodium-rich plagioclase, and often minor dark-colored minerals. Shallow-focus earthquakes result in these areas from the tension produced as the continental crust is stretched apart, much as taffy is pulled.

Many rift valleys never become very large. Others grow and may actually rip apart the continents to expose the underlying oceanic crust and rise, as is occurring in the Red Sea. There the oceanic crust is near enough to the continents that it is covered with sediment. Eventually, the continents on both sides of the Red Sea may be pulled apart so far that the underlying oceanic floor will be exposed, with no

sediment cover. About 240 million years ago, the continents of North and South America, Europe, and Africa were joined in an ancient landmass called Pangaea. They slowly broke apart along the north-south Mid-Atlantic Ridge from about 240 to 70 million years ago. At first, only a rift valley similar to the East African Rift was formed. Later it opened, much like the area of the Red Sea today. Finally, the continents drifted far enough apart during the last 70 million years to form a full-fledged ocean basin, the Atlantic Ocean.

HOT SPOTS

As the lithosphere moves slowly across the ocean floor, minor volcanic activity is generated over hot spots on the ocean floor. The Hawaiian Islands are situated over one of these hot spots. The basalts produced there are much richer in potassium than are those formed over oceanic rises. The Hawaiian Islands are part of a linear, northwest-trending chain of islands, about 2,000 kilometers long, that extends to the island of Midway. The volcanic rocks become progressively older from the Hawaiian Islands to Midway Island. Presumably, Midway Island formed first as the plate slid over the hot spot. As the plate moved to the northwest, the source of magma was removed from Midway, and newer volcanoes began progressively to form over the same hot spot.

SUBDUCTION ZONES

The lithospheric plate with oceanic crust eventually reaches a compressional plate boundary and may be subducted below other oceanic crust. One result is the island arcs in the western Pacific Ocean, such as Japan. Alternatively, the plate may be pulled below continental crust, often at an angle of 20-60 degrees to the horizontal (the Andes in western South America are the result of such movement). The intersection of the two colliding plates is marked by a sinuous, deep trench forming the deepest portions of the ocean floors. Sediment collects along the slopes of the trench, carried down from the topographic highs of the upper plate. Mountain belts are built up on the nonsubducted plate, as a result of the tremendous amounts of igneous rock that form and of the compressional forces of the plate collision, which throw much sediment and metamorphic rock in the nonsubducted plate to higher elevations.

414

The subducted plate can be traced to depths as great as 700 kilometers. Some of the sediments collecting along the trench are carried rapidly to great depths, where they undergo a very high-pressure and low-temperature metamorphism. (Metamorphism is the transformation of minerals in response to high temperatures and pressures deep within the earth.) Studies in 2004 indicated that large lithospheric slabs could be subducted as deep as 2,900 kilometers. Some rocks are carried more slowly to great depths and have a more normal, higher-temperature metamorphism. During metamorphism, many minerals containing water along the subducted plate gradually break down and give off water vapor, which moves up into the overlying plate. The water vapor is believed to lower the melting point of these rocks within the subducted and overlying plates so that widespread melting takes place, producing the abundant basalts and andesites that build up island arcs or continental masses above the subducted plate. In addition, much rhyolitic magma is formed in the continental crust, presumably through the melting of some of the higher-silica rocks in the continents.

EARTHQUAKE ZONES

Sometimes a continent is carried by an oceanic plate into another continent at a subduction zone, which is what happened when India collided with the Asian continent. Such a collision crumples the continents into very high mountains; the Himalaya Mountains were formed in this way. This process produces an earthquake zone that is more diffuse

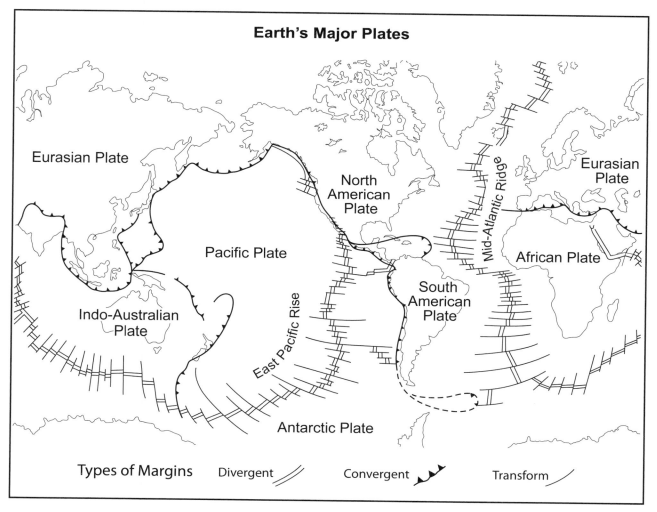

Earth's Major Plates

Eurasian Plate

North American Plate

Pacific Plate

Indo-Australian Plate

East Pacific Rise

Mid-Atlantic Ridge

South American Plate

Eurasian Plate

African Plate

Antarctic Plate

Types of Margins Divergent Convergent Transform

(with foci to depths up to 300 kilometers) than are those along subducted plates. No volcanic rocks are produced in these continental-continental plate collisions. Instead, abundant granites crystallize below the surface. Granites contain the same minerals as rhyolites. Rhyolites form small crystals by quick cooling when they crystallize rapidly in volcanic rocks; granites form larger crystals from magma of the same composition by slow cooling below the earth's surface.

DEVELOPMENT OF THE PLATE TECTONIC MODEL

Plate tectonics is a major, unifying theory that clarifies how many large-scale processes on the earth work. These include the formation of volcanoes, earthquakes, mountain belts, and many types of ore deposits, as well as the growth, drift, and fracturing of continents. The major concepts to support the theory were put together only in the late 1950's and the 1960's, yet many of the keys to developing the theory had been known for many years. Beginning in the seventeenth century, a number of scientists noticed the remarkable "fit" in the shape of the continents on opposing sides of the Atlantic Ocean and suggested that the continents could have been joined at one time. It was not until the early twentieth century that Alfred Wegener put many pieces of this puzzle together. Wegener noticed the remarkable similarity of geological structures, rocks, and especially fossils that were currently located on opposite sides of the Atlantic Ocean. Most notably, land plants and animals that predated the hypothesized time of the breakup of the continents, at about 200 million years before the present, were remarkably similar on all continents. Subsequently, their evolution in North and South America was quite different from their development in Europe and Africa. Climates could also be matched across the continents. For example, when the maps of the continents were reassembled into their predrift positions, the glacial deposits in southern Africa, southern South America, Antarctica, and Australia could be explained as having originated as one large continental glacier in the southern polar region.

One of the biggest problems with the concept of continental drift at that time was the lack of understanding of a driving force to explain how the continents could have drifted away from one another. However, in 1928, British geologist Arthur Holmes

proposed a mechanism that foreshadowed the explanation that was later adopted by science. He suggested that the mantle material upwelled under the continents and pulled them apart as it spread out laterally and produced tension. The basaltic oceanic crust would then carry the continents out away from one another much like rafts. When the mantle material cooled, Holmes believed, it descended back into the mantle and produced belts along these areas.

From the 1920's to the early 1960's, continental drift theories had no currency, for there was no real evidence for driving forces that might move the continents. It was not until the ocean floors began to be mapped that evidence was found to support a plate tectonic model. The topography of the ocean floor was surveyed, and large mountain ranges, such as the Mid-Atlantic Ridge with its rift valleys, and the deep ocean trenches were discovered. Harry Hess suggested in the early 1960's that the oceanic ridges were areas where mantle material upwelled, melted, and spread laterally. Evidence for this seafloor spreading hypothesis came from the mirror-image pattern of the periodically reversed magnetic bands found in basalts on either side of the ridges. The symmetrical magnetic bands could be explained only by the theory that they were originally produced at the ridges, as the earth's magnetic field periodically reversed, and then were spread laterally in both directions at the same rate.

Supporting evidence for plate tectonics began to accumulate during the 1960's. Further magnetic pattern surveys on ocean floors confirmed that the symmetrical pattern of matching magnetic bands could be found everywhere around ridges. Also, earthquake, volcanic rock, and heat-flow patterns were discovered to be consistent with the concept of magma upwelling along rises and seafloor material being subducted along oceanic trenches. Oceanic and lithospheric plates could then be defined, and the details of the interaction of the plate boundaries could be understood. With this overwhelming evidence, most geologists became convinced that the plate tectonic model was valid.

ECONOMIC APPLICATIONS

Plate tectonics is important economically because of the theory's usefulness in predicting and explaining the occurrence of ore deposits. Plate boundaries such as the mid-oceanic rises are areas of high temperature in which hot waters are driven up toward the surface. These hot waters are enriched in copper, iron, zinc, and sulfur, so sulfide minerals such as pyrite (iron sulfide), chalcopyrite (copper and iron sulfide), and sphalerite (zinc sulfide) form along oceanic rises. One such deposit in Cyprus has been mined for many centuries. Tensional zones sometimes created in basins behind subduction zones may form deposits similar to those at oceanic rises. In addition, ferromanganese nodules form in abundance in some places by chemical precipitation from seawater. These nodules are enriched in cobalt and nickel, as well as in iron and manganese as complex oxides and hydroxides. They could potentially be mined from ocean floors.

Deposits enriched in chromium occur in folded and faulted rocks on the nonsubducted plate next to the oceanic trench in subduction zones. This deposit is found in some peridotites (olivine, pyroxene, and garnet rocks) or dunites (olivine rock) that have been ripped out of the upper mantle and thrust up into these areas. The ore mineral chromite (magnesium and chromium oxide) is found in pods and lenses that range in size from quite small to massive. Many intrusions of silica-rich magma above subduction zones contain water-rich fluids that have moved through the granite after it solidified. The water-rich fluids deposit elements such as copper, gold, silver, tin, mercury, molybdenum, tungsten, and bismuth throughout a large volume of the granite in low concentrations. Hundreds of these deposits have been found around subduction zones in the Pacific Ocean.

Robert L. Cullers

FURTHER READING

Condie, Kent C. *Plate Tectonics and Crustal Evolution*. 4th ed. Oxford: Butterworth Heinemann, 1997. An excellent overview of modern plate tectonics theory that synthesizes data from geology, geochemistry, geophysics, and oceanography. A very helpful tectonic map of the world is enclosed. The book is nontechnical and suitable for a college-level reader. Useful "suggestions for further reading" follow each chapter.

Frisch, Wolfgang, Martin Meschede, and Ronald C. Blakey. *Plate Tectonics: Continental Drift and Mountain Building*. New York: Springer, 2010. This textbook provides a basic overview of continental drift and plate tectonics focusing on the resulting changes in the earth's surface.

Grotzinger, John, et al. *Understanding Earth*. 5th ed. New York: W. H. Freeman, 2006. This comprehensive physical geology text covers the formation and development of the earth. Readable by high school students, as well as by general readers. Includes an index and a glossary of terms.

Kearey, Philip, Keith A. Klepeis, and Frederick J. Vine. *Global Tectonics*. 3rd ed. Cambridge, Mass.: Wiley-Blackwell, 2009. This college text gives the reader a solid understanding of the history of global tectonics, along with current processes and activities. The book is filled with colorful illustrations and maps.

Kious, Jacquelyne W. *This Dynamic Earth: The Story of Plate Tectonics*. Washington, D.C.: U.S. Department of the Interior, United States Geological Survey, 1996. Kious is able to explain plate tectonics in a way suitable for the layperson. The book deals with early development of the theory. Illustrations and maps are plentiful.

Motz, Lloyd M., ed. *The Rediscovery of the Earth*. New York: Van Nostrand Reinhold, 1979. An unusual book, as it is written by many of the experts who developed the plate tectonic model. Begins at an elementary level so that someone without much background in plate tectonics should be able to understand the discussion; the discussion progresses, however, to an advanced level. Beautifully illustrated with photographs and diagrams.

Seyfert, Carl K., and L. A. Sirkin. *Earth History and Plate Tectonics*. New York: Harper & Row, 1973. This book integrates the plate tectonic concept with the evolution of plants and animals through geologic time. Some understanding of plate tectonics, rocks, and minerals would be helpful before using this source. Written as an introductory text in historical geology, so important concepts are reviewed. Good illustrations.

Skinner, Brian J., et al. *Resources of the Earth*. 3rd ed. Englewood Cliffs, N.J.: Prentice-Hall, 2001. A good book for the layperson who is interested in the history, use, production, environmental impact, and geological occurrence of ore deposits. Technical terms are kept to a minimum. Well illustrated; contains a glossary. Suitable for someone who is taking a course in geology.

van Hunen, Jeroen, and Arie P. van den Berg. "Plate Tectonics on the Early Earth: Limitations Imposed by Strength and Buoyancy of Subducted

Lithosphere." *Lithos.* 103 (June 2008): 217-235. This article provides new evidence supporting the theory of plate tectonics.

Windley, Brian F. *The Evolving Continents.* 3rd ed. New York: John Wiley & Sons, 1995. A more advanced source than the others listed. Summarizes how the continents have evolved through geologic time. The reader should understand plate tectonic processes well before attempting to read this book.

See also: Continental Drift; Creep; Earthquake Distribution; Earthquakes; Earth's Core; Earth's Differentiation; Earth's Lithosphere; Earth's Magnetic Field; Earth's Mantle; Faults: Transform; Heat Sources and Heat Flow; Lithospheric Plates; Magnetic Reversals; Mantle Dynamics and Convection; Mountain Building; Plate Motions; Plumes and Megaplumes; Subduction and Orogeny; Tectonic Plate Margins; Volcanism.

PLUMES AND MEGAPLUMES

A plume is a pipe that extends into the mass of hot rocks that exist in the mantle of the earth and brings them to the surface, forming a "hot spot." A megaplume is a supermass of extremely hot rocks that moves very slowly under the surface of the earth and influences the breakup of tectonic plates.

PRINCIPAL TERMS

- **crust:** the rock and other material that make up the earth's outer surface
- **guyot:** a formation made by plume activity in the ocean that has a flat top wholly under water
- **hot spot:** a heat source fed by a plume that reaches deep into the earth and produces molten rock
- **magma:** molten rock generated deep within the earth that is brought to the surface by volcanoes and plumes
- **mantle:** the part of the earth below the crust and above the core composed of dense, iron-rich rocks
- **plate tectonics:** the theory that accounts for the major features of the earth's surface in terms of the interactions of the continental plates that make up the surface
- **seamount:** an isolated dome formed under the sea by plumes reaching a height of at least 2,300 feet
- **tectonics:** the history of the larger features of the earth—rocks and mountains, islands and continents—and the forces and movements that produce them

Hot Spots

There are more than one hundred regions of the world known as "hot spots," which are fed by plumes of hot rock rising from deep in the earth's mantle. These hot spots are responsible for a particular type of volcanic activity, which, unlike other active volcanoes, has its origins deep in the interior of the earth. Plumes are found far away from the most active centers of volcanic activity and are usually most active in flat landscapes or at the bottom of oceans rather than in mountainous regions, as is true of more-typical volcanoes.

The hot spots come from material found deep within the earth's mantle, the solid layer of rocks that extends to more than 3,000 kilometers below the earth's surface, just above the core. Plumes apparently arise in regions of the mantle that are stirred about as the large continental plates that cover part of the earth move slowly across the surface. This

movement of plates has been going on since early in the earth's history, beginning at least 4.6 billion years ago, when the earth's crust was just forming. As the huge plates of rock that make up the continents formed, they were originally one giant mass, but they began breaking up and moving apart at a fraction of an inch per year. "Plate tectonics," as the study of these movements is called, describes how the continents reached their present locations; they are still moving apart, and still only by fractions of an inch every year. The plate movement helps explain the building of mountain ranges, for as the plates crash into one another, they push their margins up into mountains such as the Himalayas and the Andes. Plate movement also helps scientists to understand the activities of volcanoes. Most volcanic activity occurs in those areas where the major plates that make up the earth's surface (including the Eurasian, American, African, Pacific, Indian, and Antarctic plates) come together. At these margins of plate contact, the pressure of the plates pressing against one another creates fissures and breaks in the earth's surface, through which magma—the hot, molten material coming from deep in the mantle—can flow. The plates that form the continents are called "continental plates." Other plates are found at the bottom of the oceans; these are called "oceanic plates."

Hot Spot Volcanoes

As the plumes of mantle material move upward toward the surface, they feed and create what are called "hot spot" volcanoes. These range in size according to how deep the plume has reached into the depths of the earth. The deepest plumes create the largest volcanoes. The material coming up through the plumes (magma) consists of gigantic blobs of melted rock. Plumes and the hot spots connected to them move much more slowly than the continents above them. When one of the continental plates crosses over a plume, the magma flowing upward creates a large structure that looks like a dome. Such domes are usually about 125 miles wide and can be hundreds of miles long. Approximately 10 percent of the earth's surface is covered with these domelike structures. As the magma continues to burst upward, the dome increases in size; as it does, cracks and small openings appear, and the hot magma flows through these openings onto the surface. The most well-known domes created by plumes are the Hawaiian Islands. Geologists believe that all the islands in the Hawaiian chain were created from a single plume. As the Pacific plate passed over the hot spot, the islands popped up and out from the ocean floor, and the plume pumped huge quantities of magma into and out of the resulting dome.

In the Atlantic, volcanic islands formed from plumes are found along the mid-ocean ridge. The Azores and Ascension Islands appear on this ridge. The location of these island chains and hot spots can help scientists understand the movement of tectonic plates. The plumes appear to be fixed in relationship to one another and move at velocities of only a few millionths of an inch each year. Sophisticated measuring instruments can account for this movement, however. They also move in what appear to be well-established tracks, and as they move, their heat weakens the rocks above. Over time, these weakened surfaces begin to crack, causing rifts, or giant cracks in the earth. Some of these rifts become huge valleys, such as that found in the East African nation of Ethiopia.

HOT SPOT DISTRIBUTION

Of the hundred or so hot spots, more than half are found on the continental plates, with about twenty-five found in Africa. The African plate has remained over these hot spots for millions of years. The shape

A plume containing sulfur dioxide gas and ash particles rising from the floor of Halemaumau at the summit of Kilauea volcano in Hawaii Volcanoes National Park. (AP Photo)

of the continent, which is covered by hundreds of basins, domes, and ridges, was greatly influenced by the slow movement of the continental plate over these plume-fed hot spots. Hot spots are also found in great numbers under the Antarctic and Eurasian plates. Hot spots are more likely to be found under slow-moving plates, since those continents that are moving more rapidly, such as North and South America, have only a very few areas of volcanic activity caused by hot spots.

One area in North America located over a hot spot is Yellowstone National Park. The hot spot below the park creates the many geysers, including Old Faithful, that are found in the region. Geysers are created when surface water seeps into the ground. When it comes in contact with boiling magma, the water is heated rapidly; it then boils upward until it explodes through cracks in the earth's crust.

Hot spots do have a limited life span. Typically, a plume feeding a hot spot cools off and disappears after about 100 million years. Their positions also change. The plume feeding the Yellowstone geysers originated farther to the north, around the Snake River in Idaho, 400 miles away, around 15 million years ago. Over time, the North American plate has slipped across it, putting the hot spot in its present location. Yet Yellowstone, too, is only a temporary home for this hot spot, and its slow movement to the southwest continues. The slow drift accounts for the volcanic activity in the area. Scientists believe that at least three major volcanic eruptions have taken place in this region during the last two million years. They predict that another massive explosion, hundreds of times greater than the huge Mount St. Helens eruption in 1981, will hit the area sometime in the next few thousand years.

ICELAND

The hot-spot theory is an important contribution to the science of plate tectonics. Few geologists doubt the existence of these hot spots and the plumes that create them. One of the most intensely studied hot spots is the dome that makes up the North Atlantic island of Iceland. This megaplume, which was raised above the ocean floor more than 16 million years ago, lies across a formation known as the Mid-Atlantic Ridge. The dome is actually about 900 miles long, but only about 350 miles of it—Iceland—lies above sea level. To the south of the island, the dome

tapers off gradually and dips below the sea. In 1918, a volcanic eruption under a glacier to the north of Iceland melted enough ice to create an oceanic flood of water that was twenty times greater than the yearly flow of the Amazon River, the world's largest river. Had the floodwater hit land many islands and the European coastline would have been devastated.

Within the next few million years, a very short period in geologic time, the Mid-Atlantic Ridge will have moved away from the hot spot, carrying Iceland with it. This will dry up the source of magma supplying the volcanoes on Iceland, and they will no longer erupt.

SEAMOUNTS AND GUYOTS

Most volcanic hot spots never rise above sea level and remain underwater volcanoes. Magma erupting from these plumes forms structures called "seamounts." These are isolated, though they form into long chains along the surface of the oceanic plate. A few seamounts are found with extensive fissures and cracks. Within these cracks, magma has cooled over hundreds of thousands of years, piling up thousands of magma flows, one on top of the other. A few of these are high enough to break through the ocean's surface. These seamounts become volcanic islands and dot various ridges in the crust, forming island chains such as the Galapagos Islands off the west coast of Ecuador. The tallest seamounts rise more than two and one-half miles above the sea floor and are found to the east of the Philippine Islands, where the crust is over 100 million years old. Generally, the older the crust, the larger the number of undersea volcanoes. The majority of seamounts are found in the Pacific Ocean, where there are between five and ten volcanic hot spots in every 5,000 miles of ocean floor. The Hawaiian Islands were created as the oceanic plate passed over a plume and formed the Emperor Seamounts, the name used by students of plate tectonics to refer to the Hawaiian Island chain.

Other plumes, in ancient geological times, formed undersea volcanoes called guyots (pronounced GHEE-ohs). Dozens of these once rose high above the Pacific Ocean. Over millions of years, however, constant wave action eroded the tops of these guyots below the sea's surface. Guyots, over time, moved away from their sources of magma, the hot spots, and this also helps to account for their disappearance beneath the ocean.

MID-CRETACEOUS MEGAPLUME ACTIVITY

The plumes described here are moderate in size and are considered to be a normal part of the earth's mantle. Some geologists are convinced that millions of years ago the earth went through an extremely intense period of volcanic eruptions. During this period, giant megaplumes exploded from deep within the earth, expelling huge quantities of molten material. These structures spread across the earth's surface, becoming ten times larger than average plumes. These "super-plume" explosions were responsible for the volcanic activity that affected the ocean floor during the mid-Cretaceous period about 90 million to 100 million years ago. One result of this unusually violent activity was the creation of hundreds of seamounts in the western Pacific. Another area affected by these megaplumes was the Parana River basin in Brazil, where hundreds of rift valleys were created. It was also during this time that the Andes Mountains in South America and the Sierra Nevada in the western United States were formed.

Megaplume activity during the mid-Cretaceous led to a 100-foot rise in sea level and a 10-degree increase in the temperature of the earth's air. This increase was caused by the release of huge amounts of carbon dioxide into the air during volcanic explosions. A key result of this activity was an enormous increase in plankton, the microscopic organisms that drift in the oceans and are the first link in the ocean's food chain. As the plankton died, they devolved into huge deposits of oil. Perhaps 50 percent of the world's known oil supply dates to this period of megaplume activity. The volcanic activity of the giant plumes also brought large quantities of diamonds from the earth's interior closer to the surface, from which they are mined.

ROLE IN EARTH'S HISTORY

Plumes and megaplumes have had a dramatic influence on the history of the earth. Plume activity has created islands, volcanoes, valleys, and mountains. The superheated rocks brought from far within the interior of the earth have helped to form geysers, oil deposits, and diamonds. Tracking the slow movement of hot spots can help scientists to predict future events, such as the possibility of volcanic eruptions or the creation of new rift valleys. The study of past volcanic explosions can help scientists to keep the public informed about the potential harm that can be expected from future eruptions of plumes.

Volcanic eruptions have occurred periodically throughout the earth's history. From a geological point of view, periods of intense volcanic activity last for relatively brief spans of time, perhaps from 2 million to 3 million years. There are particularly intense periods of major activity about every 30 million years due to a combination of causes, including volcanic activity, impact events, and sea-level falls. These latter periods—and the mid-Cretaceous period might have been one of them—coincide with mass extinctions of life, as volcanic gases flow into the atmosphere, releasing thousands of tons of sulfur and other dangerous chemicals. Some of the released gases are converted into acids that also have a devastating impact on living things. Hot spots and megaplumes expel huge amounts of ash, dust, and molten rock from their cracks and fissures. These materials absorb the sun's radiation and can cause intense heating or cooling of the atmosphere. The dust can also shade out the sun's light for long periods of time. The reduced sunlight can cause mass extinctions of plants and animals because of the extreme cold produced. Intense volcanic activity can also produce acid rain, which could kill the leaves of plants and make the oceans and lakes unlivable. Scientists believe that the earth has been victimized by such violent activity at least three times in the past, the last time being about 65 million years ago. Hot spots and megaplumes were responsible for much of this violent volcanic activity.

Leslie V. Tischauser

FURTHER READING

Ballard, Robert D. *Exploring Our Living Planet*. Washington, D.C.: National Geographic Society, 1983. A well-illustrated guide to modern theories of continental drift, plate tectonics, and the activities of volcanoes. A good place to begin an investigation of the history of the earth's formation and the various forces that have created the earth's features. Includes pictures, maps, and an index.

Condie, Kent C. *Plate Tectonics and Crustal Evolution*. 4th ed. Oxford: Butterworth Heinemann, 1997. An excellent overview of modern plate tectonics theory that synthesizes data from geology, geochemistry, geophysics, and oceanography. A very helpful tectonic map of the world is enclosed. The book is nontechnical and suitable for a college-level reader. Useful "suggestions for further reading" follow each chapter.

Davies, Geoffrey F. *Mantle Convection for Geologists.* New York: Cambridge University Press, 2011. Begins with strong foundational material upon which to build convection concepts. Chapter 7 covers the plumes and hot spots. Although the title implies technical writing, the author's intended the text be for anyone studying geological processes or university level students.

Eicher, Don L., A. Lee McAlester, and Marcia L. Rottman. *The History of the Earth's Crust.* Englewood Cliffs, N.J.: Prentice-Hall, 1984. A brief introduction to plate tectonics and geological history. A good beginning for those unfamiliar with the topic. Useful illustrations, charts, and an index.

Erickson, Jon. *Plate Tectonics: Unraveling the Mysteries of the Earth.* New York: Facts on File, 2001. An excellent, well-written, easily understandable description of the forces shaping the earth's geology, including a detailed and illustrated discussion of plumes and hot spots. Megaplumes, however, are not described. A very good introduction to the subject. Illustrations, bibliography, index.

Foulger, G. R. *Plates vs. Plumes: A Geological Controversy.* New York: Wiley-Blackwell, 2010. Discusses volcanism, seismology, and other topics related to plumes. Constantly returns to plate and plume hypotheses as each new topic is presented. Provides a website for discussion of the plume model and non-plume models in volcanism.

Kearey, Philip, Keith A. Klepeis, and Frederick J. Vine. *Global Tectonics.* 3rd ed. Cambridge, Mass.: Wiley-Blackwell, 2009. A textbook written in somewhat technical language; nevertheless contains some good illustrations and a detailed discussion of megaplumes. Designed for college courses in geology. Index and bibliography.

Olsen, Kenneth H., ed. *Continental Rifts: Evolution, Structure, Tectonics.* Amsterdam: Elsevier, 1995. The various essays provide good explanations of plate tectonics and continental rifts. Slightly technical but suitable for the careful reader. Illustrated.

Reynolds, John M. *An Introduction to Applied and Environmental Geophysics.* 2d ed. New York: John Wiley, 2011. An excellent introduction to seismology, geophysics, tectonics, and the lithosphere. Appropriate for those with minimal scientific background. Includes maps, illustrations, and bibliography.

Seyfert, Charles K., and L. A. Sirkin. *Earth History and Plate Tectonics: An Introduction to Historical Geology.* New York: Harper & Row, 1973. A textbook for geology students, but easy to understand and well illustrated. Read this book after consulting some of the briefer descriptions of the formation of the earth's mantle and crust.

Sullivan, Walter. *Continents in Motion: The New Earth Debate.* 2d ed. American Institute of Physics, 1993. A somewhat dated but still useful summary of the differing points of view of various theorists of the earth's formation and how such views have changed over time. Popularly written; easily understood without a technical background in geology.

See also: Earth's Core; Earth's Differentiation; Earth's Lithosphere; Earth's Mantle; Heat Sources and Heat Flow; Lithospheric Plates; Mantle Dynamics and Convection; Plate Motions; Plate Tectonics; Volcanism.

POLAR WANDER

Evidence from several of the earth sciences clearly demonstrates that the earth's magnetic and geographic poles have been located at widely separated places relative to its surface during the planet's geological history.

PRINCIPAL TERMS

- **asthenosphere:** a hypothetical zone of the earth that lies beneath the lithosphere and within which material is believed to yield readily to persistent stresses
- **ice ages:** periods in the earth's past when large areas of the present continents were glaciated
- **lithosphere:** the outer layer of the earth
- **north geographic pole:** the northernmost region of the earth, located at the northern point of the planet's axis of rotation
- **north magnetic pole:** a small, nonstationary area in the Arctic Circle toward which a compass needle points from any location on the earth
- **paleomagnetism:** the intensity and direction of residual magnetization in ancient rocks
- **plate tectonics:** the study of the motions of the earth's crust

CONTINENTAL DRIFT

Shortly before World War II, geophysicists discovered a method of determining the location of rocks on the earth's surface at the time they were formed, relative to the north magnetic pole. Thus began the study of paleomagnetism. Paleomagnetic studies quickly yielded very puzzling and often contradictory results. The new science produced evidence that the north magnetic pole has changed its location by thousands and even tens of thousands of miles hundreds of times during the earth's geologic history. Since earth scientists are generally agreed that the north magnetic pole has always corresponded closely with the north geographic pole, this evidence seemed to indicate that the earth's axis of rotation had changed—a highly unlikely occurrence.

As the paleomagnetic evidence for different locations of the poles in the past accumulated through measurements of rock formations from around the world, more and more earth scientists began to accept the theory of continental drift. This theory offered an explanation of the paleomagnetic evidence without the necessity of postulating that the earth's axis of rotation had changed in the past. Early in the twentieth century, German scientist Alfred Wegener had drawn attention to the theory that the continents moved in relation to one another. Most geologists initially greeted his theories with derision, but many others agreed with him, causing an often bitter controversy in the earth sciences that lasted almost half a century. The ever-growing body of paleomagnetic evidence could be explained by postulation that the surface areas of the earth move in relationship to the planet's axis of rotation. This explanation proved to be more acceptable to geologists than the idea that the axis of rotation changed.

With the growing acceptance of the theory of continental drift in the 1940's, geologists began trying to explain the mechanism that caused it. They postulated that the earth has a stable and very dense core overlain by an area called the asthenosphere, which is made up of rock rendered plastic by heat and pressure. Floating on the asthenosphere is the earth's outer crust, or lithosphere. Dislocation within the earth caused by the action of heat and pressure result in the movement of the lithosphere relative to the core and to the axis of rotation. The initial attempts to explain continental drift have been considerably revised and refined into the modern theories of plate tectonics and ocean-bed spreading, but the basic premise remains the same: The surface areas of the earth move in relationship to its core and to its axis of rotation. The result of the movement of the earth's lithosphere is that the surface area located at the axis of rotation does not remain the same over long periods of time. This shifting accounts for the apparent "wandering" of the poles as well as for several other puzzling aspects of earth's geologic history.

EVIDENCE FROM GLACIERS AND PALEOCLIMATOLOGY

Striking evidence that the surface areas of the earth have moved enormous distances during geologic history relative to its axis of rotation comes from the study of glaciers. Observations from around the globe show that almost all the land areas of the earth have been glaciated at some time in the past, including parts of Africa, India, and South America presently located on or near the equator. Without postulating either a substantial shifting of the earth's surface relative to its axis of rotation or a change in the axis, equatorial glaciation is inexplicable. If global temperatures dropped to levels sufficient to glaciate even the equator at some time in the past, all life on earth would have been destroyed. If the areas of Africa, India, and South America, which are presently located in tropical locales, once shifted to the polar regions and shifted from there to their present locations, their ancient glaciation is not at all mysterious.

The study of paleoclimatology has also produced evidence supporting the proposition of the shifting of the earth's crust relative to its axis of rotation. Paleoclimatologists study the climates of past ages on the various parts of earth's surface. They have found that Antarctica once supported rich varieties of plant and animal life, many of which could only have lived in temperate and even subtropical climates. Explorations in the far northern regions of Canada, Alaska, and Siberia have revealed that those areas also supported multitudes of animals and luxurious forests in the past, as did many of the islands presently located within the Arctic Circle. Obviously, those regions must have had much warmer climates at the times when the plant and animal life flourished there, which can be explained in only one of two ways: Either the climate of the entire world was much warmer in the past, or those areas now located near the poles were once located in much more temperate latitudes. If the entire world had warmed to the point that the polar areas had temperate climates, the tropical and subtropical areas of the earth would have been much too hot to support life, which is demonstrably untrue according to the fossil record. Thus, the areas now near the poles must have been located in temperate climatic latitudes in the past.

APPROXIMATE CHRONOLOGY

Earth scientists, using the evidence discussed above and paleomagnetism, have established an approximate chronology showing which areas of the earth's surface were located at its north rotational axis during past ages. At the beginning of the Cambrian period (roughly 600 million years ago), the area of the Pacific Ocean now occupied by the Hawaiian Islands was at or near the earth's north rotational axis. By the Ordovician period 100 million

years later, the surface of the earth had shifted in such a manner that the area approximately 1,000 miles north and east of modern Japan was on or near the North Pole. Fifty-five million years later, during the Silurian period, modern Sakhalin Island north of Japan was within the Arctic Circle. During the next 20 million years, the area of modern Kamchatka in eastern Siberia shifted to a position very near the pole. Earth scientists have identified ninety-nine separate locations that occupied the polar regions at one time or another during the ensuing 395 million years from the Silurian to the Pleistocene. During the past million years, forty-three different areas of the earth's surface have been on or near the north geographic poles, averaging more than 1,500 miles distance from each other.

REMAINING CONTROVERSIES

Although contemporary earth scientists have reached a consensus that the surface of the earth has shifted relative to the planet's axis of rotation many times in the past, several problems remain. One area on which there is no unanimity of opinion is the mechanism responsible for crustal shift. The answer most likely lies in high-pressure physics and the nature of the asthenosphere. Another, more controversial, problem concerns the speed of crustal shifts. During most of the twentieth century, almost all the geologists who were daring enough to accept the theory of continental drift assumed that the movement of surface features of the earth relative to the axis of rotation and relative to one another was very slow, on the order of a few inches per year at most. Then an increasing number of earth scientists began arguing for short periods of relatively rapid movement of the earth's crust and long periods of stability.

These problems notwithstanding, there can no longer be any doubt that the surface of the earth has shifted many times relative to its rotational axis. The phenomenon has led to the mistaken assumption that the rotational axis has moved relative to the earth's surface—thus the term "polar wander." The rotational axis of the earth has remained constant throughout its history; apparent polar wander is caused by the shifting of the earth's crust. There is growing evidence that the moon might have formed following Earth's impact with a Mars-sized object, which would have changed Earth's rotational axis early in its history.

STUDY OF PALEOMAGNETISM

The study of paleomagnetism has yielded irrefutable evidence that many different areas of the earth's surface have occupied polar positions during the history of the planet. Scientists studying paleomagnetism measure the weak magnetization of rocks. Virtually all rocks contain iron compositions that can become magnetized. In the study of paleomagnetism, the most important of these compositions are magnetite and hematite, which are commonly found in the basaltic rocks and sandstones. Paleomagnetism may also be measured in less common rocks that contain iron sulfide. In igneous rocks, magnetization takes place when the iron compositions within the rocks align themselves with the earth's magnetic field as the rocks cool. In sedimentary rocks, small magnetic particles align with the magnetic field as they settle through the water and maintain that alignment as the sediments into which they sink solidify.

Magnetized rocks not only indicate the direction of the north magnetic pole at the time they were formed but also show how far from the pole they were at formation by the angle of their dip. Scientists call their horizontal angle the variation and their dip the inclination. Variation reveals the approximate longitude of the rock sample at the time of its formation, relative to the north magnetic pole, and inclination gives its approximate latitude. By ascertaining the date at which the rock sample being examined was formed, using well-known dating methods, scientists are able to establish the area of the earth's surface relative to the north magnetic pole that was occupied by the rock at the time of its formation.

PROBLEMS IN PALEOMAGNETIC STUDIES

There are, however, many pitfalls for the unwary scientist investigating paleomagnetism. A rock whose magnetism is being studied may have moved considerable distances from its place of formation by glacial action or by crustal movement along a major fracture in the earth's surface, such as the San Andreas fault on North America's west coast. High temperatures, pressure, and chemical action can distort or destroy the magnetization of a rock. Folding and the movement of the continents relative to one another may also alter the original orientation of the rocks whose magnetism is being studied. All these pitfalls may be avoided through the expedient of basing estimates of the relative position of the north magnetic pole

on a great number of rock samples of the same age, gathered from many different locations on all the continents.

Another problem in paleomagnetic studies involves the constant movement of the north magnetic pole relative to the north geographic pole. Recent studies show that the north magnetic pole moved from 70 degrees to 76 degrees north latitude (approximately 345 miles, or 576 kilometers) during the period 1831-1975. This phenomenon might accurately be called true polar wander, though it does not involve any alteration either of the earth's axis of rotation or of the surface of the planet relative to its axis of rotation. Most geophysicists studying this movement have concluded that over a period of several thousand years, the average position of the north magnetic pole coincides closely with that of the north geographic pole. Thus, when scientists learn that the north magnetic pole was located near Hawaii 600 million years ago, it is a virtual certainty that modern Hawaii was at that time located near the north geographic pole.

RECONSTRUCTION OF CONTINENTAL POSITIONS

The position of the magnetic field as determined for rocks of different ages in North America yields a wandering path from the eastern Pacific Ocean, looping across the ocean basin, through western Asia, to its present position. The apparent path of polar wandering as determined from Europe matches that of the North American path in general shape but does not coincide except at the present pole position. The curves are separated by about 40 degrees of longitude, which is about the width of the Atlantic Ocean. These paths do coincide if the Atlantic Ocean is removed and the continents are shifted to allow a fit along opposing coastlines. Analysis of apparent pole positions from all of the continents indicates that each continent has its own series of magnetic poles.

Recognition of polar wander supports the basic concept of continental drift, which is now imbedded in the phenomenon of seafloor spreading. Repositioning landmasses so that rocks of the same age agree in their apparent North Pole position allows reconstruction of continental positions through time. Thus, paleomagnetic data indicate that the continents were once assembled into a single landmass, known as Pangaea, which was split apart as the Atlantic Ocean opened. Once the position of landmasses throughout time is fully understood, paleomagnetic data from deformed or displaced rocks within that landmass may be used to reconstruct the terrain prior to deformation.

Paul Madden

FURTHER READING

Besse, J., and V. Courtillot. "Apparent and True Polar Wander and the Geometry of the Geomagnetic Field Over the Last 200 Myr." *Journal of Geophysical Research.* 107 (2002): 2300. A brief discussion of the concept of true polar wander. Provides evidence to support the need to differentiate apparent from true polar wander.

Brooks, C. E. P. *Climate Through the Ages.* New York: McGraw-Hill, 1949. This work synthesizes data from many earth sciences to demonstrate clearly that every surface area of the earth has at different times in the planet's history been subjected to every extreme of climate, from arctic to equatorial and everything in between. A classic work that has recently been republished (Brooks Press, 2007) with the original text and art.

Daly, R. A. *Our Mobile Earth.* New York: Scribner's, 1926. Daly was one of the first earth scientists to propose that the earth's surface has shifted over long distances relative to its axis of rotation and over relatively brief periods of time; this book explains Daly's views and his theory on the mechanism that causes the shifts. Suitable for anyone with a high school education.

Doell, Richard R., and Allan Cox. *Paleomagnetism.* Vol. 8, *Advances in Geophysics.* New York: Academic Press, 1961. Despite its extensive use of technical terms, this book can prove informative to the layperson interested in the scientific underpinnings of paleomagnetism and associated problems.

Evans, D. "True Polar Wander and Supercontinents." *Tectonophysics* 362 (2003): 303-320. Evans provides a model of true polar wander and supporting arguments from the paleomagnetic database.

Hapgood, Charles H. *Earth's Shifting Crust.* Philadelphia: Chilton, 1958. Revised and reissued by Chilton in 1970, and again by Adventures Unlimited Press in 1999, as *The Path of the Poles.* Hapgood's pioneering work presents a sound and clear argument that the earth's surface has shifted rapidly many times relative to its axis of rotation.

Includes a foreword by Albert Einstein. Readily understandable to the layperson.

Hibben, Frank C. *The Lost Americans.* New York: Thomas Y. Crowell, 1946. Hibben provides a wealth of information concerning the great animal extermination at the end of the Pleistocene. His evidence shows clearly that the extermination was the result of one or more natural catastrophes of enormous proportions. Material is presented in nontechnical language accessible to any reader with a high school education.

Hooker, Dolph Earl. *Those Astonishing Ice Ages.* New York: Exposition Press, 1958. Hooker's book is designed to make information concerning past ice ages accessible to a general audience. Includes evidence that areas now on or near the equator were once glaciated.

Irving, E. "Pole Positions and Continental Drift Since the Devonian." In *The Earth: Its Origins, Structure, and Evolution,* edited by M. W. McElhinny. New York: Academic Press, 1980. Irving uses the results of half a century of magnetic measurements to establish which areas at the earth's surface were located at its magnetic pole during the various geological periods. A layperson can follow the gist of his arguments.

King, Lester C. *Wandering Continents and Spreading Sea Floors on an Expanding Earth.* New York: John Wiley & Sons, 1983. Most of this book is written in language much too technical for the general public. It does contain a chapter on paleomagnetism with a good explanation of the techniques and pitfalls of that science and a chapter of plate tectonics with considerable evidence that shiftings of the earth's crust have occurred rapidly and at irregular intervals over geological history.

Merrill, R. T., and M. W. McElhinney. *The Magnetic Field of the Earth: Paleomagnetism, the Core, and the Deep Mantle.* San Diego: Academic Press, 1998. The authors cover the basic material associated with the earth's magnetic field. Chapters deal with the origin of the magnetic field, as well as the origin of secular variation and field reversals. A strong background in mathematics is recommended. Bibliography and index. Numerous tables, figures, and mathematical equations.

Munyan, Arthur C., ed. *Polar Wandering and Continental Drift.* Tulsa, Okla.: Society of Economic Paleontologists and Mineralogists, 1963. The articles contained in this publication range in topic from ancient climates through the study of paleomagnetism. Although a background in geology is necessary to understand all the nuances in the articles, most of them can be followed by the general reader.

Opdyke, Neil D., and J. E. T. Channell. *Magnetic Stratigraphy*. San Diego: Academic Press, 1996. Intended for someone with little scientific background, *Magnetic Stratigraphy* examines the magnetic fields of the earth, focusing largely on paleomagnetism. Contains fifty pages of bibliographical resources, as well as an index.

Tarbuck, Edward J., Frederick K. Lutgens, and Dennis Tasa. *Earth: An Introduction to Physical Geology*. 10th ed. Upper Saddle River, N.J.: Prentice Hall, 2010. This college text provides a clear picture of the earth's systems and processes that is suitable for the high school or college reader. It has excellent illustrations and graphics. Bibliography and index.

Tarling, Donald H., et al., eds. *Paleomagnetism and Diagenesis in Sediments*. London: Geological Society, 1999. This collection of essays, written by leading scientists in their respective fields, deals with key paleomagnetic concepts and principles. Although technical at times, there are many charts and illustrations that help to clarify complicated subjects. Bibliography and index.

Tauxe, Lisa. *Paleomagnetic Principles and Practice*. Norwell, Mass: Kluwer Academic Publishers, 2002. Tauxe offers a clear definition of paleomagnetism, then explores its causes and its effects on the earth and the earth's systems. The book comes with a CD-ROM that complements the material covered in the chapters.

Whitley, D. Gath. "The Ivory Islands of the Arctic Ocean." *Journal of the Philosophical Society of Great Britain* 12 (1910). Whitley describes in detail the myriad bones of large land mammals which lived during the Pleistocene, stacked to heights of more than 100 feet on many of the islands within the present Arctic Circle. That these animals could not have lived in those areas given present climatic conditions is axiomatic.

See also: Earth's Magnetic Field; Fossil Record; Geobiomagnetism; Magnetic Reversals; Magnetic Stratigraphy; Mass Extinction Theories; Milankovitch Hypothesis; Plate Tectonics; Rock Magnetism.

POTASSIUM-ARGON DATING

Radioactive decay of potassium-40 into argon-40 is used by earth scientists as a natural clock to determine the age of rocks. A wide range of ages can be measured using this technique. Moon rocks brought back by the Apollo astronauts were shown to be more than 2 billion years old, and the volcanic lava that formed the island of Hawaii has been dated at less than 1 million years old.

PRINCIPAL TERMS

- **atomic spectroscopy:** a method to identify various elements by the unique spectrum of light waves that each one emits
- **half-life:** the time required for half of the atoms in a radioactive sample to decay, having a constant value for each radioactive material
- **igneous rocks:** rocks formed by solidification of molten magma from within the earth
- **isotopes:** atoms of the same element but having different masses as a result of extra neutrons in the nucleus
- **mass spectrometer:** an apparatus that is used to separate the isotopes of an element and to measure their relative abundance
- **photometer:** a device to measure light intensity, using a light meter with a numerical output reading
- **radiogenic:** an isotope formed by radioactive decay

DISCOVERY OF ISOTOPES

The idea of using radioactivity as a clock for geology was first suggested by the British physicist Ernest Rutherford in 1905. Rutherford and co-workers had shown that uranium decays into lead in a radioactive series, with helium gas formed as a by-product. If the lead and helium are retained in the pores of a rock that contains uranium, the ratio of helium to uranium or lead to uranium can be used to calculate an age. The older the rock, the more lead and helium it should contain. There were some problems, however, that could produce incorrect results. An unknown fraction of helium might have escaped over the centuries, or extra lead might be present in a rock as the result of natural lead deposits not coming from uranium decay. Experimental uncertainties were undeniable.

An important development in 1914 was the invention of the mass spectrometer by J. J. Thomson. When neon gas was analyzed with this apparatus, Thomson was able to show that it actually consisted of a mixture of two different kinds of atoms with masses of 20 and 22 units (relative to hydrogen). His experiment constituted the discovery of isotopes. Over the next several decades, all the elements in the periodic table

Sir Joseph John Thomson (1856-1940), the English physicist who discovered the electron in 1897 and determined its charge and mass. He is considered one of the founders of modern physics. In 1906, Thompson was awarded the Nobel Prize for physics. His achievements include the development of a means for separating atoms and molecules according to their atomic weights and a mathematical theory of electricity and magnetism that contributed to the discovery of isotopes of neon. As director of Cambridge University's Cavendish Laboratory, Thomson made it a center for major developments in modern physics. (Photo Researchers)

were analyzed. The sensitivity of the apparatus was greatly improved so that even isotopes whose abundance is less than 1 percent could be measured with precision. In particular, Alfred Otto Carl Nier showed in 1935 that the element potassium (K) has three isotopes, of which potassium-40 is radioactive and has an abundance of less than 0.01 percent.

POTASSIUM DECAY

The atmosphere of the earth is known to consist of about 99 percent nitrogen and oxygen, nearly 1 percent argon, and very small amounts of other gases. It was a mystery why so much argon is present in the air. Mass spectrometer data for argon (Ar) showed that argon-40 was by far the most abundant isotope. The German physicist C. F. von Weizsäcker made the suggestion in 1937 that the unexpectedly large amount of argon in the air could have come from radioactive decay of potassium in rocks. A test of this idea would be to analyze old rocks that contain potassium to see if they also contain a higher percentage of argon-40 than that found in the air. In 1948, Nier conducted this experiment on several geologically old minerals. He showed that the argon-40 isotope was indeed greatly enriched in the old rocks. The mystery of excess argon in the air had been solved: It came from potassium decay.

Many rocks contain the element potassium. All these rocks show a slight radioactivity as the potassium slowly decays into argon. The key idea of K-Ar dating is to measure accurately the relative amounts of potassium and argon. Very old rocks contain a larger amount of argon because more time has elapsed for the argon atoms to accumulate. The point at which the rock cooled to its solid form sets the starting time for the K-Ar clock. Before the rock crystallized, the argon gas could escape, but after the rock became solid, the argon gas would be retained. In some cases, corrections have to be made in the measurements because argon may have been gained or lost over long periods of time. Some minerals retain argon gas much better than others. With experience, geologists have learned to be selective in finding those applications where the K-Ar technique has its greatest validity.

Radioactivity measurements have established that potassium-40 decays into two possible end products. About 11.2 percent of the decays become argon-40, and the other 88.8 percent of the decays become

an isotope of calcium, calcium-40. The calcium-40 cannot be used in dating a rock because there is too much calcium in the rock already. Only ratios are used in actual calculations. For example, after 200 million years, the ratio of potassium-40 to argon-40 is about 80 to 1. Ratios can be calculated for any age. For meteorites with an age of 4 billion years, nearly 90 percent of the original potassium-40 has decayed. For time intervals shorter than 3 million years, the amount of argon-40 becomes very small and hard to measure, even with a mass spectrometer. As the sensitivity of the apparatus has been improved, however, it has become possible to date rocks with an age as low as 100,000 years or even less.

DETERMINING POTASSIUM AND ARGON CONTENT

To understand how the potassium-argon clock is used in geology in a quantitative way, it is necessary to look at the methods by which K and Ar are determined. For potassium, the general techniques of chemical quantitative analysis can be used. A rock sample may contain as much as 5 percent potassium or more. The sample is crushed and dissolved in acid; after unwanted elements are removed by heating and precipitation, the potassium is converted to an insoluble salt and the precipitate is collected by centrifuging. The amount of potassium can then be determined by weighing. A second method, which gives better precision (typically ± 1 percent), makes use of atomic spectroscopy. Potassium atoms emit a characteristic purple light with a particular wavelength. A standard solution with a known amount of potassium is prepared and vaporized in a burner flame. A photometer set for the proper wavelength measures the light intensity from the standard. The same process is repeated for the rock sample solution. The ratio of light intensities from sample and standard is used to calculate the potassium content of the rock. After the total amount of potassium in a rock sample has been determined, it is an easy step to calculate how much radioactive potassium-40 it contains because the relative isotope abundance is known.

For argon analysis, the mass spectrometer has to be used because the quantity is so small. A 5-gram rock sample with 5 percent potassium would contain one-fourth of a gram of potassium, but perhaps only a few billionths of a gram of argon-40. The older the rock, the more time has elapsed to allow a larger amount of argon-40 to accumulate. The rock is crushed and heated in a vacuum to collect the gases. The small amount of argon is separated; when it is run through a mass spectrometer, an electrical current is observed at mass number-40. How can one know the amount of argon that caused this current? A calibration standard using a known amount of another argon isotope, argon-38, is added. (The technique of adding another isotope is called isotope dilution.) A known mass of argon-38 is mixed with the argon sample from the rock, the gas mixture is run through the mass spectrometer, and the ratio of the electrical currents at mass numbers 38 and 40 is measured. Since the amount of argon-38 is known, the amount of argon-40 can then be calculated by simple proportion.

In the argon-40 determination, one possible source of error needs to be considered. Suppose the mass spectrometer shows the presence of another argon isotope at mass number 36, which cannot be caused by radioactive decay of potassium. Where did it come from? The most likely source of this contamination is the argon contained normally in the atmosphere. It could have gotten into the rock sample or it might be the result of residual air in the vacuum system used for analysis. In either case, the total argon-40 measured by the mass spectrometer would be the sum of radiogenic argon-40 (which comes from radioactive decay of potassium-40) plus atmospheric argon-40. Fortunately, the ratio of argon-36 and argon-40 in the atmosphere is well known, so the measured argon-36 can be used to subtract the atmospheric argon-40 from the total. The radiogenic argon-40 alone should be used for calculating the age of the rock.

AR-AR METHOD AND APPLICATION

Once the potassium-40 and the argon-40 content of a rock have been determined, all the essential data for an age calculation are available. A relatively new development in K-Ar dating is the so-called Ar-Ar method. If a rock sample is irradiated with neutrons, some of the stable potassium atoms will be converted to a new argon isotope, argon-39. With a mass spectrometer, the ratio of argon-40 to argon-39 is then determined. The argon-39 is an indirect measure of the potassium content. It is no longer necessary to do a separate potassium analysis of the rock sample. This procedure has been used to investigate possible loss of argon from the outer layers of a rock fragment during metamorphism. The inner part of a rock will

show a larger argon-40/argon-39 ratio than the outer part because the argon-40 will be retained most effectively in the interior. In favorable cases, it is possible to reconstruct a history of the rock fragment since its time of crystallization.

An application of the Ar-Ar method of dating will be described, to show its specialized applications. The astronauts brought back a remarkable orange, glass-like rock from the moon. At first it was thought possible that it might be the product of recent volcanic activity. A sample of less than one-tenth of a gram of this glass was selected for analysis. The sample was irradiated for several days in a nuclear reactor to convert some of the stable potassium-39 into argon-39. The sample was then heated to a moderate temperature of about 650 degrees Celsius, releasing argon gas only from the outermost part of the rock. This gas was collected and analyzed in a mass spectrometer. The rock sample was heated in successive steps of 100 degrees to about 1,350 degrees Celsius, each time collecting the argon gas that was released from the more interior parts of the rock. The argon-40/argon-39 ratio was plotted against the temperature of release. The resulting graph showed that the ratio varied but reached a constant plateau at higher temperatures. Using the argon ratio where it became constant, an age of 3.7 billion years was calculated. Variation in the argon ratio at lower release temperature was attributed to gain or loss of argon in the outer layers of the sample. The Ar-Ar method made it possible to discard erroneous data from the outer part of the rock when determining the age.

SUCCESSFUL K-AR APPLICATIONS

One successful application of K-Ar dating was to determine the ages of the individual Hawaiian Islands. The islands were formed by volcanic activity, and small rocks in the lava would have cooled rather rapidly after an eruption. With rapid cooling, it is possible that some atmospheric argon might have become trapped in the rocks, so adjustments had to be made in the calculations. The K-Ar measurements gave the following results: The island of Kauai is the oldest, at about 5 million years; Hawaii is the youngest, at less than 1 million years; and the other islands show a regular sequence of ages in between. Scientists concluded that the volcanic activity started in Kauai and gradually migrated to Hawaii, about 300 miles to the southeast, forming a chain of islands at regular time intervals over the 5 million years.

As another example, several hundred ages have been measured for granitic rock samples from the Sierra Nevada region (near Yosemite National Park) using K-Ar dating. The time when these igneous rocks formed, before they rose to the surface, is of significance for a geological understanding of the region. Samples from Half Dome, Cathedral Peak, and other sites gave initial results which were in the range of 80 million years. The data, however, had to be corrected for loss of argon as a result of reheating from later molten rock intrusions that moved upward toward the surface. With corrections, the age of the Sierra Nevada is estimated to be in the range of 140-210 million years.

Age measurements of a much longer time span have been taken for some rock samples brought back from the surface of the moon by the Apollo astronauts in the 1970's. Radioactive dating by K-Ar, as well as uranium and rubidium decay, gave ages of about 4 billion years for some of the rocks from the lunar exploration. Four billion years is equal to about three half-lives for potassium, so almost 90 percent of the original potassium 40 would have decayed. It has been suggested that some moon rocks may be even older than the measured 4 billion years, because a portion of the argon may have been lost as a result of a high-temperature episode in the early history of the solar system.

As a final example of the K-Ar dating method, consider an application to archaeology. In 1959, L. S. B. Leakey and his wife, Mary, discovered the fossil of a humanoid skull in the east African nation of Tanzania. The remains were found in an area of volcanic deposit suitable for dating by radioactivity, and rock samples lying in strata near the fossil remains were dated. At first, there was controversy about the results because some samples selected for analysis came from rock strata that were not accurately correlated with the fossils. In addition, some data had to be discarded because weathering and possible contamination by water were a problem. General agreement has now been reached that the fossil remains, dated by K-Ar of properly chosen nearby rock samples, are about 1.75 million years old. This age has been confirmed by other methods of radioactive dating. When close correlation is obtained by independent methods, one can conclude with confidence that the age determination is valid.

SCIENTIFIC AND PRACTICAL VALUE

Dating with potassium-argon has been applied to many geologically interesting questions. Among them are the early history of the solar system, the age of meteorites, the exploration of the surface of the moon, the dates when the earth's magnetic field reversed, the general geological history of various mountain ranges, the eruption of volcanoes, and the dates of fossil remains for archaeology. In some cases, the Ar-Ar technique using two different isotopes can be used to analyze the history of a rock sample through periods of cooling and reheating.

The problem of storing high-level radioactive wastes began to receive increased attention as more electricity was generated by nuclear power plants worldwide. Nuclear power has an advantage over coal because it does not generate acid rain or carbon dioxide. Some environmental groups now find nuclear power plants less damaging to the environment than coal or oil plants. The disadvantage of nuclear power is the radioactive waste that is produced. The information obtained by geologists from K-Ar dating has made a contribution to the development of waste storage technology because it has been shown that certain rocks can retain radioactivity for very long periods of time. Research is being done on embedding radioactivity in synthetic rocks (the SYNROC process): The radioactive waste is to be incorporated in the rock material itself, where it would be relatively impervious to water and weathering. Geological study of K-Ar dating on very old rocks can thus help scientists to find an acceptable solution to the very practical problem of nuclear waste storage.

Hans G. Graetzer

FURTHER READING

Burchfield, Joe D. *Lord Kelvin and the Age of the Earth.* Chicago: University of Chicago Press, 1990. Lord Kelvin (1824-1907) was widely regarded as the greatest physicist of his era. He published some articles in the 1860's that suggested an age of 100 million years or less for the earth. The scientific basis for and the general acceptance of this estimate by most geologists before 1900 are documented. Eventually, Kelvin's conclusion was overthrown by the new technique of radioactive dating. Many original documents are cited after each chapter.

Criss, Robert E. *Principles of Stable Isotope Distribution.* New York: Oxford University Press, 1999. Criss describes isotopes and their properties with clarity. In addition to well-written text, the book features diagrams and illustrations that present a clear picture of the different phases of isotopes and isotope distribution. Includes a bibliography and index.

Dalrymple, G. Brent; and Marvin A. Lanphere. *Potassium-Argon Dating: Principles, Techniques, and Applications to Geochronology.* San Francisco: W. H. Freeman, 1969. The two authors are scientists with the U.S. Geological Survey, writing from their extensive personal experience with K-Ar measurements. Sample preparation, instrumentation, sources of error, and useful applications are discussed with careful attention to detail. The best technical overview of K-Ar dating, compiled into a compact volume of about 250 pages.

Durrance, E. M. *Radioactivity in Geology.* New York: Halsted Press, 1986. The author shows the wide scope of radioactivity measurements in geological investigations. Up-to-date information is presented on environmental radioactivity (including the radon hazard), heat generation, and various isotope-dating procedures. A bibliography of articles published in professional as well as popular journals follows each chapter.

Emsley, John. *The Elements.* 3rd ed. Oxford: Oxford University Press, 1998. Emsley discusses the properties of elements and minerals, as well as their distribution in the earth. Although some background in chemistry would be helpful, the book is easily understood by the high school student.

Faure, Gunter. *Isotopes: Principles and Applications,* 3rd ed. New York: John Wiley & Sons, 2004. Originally titled *Principles of Isotope Geology.* An excellent, though technical, introduction to the use of radioactive and stable isotopes in geology, including a thorough treatment of the Rb-Sr technique. The work is well illustrated and well indexed. Suitable for college-level readers.

Fleischer, Robert Louis. *Tracks to Innovation: Nuclear Tracks in Science and Technology.* New York: Springer, 1998. The author explains the method of fission track dating and describes experiments done to compare it with other dating techniques. The book also emphasizes the mechanism of track formation and the use of solid-state track detectors to determine the charge and energy of each particle. Designed to acquaint geologists with the technique of fission track dating, it is a suitable introduction for general readers.

Kerr, Richard A. "Two Geologic Clocks Finally Keeping the Same Time." *Science* 320 (2008): 434-435. This article compares argon-argon radiometric dating with potassium-argon dating. It discusses the recalibration of Ar-Ar dating.

Levin, Harold L. *The Earth Through Time.* 9th ed. Philadelphia: Saunders College Publishing, 2009. This college-level text contains a brief, clear description of the Rb-Sr method. A diagram of a whole-rock isochron is included. Five other radiometric dating techniques are discussed, and background information on absolute age and radioactivity is provided. Includes review questions, a list of key terms, and references.

Parker, Sybil P., ed. *McGraw-Hill Encyclopedia of the Geological Sciences.* 2d ed. New York: McGraw-Hill, 1988. This source contains entries on radioactivity and radioactive minerals. The entry on dating methods includes a brief section on the Rb-Sr method. Includes the formula for radioactive decay and a table of principal parent and daughter isotopes used in radiometric dating. The entry on rock age determination has a longer discussion of the Rb-Sr dating and includes an isochron diagram. For college-level audiences.

Skinner, Brian J., and Stephen C. Porter. *Physical Geology.* New York: John Wiley & Sons, 1987. A widely used college-level textbook for an introductory course in geology. One chapter deals with geological time and its determination, using radioactivity and other methods. Both the K-Ar method and the more recent argon-40/argon-39 ratio are described in a readable way.

Tuniz, Claudio, et al., eds. *Accelerator Mass Spectrometry: Ultrasensitive Analysis for Global Science.* Boca Raton, Fla.: CRC Press, 1998. This book looks at the processes involved with accelerator mass spectrometry and the instruments required. There is also a substantial amount of care given to radioactive dating and its protocols, principles, and usefulness. Bibliographic references and index.

Wagner, Gunther A., and S. Schiegl. *Age Determination of Young Rocks and Artifacts: Physical and Chemical Clocks in Quaternary Geology and Archaeology.*

New York: Springer, 2010. The authors cover various materials and dating methods. Well organized, accessible to advanced undergraduates and graduate students.

Walker, Mike. *Quaternary Dating Methods.* New York: Wiley, 2005. This text provides a detailed description of current dating methods, followed by content on the instrumentation, limitations, and applications of geological dating. Written for readers with some science background, but clear enough for those with no prior knowledge of dating methods.

See also: Earth's Age; Earth's Oldest Rocks; Elemental Distribution; Environmental Chemistry; Fission Track Dating; Mass Spectrometry; Nucleosynthesis; Radioactive Decay; Radiocarbon Dating; Rubidium-Strontium Dating; Samarium-Neodymium Dating; Uranium-Thorium-Lead Dating.

R

RADIOACTIVE DECAY

Radioactive decay is the release of energy by nuclei through the emission of electromagnetic energy or several types of charged particles. In geology, radioactive decay is important not only as the basis for most of the standard dating techniques and as a tracer for fluid flows and chemical reactions but also for its role in heating the interior of the earth and changing the character of minerals.

PRINCIPAL TERMS

- **alpha particle:** the nucleus of a helium atom, which consists of a tightly bound group of two protons and two neutrons
- **atom:** the smallest piece of an element that has all the properties of the element
- **electron:** a negatively charged particle that forms the outer portion of the atom and whose negative charge is equal in magnitude to the positive charge of the proton
- **gamma radiation:** high-energy electromagnetic radiation emitted when a nucleus emits excess energy
- **half-life:** the time during which half the atoms in a sample of radioactive material undergo decay
- **neutron:** the uncharged particle that is one of the two particles of nearly equal mass forming the nucleus
- **nucleus:** the central portion of the atom, which contains all the positive charge and most of the mass of the atom
- **positron:** a positively charged electron, a form of antimatter
- **proton:** the positively charged particle that is one of the two particles of nearly equal mass forming the nucleus

BEHAVIOR OF ATOMIC NUCLEI

Radioactive decay is the release of energy by the nucleus of an atom. Nuclei discharge energy either through the emission of electromagnetic radiation—a form of pure energy that does not alter the chemical nature of the atom—or through the emission of a particle that changes the atom into an atom of a different chemical element. To understand radioactive decay, it is necessary to understand the behavior of atomic nuclei.

Atomic nuclei occupy a very tiny central portion of the atom. If the atom were the size of a two-story house, the nucleus would be the size of the head of a pin. In spite of its small size, the atomic nucleus contains nearly all the mass of the atom. Nuclei are composed of two particles with nearly identical masses: the proton, which carries one unit of positive electric charge; and the neutron, which is uncharged. The nucleus is orbited by the electrons. Each electron carries one unit of negative electric charge equal in size to the positive charge of the proton—even though the mass of the electron is only five-hundredths of a percent of that of a proton or neutron. The atom is electrically neutral. Therefore, the number of electrons orbiting the nucleus under normal conditions equals the number of protons contained in the nucleus. The number of protons or electrons determines the chemical element to which the atom belongs.

The particles in the nucleus are held together by the nuclear force, which is strong enough to overpower the electrical repulsion of the protons at very short distances. Just as the atomic electrons orbit the nucleus in patterns with definite energies, the nuclear particles fill states of definite energy within the nucleus. The energies of the neutron states are a little lower than those of the protons, because the neutrons are not forced apart by electrical repulsion. For elements with few protons in the nucleus, the numbers of protons and neutrons in the nucleus are nearly equal. For elements with larger numbers of protons, the electrical repulsion becomes strong enough so that neutrons in heavier atoms considerably outnumber protons to help stabilize the nucleus. For most elements, there are several types of nuclei that contain different numbers of neutrons but have the same number of protons. Such atoms with equal numbers of protons but different numbers of neutrons are called isotopes of an element.

435

HALF-LIFE OF THE DECAY

If a nucleus has extra energy, it will seek to rid itself of that extra energy by emitting either electromagnetic radiation or a particle. This emission is called radioactive decay. The time at which an excited nucleus (one with extra energy) will decay is not predictable except as a probability, which depends on the time elapsed since the nucleus was formed. This probability is described in terms of the half-life of the decay, which is the time it takes for half the excited nuclei in a sample to decay. For example, if there are originally four hundred excited nuclei in a sample, two hundred of them will undergo radioactive decay during the first half-life, one hundred of them in the second, fifty in the third, and twenty-five in the fourth. After the first half-life, there will be two hundred excited nuclei left in the sample; one hundred will be left after the second, fifty left after the third, and twenty-five excited nuclei left after the fourth.

Nuclei with short half-lives decay rapidly and disappear quickly, but the large number of energized particles they emit may do much damage to their surroundings. Nuclei with longer half-lives exist much longer. Their radioactive decay will do less immediate damage to their surroundings but will continue to do damage over a considerable period of time.

TYPES OF DECAY

Three major types of radioactive decays are found in nature. They are called alpha, beta, and gamma decay (named for the first three letters of the Greek alphabet). The particles emitted in the decay are called alpha, beta, and gamma particles or rays. The mechanisms for the decays, their half-lives, and their effect on their surroundings differ widely.

Emission of high-energy electromagnetic radiation consisting of higher-frequency X-rays, called gamma decay, allows the protons and neutrons to settle into lower energy states without changing the number of protons in the nucleus that decayed. Gamma decay particles carry no charge and have no mass. Because they are electromagnetic radiation, they travel long distances through matter and do little damage to the atoms through which they pass compared to the damage done by the passage of charged particles. The half-lives of gamma decays are usually very short. Common half-lives are about a billionth of a second; it is rare to find a gamma half-life as long as a second. The energy of the gamma radiation is characteristic of the energy levels of the protons and neutrons in the nucleus that emitted it. The pattern of emitted gamma rays can be used to identify a particular nuclear species.

Alpha decay occurs mostly in heavy nuclei that emit an alpha particle—the nucleus of a helium atom, consisting of two protons and two neutrons. This massive particle is believed to form a tightly bound unit inside these heavy nuclei. It takes advantage of a unique phenomenon of quantum mechanics to escape far enough from the vicinity of the nucleus that the positive electrical repulsion of the nucleus acts on the alpha's positive charge to drive it out of the atom. Because they are very massive and carry two units of positive charge, alpha particles heavily damage their surroundings even though they travel very short distances in matter and are easily stopped by a thin sheet of paper. Like gamma decay, each alpha decay is characterized by a unique energy determined by the energy structure of the nucleus from which it has escaped. The remaining nucleus now forms an atom of a different element, with two fewer protons and therefore two fewer electrons in the neutral atom. The half-lives of alpha decays are usually very long. Uranium-238, for example, has a half-life of 4.51 billion years, which is believed to be approximately the age of the earth.

BETA DECAY

There are two types of beta decay: negative (the emission of an electron and changing of a neutron to a proton in the nucleus) and positive (the emission of a positron—a positively charged electron—and changing of a proton to a neutron in the nucleus). Both types of beta particles lie between alpha and gamma decay in terms of the damage they do to their environment. They are typically stopped by thin blocks of aluminum and do less damage than alpha particles. Positrons—a form of antimatter—destroy themselves by uniting with an electron and annihilating themselves with the emission of two gamma rays that damage their surroundings. The half-lives of beta decays range from a few seconds to thousands of years.

One of the more important beta decays is the decay of the carbon-14 isotope of carbon with six protons and eight neutrons to the isotope of nitrogen with seven protons and seven neutrons by the emission of an electron. This decay has a half-life of 5,730 years. If a positron is emitted, the remaining nucleus has one proton fewer than it did before. Several varieties of beta decay involve phenomena such as the capture of one of the atomic electrons by a proton to turn itself into a neutron. In this case, there is no emitted positron, but an observer sees an X-ray as other electrons fall close to the nucleus to replace the electron that was captured.

When they were first discovered, beta decays puzzled researchers because they did not exhibit the definite energies that characterized alpha and gamma decays. It was finally realized that nuclei undergoing beta decay emit not only an electron or a positron but also a tiny uncharged particle called a neutrino. Neutrinos carry off part of the definite energy of the nuclear decay so that the beta particles from a particular nuclear transition exhibit a statistical energy distribution. Because they are uncharged and interact very little with other atomic particles, neutrinos can pass through the mass of several earths with less than a 50 percent chance of interacting. Consequently, they are very difficult to detect. (An example of a neutrino detector is a hole in a mine the size of a ten-story building, filled with water and surrounded by detectors.) Despite their small chance of interacting, neutrinos are important to scientists' understanding of the structure of the universe, as they are believed to be produced in the nuclear reactions at the core of the sun and other stars. Thus, they may constitute a large portion of the mass of the universe. A debate rages over whether the neutrino is massless or has a very tiny mass that has not yet been measured. By 2010 several studies pointed to a tiny mass for the neutrino, including analysis of cosmic background radiation and evidence for neutrino oscillations requiring mass.

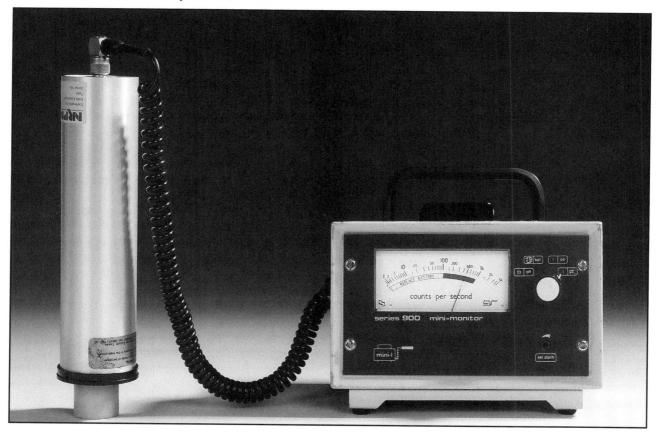

Geiger counter detecting a source of cesium-137 radiation. It is showing a reading of 500 counts per second. This radioactive isotope is extremely toxic to humans, even in minute amounts, and remains so for 300 years. Small amounts of cesium-137 have been released into the environment during the testing of nuclear weapons and by the Chernobyl disaster. (Radiation Protection Division/Health Protection Agency/Photo Researchers, Inc.)

RADIATION DETECTORS

Radioactive decay was discovered by accident when Antoine-Henri Becquerel (1852-1908) accidentally left a piece of uranium-bearing rock on top of a photographic plate in a darkened drawer. The rock left its image on the film. The first studies of radioactive decay used film as a detector for radiation. The next generation of radiation detectors used fluorescent screens which would glow when struck by a decay particle. The screens could not be connected to electronic timers, and the flashes had to be counted through a microscope.

The Geiger-Müller counter is a gas-filled tube with a charged wire running up its center. When a gamma or a beta particle enters the tube, it knocks electrons off the gas atoms and makes the gas a conductor. The electrons are collected by the center wire and can be counted electronically or used to activate a speaker and make the characteristic click of a counter in the presence of radiation. Geiger-Müller counters often are not sensitive to alpha particles, as the metal used to keep the gas inside the tube also keeps alpha particles from entering the counter.

The proportional counter, also a gas-filled tube, works on the same principle as does a Geiger-Müller counter except that it carefully measures the number of electrons that reach the central electrode. Because the number of electrons knocked off gas atoms is directly related to the energy of the particle that knocked them off, the number of electrons reaching the central electrode is directly related to the energy of the particle that produced them.

Scintillation counters utilize transparent materials that emit a flash of light when a particle passes through them. The amount of light is proportional to the kind of particle that passes through and its energy. The light is collected by a special tube, called a photomultiplier tube, that converts the light into a current pulse whose size is proportional to the amount of light emitted. The pulse can be used to drive an electronic counting system.

MODERN DETECTORS

Many modern studies of gamma decay use solid-state detectors which take advantage of the fact that silicon and germanium crystals can be grown with very small amounts of impurities. If the crystals are carefully prepared, they will become conductors when radiation knocks electrons off the atoms in the regions containing the impurities. Once again, the number of electrons produced depends on the energy of the gamma that originated them. The crystal is placed between electrodes with a large voltage across them, and the electrodes collect the electrons produced by the gamma.

Modern detectors produce a pulse of electrons—an electric current—the strength of which is proportional to the energy of the radioactive decay particle that produced it. This current pulse is amplified and its size determined to study the energy of the decay particle that produced it. Such studies typically involve several stages of amplification during which the researcher must be careful not to alter the shape of the current pulses he or she is studying. The pulses are then fed into a device called a multichannel analyzer, which sorts them according to their strength and which stores each pulse as a count in a series of electronic bins. The bins can be displayed to show the number of counts received at each energy level. Called a spectrum of the decay, the result is used to identify the nucleus that emitted the decay product. In modern systems, the multichannel analyzer is replaced by a dedicated computer that automatically identifies the energy of the decay and, in many cases, can tell the researcher which nucleus produced it. Such systems have memories stored with data on energies and half-lives of numerous radioactive decays that have been accumulated since World War II and carefully tabulated.

Radiation detectors and their associated counting systems have become smaller and more rugged. Studies of radioactive decays were once conducted only in laboratories; however, portable systems can now be taken into the field and are found, for example, at petroleum drilling sites. Detection systems are frequently carried into space and have been used in studies of the cosmic radiation and numerous other phenomena. The more sensitive systems used in radioactive dating are still confined to the laboratory, as they must be protected from the radiation in the environment produced by cosmic radiation and by radiation from common minerals.

RESEARCH AND EXPLORATION TOOL

In the decades following World War II, radioactive decay ceased to be a laboratory curiosity and became a widely used tool. The understanding of radioactive decay has led to the development of the means to date geological and archaeological specimens. Radioactive-dating techniques take advantage of the fact that each species of excited nucleus will decay so that half of it disappears after a particular amount of time elapses. If no new excited nuclei have been added to the sample since it was formed, one can compare the number of excited nuclei remaining in the sample to the number of nuclei in the sample formed by the radioactive decay and thus determine how long it has been since the sample was formed. The understanding of geologic time is based largely on radioactive dating.

In addition to its importance as a dating tool, radioactive decay is believed to be a major source of the heating of the earth. This heat flow is important to the overall heat budget of the earth and may be partially responsible for present temperatures on the earth's surface. Radioactive decay may also contribute to driving convection currents in the earth's interior, and it is probably at least partially responsible for the heat extracted as geothermal energy. Radioactive decay is also responsible for changing the nature of certain minerals, as in metamictization.

The presence of a large number of nuclei undergoing radioactive decay makes nuclear wastes hazardous. Although short-lived nuclei decay within a year and disappear, wastes from nuclear reactors are characterized by the presence of nuclei with long half-lives. They must therefore be stored in such a manner that they will not come in contact with the environment for tens of thousands of years. Only very stable geologic formations where wastes cannot be reached by groundwater will permit such storage. The search for suitable areas has covered many states. Radioactive decay has become a concern of many home owners with the discovery that seepage of radioactive radon gas, produced by the decay of minerals in the earth, has raised radiation levels in some homes above levels deemed healthy.

Despite its hazards, radioactive decay has become a useful tool in many types of geological research. For example, small amounts of short-lived radioactivity have been injected into geothermal systems along with cooled water to see how long it takes the reinjected water to reach the production end of the system. Radiation detectors are inserted into boreholes during petroleum exploration to map the presence of radioactive minerals along the walls of the borehole, which assists in the identification of shale layers within a sandstone formation. Radioactive decay has proved to be a useful tool for scientific research and exploration.

Ruth H. Howes

FURTHER READING

Adolff, Jean Pierre, and Robert Guillaumont. *Fundamentals of Radiochemistry*. Boca Raton, Fla.: CRC Press, 1993. A thorough approach to radiochemistry, this book describes all aspects and applications associated with the field. An excellent source of information for someone without much knowledge or background in radiochemistry, this book is well illustrated with diagrams and charts. Bibliography and index.

Choppin, Gregory R., Jan Rydberg, and Jan-Olov Liljenzin. *Radiochemistry and Nuclear Chemistry*. 3rd ed. Oxford: Butterworth-Heinemann, 2001. This widely used college text introduces the reader to the basics of nuclear chemistry and radiochemistry. It explores the theories surrounding those fields and their applications. Well illustrated with clear diagrams and figures, this is a good introduction for someone without a strong background in chemistry.

Dosseto, Anthony, Simon P. Turner, and James A. Van-Orman, eds. *Timescales of Magmatic Processes: From Core to Atmosphere*. Hoboken, N.J.: Wiley-Blackwell, 2010. Covers many aspects of the earth's history, from the formation and differentiation of the earth, to magma ascent, cooling, and degassing. Uranium series isotopes are referenced multiple times in evaluating the time scales of multiple concepts.

Durrance, Eric M. *Radioactivity in Geology: Principles and Applications*. New York: John Wiley & Sons, 1987. This monograph presents a very thorough review of the principles of radioactive decay and its many applications in geology. In addition to its thoroughness, this book is written for a nonspecialist in nuclear physics and is fairly readable as well as complete, discussing not only dating methods but also exploration methods, the use of radioactivity in petroleum studies, environmental radioactivity, and the role of radioactivity in heating the earth's interior.

Keller, C. *Radiochemistry*. New York: John Wiley & Sons, 1988. In addition to covering the basic theory of radioactive decay, this volume stresses the applications of radioactive decay in the study of problems of geological interest. Unlike the preceding reference, it emphasizes radioactivity produced by humans for their own purposes and is of less general interest than is Durrance's monograph.

Lieser, Karl Heinrich. *Nuclear and Radiochemistry: Fundamentals and Applications*. 2d ed. New York: Wiley-VCH, 2001. Lieser's book gives the reader a basic understanding of the practices and principles involved in the fields of radiochemistry and nuclear chemistry. Although there is quite a bit of chemistry involved in the author's explanations, someone without a chemistry background will still find the book useful. Illustrations, charts, and diagrams help clarify difficult concepts and theories. Bibliography and index.

Mares, S., and M. Tvrdy. *Introduction to Applied Geophysics*. New York: Springer, 2011. Provides a good overview of the methods used in studying geophysics. Chapter 3 covers radioactivity and radiometric methods.

Mozumder, A. *Fundamentals of Radiation Chemistry*. San Diego: Academic Press, 1999. Mozumder presents a concise introduction to radiochemistry, examining the procedures and theories associated with it. Some background in chemistry would be helpful but is not necessary. Bibliography, illustrations, and index.

Rhodes, Richard. *The Making of the Atomic Bomb*. New York: Simon & Schuster, 1986. The early chapters of this excellent history of nuclear physics are devoted to a detailed description of the discovery of nuclear decay and the development of a theory of the nucleus. The historically important experiments and ideas are presented in detail. Because the emphasis is on people and the development of a theory, Rhodes has written a very readable as well as technically accurate description of the theory of radioactive decay.

Wagner, Gunther A., and S. Schiegl. *Age Determination of Young Rocks and Artifacts: Physical and Chemical Clocks in Quaternary Geology and Archaeology*. New York: Springer, 2010. The authors cover various materials and dating methods. Well organized, accessible to advanced undergraduates and graduate students.

Walker, Mike. *Quaternary Dating Methods*. New York: Wiley, 2005. This text provides a detailed description of current dating methods, followed by content on the instrumentation, limitations, and applications of geological dating. Written for readers with some science background, but clear enough for those with no prior knowledge of dating methods.

Walther, John Victor. *Essentials of Geochemistry*. 2d ed. Jones & Bartlett Publishers, 2008. Contains chapters on radioisotope and stable isotope dating, and radioactive decay. Geared more toward geology and geophysics than toward chemistry, this text provides content on thermodynamics, soil formation, and chemical kinetics.

See also: Earth's Age; Earth's Oldest Rocks; Environmental Chemistry; Fission Track Dating; Heat Sources and Heat Flow; Neutron Activation Analysis; Nucleosynthesis; Potassium-Argon Dating; Radiocarbon Dating; Rubidium-Strontium Dating; Samarium-Neodymium Dating; Uranium-Thorium-Lead Dating.

RADIOCARBON DATING

Radiocarbon dating is a means of determining the approximate time at which biological processes ceased in a once-living organism or in related organic substances. It allows scientists to estimate the ages of organic materials and the formations in which they occur.

PRINCIPAL TERMS

- **dendrochronology:** the study of tree rings; it provides a means of calibrating radiocarbon dates with absolute chronology
- **half-life:** the period required for half of any given quantity of a radioactive element to revert to a stable state
- **isotopes:** forms of the same element with identical numbers of protons but different numbers of neutrons in their atoms' nuclei
- **Maunder minimum:** the period from 1645 to 1715, when sunspot activity was almost nonexistent
- **photosynthesis:** the process of fixing atmospheric carbon in organic compounds in plants with free oxygen as a by-product
- **radioactive isotope:** an unstable isotope that decays into a stable isotope
- **thermoluminescence:** the process by which some minerals trap electrons in their crystal structures at a fixed rate and release them when heated

RADIOACTIVE CARBON-14

Carbon compounds are among the most abundant in nature. Carbon, in the form of atmospheric carbon dioxide, is continuously cycled through major environmental systems. The photosynthesis of carbon dioxide establishes carbon in various compounds in plants, which in turn may be consumed by animals. Most carbon is absorbed as carbon dioxide by the oceans or appears there as dissolved carbonate or bicarbonate compounds. Carbon occurs naturally in three isotopes: carbon-12, the most common; carbon-13, also a stable isotope; and the radioactive isotope carbon-14, which exists in minute quantities. Radiocarbon dating draws on several assumptions concerning the natural production of carbon-14, its presence in various environmental cycles, and its rate of decay, or half-life.

Radioactive carbon-14 is formed in the upper atmosphere as the result of bombardment by energetic cosmic radiation emanating from deep space. Statistically, it is most likely that free neutrons, which result from collisions between proton cosmic radiation and atmospheric gases, will shortly collide with molecules of stable nitrogen, nitrogen-14, by far the most abundant gaseous element in the atmosphere. The resulting reaction normally expels a proton from the nitrogen nucleus, producing an atom which now behaves chemically like carbon but is heavier than the stable carbon isotopes. Free carbon does not remain long in the upper atmosphere. It quickly combines with oxygen molecules to form carbon dioxide, whereupon it enters into various geological and biological processes.

BETA DECAY

The half-life of carbon-14 usually is expressed as 5,568 ± 30 years, though it is more likely to be on the order of 5,730 ± 30 years. (Once large numbers of dates had been calculated on the basis of the former figure, leading scientific journals generally preferred to stay with it.) The particular process of radioactive decay of carbon-14 is called beta decay, in which the "extra" neutron in the nucleus emits a beta particle (essentially an electron) and a neutrino (an uncharged particle), thus changing itself, in effect, from a neutron into a proton. The result is once again an atom of stable nitrogen.

As long as an organism is alive and normal biological processes are occurring, the rate of accumulation of carbon-14 is in approximate equilibrium with the rate of radioactive decay of the carbon-14 already in the organism. This level of equilibrium is extremely small, on the order of one atom of carbon-14 to every 1 trillion atoms of carbon-12. The moment that biological processes cease, however, the equilibrium is broken, and the quantity of carbon-14 in the once-living organism begins to decrease at the predictable rate of radioactive decay. By measuring the rate of beta radiation from the residual carbon-14, one may estimate the age of a material, or, more precisely, when the biological processes of the organism from which the material derives ceased to operate. In general, the age of material which gives off only half as much beta radiation as a living organism would be

5,730 years; the age of material emitting only 25 percent as much radiation would be 11,460 (5,730 × 2) years; and so on.

EVOLUTION OF MEASUREMENT TECHNIQUES

American physicist Willard F. Libby introduced the radiocarbon dating method in 1949 after fifteen years of research. Since Libby's pioneering work, radiocarbon dating has evolved to include several counting techniques. Normally the substance under study must be destroyed by combustion or other processes to produce gaseous carbon dioxide or a hydrocarbon gas, whose carbon-14 content is measured by a gas counter. Liquid counting techniques have also been developed, in which the carbon dioxide from the substance under study is synthesized into more complex liquid hydrocarbons, such as benzene. After 1980, direct measurement techniques, using particle accelerators and mass spectrometers, increasingly replaced gas and liquid counting.

The evolution of measurement techniques has greatly enhanced the usefulness of radiocarbon dating. The minimum acceptable mass of a sample substance is much smaller than that required in the years immediately following Libby's introduction of the method. Refined techniques and improved instrumentation also have extended the effective chronological limits of the method. At first, only materials less than about 20,000 years old could be dated with any confidence, but by the fortieth anniversary of Libby's initial dating experiments, there was wide agreement that the effective limit was about 40,000 years and, when supporting data could be gleaned from other dating methods, possibly 70,000 years.

RISK OF CONTAMINATION

The extremely small quantities of carbon-14, even in living organisms, demand that substances undergoing counts be prepared most carefully and that every possible source of contamination be considered. For example, fallout from atmospheric testing of thermonuclear weapons in the 1950's interfered with some of Libby's early experiments. Numerous chemical processes affect archaeological artifacts and other substances which might be candidates for radiocarbon analysis. Carbon compounds from associated soil layers find their way into dating samples. Carbonate encrustations may develop around other samples, especially in locations with abundant

groundwater. In the 1980's, some radiocarbon laboratories had to consider their local environments, where, in many cases, the buildup of carbon dioxide and other carbon compounds in the atmosphere from fossil fuel combustion threatened to interfere with sample integrity and analysis. (Fossil fuels, because of their great geological age, contain virtually no carbon-14.)

Some substances are at a greater risk of contamination than others. Woody plant remains, such as charcoal or building materials carbonized by fire, produce the best results. Less dependable are textiles, fibers, and the remains of nonwoody plants, which do not live as long as woody species and, as a result, may reflect short-term or local carbon-14 anomalies. Bone is notorious for its ability to absorb carbon compounds from its surroundings. Inorganic substances which contain carbon compounds, such as eggshell and marine shell, also present serious problems of carbonate contamination. Of these substances, marine shell is preferred, since the extraneous carbonate levels can be determined on the basis of relatively constant values present in ocean water.

ADJUSTING FOR VARIABILITY

In reality, radiocarbon dating is subject to numerous variable factors and is not nearly so accurate. Radiocarbon dates are expressed as years B.P. or "before present." For example, increases in atmospheric carbon dioxide since the Industrial Revolution have diluted the concentration of carbon-14 in what is known as the Suess effect, named after Austrian chemist Hans Suess. Observations by the THEMIS spacecraft fleet have shown that twenty times more solar particles cross the earth's magnetic shield when it is aligned with the sun's magnetic field, compared to when the two magnetic fields are oppositely directed. "Present" is actually a zero base date of 1950. Scientists chose this year because it was close to the first experimental application of the method and because the buildup of atmospheric radioactivity from nuclear weapons tests in later years introduced complications into the measurement process. Each date is expressed with a standard error, or standard deviation—in essence, a "confidence level"—indicated by a plus-or-minus sign. A typical radiocarbon date of, say, 2750 B.P. ± 60 indicates that there is a 66 percent probability that the true date falls within sixty years of 2750 B.P. and a 95 percent probability that the true date falls within twice the standard error—in this case, 120 years.

Another source of variability in test results was only dimly suspected before about 1970. Libby assumed that the formation rate of carbon-14 in the upper atmosphere had a constant value over time, since the process depended on cosmic radiation. Scientists now know that the formation rate varies over time and space, in part according to the strength and contours of the earth's magnetic field. The field deflects a certain portion of cosmic radiation, so some less energetic particles never reach the atmosphere. At any time, for example, more carbon-14 is formed in the atmosphere near the poles, where the magnetic field is relatively weak, than near the equator, where it is stronger.

The strength of the magnetic field itself is affected by cyclical changes in solar activity. The sun has roughly an eleven-year cycle of sunspot activity; the earth's magnetic field is most energetic when sunspots are most numerous. Longer-term and even more significant changes apparently have occurred in more recent decades of solar history, as in the case of the so-called Maunder minimum (1645-1715), when sunspot activity may have been almost nonexistent. During such a period, the earth's magnetic field would be less energetic, carbon-14 formation levels would be abnormally high, and materials from the era would appear "younger" in radiocarbon terms than they otherwise would.

Dendrochronology (the study of tree rings) provides a means of identifying these anomalies and thus calibrating radiocarbon dates more closely with absolute chronology. In principle, each growth ring in a tree lives only for one year. By comparing the known age of tree rings with their radiocarbon ages, scientists can locate anomalies in the carbon-14 absorption rate and, therefore, the formation rate. Wood samples from the extremely long-lived bristlecone pine provide invaluable data. The oldest living bristlecone is more than 4,500 years old, and the remains of dead specimens have yielded a calibration matrix extending almost nine thousand years into the past.

Several other dating systems have been developed which utilize the half-life of radioactive isotopes, and some provide corroboration of radiocarbon chronologies. Further support comes from closely related techniques, such as thermoluminescence. These methods, as well as the established practices

of classical archaeology, strengthen the credibility of the radiocarbon technique.

APPLICATION TO ARCHAEOLOGY

Through radiocarbon dating applications, scientists have amassed a wealth of information on such matters as sea-level fluctuation, climatic change, glaciation, habits of marine life, volcanic activity, and early forms of atmospheric pollution from the combustion of fossil fuels. From time to time, results of carbon-14 analysis also attract attention by establishing the authenticity of artistic works or the ages of religious relics.

By far the most significant application of radiocarbon dating, however, has been in providing archaeologists with the means of constructing chronological relationships independently of traditional archaeological assumptions about cultural processes. The technique is especially valuable in prehistoric archaeology, for which little or no documentation is available and artifact interpretation previously could allow only relative chronological sequences. In the Western Hemisphere, Africa south of the Sahara, and other parts of the world where "prehistory"—in the sense of absence of documentation—extends nearly to the present and chronologies had been mere guesswork, carbon-14 dating was a revolutionary technological breakthrough.

The results obtained from applying carbon-14 dating to prehistoric materials, however, made the technique immediately controversial. For example, early dates for the origins of agriculture in the Near East—generally referred to as the Neolithic Revolution—were dramatically earlier than archaeologists had expected. At Jericho, one of the first sites to provide such material, scholars previously had placed the start of the Neolithic Revolution at around 4000 B.C.E., but carbon-14 dates were on the order of four thousand years earlier. Further research in the Near East has established a chronological frontier for the Neolithic Revolution at 10,000 B.C.E.. or even earlier.

A second major achievement of radiocarbon dating was the establishment of the first outlines of an absolute chronology for developments similar to the Neolithic Revolution in Mexico and Central America. Supported by pollen analysis, carbon-14 dating of the remains of early domesticated maize suggested an astonishingly long continuum for agriculture in the region extending back to 5000 B.C.E..

This finding was the first real indication of the depth and complexity of Mesoamerican cultural heritage.

In sub-Saharan Africa, the immense potential of charcoal for providing carbon-14 dates led to reconstruction of the prehistory of a region which, before 1950, was unknown. Charcoal is especially plentiful at the large numbers of iron-smelting sites in Africa. Slag from the smelting operation, sometimes even iron remains, may be dated using carbon-14; carbon migrates as an impurity into the iron. Before 1950, conventional wisdom attributed the presence of iron-smelting technology in sub-Saharan Africa to cultural borrowing, either from Phoenician traders or from the Egyptians. Carbon-14 dates clearly established a thriving iron technology in West Africa as early as 1000 B.C.E., several hundred years before the earliest known incidence of smelting in Egypt itself.

CONTROVERSIAL RESULTS

Perhaps the most disturbing revelation of the first decade or so of carbon-14 dating was its support for an extended Neolithic chronology in Europe. Although archaeologists already were debating the merits of a "long" versus "short" European Neolithic, most scholars and cultural opinion favored the latter, since it accommodated the notion of cultural "diffusion" from the ancient Near East, relegating Europe to a state of barbarism until stimulated by classical civilizations. The famous megalithic structure of Stonehenge in southern England, for example, was thought to have been inspired by Mycenaean influence and was therefore dated to around 1400 B.C.E. Yet, carbon-14 results obtained from reindeer bone fragments at the site—the bones probably were digging implements used in construction—suggested dates for vital parts of the Stonehenge complex that were several centuries earlier, making it, in effect, pre-Mycenaean. These findings threatened to sever the "diffusionist" link with the Near East and forced a reassessment of the state of prehistoric European culture.

The most severe test of the radiocarbon technique occurred when it was used to corroborate dates for Egyptian historical artifacts that already had been fairly accurately dated with conventional archaeological methods. Early results were not reassuring. The method repeatedly generated dates for materials from the second millennium B.C.E. that were several centuries too recent. Since these discrepancies were more or less consistent in scope and did not alter the *sequences* of Egyptian history, they raised serious questions about the integrity of the whole chronological framework of ancient history patiently constructed over decades by archaeological research. Conversely, many traditional scholars, confident of their work and suspicious of a technique derived from nuclear physics, preferred to reject the radiocarbon system out of hand.

By the early 1970's, these difficulties, together with the developing precision of sample handling techniques and instrumentation, had led to the realization that there were variables in the carbon-14 cycle unaccounted for by Libby's earlier formulations—principally the magnetic field fluctuations, which, as noted earlier, result in periodic changes in the carbon-14 formation rate. Once radiocarbon dates could be calibrated on the basis of tree-ring data, they agreed very closely with previously established dates for Egyptian artifacts and therefore confirmed the work of traditional archaeologists. With that, the last serious barrier to acceptance of the technique disappeared. Some of the prehistoric dates for Europe, however, when corrected according to tree-ring data, actually pushed the Neolithic horizon even further into the past. The earliest portions of Stonehenge are now believed to be separated from what once were thought to be their Mycenaean origins by what one archaeologist has called a "yawning millennium."

In the 1980's, investigators from England, Switzerland, and the United States applied radiocarbon dating techniques to fibers from the "Shroud of Turin." Held in reverence by many as the burial shroud of Jesus, this piece of linen retains the image of a bearded man with marks consistent with crucifixion. Measurements by separate laboratories agreed that the flax from which the linen was produced grew sometime in the thirteenth or fourteenth centuries—far too recent to have been the burial shroud of Jesus.

INTERFACE BETWEEN SCIENCE AND TRADITION

Radiocarbon dating, together with similar techniques using isotopes of other elements and a variety of methods drawn from the physical and life sciences since 1950, has elaborated a picture of human history and earth history to a degree that could not have been conceived by earlier scholars. Among the method's practical benefits is a much more sophisticated knowledge of the scope of natural climatic change, without which it would not be possible to make useful scientific or political decisions on matters that may affect future, human-induced climatic and environmental change. Radiocarbon dating also has shattered many long-standing notions about European prehistory, Europe's historical relationship with the ancient Near East, and the antiquity and complexity of non-European civilizations, thereby undermining fundamental assumptions about the centrality of Western civilization in the human saga.

The history of the application and results of carbon-14 dating require that one look carefully at the conditions and assumptions associated with certain dates and at the stage in the technique's development from which those dates derive. A prime example of how these matters can fuel controversy was the confusion generated by some carbon-14 dates which seemed to suggest that established dates for Egyptian artifacts of the second millennium B.C.E.. were several centuries too early. Biblical scholars had long worried about an unexplained gap between the dates, derived from scientific genealogical and textual studies, for the Hebrew Exodus from Egypt and the establishment of the ancient kingdom of Israel. In terms of regarding the Old Testament as historically accurate, it would be convenient if several centuries of Egyptian history could be "erased" and earlier events moved up to fill the gap. Ironically, that is just what early carbon-14 dating of Egyptian artifacts suggested. Some biblical scholars were ecstatic that science seemed to verify the accounts of the Bible. However, later calibration of these Egyptian dates using tree-ring data reinstated traditional Egyptian chronology and nullified this temporary congruence. Clearly, the lay reader who encounters results of radiocarbon dating encounters a complex interface between modern science and the most fundamental issues in the Western religious and historical tradition.

Ronald W. Davis

FURTHER READING

Agrawal, D. P., and M. G. Yadava. *Dating the Human Past.* Pune: Indian Society for Prehistoric and Quaternary Studies. 1995. An interesting look at the use of radiocarbon dating technology in relation to anthropology. Suitable for the nonscientist. Illustrations.

Chemist Dr. Willard Libby (1908-1980) is known for his development of radiocarbon or carbon-14 dating. Dr. Libby also worked as a member of the Manhattan Project before winning the Nobel Prize in Chemistry in 1960. (Jack Fields/Photo Researchers, Inc.)

Bard, E., F. Rostek, and Guillemette Menot-Combes. "A Better Radiocarbon Clock." *Science* 303 (2004): 178-179. A strong discussion of the need for a radiocarbon curve that extends beyond that of INTCAL98. This article is written in a manner that is accessible to undergraduates as well as graduate students.

Bard, Edouard, and Wallace S. Broecker, eds. *The Last Deglaciation: Absolute and Radiocarbon Chronologies.* Berlin: Springer-Verlag, 1992. This book deals with the radiocarbon dating processes and techniques that have been used to unravel the mysteries of glaciers and massive changes in climate. This book provides a simple explanation of radiocarbon dating and illustrates its applications.

Currie, L. A. "The Remarkable Metrological History of Radiocarbon Dating." *Journal of Research of the National Institute of Standards and Technology* 109 (2004): 185-217. This article accounts the discovery and progression of radiocarbon dating as a tool in geological and archaeological studies. It is extremely detailed and provides an excellent background for anyone using carbon dating techniques.

Fleming, Stuart. *Dating in Archaeology: A Guide to Scientific Techniques.* New York: St. Martin's Press, 1976. Places the methods and results of radiocarbon dating in the context of other dating techniques, such as dendrochronology, thermoluminescence, fission track dating, pollen analysis, and chemical methods. Discusses how several of these techniques may corroborate one another in establishing chronologies. Excellent bibliography.

Lowe, J. John, ed. *Radiocarbon Dating: Recent Applications and Future Potential.* Chichester, N.Y.: John Wiley and Sons, 1996. This college-level book offers a comprehensive overview of the techniques and protocols of radiocarbon dating. Several of the essays explore possible future usage and applications. Illustrations and maps help to clarify difficult concepts. Bibliographical references.

Reimer, Paula J., et al. "IntCal104 Terrestrial Radiocarbon Age Calibration, 0-26 cal kyr BP." *Radiocarbon.* 46 (2004): 1029-1058. This article provides a technical review of the past state of the radiocarbon calibration curve. More recent data are used to update and refine the curve. Written in a highly technical manner for practicing geologists and graduate students.

Renfrew, Colin. *Before Civilization: The Radiocarbon Revolution and Prehistoric Europe,* 2d ed. London: Penguin Books, 1990. A comprehensive discussion of the development of radiocarbon dating, the early discrepancies, and their correction using tree-ring data. Offers a thorough analysis of some alternative approaches to prehistory based on carbon-14 results from Europe.

Rick, T. C., R. L. Vellanoweth, and J. Erlandson. "Radiocarbon Dating and the 'Old Shell' Problem: Direct Dating of Artifacts and Cultural Chronologies in Coastal and Other Aquatic Regions." *Journal of Archaeological Sciences* 32 (2005): 1641-1648. An interesting and easy-to-follow discussion of a common problem in sample selection for carbon dating. This article addresses the problem of "old wood" and compares it to the sampling of shells at archaeological sites. Provides the basic principles of radiocarbon dating in a manner accessible to the layperson with some science background.

Scott, E. M., M. S. Baxter, and T. C. Aitchison. "A Comparison of the Treatment of Errors in Radiocarbon Dating Calibration Methods." *Journal of Archaeological Science* 11 (1984): 455-466. Discusses several methods of calibration with respect to the varying degrees of accuracy required in specific applications.

Taylor, R. E. "Fifty Years of Radiocarbon Dating." *American Scientist* 88 (January/February 2000): 60-67. This paper reviews the entire five decades of development of this remarkable technique. Taylor reviews the basic technique, deviation of carbon-14 dates from true dates due to variation in the magnetic field, means of correction, and attaining increased sensitivity by the application of mass spectrometry.

_____. *Radiocarbon Dating: An Archaeological Perspective.* New York: Academic Press, 1987. An excellent treatment of the procedures and complexities involved in measuring radiocarbon content. Discusses instrumentation, sources of contamination, case studies using various substances, and the historical development of radiocarbon methodology. The bibliography covers a broad range of sources.

Vita-Finzi, Claudio. *Recent Earth History.* New York: Halsted Press, 1974. An account of the physical changes undergone by the earth during the Holocene (modern) geologic era, presented in the form of a stratigraphical narrative based throughout on radiocarbon dates.

Wagner, Gunther A., and S. Schiegl. *Age Determination of Young Rocks and Artifacts: Physical and Chemical Clocks in Quaternary Geology and Archaeology.* New York: Springer, 2010. The authors cover various materials and dating methods. Well organized, accessible to advanced undergraduates and graduate students.

Walker, Mike. *Quaternary Dating Methods.* New York: Wiley, 2005. This text provides a detailed description of current dating methods, followed by content on the instrumentation, limitations, and applications of geological dating. Written for readers with some science background, but clear enough for those with no prior knowledge of dating methods.

Walther, John Victor. *Essentials of Geochemistry,* 2d ed. Jones & Bartlett Publishers, 2008. Contains chapters on radioisotope and stable isotope dating and radioactive decay. Geared more toward geology and geophysics than toward chemistry, this text provides content on thermodynamics, soil formation, and chemical kinetics.

Wilson, David. *The New Archaeology.* New York: New American Library, 1974. Perhaps the best account for the general reader of the development and subsequent refinement of radiocarbon dating techniques. Discusses their enormous impact on the field of prehistoric archaeology and perceptions of prehistory.

See also: Carbon Sequestration; Earth's Age; Earth's Magnetic Field; Earth's Oldest Rocks; Environmental Chemistry; Fission Track Dating; Potassium-Argon Dating; Radioactive Decay; Rubidium-Strontium Dating; Samarium-Neodymium Dating; Uranium-Thorium-Lead Dating.

RARE EARTH HYPOTHESIS

The rare Earth hypothesis states that complex life on any planet other than Earth is highly improbable, as it requires just the right combination of favorable conditions. These conditions include location in the galaxy; type of star and distance from the star; planetary mass, tilt, and orbit; size and orbit of a large moon; amount and kind of atmosphere and oceans; plate tectonics; and magnetic iron core.

PRINCIPAL TERMS

- **complex life:** multicellular organisms with integrated organ systems comprising animals and plants
- **dynamo theory:** an attempt to explain the magnetic field of some celestial objects in terms of rotation, heat convection, and electrical conduction in a fluid, metallic inner core
- **escape velocity:** the minimum speed required for an object to escape from the gravity of the earth or other celestial object
- **extremophiles:** microbial forms of life that live in extreme environments above or below the temperature range for liquid water
- **giant impactor theory:** a lunar origin theory suggesting that a Mars-size planetesimal collided with the earth, blasting enough debris into orbit to form Earth's moon
- **greenhouse gases:** atmospheric gases such as water vapor, carbon dioxide, and methane that trap heat by absorption of solar radiation, causing an increase in temperature
- **habitable zone:** the region around a star where a planet with an atmosphere similar to that of the earth would have a surface temperature that permits liquid water
- **plate tectonics:** the geological processes involving the large-scale motions of the earth's crust that account for continent formation, mountain building, and continental drift
- **snowball Earth:** conditions that prevailed in the earlier periods of Earth's history when the planet was covered with ice, perhaps extending to the equator
- **solar wind:** the stream of high-energy charged particles flowing out from the upper atmosphere of the sun, consisting mostly of protons and electrons
- **subduction:** a plate tectonics process in which an oceanic plate is forced under an adjacent continental plate, often resulting in volcanic eruptions

ORIGIN OF THE RARE EARTH HYPOTHESIS

The rare Earth hypothesis was proposed by paleontologist Peter Ward and astronomer Donald Brownlee, both of the University of Washington, in their book *Rare Earth: Why Complex Life Is Uncommon in the Universe* (2000). Contrary to prevailing opinion, the two scientists proposed that complex life is extremely rare in the universe, although they conceded that single-cell life might be quite common. Thus, the rare Earth hypothesis combined two predictions: the commonness of microbial life and the rarity of complex life in the universe, leading to the conclusion that Earth might be the only planet in the Milky Way galaxy with complex life.

The hypothesis challenged the accepted view of most of the scientific community that Earth was not special or unique, but was just one of billions of other similar planets in the galaxy of some 200 billion stars. This view is called the principle of mediocrity, or the Copernican principle, and it suggests that complex life is to be expected throughout the universe. A famous prediction by popular astronomer and astrophysicist Carl Sagan in the 1960s, and based on a proposal by astronomer and astrophysicist Frank Drake, concluded that one million intelligent civilizations exist in the galaxy.

The rare Earth hypothesis recognizes that microbial life exists throughout the earth under nearly all conditions, and may even be common on planets quite different from Earth. This view was reinforced with the discovery in the 1970s of new microbes called extremophiles, which live under extreme conditions of temperature and pressure in deep-sea hydrothermal vents and in volcanic hot springs and geysers. However, complex life appears to require much more limited environments, in which can be found liquid water, minimal radiation, and a favorable atmosphere. These conditions greatly restrict the location, size, and type of planet that can support complex life. Analysis of these requirements leads to a low probability that complex life as known on Earth exists elsewhere in the universe.

The rare Earth hypothesis provides a solution to the Fermi paradox, in which Italian-born American physicist Enrico Fermi's answer to the question of intelligent life on other planets was "Where are they?" He assumed that if intelligent life was common on other planets, some of them would have developed technology over a few million of the last few billion years and would have made contact with Earth by now.

In the following sections, some life-supporting requirements are reviewed in terms of astronomical conditions of location in the galaxy, geological conditions of size and kind of planet, and biological conditions imposed by the nature of evolutionary processes.

ASTRONOMICAL REQUIREMENTS FOR LIFE

According to the rare Earth hypothesis, the earth is a "Goldilocks planet" with just the right conditions for life: It exists in the right kind of galaxy and in the right location in that galaxy, it has the right kind of sun and is located the right distance from that sun, and it is the right kind of planet with the right kind of moon. Only spiral galaxies contain stars with enough of the heavier elements needed for life. These are second- or third-generation stars that result from supernova explosions, which are the only known processes for producing elements heavier than lithium. Elliptical galaxies, small galaxies, and globular clusters contain mostly first-generation stars with few heavy elements; most of these stars of a suitable mass have evolved into giant stars that are too hot for Earth-like, rocky, inner planets to retain liquid water and sustain complex life.

Even within spiral galaxies containing metal-rich stars (heavy elements), only a narrow galactic habitable zone exists with the right conditions for life. Near the galactic center and inside the spiral arms, energetic processes produce too much radiation for life to develop; in the outer edges of a galaxy most stars are metal-poor. Thus the habitable zone in a spiral galaxy is a narrow ring around the center of the galaxy, which must be nearly circular for a star to maintain the right distance from the center as it orbits around the galaxy. Only about 5 to 10 percent of stars reside in this habitable zone, limiting life-supporting stars in a spiral galaxy to about ten or twenty billion stars.

Most stars are too hot or too cold to allow for liquid water on any planets these stars might accompany. Hot stars have shorter lives and eventually expand to become red giants that would envelop any rocky inner planets. Cooler stars might provide sufficient warmth for nearby planets, but these planets would be subject to dangerous radiation from solar flares and would tend to have one side gravitationally locked toward the star, making that side too hot and the other side too cold. Various estimates suggest that less than 10 percent of stars are of a suitable size and temperature for supporting life. Furthermore, any rocky planets orbiting a suitable sun-like star must be at the right distance from the star in a planetary habitable zone.

Estimates for the habitable zone for Earth indicate that a 5 percent decrease in orbital radius would be too hot, like Venus, and a 15 percent increase would be too cold, like Mars. Stable orbits are also required, which would preclude most multiple-star systems or highly elliptical orbits that allow for too much interaction with neighboring planets. Most of these astronomical requirements for life were not taken into account in the original optimistic estimates by Drake and Sagan of the existence of one million intelligent civilizations in the galaxy.

GEOLOGICAL REQUIREMENTS FOR LIFE

Several characteristics of the earth are unique among the planets and provide its life-supporting capability. These characteristics include being the right size, supporting liquid water, having a favorable atmosphere, having surface chemicals, undergoing plate tectonics, having extended continents, having a strong magnetism, and having a large moon. The size and mass of the earth are critical in retaining its atmosphere and oceans. A smaller mass would not have sufficient gravity to hold water vapor long enough to form the oceans. A larger mass would trap larger amounts of heavier gases such as carbon dioxide and retain too much water to allow for continents to rise above the oceans. A denser atmosphere would have produced a severe greenhouse effect and high temperatures, boiling away the oceans, as seen on Venus, whose atmosphere is one hundred times denser than that of Earth. A thin atmosphere would cause freezing temperatures, as seen on Mars, which has an atmosphere one hundred times less dense than that of Earth.

Many of the more volatile elements needed for life did not readily condense from the solar disk except in the cooler regions farther from the sun, where the giant gas planets formed. Some of this material was returned to Earth's surface by the gravity of the giant gas planets in a heavy bombardment of comets and asteroids, seeding the earth with carbon, water, and other materials needed for life.

There is some evidence from computer simulations to suggest that Jupiter and Saturn may have played a different role (after this initial heavy bombardment decreased) in helping to trap the larger and more energetic comets and asteroids before they could reach the earth, reducing the frequency of giant impacts and the extinctions of life they produce. In either case, it appears that giant gas planets such as Jupiter are needed at the right location relative to an Earth-like planet (but not so close as to destabilize its orbit) to provide chemicals needed for life and to protect from mass extinctions.

The rare Earth hypothesis also recognizes the important role of plate tectonics in supporting life and the unusual conditions on Earth that made it possible. Plate tectonics is the process that forms the continents through the slow motions of large crustal plates. Without the continents, Earth would be covered with water and could support only marine life.

Plate tectonics also helps to regulate long-term climate by acting like a global thermostat. Carbon dioxide in the atmosphere helps to raise temperatures through the greenhouse effect. As temperatures increase, weathering increases and removes carbon dioxide from the atmosphere into the oceans, lowering temperatures. The carbon dioxide becomes trapped in limestone deposits, which are eventually recycled by the subduction of oceanic plates beneath continental plates and released into the atmosphere by volcanic eruptions. This raises the temperature and begins the process again. Earth is the only known planet with just the right combination of crustal thickness, oceans, core heating, and internal heat convection to produce plate tectonics and its resulting continent building and climate regulating results.

Earth is also unique among the known rocky planets with its large moon and large magnetic field, which protects the earth by deflecting high-energy particles in cosmic rays and solar wind. Without this protection, early life forms would have had little chance to survive, and Earth's atmosphere would

have been slowly depleted. The dynamo effect that produces Earth's magnetism requires strong heat convection, which is enhanced by plate tectonics. Earth's large moon also appears to have contributed to a strong magnetic field and several other beneficial results. Computer simulations support the giant impactor theory for the formation of the moon. This theory also maintains that there was a substantial increase in Earth's iron core and internal heat, both of which are required for a strong dynamo effect.

The giant impact also increased the earth's size, spin, and tilt, making them more favorable for life. The impact probably thinned the earth's crust as well, making plate tectonics more likely. The moon also stabilizes the tilt of the earth, giving it regular seasons over long periods of time that provide better support for the development of life.

BIOLOGICAL REQUIREMENTS FOR LIFE

Several other unusual biological features of planet Earth made the evolution of life possible, including early conditions that nurtured complex molecules, the right kind of atmosphere for plants and animals, the emergence of photosynthesis, the right amount of oxygen, and certain trigger events that stimulated evolutionary diversity and opened new niches for complex life. The giant impactor theory of the moon implies that the moon was once closer and that it produced huge Earth tides, sweeping far inland and enriching the oceans with minerals needed for life. These tides would also produce tidal pools that could have concentrated nutrients by evaporation for emerging life forms. Tidal cycling of wetting and evaporation in these intertidal pools might have provided the kind of environment in which proto-nucleic acid fragments could begin to associate and assemble molecular strands, leading to the origin of life.

Computer simulations of the giant impact formation of the moon suggest that the impact removed much of the primordial atmosphere of the earth, thus avoiding a greenhouse effect like that on Venus. As the earth cooled after the giant impact, a new atmosphere was produced through outgassing and comet collisions. Eventually, a new crust formed and water vapor condensed to form oceans, which then began to absorb carbon dioxide. The reformulated atmosphere on Earth after the collision, and subsequent water condensation, was thin enough to prevent a runaway greenhouse effect and was sufficiently transparent to eventually allow photosynthesis to occur (with its associated production of oxygen). Plate tectonics increased the rate of oxygen production by enhancing biological activity through the recycling of nutrients such as phosphates and nitrates.

Rare-Earth proponents also argue that evolutionary diversity was greatly enhanced by unusual events on Earth that are highly unpredictable and perhaps unique to Earth in their magnitude and timing. Two snowball Earth periods of nearly global glaciations coincided and probably triggered mass extinctions that were followed by explosions of new species. Single-celled (prokaryotic) life on Earth began about 3.8 billion years ago with little development until approximately 2.4 billion years ago, when falling temperatures led to the first snowball Earth; this was followed by the emergence of the first nuclear-celled (eukaryotic) life. Through the next billion years came many new multicellular species.

The second snowball Earth occurred between 800 and 600 million years ago, followed by the Cambrian explosion of new species and the great diversification of animal phyla, with no known phyla appearing since. Apparently, the great stress of environmental changes produced new niches and stimulated rapid evolutionary changes. Similar mass extinctions followed by new evolutionary developments have been associated with giant impacts, such as the one that appears to have ended the age of dinosaurs 65 million years ago. If these events had involved larger colliding objects, they could have sterilized the earth and ended all complex life.

THE IMPROBABILITY OF COMPLEX LIFE BEYOND EARTH

The rare Earth hypothesis greatly constrains the probability of life in the universe. The Drake equation that predicted one million intelligent civilizations in the galaxy requires many changes to its original fractional factors and many new factors. These factors include the fraction of stars in the galactic habitable zone and the fraction of those stars with planets; the fraction of rocky planets, of planets in habitable zones that develop microbial life, and of planets where complex life evolves; the fraction of the lifespan of a planet during which complex life is present; the fraction of planets with a large moon; and the fraction of those planets with large gas planets in the right location, and of planets with

a critically low number of extinction events. If each of these nine fractions averaged one-tenth, the probability of complex life in the galaxy would be reduced by one billion, even without other rare-Earth factors such as plate tectonics and snowball Earth events. Thus the rare Earth hypothesis implies that there is less than one chance in one thousand that another planet in the galaxy can support complex life, and even less for one with intelligent civilizations.

Although the requirements for simple microbial life are fairly lenient and may be common on many planets or moons, requirements for complex life are much more stringent. During nearly four billion years of life on Earth, complex life has existed for only about one billion of those years. Many biologists agree with the rare Earth hypothesis that the evolutionary path from microbial life to complex life, and especially to human life, was so highly improbable that it is unlikely to occur again, even given a "Goldilocks" environment like that of Earth. The biologists also emphasize the many adaptive disadvantages in the high metabolism required for a large brain and the long gestation period and childhood for humans. Other unusual characteristics such as hand-eye coordination and vocal apparatus allowing for speech and cooperative efforts also seem highly improbable. If the rare Earth hypothesis is correct, then Earth is not just a mediocre planet in a far corner of the galaxy, but a very special place indeed.

Joseph L. Spradley

FURTHER READING

Dartnell, Lewis. *Life in the Universe: A Beginner's Guide.* Oxford, England: One World, 2007. This book describes the processes and possibilities of evolution and how it came to occur on Earth, and speculates on various possibilities for the evolution of life elsewhere in the universe.

Davies, Paul. *The Eerie Silence: Searching for Ourselves in the Universe.* Boston: Houghton Mifflin Harcourt, 2010. The author responds to issues raised by the rare Earth hypothesis, suggesting new approaches in the search for extraterrestrial intelligence and the significance of the possibility that humans are alone in the universe.

Gonzalez, Guillermo, and Jay Richards. *The Privileged Planet: How Our Place in the Cosmos Is Designed for Discovery.* Washington, D.C.: Regnery, 2004. This book presents a comprehensive review of evidence for a rare Earth, with the added insight that the place and time of humans in the universe appear to be uniquely adapted for observing and understanding that universe.

Morris, Simon Conway. *Life's Solution*. New York: Cambridge University Press, 2003. The author, a leading paleontologist who believes that evolution converges toward intelligent life, is a proponent of the rare Earth hypothesis. He discusses the hypothesis in Chapter 5.

Ward, Peter, and Donald Brownlee. *Rare Earth: Why Complex Life Is Uncommon in the Universe*. New York: Copernicus Books, 2000. This groundbreaking book describes many unusual life-sustaining features of the earth in a readable format. The authors were the first scientists to use the term "rare Earth hypothesis."

Webb, Stephen. *If the Universe Is Teeming with Aliens . . . Where Is Everybody? Fifty Solutions to Fermi's Paradox and the Problem of Extraterrestrial Life*. New York: Copernicus Books, 2002. Webb considers the possibility that aliens are already among humans, or that they have not yet communicated with humans. In Chapter 5, Webb reviews evidence from the rare Earth hypothesis and concludes that aliens probably do not exist.

See also: Earth-Moon Interactions; Earth's Differentiation; Environmental Chemistry; Importance of the Moon for Earth Life; Oxygen, Hydrogen, and Carbon Ratios; Plate Tectonics.

RELATIVE DATING OF STRATA

The successive deposition of identifiable layers of geologic materials provides a way to ascribe relative age to the contents of those layers. The dating is based on the assumption that the closer a layer is to the surface, the more recently it was deposited. It is also assumed that unconnected layers in different locations are of the same age if they have the same composition and the same place within a pattern of layers. Sedimentation, fluvial processes, and cataclysmic or volcanic processes are the most common depositors of strata. Stratigraphic dating can only be relative to other stratigraphic layers; absolute dating requires a "clock" mechanism such as radioactive decay that effectively counts time back from the present.

PRINCIPAL TERMS

- **arenaceous:** rocks or sediments having a sandy composition, composed of grains of sand
- **argillaceous:** rocks or sediments formed principally from clay or clay-mineral particles
- **dip:** the angle between the horizontal and the apparent plane between stratigraphic layers
- **facies:** the combination of characteristic lithology and paleontology of a sediment from which the environmental conditions at the time the sediment was deposited can usually be inferred
- **fluvial:** having to do with or being the result of flowing water or other liquids
- **mudstone:** sedimentary rock type formed from mixtures of particles ranging from fine clay to coarse sand grains
- **orogeny:** the process of mountain building by tectonic movement
- **sandstone:** sedimentary rock type formed from sand-sized silicate grains agglomerated and consolidated usually with carbonate minerals
- **seat earth:** a fossil soil layer generally found directly beneath coal beds, often with plant rootlets still in place
- **siltstone:** sedimentary rock type, typically shale, formed from fine clay and silt particles less than four microns in size
- **strata:** defined layers in sedimentary rock, typically separated from each other by identifiable bedding planes

STRATIGRAPHIC LAYERS

British geological engineer William Smith determined in 1793 that there is a direct correlation between the fossil content of rock layers and the layers themselves. Although stratigraphic layers of rock and soil always have been apparent to the human eye, as have fossils, no combination of the two observations had ever occurred. Smith came to understand that the fossil content of a layer was characteristic of that layer and, hence, of the age in which the layer was formed. His discovery permitted, for the first time, the mapping of subsurface terrains and has since proved invaluable for mining, engineering, resource exploration, hydrology, construction, geology, paleontology, and every other branch of subterranean endeavor.

Smith's realization came from the deceptively simple observation that the same kinds of fossils were always found in the same kind of rock layers in excavations carried out for the construction of canals, bridges, and other structures. By identifying the fossils that were unearthed, he found he could tell exactly what the stratigraphy would be both above and below the location of the fossils, regardless where they were located. In many cases, the layers were discontinuous, having been incised by erosion or other features. This did not seem to alter the stratigraphic ordering of the layers, however, leading to the conclusion that any particular layer must have been laterally continuous when it was formed. Any variations and interruptions in a particular layer were thus caused by erosive and geological processes that occurred in the time since the layer was formed.

Layer Formation and Rock Types

Earth is a dynamic planet. It has an active core and mantle that exert forces on the relatively eggshell-thin crust, causing it to split and reorganize, bring new material to the surface through volcanic activity, subduct older material back into the mantle, and deform existing structures through the actions of heat and pressure. Each of these processes leaves its imprint in the geological structure of the crustal material in a variety of ways. In addition, fluid processes caused by the movement of air and water over the surface play significant roles in the development of surface structures and stratigraphic layers.

Chronostratigraphy is a practice that allows geologists to assign the relative ages and general dates at which specific processes occurred according to the characteristics of particular stratigraphic layers. When tied to an absolute dating method such as radiometric dating, the assignment of age according to stratigraphic location can be quite accurate.

Stratigraphic layers are formed by two primary methods: volcanic activity and sedimentation. Through volcanic activity, various rock and mineral formations can be formed as stratigraphic layers. Molten lava spreading and solidifying over existing structures forms layers of igneous rock, while volcanic ash and other ejecta fall to the ground and form a covering layer.

Strata (horizontal bands) are parallel layers of different-colored or differently structured rock or soil lain down by a process of sedimentation. Individual layers may vary from millimeters to a kilometer or more. Limestone and shale are examples of sedimentary rocks, formed by the compaction of water deposits. Photographed in the Negev Desert, Israel. (PhotoStock-Israel/Photo Researchers, Inc.)

Sedimentary processes, in which particulate mineral matter settles out of a fluid medium, deposit layers of sediment, dust, sand, and larger particles on top of existing surface structures. Sedimentation occurs in order of relative density, with the most dense particles settling out first and the least dense last. Typically this means that light organic materials are deposited at the top of each particular deposition layer, over time forming bedding planes between successive layers. In many sedimentary formations, bedding planes permit the cleavage and separation of layers to reveal physical details of materials and objects that have been captured in the formation. Sedimentary rock formations are found to contain fossil formations, while igneous rock formations do not.

Between igneous and sedimentary rock types are the metamorphic rocks. These may be either igneous or sedimentary in origin, but both have been altered through heat and pressure from their original form. An example is the argillaceous shale known as slate. Shale is formed by the deposition of fine clay particles less than four microns in size. Through subjection to prolonged heating and pressure the shale structure becomes baked and compressed, or tempered, into the much harder and more glass-like structure of slate. Metamorphic rock that is sedimentary in origin may still contain fossils, but the metamorphosis tends to destroy fossil remains, and at the very least makes it much more difficult to release them.

FOSSILS AND FOSSILIZATION

Essentially all fossils are formed as the result of fluvial processes. Typically, fossilization occurs after an organism dies and its remains become covered by sediments before they are able to decompose or be ravaged by scavengers. As successive layers of sediment build up over the remains, decomposition slowly occurs, as the surrounding material is compressed by the weight of material above it. The extra weight eventually compresses the sediments sufficiently to bring about chemical alterations that cement together its component particles. At the same time, minerals dissolved in water percolating through the sedimentary layer replace the original mineral content of the remains to produce an exact replica of the remains in stone.

To appreciate the formation and importance of fossil formations, it is necessary to understand the length of time over which such processes occur. While it is possible for the entombment of remains to occur very quickly, as when an animal has perished in a flash flood or the collapse of a sand dune, it is far more common that the deposition of sedimentary layers has occurred slowly over many years. Geologists use present-day rates of sedimentation as model data for past processes.

The deposition of silt in a slow-moving stream or river is as little as one millimeter, essentially no more than the layer of clay that coats rocks in slow-moving waters. Such a layer compresses to a considerably smaller thickness, indicating that a stratigraphic layer one meter in thickness and consisting of a single type of sediment could have taken tens of thousands of years to accumulate. In early marine environments that held only simple mollusks and crustaceans with calciferous remains, the life cycle of those organisms would have produced an almost constant deposition of shells and other remains to the sea floor over long periods of time. As the remains, consisting mainly of calcium carbonate, accumulated, they eventually compressed to form thick, uniform deposits of chalk. Such deposits, such as the white cliffs of Dover in England and extending into western Europe, are known to have thicknesses measured in hundreds of meters. It was the consistency of such layers occurring in recognizable patterns, along with their characteristic fossil content, which provided the basis for Smith's observations.

The Suppositions of Chronostratigraphy

Relative dating according to position within stratigraphic layers, or chronostratigraphy, is based on three very basic, but important, suppositions. The first of these is the principle of superposition. This is the principle by which every stratigraphic layer that forms must do so on top of existing stratigraphic layers. That is to say, each layer is correspondingly older than the layer that is immediately above it. The basic premise here is that layers form according to the succession of time; a layer of sediment that forms in any given year can never become situated below the layer that formed in any previous year. This provides the basis for the vertical ordering of layers.

The second supposition is the principle of lateral continuity. According to this principle, stratigraphic layers having identical mineralogical content and structure, containing identical distributions of fossil or other extraneous content, must have formed at the same time and as a single continuous layer. Various erosive and geological processes that operate in the time between the formation of the layer and some more recent date

function to form discontinuities in the layer that were not present when it was formed. These may be such actions as water erosion, seismic, tectonic and orogenic activity, and wind erosion.

Each of these processes has the potential to remove over time some portion of a laterally continuous stratigraphic layer, or to induce some form of discontinuity in the layers of a specific region in which the activity takes place. Orogenic processes raise and fold rock formations, producing angled beds that, when coupled with the results of erosive actions, mask the innate continuity of the stratigraphic layers within the affected formation. This is the feature that makes Smith's deductions the stroke of genius rather than mere observation.

The third basic principle is horizontality, in which identical layers are tied to the same period or absolute date regardless of their relative elevations and orientations. For example, a layer formation that has been elevated by orogenic activity is nevertheless the product of the same period as the identical layer at a lower elevation. The dating of a layer is therefore transferable from one location to another only if the two layers are otherwise identical in composition, content, and position within the pattern of the encompassing layers.

Given these three basic principles, the age of any particular layer relative to other layers can be ascribed with a high degree of accuracy. The pattern of stratigraphic layers, however, forms a self-consistent internal system with the condition that such a system might reflect a span of time that could have occurred at any time within the age of the planet. That is to say, it is not possible to know precisely when the span of time reflected in the pattern of stratigraphic layers actually came to pass. This requires the correlation of some point within that span of time to a known point in time, or absolute date.

Absolute or Radiometric Dating

To know precisely when a particular stratigraphic layer in a geological formation came into being, it must be possible to relate that unknown point in time to some known point in time. The most obvious point in time to which the past can be tied is the present. To achieve this, some kind of "clock" process that can be used to count time back to some specific starting point is required. Nature provides just such a clock mechanism in the structure of atoms.

Chemical elements are known to occur in isotopic forms in which certain atoms containing the same number of protons in the nucleus, by which they are atoms of the same element, have different numbers of neutrons. Some isotopes are radioactive. That is, they decay into atoms of different elements by ejecting specific portions of their nuclear mass. The process has a specific starting point and continues through exact steps until a stable atom is formed. Accordingly, a specific radioactive parent element produces only one specific stable daughter element.

Equally important is the rate at which the exponential radioactive decay process takes place. The rate is directly dependent on the amount of radioactive material that is present; this rate is described by the mathematical equation $A = A_o\, e^{Kt}$, where A is the amount of the element at time t, A_o is the original amount of the element, t is the time that has elapsed form the starting condition, and K is a constant. A special relationship exists for such a system when the amount remaining is exactly one-half of the original amount. According to the mathematical relationship, it takes exactly the same amount of time for any quantity of the material to break down to one-half of that quantity. Thus it requires the same amount of time for one kilogram of a radioactive element to decay to 500 grams as it does for one gram to decay to 500 milligrams. This amount of time is called the half-life of the process. By determining the amounts of starting radioactive element and stable product element in a sample, it is then easy to determine the number of half-lives that have elapsed since the initial condition of the system was established. Essentially, the equation can determine how long ago the initial condition was established.

One of the most-used relationships for determining geological time is the ^{40}K–^{40}Ar decay system, in which radioactive potassium atoms of mass 40 decay into stable argon atoms of mass 40. Because this occurs within the structure of potassium-containing rocks, the argon that is produced is trapped within the same rocks. Mass spectrometric analysis of the potassium and argon recovered from a particular rock formation can determine the relative amounts of ^{40}Ar and ^{40}K that remain. Measurement of the rate of decay for the process can determine the value of K. These values provide the data needed to determine the time t that has elapsed.

There are, of course, practical limits based on the error limits of detection of the isotopes. The half-life of the K-Ar transition is 1.28×10^9 years. An error of just 0.01 percent on this determination translates to a temporal error of 128,000 years, an error range of 256,000 years altogether. More precise measurement enables a more precise determination of absolute age, but it must be remembered that other sources of error also are involved. Thus, geological ages are always rounded off or expressed as a range.

An excellent example of these limits is the so-called K-T boundary event, which terminated the Cretaceous period and ended the age of dinosaurs. In that event, a very large meteorite collided with Earth near the Yucatan Peninsula, initiating a globe-spanning firestorm and depositing a recognizable layer of iridium-rich material over the entire planet. This layer delineates the Cretaceous-Tertiary boundary and is dated radiometrically to 65 million years ago. A more precise date is not yet knowable, even though the actual event occurred at one specific instant in time.

RELATING AGES

With an appropriate absolute-dating method, the absolute age of specific contents (such as a fossil) of a stratigraphic layer can be determined. From that point on, the presence of fossils identified as the same species in a stratigraphic layer at any other location also identifies the age of that layer. Similarly, the presence of a specific mineral type in one layer can be used to assign an age to another layer in which that same mineral type appears. This is somewhat less reliable method, however, because Earth is a dynamic planet and because the chemical processes that create specific minerals are constantly in operation, permitting the formation of the same mineral type at different times. In unique cases, such as that of the K-T boundary, however, the identification of relative age is absolute, and it is certain that that stratigraphic layer identifies the same point in time wherever it is found.

Richard M. Renneboog

FURTHER READING

Bennett, Sean J., and Andrew Simon, eds. *Riparian Vegetation and Fluvial Geomorphology.* Washington, D.C.: American Geophysical Union, 2004. This book closely examines the various processes that take place along flowing water, such as how erosion and sedimentation are related and how they are affected by vegetation and geological structures.

Koutsoukos, Eduardo A. M., ed. *Applied Stratigraphy.* New York: Springer, 2007. This book presents the historical background of stratigraphy, and builds on that foundation to present the theory and practice of chronostratigraphy as a research tool.

Petersen, James F., Dorothy Sack, and Robert Gabler. *Fundamentals of Physical Geography.* Belmont, Calif.: Cengage Learning, 2011. Chapter 14 is dedicated to fluvial processes and how they relate to various landforms, an important factor in the formation of sedimentary stratigraphic layers.

Rey, Jacques, and Simone Galeotti, eds. *Stratigraphy Terminology and Practice* Paris: Technips, 2008. This book focuses on the essential theory and practices of the five major fields of chronology based stratigraphic methods, relating them to each other and to the fields of study in which each is best employed.

Wicander, Reed, and James S. Monroe. *The Changing Earth:. Exploring Geology and Evolution.* 5th ed. Belmont, Calif.: Cengage Learning, 2009. This basic college-level geology textbook provides a thorough discussion of geological processes before discussing how they provide the stratigraphic basis for relative dating.

Winchester, Simon. *The Map That Changed the World:. William Smith and the Birth of Modern Geology.* New York: HarperCollins, 2001. This well-researched book details the history of William Smith's realization of stratigraphy and the lateral continuity of those layers.

See also: Carbon Sequestration; Earth's Interior Structure; Earth's Oldest Rocks; Elemental Distribution; Environmental Chemistry; Experimental Petrology; Freshwater Chemistry; Isotope Geochemistry; Magnetic Stratigraphy; Mountain Building; Plate Tectonics; Radioactive Decay; Radiocarbon Dating; Rock Magnetism; Soil Liquefaction; Subduction and Orogeny; Volcanism.

REMOTE-SENSING SATELLITES

Remote-sensing satellites use instruments for gathering data about Earth's atmosphere, climate, weather, and geography, and also for exploring celestial bodies.

PRINCIPAL TERMS

- **aerial photography:** geospatial observation from above; in some respects, a fairly limited predecessor of modern remote sensing
- **altimetry:** a determination of the topography of geographic features; with satellites, it involves sending a signal and determining its time of return
- **bathymetry:** a determination of ocean depths, previously measured through depth sounding but now measured through satellite altimetry
- **geographic information system (GIS):** a computer system designed to store and handle the large quantity of data generated by geographic research
- **geospatial:** pertaining to the location of objects on the surface of the earth and their relationship to each other
- **light detection and ranging (LIDAR):** a system in which light signals are pulsed to the ground, detecting range to map an area's topography
- **multispectral photography:** photography generated by utilizing the varying emission wavelengths of different rocks and vegetation
- **remote sensing:** the use of instruments not in contact with an object to gain information about that object
- **satellite:** a human-made object put into orbit around a celestial body, typically equipped with instruments for various measurements or observations
- **topography:** a representation of the geographic features of an area, such as on a map
- **topology:** the study of the topography of a certain area over time

THE MEANING AND HISTORY OF REMOTE SENSING

While remote sensing has applications for greater discovery on Earth and the celestial bodies beyond, a person can be engaged in remote sensing, at its most basic level, at any given moment. Remote sensing is utilizing instruments to obtain information about objects without coming into contact with those objects. (By reading this text, one engages in remote sensing, using an "instrument" of the body.)

The history of remote-sensing satellites significantly precedes that of space travel. Early attempts at aerial surveillance go as far back as the eighteenth century. In 1794, French troops utilized hot air balloons to obtain information on enemy troop formations at the Battle of Fleurus, during the French Revolutionary Wars.

In the nineteenth century, with the invention of photography, balloons were used to assist in mapping. Later, aerial reconnaissance was used by the Union and Confederacy forces during the American Civil War to learn more about enemy troop positions. Similar balloons were used for mapping in the Spanish-American War at the close of the 1890s.

The twentieth century saw a boom in aerial photography. In 1903, German engineer Julius Neubronner invented a camera to be placed on the breast of a pigeon to take photographs at set intervals. In 1911, Italian airplane racer and air force commander Carlo Piazza flew probably the first reconnaissance mission using photography during the Italo-Turkish War. Aerial photography accompanied the widespread use of airplanes in World War I. The technology spread afterward, allowing pilots to photograph difficult-to-access areas like Antarctica from above.

During World War II, Nazi Germany developed V-2 rockets, missiles with a range of about 305 kilometers (190 miles) and designed to strike land targets. Following the war, Wernher von Braun, the physicist who headed the team that had developed the V-2, came to the United States and worked with the U.S. Army and later with the National Aeronautics and Space Administration (NASA) in developing rockets. His research at White Sands, New Mexico, involved the use of rockets, often outfitted with cameras or video recorders.

A key step from aerial photography and surveillance to remote sensing took place with the launch of the first satellite, Sputnik I, by the Soviet Union on October 4, 1957. While later satellites would conduct remote sensing to gather data about the planet, Sputnik I was effectively different. It sent out a radio signal that could be picked up on Earth; remote sensing of space, as it applied to this satellite, was conducted by operations on Earth.

The launch of Sputnik triggered strong public opinion and is cited as a primary reason for the founding of NASA in 1958. A number of Russian and later U.S. satellites followed, many of which were equipped with instruments to gather data. Some were sent to make measurements of outer space, but others were designed to observe the earth.

Researchers recognized the limits of conventional photography when it came to observing the earth and reached out for newer tools, including multispectral, infrared, and ultraviolet photography. Recognizing a change in the field, Evelyn Pruitt, a geographer working for the U.S. Office of Naval Research, coined the term "remote sensing" in 1960. Pruitt explained that a new term was needed to change the notion of what the field involved.

Beyond aerial photography, remote sensing incorporated photography methods outside the visual light spectrum, ultimately "seeing" the earth in different ways. The great amount of data being collected by remote-sensing satellites also required a better processing system for the data. Geographic information systems (GISs) were used beginning in the 1970s to synthesize all the data, enabled by the increasing memory and processor speeds of computers becoming available.

GIS allows for the development of many different "maps," including, for example, a map that would include details of a location's topography, soil, vegetation, and temperature. Maps not generated by satellite data also can be incorporated into such a system (for example, a census of a region's population). GIS, like other remote-sensing satellite technologies, is now available to consumers. Starting in the early twenty-first century, Web-based programs such as Google Earth have allowed for the observation of satellite data on personal computers.

Remote-sensing satellites are used in a wide variety of applications. These applications include determining the location of objects on the earth's surface and monitoring pollution, weather systems, oceanic activity, and the changing geographic terrain of areas such as glaciers, rainforests, and crop lands. These satellites also are used for space exploration.

OBSERVING THE EARTH FROM ABOVE

In 1972, NASA and the U.S. Geological Survey started the Landsat program. This program, originally called the Earth Resources Technology Satellite I, was designed to monitor the natural resources of the planet. The program is ongoing, and seven Landsat satellites have been launched, the most recent in 1999. However, only Landsat 5 and Landsat 7 are operational.

Landsat satellites photograph the earth's surface, and the images are stored to give a picture of changing geography over time. Landsat's imaging is at a medium resolution, which allows the viewer to see major geographic features and larger roads but not individual houses.

The European Space Agency also monitors Earth's climate and natural resources, with programs such as Envisat, which launched a satellite in 2002. Envisat provides data on Earth's land surface, oceans, and atmosphere.

Many satellites also are equipped with advanced instruments that can obtain data beyond visible light frequencies. The National Oceanic and Atmospheric Administration, for example, launched the Advanced Very High Resolution Radiometer satellite program in 1998. The satellite captures images at infrared and near-infrared frequencies.

Infrared is key in measuring polar ice, glaciers, and snow fall, all of which are crucial to scientific inquiry into global climate trends. Infrared sensing allows for gauging snowfall and the density of ice. The earth's reduced ice density indicates a warming trend for the planet, by which ice melts and does not refreeze as the temperature cools. This reduced density is evident not only in the polar regions but also in areas where glaciers have been shrinking with the passing years.

Other monitoring issues are more localized. Many of the satellites used to gauge land usage are accurate to within about 1 kilometer (0.6 mile). This accuracy is enough to gauge the effect of human activity on local ecosystems. A common issue is the conversion of other types of land into cropland. Two problems that often accompany cropland development are deforestation and soil erosion. Satellite observation has shown that tens of thousands of square kilometers of rainforest are lost each year in Southeast Asia, Africa, and Latin America. In addition to a loss of wildlife and other parts of the rainforest ecosystem, the loss of rainforest can lead to soil erosion.

Monitoring the earth through satellites goes beyond landforms. One method of mapping that has been distinctly changed by use of satellites is bathymetry. From ancient times, sailors have "sounded" the ocean floor by using poles to determine water depth. Over time, instruments such as sonar were developed to help measure the depths of the ocean floor—measurements that could provide three-dimensional maps.

In the 1850s, experts determined that height analysis at the water's surface could indicate depth. Height analysis was further tested in the late 1970s. Later, altimetry from orbiting satellites was used to gauge the depths of the oceans.

Another important measurement of altimeters on satellites is that of sea levels. Rising global temperatures are leading to a rise in ocean levels, which is caused by melting ice caps in the polar regions. Satellite altimetry, which is thought to be more accurate than land-based measurement, in which movement of the land may confound the measurements, has shown a rise in ocean levels of 1 to 2 millimeters per year. The small changes involved in detecting changes in sea level require a fairly sensitive measurement.

MONITORING THE ATMOSPHERE

A key satellite component in measuring substances in Earth's atmosphere is NASA's Moderate-Resolution Imaging Spectroradiometer (MODIS), which is included on several satellites. MODIS instruments observe visible and infrared radiation. Although designed to make surface observations, MODIS also can collect data on precipitation, clouds, and surface temperatures.

Other satellites involved in monitoring the atmosphere use similar technologies to observe the levels of pollutants. Some of the molecules monitored are carbon dioxide, methane, and ozone. Carbon dioxide and methane are special concerns because they can absorb heat, contributing to increased temperatures on Earth. Ozone, a component of the atmosphere, is reactive. Certain pollutants, when released into the atmosphere, will break the bonds of the ozone. This effect is leading to concerns about increased greenhouse gases with lower levels of ozone in the atmosphere.

Weather systems also can be tracked by satellite. However, for local weather, a key component of remote sensing is on the ground. Doppler radar utilizes the Doppler principle, according to which radiation from incoming objects arrives at a higher frequency. The stationary towers with Doppler radar are thus able to predict incoming storms and other systems. For this reason, they are often located near airports, where knowledge of local wind patterns is critical for planes departing and landing.

REMOTE SENSING AND SPACE EXPLORATION

While humans traveling to other planets has been a regular theme of science fiction, for the foreseeable

future satellites will be doing the majority of space exploration. The same instruments used to gather information about the topographic features of Earth have been utilized for gathering data about the geographic features of other planets.

Europa, Jupiter's smallest moon, has been found to be covered in a layer of ice, beneath which there is flowing water (the only known extraterrestrial ocean in the solar system). This ocean was discovered through remote sensing—specifically, through infrared photography from the Galileo spacecraft, which was launched by the United States in 1989. The Hubble Space Telescope, launched by NASA in 1990, is perhaps the best-known satellite for space exploration. Orbiting Earth, its cameras capture previously unseen images of outer space.

OTHER REMOTE-SENSING APPLICATIONS

The products of remote sensing have had perhaps their greatest influence in the consumer marketplace. The U.S. Department of Defense has a series of more than thirty satellites orbiting Earth, satellites that communicate with one another about their respective position. Although this system was originally created to help the U.S. military with global positioning, a second, civilian signal was introduced in 1983. Initially much less accurate than the encrypted military signal, it has improved in quality, with positioning accurate to within a few meters of a location.

In addition to helping determine the current state of the planet, remote-sensing satellites play a role in discovering more about Earth's past. Remote sensing has long been used in archaeology. Some of the earliest aerial photographs of archaeological sites came just after the first aerial photographs were taken. Satellites allow for a wider survey of archaeological sites, assisting in the understanding of the topology of a region and whether certain changes over time are caused by human activity or by natural, nonhuman processes. Multispectral photography is one tool that can display human-caused disturbances in the soil or other geographic features. Remote-sensing satellites also allow for the study of areas, such as rainforests, that are remote and difficult for humans to reach.

THE POLITICS OF REMOTE SENSING

Remote sensing and its products have played a part in two major political debates in recent years. While GPS technologies were originally developed for use by the U.S. military, GPS has also been used as a weapon against the military. After capturing insurgents following battles, soldiers have recovered GPS units used to gather information about terrain and potential troop positions.

Perhaps greater public concern has surrounded remote sensing's involvement in acts of terrorism. Following the terrorist attacks on the city of Mumbai, India in 2008, it was revealed that terrorists had used Google Earth to map their attack route. These concerns may influence scientific investigations. Researchers have noted that in certain parts of the world, the use of some of the more sensitive remote-sensing technologies (such as LIDAR) is restricted for military use only.

Privacy also is a concern. GPS is feared by some as a tracking device. It should be noted, however, that many of these concerns arise out of a misunderstanding of GPS technology. With GPS, the receiver tracks the satellites; the satellite does not track the receiver. Privacy advocates should not be concerned with GPS and remote-sensing satellites; rather, they should focus their concerns on service providers, cell phone towers, closed-circuit television cameras, and other remote-sensing devices much closer to earth.

Climate change is another major political issue associated with remote-sensing satellites. The proliferation of satellite technologies in the late twentieth century has enabled the widespread study of Earth's climate. In addition to land-use and ecosystem changes, the overall climate of the planet is changing. Data suggest that the planet is becoming warmer, which could spell catastrophe for life on Earth. The scientific consensus is that much of this climate change is caused by humans.

Finding a solution to this issue is likely to take some time. While local environmental issues can be fixed more quickly and easily, harmful global climate change will require worldwide efforts to alleviate. Modifying human habits will likely prove difficult, and the international diplomacy required will likely lead to roadblocks, as developing nations may face conflicts between economic development and reducing pollution levels.

Joseph I. Brownstein

FURTHER READING

Chuvieco, Emilio, and Chris Justice. "NASA Earth Observation Satellite Missions for Global Change Research." *Earth Observation of Global Change: The Role of Satellite Remote Sensing in Monitoring the Global Environment*, edited by Emilio Chuvieco. Berlin: Springer, 2008. The authors enumerate the various methods, technologies, and specific missions NASA has undertaken to monitor global environmental change.

Garfinkel, Simson. "Google Earth: How Google Maps the World." *Technology Review* 110, no. 6 (November/December 2007): 20-21. A brief overview of how Google uses a satellite network to provide high-resolution images of the earth and allows users to view specific areas using only the power of a home computer.

Kramer, Herbert J., and Arthur P. Cracknell. "An Overview of Small Satellites in Remote Sensing." *International Journal of Remote Sensing* 29, no. 15 (August 10, 2008): 4285-4337. This article provides a detailed history of satellites, listing the various missions and countries of origin of satellites that have been launched since Sputnik and the technological changes that have taken place along the way.

Mayaux, Philippe, et al. "Remote Sensing of Land-Cover and Land-Use Dynamics." In *Earth Observation of Global Change: The Role of Satellite Remote Sensing in Monitoring the Global Environment*, edited by Emilio Chuvieco. New York: Springer, 2008. The authors provide a view of how recent developments enable remote-sensing satellites to give a picture of resource availability. The book discusses problems such as deforestation and how the use of land for crops has affected other ecosystems.

Norris, Pat. *Watching Earth from Space: How Surveillance Helps Us, and Harms Us.* New York: Springer, 2010. This book provides an overview of some of the benefits of remote sensing, but also discusses some of the incidents where this information has been used to plan terrorist acts.

Pruitt, Evelyn L. "The Office of Naval Research and Geography." *Annals of the Association of American Geographers* 69, no. 1 (March 1979): 103-108. In this overview, Pruitt explains, among other developments, the reasons for the changes in the field of remote sensing and why she developed a new term for this field of study.

Verbyla, David L. *Satellite Remote Sensing of Natural Resources.* Boca Raton, Fla.: Lewis, 1995. This short book provides an overview of how many of the technologies available in remote sensing are used to map the earth's surface.

See also: Climate Change: Causes; Geodetic Remote-Sensing Satellites; Geodynamics; Geologic and Topographic Maps.

ROCK MAGNETISM

Rock magnetism is the subdiscipline of geophysics that has to do with how rocks record the magnetic field, how reliable the recording process is, and which conditions can alter the recording and therefore raise the possibility of a false interpretation being rendered by geophysicists.

PRINCIPAL TERMS

- **basalt:** a very common, dark-colored, fine-grained igneous rock
- **blocking temperature:** the temperature at which a magnetic mineral becomes a permanent recorder of a magnetic field
- **Curie temperature:** the temperature above which a permanently magnetized material loses its magnetization
- **daughter product:** an isotope that results from the decay of a radioactive parent isotope
- **detrital remanent magnetization:** the magnetization that results when magnetic sediment grains in a sedimentary rock align with the magnetic field
- **ferromagnetic material:** the type of magnetic material, such as iron or magnetite, that retains a magnetic field; also called a permanent magnet
- **granite:** a low-density, light-colored, coarse-grained igneous rock
- **magnetite:** a magnetic iron oxide composed of three iron atoms and four oxygen atoms
- **radioactivity:** the spontaneous disintegration of a nucleus into a more stable isotope
- **thermal remanent magnetization:** the magnetization in igneous rock that results as magnetic minerals in a magma cool below their Curie temperature

MAGNETIC FIELD PRODUCTION

The direct study of the earth's magnetic field began in the 1600's. This study involves the measurement of the field with scientific instruments and subsequent analysis of the resulting data. Four centuries is a very small fraction of the 4.6 billion years that the earth has existed; thus, direct study affords scientists very little understanding of the nature of the field over long periods of time. It is useful to know what happened to the earth's magnetic field in those billions of years before the present, because the field can be a source of information about conditions on the earth's surface and its interior. Magnetic minerals in rocks serve as recording devices, giving scientists

clues regarding the nature of the ancient magnetic field.

A moving electric charge, such as an electron, produces a magnetic field that is the ultimate source of any larger magnetic field. An atom is composed of a nucleus, with its protons and neutrons, and the electrons that surround the nucleus. The protons do not orbit within the nucleus, but their spinning does produce a small magnetic field, which is cancelled out if there is an even number of protons. The electrons, however, orbit the nucleus, and this movement produces a weak magnetic field. In addition, the electrons spin on their axes, and this activity also gives rise to a small magnetic field.

TYPES OF MAGNETISM

Because all atoms have electrons orbiting and spinning, one might think that all materials should have a permanent magnetic field, but the situation is more complicated. Strictly speaking, every material is magnetic, but there are different types of magnetism. Some materials are paramagnetic: When they are placed in an external magnetic field, the atoms align with the field. The atoms act as small compasses, orienting with the field, and the material is magnetized; the magnetic fields produced by the atom's electrons add to the intensity of the external field. When the external field is removed, however, the atom's orientation becomes randomized because of vibrations caused by heat, and the material is consequently demagnetized. Many materials, such as quartz, are paramagnetic and are not able to record the earth's magnetic field.

A much smaller number of minerals are ferromagnetic. There are various types of ferromagnetism, but the underlying principle is the same. In ferromagnetic materials, an external magnetic field again aligns the atoms parallel to the field, and the material is magnetized. When the field is removed, however, the atoms remain aligned, and the substance retains its magnetization; it is "permanently" magnetized. Actually, the substance can be demagnetized by heating or stress. Dropping a bar magnet on the floor or striking it with a hammer will demagnetize it slightly. The shock randomizes some of the atoms so that they cease to contribute to the overall magnetic field. The heating of a magnet above its Curie temperature also destroys its magnetization by randomizing the atoms and making the material paramagnetic. As the temperature drops below the Curie point, the material becomes slightly re-magnetized, because the weak field of the earth aligns some of the atoms.

In ferromagnetic materials, atoms are not all aligned in one direction; rather, they are found in aligned groups, called domains. Under a microscope, the domains are barely visible. Within a particular domain, the atoms are aligned, but all the domains are not aligned in the same direction. A "permanent" magnetic material that is unmagnetized has all the domains randomly aligned, and the overall field cancels to zero. When placed in a magnetic field, some of the domains realign parallel to the direction of the field and stay aligned after the field is removed. It is these domains that give the material its overall magnetization. If a high enough magnetic field is applied, all the domains align with the field, and the magnetization has reached its saturation point; the strength of the material's magnetic field is at a maximum. One of the areas of research for physicists is the quest for materials that have high magnetic field strengths but with less material. Such materials are useful in making small, but powerful, electric motors.

MAGNETIC MINERALS

Rocks are classified into three main groups: igneous, formed from crystallized molten rock; sedimentary, formed from weathered rock material; and metamorphic, produced when other rock is modified with heat, pressure, and fluids. Most magnetic minerals occur in igneous and sedimentary rocks.

Materials such as iron, cobalt, and nickel are ferromagnetic. For this reason, they are used in making various permanent magnets. These metals are not found naturally on the earth's surface in the uncombined state, so they do not contribute to rocks' recording ability. Most of the minerals that make up rocks, such as quartz and clay, are not ferromagnetic. These minerals are useless as recorders, but many rocks contain magnetite or hematite, which are good recorders. These common magnetic minerals are oxides of iron.

Hematite is Fe_2O_3, which means that there are two iron atoms for every three oxygen atoms. Hematite is red in color, similar to rust on a piece of iron. Most reddish-brown hues in sedimentary rock are caused by hematite. This magnetic mineral is not a very strongly magnetized compound, but it is a very stable recorder in sedimentary rocks. Unfortunately, in many cases, its formation postdates that of the rock in which it occurs, so it does not necessarily record the magnetic field at the time of the rock's formation. Magnetite (Fe_3O_4) has been known as lodestone for several millennia. It is a strongly magnetized iron compound that makes some igneous rocks very magnetic and supplies some of the recording ability of sedimentary rocks. The magnetite in rocks can record the field direction by one of several methods.

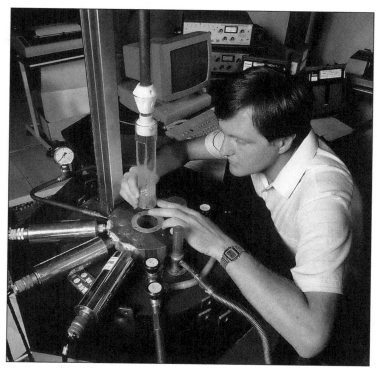

Laboratory study of a rock core sample. The sample is being loaded into a cryogenic magnetometer, to determine the strength and alignment of its magnetic field. This is part of research into the phenomenon of geomagnetic reversal—the occasional reversal of the earth's magnetic field polarity over geological time scales. (Geoff Lane/CSIRO/Photo Researchers, Inc.)

THERMAL REMANENT MAGNETIZATION

In igneous rocks, magnetite crystals form as the magma cools. As the crystals grow, they align themselves with any magnetic field present. This process is called thermal remanent magnetization (TRM). If the crystals are quite small or quite large, they cannot permanently record the field direction; after a short time the recording fades and becomes unreadable. The magnetism of such small grains is called superparamagnetism: They do align with a magnetic field, but they easily lose their orientation. The larger grains contain many magnetic domains that become misaligned over time so that the recording fades.

Grains the size of fine dust are good recorders. Unfortunately, not all igneous rocks have grains of the proper size. The size of the mineral crystal depends on the rate of cooling: When magma is cooled very slowly, large crystals are produced, while a rapid cooling results in smaller crystals. Granite is coarse-grained and thus is not the best recorder. The best igneous recorder is basalt, a black, fine-grained rock. Basalt is fairly common on the surface of the earth, particularly in the ocean basins, where nothing but basalt underlies the sediment on the basin floor.

A useful magnetic recorder must provide information about how old it is. Basalt again fills this requirement, as its crystallization can be dated by measuring the amount of radioactive elements and their daughter products it contains. Clearly, basalt is an ideal source of information on the magnetic field. Unfortunately, it does not occur everywhere on the earth; moreover, as a recorder, it covers only times of eruptions of magma. Some other recorder must be used to fill in the blanks.

DETRITAL REMANENT MAGNETIZATION

Sedimentary rock is formed from the products of the rock weathering that accumulate mostly in watery environments, such as rivers, lakes, and oceans. Clastic sedimentary rocks are formed from fragments of rock and mineral grains, such as grains of quartz in sandstone. Chemical sedimentary rock is derived from chemical weathering products, such as calcium carbonate or calcite, which is the major constituent of limestone. Most of the material in sedimentary rocks is not ferromagnetic, but there are a few grains of magnetite and other ferromagnetic compounds. As the grains fell through the water, they aligned with the magnetic field present at that time. When they

hit the bottom, they retained the orientation, for the most part, and were subsequently covered by more sediment. This process is termed detrital remanent magnetization (DRM).

An interesting aspect of DRM is the role that organisms play in its formation. The grains of magnetic minerals that fall through the water are oval-shaped, and when they strike the surface of the sediment they become misaligned with the field. Organisms such as worms disturb the sediment in a process known as bioturbation, which moves the sediment around and realigns the magnetic grains with the field. In the mid-1980's, it was discovered that certain varieties of bacteria have small grains of magnetite in their bodies. The bacteria use the grains like compasses to find their way down into the sediment on which they feed. The bacteria eventually die, and the magnetite grains become part of the sediment, aligned with the magnetic field; this phenomenon is known as biomagnetism.

The grain-size problem also occurs in DRM, given that a sediment particle can be the size of a particle of clay, a boulder, or anything in between. Conglomerate, a rock composed of rounded pebbles and other large particles, is not a good recorder, nor is coarse sandstone. Finer sandstones, shales, siltstones, and mudstones are much better. Most chemical rocks, such as halite (common table salt), are poor recorders; limestone may or may not be good, depending on the conditions of formation.

The magnetization in sedimentary rocks is generally between one thousand and ten thousand times weaker than is the magnetization in a basalt. Very sensitive magnetometers are needed to measure the magnetic field in these specimens. To be useful in geomagnetic studies, sedimentary rocks must be dated, but this is a difficult task, as they cannot be dated using radioactive methods. By a complex method of determination, fossils can act as indicators of the age of the rock in which they are found. If igneous rock layers are located above and below the rock layer of interest, and if these igneous rock layers can be dated, an intermediate age can be assigned to the sedimentary layer.

STUDY OF ROCK MAGNETISM

A magnetometer useful in the study of rock magnetism is the superconducting rock magnetometer (SCM). Superconductivity is the phenomenon of a material's losing its resistance to electric current at low temperatures. Liquid helium is used to cool a portion of the magnetometer, composed of a cylinder of lead closed at one end. As the lead cools, it becomes superconducting, and if done in a region of low magnetic-field intensity, this low field is "trapped" inside the cylinder. Magnetic field sensors known as SQUIDS (or superconducting quantum interference devices) are very sensitive to low-intensity magnetic fields. The sample is lowered into the device, and its electronic display shows the intensity of the sample's magnetization. Such devices are useful in studying the rock magnetism of low-intensity sedimentary rocks.

The Curie temperature is important for establishing the thermal remanent magnetization for igneous rocks. A sample of a particular ferromagnetic material in a magnetic field is heated and the temperature is measured; the sample's Curie temperature is determined when the pull of the magnetic field on the sample weakens. The Curie point for various ferromagnetic materials is established by this method. Once that is done, the procedure is reversed. A sample of an unknown ferromagnetic material can be heated in a magnetic field to determine its Curie point, which can then be compared with the established table of values to identify the magnetic mineral. This method does not establish the exact composition of the material, but it does narrow down the possibilities, which is of value because other methods for determining composition are more expensive. In addition, it has been discovered that Curie temperature is not the only factor critical to the recording process. At the Curie point, the material is ferromagnetic but the recording ability is weak. The material has to cool through the blocking temperature for recording stability. Thereafter, magnetic minerals are magnetically stable for periods of billions of years.

Another area of study is the determination of the best grain size and shape for magnetic recording. Researchers experiment with different sizes and shapes of magnetic grains in magnetic fields of various strengths and directions and measure their responses to changes. It was found that crystals of magnetic materials such as magnetite develop features known as domains. These are areas where the atoms are aligned in one direction and produce the unified magnetic field for the domain. A small crystal has only one domain that can easily shift to another

direction; therefore small crystals are poor recorders. If the crystal is quite large, it has many domains in which it is again easy to shift direction. Crystals with one large domain or several small domains are magnetically "hard" in that it is more difficult to shift the magnetic alignment. For magnetite, these are dust-sized particles, around 0.03 micron in diameter.

METHODS OF "MAGNETIC CLEANING"

Other research reveals that a rock's recording of the magnetic field is not as "neat and clean" a process as portrayed in the previous paragraphs. Many events can lead to the alteration of the magnetic alignment. If the rock is heated above the Curie point and then cooled, the magnetic alignment is that of the field present at that time, and the old alignment is erased. The rock may be changed chemically, and old magnetic minerals may be destroyed and new ones produced. This process is referred to as chemical remanent magnetization (CRM).

These secondary magnetizations can be removed in some cases, and they can even provide more information on the rock's history. One method of magnetic cleaning or demagnetization involves subjecting the rock sample to an alternating magnetic field while other magnetic fields are reduced to zero. This "cleaning" will remove that portion of the mineral's magnetization that is magnetically "softer" than the maximum alternating field. The magnetization above this level is unaffected and should represent the original magnetization. Heating a sample to a certain temperature is another method of demagnetization.

ROLE IN STUDY OF EARTH'S HISTORY

The study of the earth's magnetic field history, and all the inferences about the earth drawn from that study, depends on the ability of rocks to record information about the magnetic field at the time of the rocks' formation. That ability, in turn, is dependent upon the magnetic characteristics of a few permanently magnetized minerals, such as magnetite.

The study of rock magnetism is rather esoteric; only a few individuals worldwide are involved in this subdiscipline of geomagnetism. Yet, such study has shown that rocks can faithfully record the history of the earth's magnetic field. This record is used to infer conditions on the earth hundreds of million years ago. Such studies have lent support to the idea that the continents have actually moved over the surface of the globe—and thus the theory of plate tectonics was born, with all its implications for the formation and location of petroleum and ore deposits, the origin of earthquakes and volcanoes, and the formation of mountain ranges such as the Himalaya. Such is an example of the odd twists and turns that science can take. Seemingly inconsequential findings can lead to a theory with great potential for making the earth and its workings much more understandable.

Stephen J. Shulik

FURTHER READING

Butler, Robert F. *Paleomagnetism: Magnetic Domains to Geologic Terranes.* Boston: Blackwell Scientific Publications, 1992. Butler's exploration of the earth's magnetic fields begins with basic descriptions of what paleomagnetism is and how it occurs. The book is filled with illustrations to back up difficult concepts covered in the text.

Cox, Allan, ed. *Plate Tectonics and Geomagnetic Reversals.* San Francisco: W. H. Freeman, 1973. Cox provides fascinating introductions to chapters that are composed of seminal papers concerning magnetic reversals and their contribution to the development of the theory of plate tectonics. Information on rock magnetism is scattered throughout the book in discussions on baked sediments, magnetization of basalt, magnetic intensity, and self-reversals in rocks. The papers are advanced for the average reader, but there are many graphs, diagrams, and figures that merit attention.

Dunlop, D. J., and O. Ozdemir. *Rock Magnetism.* New York: Cambridge University Press, 2001. A valuable resource for graduate students and researchers. This text covers topics of general earth magnetism as well as putting rock magnetism into the context of sedimentary, igneous and metamorphic rock. The focus of these topics is on the mathematics and physics aspects; a strong understanding of mathematics is recommended.

Glen, William. *The Road to Jaramillo: Critical Years of the Revolution in Earth Science.* Stanford, Calif.: Stanford University Press, 1982. This book gives a history of the plate tectonics revolution of the mid-1950's to the mid-1960's. Rock magnetism is specifically covered on pages 103-109. Other aspects of rock magnetism are discussed in various portions of the book—for example, those dealing with the magnetic minerals associated with rock

magnetism and deep-sea core work. Of particular interest are the sections devoted to instruments used to measure rock magnetism. Best suited for students of geology, the writing is very technical.

Hamblin, William K., and Eric H. Christiansen. *Earth's Dynamic Systems.* 10th ed. Upper Saddle River, N.J.: Prentice Hall, 2003. This geology textbook offers an integrated view of the earth's interior not common in books of this type. The text is well organized into four easily accessible parts. The illustrations, diagrams, and charts are superb. Includes a glossary and laboratory guide. Suitable for high school readers.

Hargraves, R. B., and S. K. Banerjee. "Theory and Nature of Magnetism in Rocks." In *Annual Review of Earth and Planetary Sciences*, Vol. 1, edited by F. Donath. Palo Alto, Calif.: Annual Reviews, 1973. The article covers the theories of the natural remanent magnetization of rocks (NRM). NRM is the combined magnetization of all magnetic minerals in a rock, such as those resulting from TRM. The various carriers of remanence are also covered, with a table that lists each mineral and its composition, crystal structure, magnetic structure, and other pertinent information. The paleomagnetic potential of rocks is also discussed. A few mathematical equations are included, but nothing too formidable. There are numerous figures and a long list of references.

Lapedes, D. N., ed. *McGraw-Hill Encyclopedia of Geological Sciences.* New York: McGraw-Hill, 1978. Pages 704-708, under the heading "Rock Magnetism," provide concise descriptions of how rock magnetization occurs, the present field, magnetic reversals, secular variation, and apparent polar wandering. The text is very readable, with no mathematics and a fair number of graphs, tables, and figures.

Merrill, R. T., and M. W. McElhinney. *The Magnetic Field of the Earth: Paleomagnatism, the Core, and the Deep Mantle.* San Diego: Academic Press, 1998. The authors cover the basic material associated with the earth's magnetic field. Chapters deal with the origin of the magnetic field, as well as the origin of secular variation and field reversals. A strong background in mathematics is recommended. Bibliography and index; numerous tables, figures, and mathematical equations.

O'Reilly, W. *Rock and Mineral Magnetism*. New York: Chapman and Hall, 1984. O'Reilly covers the atomic basis for magnetism, the magnetization process, the various remanent magnetizations such as TRM, the magnetic properties of minerals, and the applications of rock and mineral magnetism. Many tables, figures, and photographs of minerals are included. Some mathematics and chemistry are also included but should not be too difficult. References are included at the end of each chapter.

Plummer, Charles C., and Diane Carlson. *Physical Geology*. 12th ed. Boston: McGraw-Hill, 2007. A college-level introductory geology textbook that is clearly written and wonderfully illustrated. An excellent sourcebook of basic information on geologic terminology and fundamentals of geologic processes. An excellent glossary.

Stacey, F. D., and Paul M. Davis. *Physics of the Earth*. 4th ed. New York: Cambridge University Press, 2008. Under Section 25.2, "Magnetic Properties of Minerals and Rocks," the author provides a short, technical description of rock magnetism. Several figures show the various types of magnetic alignments. The domain structure of ferromagnetic materials is also discussed. Many other aspects of the earth's magnetic field are covered at a technical level as well. Reformatted into units to make topics more accessible.

Tarling, D. H. *Paleomagnetism: Principles and Applications in Geology, Geophysics, and Archeology*. New York: Chapman and Hall, 1983. A very good resource on the subject of paleomagnetism, or the ancient magnetic field. Chapter 2 is devoted to the "physical basis" for the magnetization of material, with a discussion of the atomic level and the resulting magnetic domains. The various remanent magnetizations are covered in detail. Chapter 3 deals with the various magnetic minerals and their identification. The magnetization of the various rock types is covered in Chapter 4. Chapter 5 discusses instruments used in paleomagnetic work. The remainder of the text deals with mathematical analysis used in paleomagnetic work and paleomagnetic applications.

Yamaguchi, Masuhiro, and Yoshifumi Tanimoto, eds. *Magneto-Science: Magnetic Field Effects on Materials: Fundamentals and Applications*. Berlin: Springer-Verlag, 2010. Provides an overview of magnetism and its applications in earth science, chemistry, and physics. An excellent source of basic knowledge of magnetic forces and magnetic fields.

See also: Earth's Age; Earth's Magnetic Field; Environmental Chemistry; Geobiomagnetism; Magnetic Reversals; Magnetic Stratigraphy; Plate Tectonics; Polar Wander; Radioactive Decay; Relative Dating of Strata.

RUBIDIUM-STRONTIUM DATING

Rubidium-strontium (Rb-Sr) dating is one of the most common methods of obtaining absolute (numerical) ages of geologic materials, especially older minerals and rocks.

PRINCIPAL TERMS

- **absolute age:** the numerical timing of a geologic event, as contrasted with relative, or stratigraphic, timing
- **geochronology:** the study of the absolute ages of geologic samples and events
- **half-life:** the time required for a radioactive isotope to decay by one-half of its original weight
- **isochron:** a line connecting points representing samples of equal age on a radioactive isotope (parent) versus radiogenic isotope (daughter) diagram
- **isotope:** a species of an element having the same number of protons but a different number of neutrons and therefore a different atomic weight
- **mass spectrometry:** the measurement of isotope abundances by separating the isotopes by mass and charge in an evacuated magnetic field
- **radioactive decay:** a natural process by which an unstable, or radioactive, isotope transforms into a stable, or radiogenic, isotope

RUBIDIUM

In 1904, British chemist and physicist Ernest Rutherford proposed that geologic time might be measured by the breakdown of uranium in uranium-bearing minerals. A few years later, American chemist Bertram Boltwood published the absolute, or numerical, ages of three samples of such minerals. The ages, which approximated 500 million years, indicated the antiquity of some earth materials, a finding developed by British geologist Arthur Holmes in his classic *The Age of the Earth* (1913). Holmes's early time scale for the earth was not immediately accepted by most of his peers, but it helped to set the stage for the eventual use of absolute ages as the prime quantitative components in the study of geology and its many subdisciplines. After the early study of the isotopes of uranium, including uranium-series transition isotopes, came the discovery of other unstable isotopes and the formulation of the radioactive decay schemes that have become essential to geochronology, including the rubidium-strontium method.

Rubidium (Rb), an alkali-group (lithium, sodium, potassium, rubidium, cesium, and francium) element with a valence (bonding value) of +1, is a trace element in terrestrial and solar system materials. It is not a necessary component in any known mineral; it substitutes for the major element potassium in common, rock-forming minerals such as the alkalic feldspars and micas. Because of their similar ionic size and geochemical behavior, the ratio of potassium to rubidium is a useful petrologic parameter.

Rubidium consists of two natural isotopes: rubidium-87 and rubidium-85. Several artificial isotopes also are known. Rubidium-87 has been known to be unstable since 1940, but its use in age dating did not begin until the advent of modern mass spectrometry in the 1950's. It decays by the emission of a beta particle to the stable nuclide strontium-87. The half-life for this decay is not precisely known, because of the low energy spectrum of the emitted beta particles. Half-life values range from 47 to 50 billion years. Although the commonly accepted value is 47 billion years, a comparison of Rb-Sr dates for samples of meteorites and lunar rocks with dates for the same samples yielded by other techniques indicates that a value of 48.8 billion years is more correct. In any event, the long half-life and the low parent-daughter ratio mean that the radiogenic accumulation of strontium-87 in most natural minerals and rocks is very slow.

STRONTIUM

Strontium—which, along with beryllium, magnesium, calcium, barium, and radium, belongs to the alkaline-earth group of elements—has a valence of +2. Though more abundant than rubidium in materials of the solar system, it also is a trace element. It forms its own mineral phases in the form of strontianite, a strontium carbonate, and celestite, a strontium sulfate, but it occurs more significantly as a substitute for the major species calcium in common, rock-forming minerals such as plagioclase feldspar, calcium-rich pyroxenes and amphiboles, apatite, and calcium carbonate minerals. Like the potassium-rubidium ratio, the calcium-strontium ratio has use in the understanding of various geologic processes.

Strontium comprises four naturally occurring isotopes: strontium-88, -87, -86, and -84. There are also several artificial isotopes, the best known being the nuclear-fission-produced strontium-89 and -90.

PURIFICATION OF SAMPLES

Concentrations of rubidium and strontium in minerals, rocks, and other natural substances can be determined by a variety of techniques, although isotopic parameters and precise elemental abundances are commonly determined by mass spectrometry. As the abundances of rubidium-85, strontium-88, strontium-86, and strontium-84 all are precise percentages of the total for each of these elements and do not vary as a function of nuclear instability, each can be calculated by taking the appropriate percentage of the total elemental concentrations, as determined by gravimetric analysis, atomic absorption spectrophotometry, X-ray fluorescence spectrophotometry, microbeam probe, or another type of analysis. Most modern work, however, focuses on determining the relevant isotopic parameters by mass spectrometry, after purification by chemical techniques.

Commonly, a preliminary determination of rubidium and strontium abundance and the Rb-Sr ratio is made or estimated for the minerals, rocks, or other materials to be dated by a reconnaissance technique, such as X-ray fluorescence. Samples selected on the basis of these determinations are chosen for dating. After selection, the samples are crushed—homogenized, if necessary—and a fraction is taken which contains enough of the rubidium and especially the more critical strontium for adequate isotopic analysis. Rarer materials, such as meteorites and lunar rocks, may not afford enough of a sample for optimal analysis. The fraction of material—if, as is most common, it is in the geologic form of aluminosilicate compounds, perhaps with some organic material—is dissolved in a mixture of hydrofluoric and perchloric acids and reduced by evaporation to a concentrated "mush" of material. Samples other than silicates may be dissolved in other, more appropriate solvents. The concentrated mush is spiked with an appropriate amount of purified liquids containing known amounts of rubidium and strontium of known, nonnatural isotopic composition. This material is dissolved in a small amount of hydrochloric acid and placed on calibrated ion exchange columns. The columns are washed with hydrochloric acid,

and purified portions of rubidium, then strontium, of mixed natural and spiked isotopic composition, are collected and evaporated. Achieving the highest accuracy and precision requires that the smallest amounts of rubidium and strontium from the laboratory environment (contamination) be included with the completed, purified elements.

MASS SPECTROMETRY

The standard method of mass spectrometry for Rb-Sr dating involves placing the purified samples as solids on metal filaments and installing the loaded filaments in solid-source mass spectrometers. The spectrometer source regions, evacuated to very low pressures, are constructed so that the metal filaments can be heated until the rubidium or strontium ionizes. The charged, ionized sample is accelerated through a series of collimating slits into a controlled magnetic field, where the beams of ions are separated by charge-mass ratios into beams of separate isotopes; as the charge of the elements is the same for each atom, the ions are separated on the basis of mass only. Specific isotopic beams, controlled by the magnetic field, are channeled through more collimating slits to the collector part of the spectrometer. Commonly, a Faraday cup is used to analyze the number of atoms of each isotope by the conversion of each atomic impact into a unit of charge, which is then amplified. The actual output is isotope ratio measurements—rubidium-87/rubidium-85, strontium-88/strontium-86, and so on—which are converted using mathematical programs into the required parameters for determining time. Because strontium-87 is closest in abundance to strontium-86 in most cases, and because it differs by only one mass unit, the standard for reporting the radiogenic component is with the ratio strontium-87/strontium-86.

By mixing precise amounts of "spikes" of strontium and rubidium, whose isotopic compositions differ markedly from natural isotopic compositions, with the strontium and rubidium in the natural sample, a combined mass spectrum is obtained. From these data can be calculated the precise abundances of rubidium and strontium in the natural material (a process known as isotope dilution) and the critical isotopic composition of the natural strontium.

ISOCHRON DIAGRAMS

Although the age of the analyzed sample can be calculated using the determined Rb-Sr parameters and the decay constant for rubidium-87, it is customary and more useful to determine the age graphically, with the use of an isochron diagram. In the diagram, the actual isotope ratios collected in the spectrometer are used as coordinates. Thus, the parent, unstable component, rubidium-87, is designated by reference to a common isotope of strontium, strontium-86. The other coordinate, the measure of the radiogenic component, is strontium-87/strontium-86. A line connecting points representing samples of equal ages, an isochron, has an age value that is represented by its slope on the figure; a horizontal isochron has an age value of zero, and positive slopes of successively greater degree have increasingly greater ages, given in terms of the isochron slope and the half-life of the parent rubidium-87 isotope. A single mineral or rock would furnish only one point on the diagram, so for an isochron to be drawn, there must be knowledge of or, more likely, an estimate of the sample's initial isotopic composition. Ages calculated in this way are termed "model ages."

Analysis of minerals of equal age but different compositions from a sample of plutonic volcanic rock would allow the construction of an isochron, whose slope would be proportional to the age of crystallization of the rock. An assumption that is justified in most circumstances is that at the time of crystallization from a magma or lava, all minerals formed have the same isotopic composition of strontium. The basis for this assumption is that, unlike isotopes of elements with a mass of less than about 40, there is no measurable fractionation of isotopes of an element at the physical-chemical conditions of the liquidus material. If the rock system has not been affected by open-system behavior—for example, by the introduction of parent or daughter species by metamorphism or weathering—the points representing these samples will define a perfect line. In practice, uncertainties in measuring each of the parent-daughter parameters, and perhaps some open-system behavior, result in imperfect isochronism and, therefore, uncertainties in the calculated age.

A benefit of the isochron method is that the isotopic composition of strontium at the time of origin of the rock, or strontium-87/strontium-86, is marked by the left-hand or lower intercept of the isochron, where rubidium-87/strontium-86 is equal to zero. As discussed above, for a model age calculated from a single mineral or rock analysis, this parameter would

have to be estimated. Another benefit is that one can readily see whether one or more points are aberrant or whether a poor fit might indicate an open-system history for the rock. A wide variety of terrestrial and extraterrestrial igneous rocks have been dated by this mineral isochron method with good precision.

As with minerals from a single rock, if a series of rocks of equal age from a common parent (such as fractionally differentiated rocks from a single magma or similar rocks which resulted from different degrees of partial melting of a common source) are analyzed, their Rb-Sr isotopic parameters should yield an isochron proportional to the unique age of the rocks, or a whole-rock isochron. Consequently, one may test for ages and for comagmatic properties in a suite of rocks. In practice, however, it is common for these properties to be unraveled only with supporting petrologic or geochemical data, if at all.

UNIQUENESS OF THE METHOD

The most spectacular benefit of the Rb-Sr isochron technique and one that is unique to the method is the ability, under some circumstances, to identify both the original time of crystallization of an igneous rock, such as a granite, and a later time of metamorphism resulting, for example, in a granitic gneiss. These events may be traced by constructing both mineral and whole-rock isochrons from several, chemically differing types of the gneiss. Points representing mineral components, along with points representative of their whole-rock mixtures, may form an isochron proportional to the age of metamorphism. These mineral isochrons should have the same slopes and therefore ages. However, a whole-rock isochron constructed only from the whole-rock points will be steeper, with a slope proportional to the earlier time of intrusion of the granite. This seemingly peculiar behavior results from a reequilibration of strontium isotopes at the time of metamorphism in the vicinity of the rock sampled; on a more regional scale, however, whole-rock parameters were not homogenized.

The method requires, in addition to reasonably closed-system behavior, that the later metamorphic event be capable of reequilibrating the local rock systems completely. In practice, that is accomplished either through fairly high-grade metamorphism, through the availability of sufficient water to effect the isotopic exchange, or, most likely, through both. Dry metamorphism, even if high-grade, may not result in reequilibration; the mineral isochron date obtained will therefore record only the magmatic event. Conversely, if the rocks are permeable, fine-grained, and wet, rehomogenization may be completed even under low-grade metamorphic conditions. Where conditions of economic mineralization have been sufficient to equilibrate Rb-Sr components, whole-rock isochron analysis can reveal this type of metamorphic event. If the event results in the formation of new minerals, these minerals may record the time of metamorphism. Unfortunately, in some studies, the multiple possibilities of incomplete rehomogenization, open-system behavior, and poor precision of measurement result in poor isochronism and age data of questionable or no value.

USE WITH MARINE COMPONENTS

Because of a lack of fossils with determinable ages in many sediments and sedimentary rocks, absolute age analysis by several isotopic methods, including Rb-Sr dating, has been tested. The technique works well only for certain minerals for which model ages can be calculated by assuming the ratio between rubidium-87 and strontium-86; for marine components, this assumed parameter may be reasonably determined, thanks to the effective homogenization of strontium in seawater during a given geologic time. The most commonly dated material is glauconite, although the closed-system requirement may be violated by the unavoidable inclusion of detrital components of varying strontium-isotopic composition. Other materials of limited usefulness are minerals such as phillipsite and some illite. Evaporitic minerals containing enough rubidium, such as sylvite, also have been dated by this technique, although omnipresent recrystallization may perturb the isotopic systematics.

In some cases, the isotopic composition of strontium by itself can be used to determine the age of carbonate rock. As previously stated, strontium dissolved in seawater of a particular geologic episode has the same isotopic composition everywhere in the ocean. That is because the mixing rate of marine strontium—about one thousand years—is short compared with the average "lifetime" for strontium atoms in the sea—about two million years. Thus, strontium of variable isotopic composition washed into the sea from rivers or other sources is well mixed. Because strontium-87 accumulates through time as a result of

rubidium-87 decay, terrestrial strontium, including marine strontium, becomes more radiogenic through geologic time. Theoretically, then, marine strontium of any geologic time should have a unique value, and the time could be identified simply by measuring the ratio of strontium-87 to strontium-86. Unfortunately, from the strictly chronologic perspective, strontium is not monotonic; strontium of some periods, for example, is less radiogenic than that for the preceding and succeeding periods. Careful work has delimited the marine growth curve for strontium, however, and the technique has met with considerable success.

ROLE IN UNDERSTANDING GEOLOGIC PHENOMENA

The absolute dating of geologic materials and events has had an unprecedented influence on the understanding of geologic events on Earth and of solar system minerals and rocks. The ability to establish events in terms of actual years, rather than in relative terms such as "older than" or "younger than," has led to a realistic estimate of the earth's age and to calibrated time scales for organic evolution, geomagnetic events, and the structural development of the earth's crust. One of the earliest and most useful chronometric schemes for the oldest rocks, the Rb-Sr technique has been of exceptional value, not only for dating but also for the use of strontium-isotopic composition as an indicator or tracer for a variety of geologic processes, such as the evolution of seawater. Additionally, the technique has had much success in the unraveling of igneous and metamorphic processes in complex, regionally metamorphosed geologic terrains.

Because of its usefulness in the precise determination of the ages of very old rocks, the Rb-Sr method will continue to be of major use in dating Earth's oldest rocks and extraterrestrial solar system materials—for example, lunar rocks and meteorites, including meteorites from the moon and possibly from Mars.

E. Julius Dasch

FURTHER READING

Faure, Gunter. *Origin of Igneous Rocks: The Isotopic Evidence.* New York: Springer, 2010. Descriptions of multiple radioactive isotope dating methods contained within this book. Principles of isotope geochemistry are explained early, making this book accessible to undergraduates. Data presented in diagrams, more than 400 original drawings, and a long list of references included at the end.

_____. *Principles of Isotope Geology*. 2d ed. New York: John Wiley & Sons, 1986. An excellent, though technical, introduction to the use of radioactive and stable isotopes in geology, including a thorough treatment of the Rb-Sr technique. The work is well illustrated and well indexed. Suitable for college-level readers.

Levin, Harold L. *The Earth Through Time*. 9th ed. Philadelphia: Saunders College Publishing, 2009. This college-level text contains a brief, clear description of the Rb-Sr method. A diagram of a whole-rock isochron is included. Five other radiometric dating techniques are discussed, and background information on absolute age and radioactivity is provided. Includes review questions, a list of key terms, and references.

Parker, Sybil P., ed. *McGraw-Hill Encyclopedia of the Geological Sciences*. 2d ed. New York: McGraw-Hill, 1988. This source contains entries on radioactivity and radioactive minerals. The entry on dating methods includes a brief section on the Rb-Sr method. Includes the formula for radioactive decay and a table of principal parent and daughter isotopes used in radiometric dating. The entry on rock age determination has a longer discussion of the Rb-Sr dating and includes an isochron diagram. For college-level audiences.

Smith, David G., ed. *The Cambridge Encyclopedia of Earth Sciences*. Cambridge, England: Cambridge University Press, 1981. Organized as a compilation of high-quality and authoritative scientific articles rather than a typical encyclopedia. Chapter 8, "Trace Elements and Isotope Geochemistry," covers mass spectrometry, igneous processes, radiogenic isotopes, and radioactive decay schemes. Rb-Sr decay is used to illustrate principles of geochronology. An Rb-Sr isochron diagram is provided and explained. A more technical discussion than the one in Levin's book. Suitable for the reader with some background in science.

Walther, John Victor. *Essentials of Geochemistry*. 2d ed. Jones & Bartlett Publishers, 2008. Contains chapters on radioisotope and stable isotope dating and radioactive decay. Geared more toward geology and geophysics than toward chemistry, this text provides content on thermodynamics, soil formation, and chemical kinetics.

Zalasiewicz, Jan. *The Planet in a Pebble: A Journey into Earth's Deep History*. New York: Oxford University Press, 2010. An easily accessible account of Earth's formation and history, written for the layperson. Summarizes many studies in geology, explaining basic physics and chemistry, and even delving in to radiometric dating. This text is indexed and also provides further readings and bibliographies for each chapter.

See also: Earth's Age; Earth's Oldest Rocks; Environmental Chemistry; Fission Track Dating; Mass Spectrometry; Potassium-Argon Dating; Radioactive Decay; Radiocarbon Dating; Samarium-Neodymium Dating; Uranium-Thorium-Lead Dating.

S

SAMARIUM-NEODYMIUM DATING

Sm-Nd dating is one of the more recent, yet most common methods of obtaining the absolute ages of geologic materials, especially older minerals and rocks. The method depends on the natural radioactivity of one of the seven isotopes of samarium, samarium-147, which decays to neodymium-143.

PRINCIPAL TERMS

- **absolute age:** the numerical timing of a geologic event, as contrasted with relative, or stratigraphic, timing
- **geochronology:** the study of the absolute ages of geologic samples and events
- **half-life:** the time required for a radioactive isotope to decay by one-half of its original weight
- **isochron:** a line connecting points representing samples of equal age on a radioactive isotope (parent) versus radiogenic isotope (daughter) diagram
- **isotope:** a species of an element having the same number of protons but a different number of neutrons and therefore a different atomic weight
- **mass spectrometry:** the measurement of isotope abundances by separating the isotopes by mass and charge in an evacuated magnetic field
- **radioactive decay:** a natural process by which an unstable, or radioactive, isotope transforms into a stable, or radiogenic, isotope

RARE-EARTH ELEMENTS

Samarium (Sm) and neodymium (Nd) are both rare-earth elements (REEs, or lanthanides). They occur, commonly in trace amounts, in many of the more widespread minerals and rocks and appear in high concentrations in some rare but economically viable minerals, such as bastnaesite and monazite.

The sixteen REEs, their abundances, and especially their abundance patterns have achieved a remarkable usefulness in geochemistry. Their abundances, usually plotted relative to their abundances in important major reservoirs, especially the average composition of the solar system—as it is estimated from primitive meteorites known as the carbonaceous chondrites—form patterns that have proved exceptionally useful as tracers for a wide variety of cosmic and geologic processes. These include the parentage of igneous, metamorphic, and sedimentary rocks. Rare-earth element concentration patterns, called tracers, are significant in a way that is similar to the importance of the isotopic ratios of radiogenic nuclides such as strontium-87 and strontium-86 or neodymium-143 and neodymium-144, described herein. REEs have a valence (bonding value) of +3 (except for europium and cerium, which may also exist in other valence states), and they have identical outer shell electronic configurations. The reason for this similarity is that the REEs, although exhibiting almost identical geochemical properties, nevertheless vary slightly because they have slightly different ionic radii. Consequently, REEs within a given mineral or other phase act somewhat like isotopes of a given element. Because so much is known about the geochemical behavior of the REEs, the usefulness of the Sm-Nd dating and tracer techniques is great. Samarium and neodymium belong to the light REE part of the lanthanide spectrum and occur next to each other in terms of atomic number and ionic size.

SAMARIUM AND NEODYMIUM

Samarium consists of seven natural isotopes as well as several artificial isotopes. The Sm-Nd chronometer is a result of the alpha decay of samarium-147 to neodymium-143, which has an exceptionally long half-life of 106 billion years. Samarium-147 has been known to be unstable for many years. Because of its long half-life and the small dispersion of Sm-Nd ratios in most materials, its use in age dating did not begin until the mid-1970's, with the advent of modern mass spectrometry, high-precision instruments, and the digital collection of data. The radiogenic accumulation of neodymium-143 in most natural minerals and rocks is very slow.

The REE neodymium consists of seven natural isotopes and several unstable species. In order of isotopic abundance, the seven natural isotopes are neodymium-142, -144, -146, -143, -145, -148, and -150. The important radiogenic nuclide is neodymium-143, which forms by the alpha decay of radioactive samarium-147. Like the isotopic composition of strontium (strontium-87 and strontium-86), the abundance of neodymium-143 is measured in a mass spectrometer relative to the neodymium isotope of closest mass and reasonable abundance, neodymium-144. Also like the isotopic composition of strontium, the isotopic composition of neodymium—specifically, the neodymium-143/neodymium-144 ratio—has been of exceptional use in helping scientists to understand a variety of geologic, especially igneous, processes.

DETERMINING SAMARIUM-NEODYMIUM CONCENTRATIONS AND RATIOS

Concentrations of samarium and neodymium in minerals, rocks, and other natural substances can be measured using a variety of techniques, although isotopic parameters and precise elemental abundances are commonly measured by mass spectrometry. As the abundances of most of the samarium and neodymium isotopes are precise percentages of the total for each of these elements and do not vary as a function of nuclear instability, each can be calculated by taking the appropriate percentage of the total elemental concentrations, as determined by various methods. These methods include gravimetric analysis, atomic absorption spectrophotometry, X-ray fluorescence spectrophotometry, or microbeam probe analysis. In practice, however, because of low abundances of the REEs in most minerals, the most useful techniques are neutron activation analysis and mass spectrometry. Most modern work consists of determining the relevant isotopic parameters by mass spectrometric isotope dilution after purification by chemical techniques.

A major attribute of the Sm-Nd system is that, thanks to the geochemical affinity of samarium for ultramafic and mafic minerals (minerals low in silica and rich in iron and magnesium), the method can be used to date even comparatively young low-silica rocks, such as peridotite and basalt, that cannot easily be dated by uranium-thorium-lead (U-Th-Pb) or rubidium-strontium (Rb-Sr) techniques. Additionally,

in principle, the unusual geochemical behavior of the Sm-Nd ratio in magmatic crystallization or partial melting of source rocks allows contrasting isotopic trends between Sm-Nd and other ratios to be traced.

For example, fractional crystallization of a magma yields residual liquid that becomes increasingly greater in terms of U-Th-Pb and Rb-Sr ratios, whereas Sm-Nd values commonly become lower. Thus, in the last-formed residual rocks, strontium and lead will be more radiogenic, but samarium perhaps less radiogenic, than they were in the beginning magma. The reverse is true for the partial melting of a mantle source rock such as peridotite. It has become useful, therefore, to plot time-of-crystallization (initial) isotopic ratios of lead and especially strontium against that of neodymium. For many rocks, these data form a trend of negative slope, a trend that has been called the "mantle array."

PRELIMINARY TECHNIQUES IN SM-ND DATING

Commonly, a preliminary determination of samarium and neodymium abundance and the Sm-Nd ratio is made for the materials to be dated, either by a reconnaissance technique or simply by an estimation from known concentrations of these elements in previously measured samples of the same type of mineral or rock. Samples selected on the basis of these determinations are chosen for the optimal conditions for dating. After selection, the samples are crushed, or homogenized if necessary, and a portion is taken which contains enough of the samarium and especially of the more critical neodymium for adequate isotopic analysis. Rarer materials, such as some lunar rocks or meteorites, may not afford enough of a sample for optimal analysis.

The sample of material, if it is the most common geologic form of aluminosilicate compounds, perhaps with some organic material, is dissolved in a mixture of hydrofluoric and perchloric acids and reduced by evaporation to a concentrated "mush." Samples other than silicates may be dissolved in other solvents. The mush is "spiked" with an appropriate amount of purified liquids containing known amounts of samarium and neodymium of known, non-natural isotopic composition. This material is dissolved in a small amount of hydrochloric acid and placed on calibrated ion exchange columns. The columns are washed with hydrochloric acid, and purified samples of samarium, then neodymium, of mixed natural and spiked isotopic composition are collected and evaporated. Achieving the highest accuracy and precision requires that the smallest amounts possible of samarium and neodymium from the laboratory environment (contamination) be included with the completed, purified elements.

MASS SPECTROMETRY

The standard method of mass spectrometry for Sm-Nd chronology involves placing the purified samples as solids (such as oxides or metals) on metal filaments and installing the loaded filaments in solid-source mass spectrometers. The spectrometer source regions, evacuated to very low pressures, are constructed so that the metal filaments can be heated until the samarium or neodymium ionizes. The charged, ionized sample is accelerated through a series of collimating slits and into a controlled magnetic field, where the beams of ions are separated by charge-mass ratios into beams of separated isotopes. The charge is the same for each atom, so the ions are separated on the basis of mass only. Specific isotopic beams, controlled by the magnetic field, are channeled through more collimating slits into the collector part of the spectrometer.

Commonly, a Faraday cup is used to analyze the number of atoms of each isotope by conversion of each atomic impact into a unit of charge, which is amplified. A digital readout is produced. The actual output of the procedure is isotope ratio measurements—samarium-150/samarium-147, neodymium-143/neodymium-144, and so on—which are converted to the required parameters for determining time by mathematical programs. Because the important quantity of neodymium-143 is relatively close to that of neodymium-144 in most cases, and because it is different by only one mass unit, the standard for reporting the radiogenic component is with the ratio neodymium-143/ neodymium-144.

By mixing known weights of "spikes" of samarium and neodymium, which differ markedly from the natural isotopic compositions of these elements, with the natural sample, a combined mass spectrum is obtained. From these data can be calculated the precise abundances of samarium and neodymium in the natural material (a process known as isotope dilution) and the critical isotopic composition of the natural neodymium.

ISOCHRON DIAGRAMS

Although the age of the analyzed sample can be calculated using the determined Sm-Nd parameters and the decay constant for samarium-147, it is customary and more useful to determine the age graphically, with an isochron diagram. In the diagram, the actual isotope ratios collected in the spectrometer are used as coordinates. Thus, the parent, unstable component, samarium-147, is designated by reference to the common isotope of neodymium, neodymium-144. The other coordinate—the measure of the radiogenic component—is neodymium-143/ neodymium-144. A line connecting points representing samples of equal ages, or an isochron, has a value represented by its slope; a horizontal isochron has an age value of zero, and positive slopes of successively greater degree have increasingly greater ages, given in terms of the isochron slope and the half-life of the parent samarium-147 isotope. A single mineral or rock would furnish only one point on the diagram, so to draw an isochron, the researcher must know or, more likely, estimate the sample's initial isotopic composition. Ages determined in this way are called "model ages."

DATING ROCK CRYSTALLIZATION

Sm-Nd dating is used to determine the time of crystallization of specific types of igneous rock, the time of formation of comagmatic igneous rocks, and the time of metamorphism of a sequence of rocks of varying composition. The analysis of minerals of equal age but different compositions from a sample of plutonic or volcanic rock would form the points for an isochron whose slope would be proportional to the rock's age of crystallization. A common example is a mafic rock, such as basalt, with minerals of an increasing samarium-147/neodymium-144 ratio, such as plagioclase feldspar and pyroxene. An assumption that is justified in most circumstances is that at the time of crystallization, all the minerals formed have the same isotopic composition of neodymium. If the rock system has not been affected by the introduction of parent or daughter species by metamorphism or weathering (open-system behavior), the points representing these samples will define a perfect line. In practice, however, uncertainties about the parent-daughter parameters, and perhaps some open-system behavior, result in imperfect isochronism and, therefore, uncertainties in the calculated age.

A benefit of the isochron method is that the isotopic composition of neodymium at the rock's time of origin is marked by the left-hand or lower intercept of the isochron. Another benefit is that it is easy to see whether one or more points are aberrant or whether a poor fit might indicate an open-system history for the rock. A wide array of terrestrial and extraterrestrial igneous rocks have been dated by this mineral isochron method with good precision.

DATING ROCK SERIES AND SEQUENCES

Not only can geologists use this method to date minerals from a single rock, but if a series of rocks of equal age from a common parent are analyzed, their Sm-Nd isotopic parameters should also yield a "whole-rock" isochron proportional to the rocks' age. Consequently, geologists may test for comagmatic properties and ages in a suite of rocks. In practice, however, these properties are often analyzed only with supporting petrologic or geochemical data, if at all.

The time of metamorphism of a sequence of rocks of varying composition also may be determined by the Sm-Nd technique, provided that, at the time of metamorphism, the isotopic composition of neodymium in the rocks is effectively homogenized. The geochemical behavior of the REEs, however, results in much more immobile transport for these elements as compared with, say, rubidium and strontium. Complete homogenization is accomplished through high-grade metamorphism, through the availability of sufficient water to effect the isotopic exchange, or, most likely, through both. Dry metamorphism, even if high grade, may not result in homogenization; the isochron date obtained in such a case would record only some mixture of original events or isotopic compositions. Conversely, if the rocks are permeable, fine-grained, and wet, homogenization may be completed even under low-grade metamorphic conditions. Where conditions have been sufficient to equilibrate Sm-Nd components, whole-rock isochron analysis can reveal this type of metamorphic event.

DATING MATERIALS PRECIPITATED
FROM SEAWATER

The isotopic composition of neodymium by itself can in principle be used to determine the age of manganese nodules, carbonate rocks, or other materials precipitated from seawater. For this method to work, enough must be known about the isotopic composition of neodymium in that sector of the sea through time—a history that may not readily be available. Strontium dissolved in seawater of a particular geologic episode has the same isotopic composition everywhere in the ocean. Neodymium, however, does not, because the mixing rate of marine neodymium, about one thousand years, is long compared to the average "lifetime" for neodymium atoms in the sea, less than one hundred years. Thus, neodymium of variable isotopic composition washed into the sea from rivers or other sources commonly is deposited on the sea floor before it has become isotopically well mixed. Because neodymium-143 accumulates through time as a result of samarium-147 decay, neodymium on earth becomes more radiogenic through geologic time. One might assume, then, that geologic time could be identified simply by measuring the ratios of neodymium-143 to neodymium-144; unfortunately for the chronologic usefulness of this parameter, however, neodymium is not uniform in the seas, as discussed. Therefore, marine neodymium is used more for an understanding of marine processes, such as water transport, than for chronologic studies. Scientists continue to investigate the potential of marine neodymium as an indicator of time and as a global tectonic tracer.

ROLE IN UNDERSTANDING GEOLOGIC EVENTS

The absolute dating of geologic materials and events has had a profound effect on scientists' understanding of terrestrial and extraterrestrial geologic events. Establishing the age of events in years rather than in relative terms has led to reliable estimates of the earth's age and to calibrated time scales for organic evolution, geomagnetic events, and the plate tectonic cycle. One of the most recent and most useful chronometric methods, the Sm-Nd technique has been extremely important not only for dating but also for tracking of a variety of geologic processes, such as the evolution of seawater. This technique has also helped geologists to understand igneous and metamorphic processes in complex, regionally metamorphosed geologic terrains.

The Sm-Nd method will continue to be of great use in dating Earth's oldest rocks and other solar system materials, including lunar and Martian rocks and meteorites.

E. Julius Dasch

FURTHER READING

Duckworth, H. E. *Mass Spectrometry.* Cambridge, England: Cambridge University Press, 1958. An older work, but it covers well the basic principles of mass spectrometry, the major measurement technique used in conjunction with Sm-Nd dating. For readers with some background in science.

Faure, Gunter. *Isotopes: Principles and Applications.* 3rd ed. New York: John Wiley & Sons, 2004. Originally titled *Principles of Isotope Geology.* An excellent introduction to radioactive and stable isotopes and their use in geology. It covers the Sm-Nd technique thoroughly. The work is somewhat technical but well illustrated and indexed. Written for a college-level audience.

_____. *Origin of Igneous Rocks: The Isotopic Evidence.* New York: Springer, 2010. Descriptions of multiple radioactive isotope dating methods contained within this book. Principles of isotope geochemistry are explained early, making this book accessible to undergraduates. Includes data presented in diagrams, more than 400 original drawings, and a long list of references included at the end.

Parker, Sybil P., ed. *McGraw-Hill Encyclopedia of the Geological Sciences.* 2d ed. New York: McGraw-Hill, 1988. This source contains an entry on rock age determination. Not much space is devoted solely to the Sm-Nd method, but other methods are reviewed and the general principles of dating are explained. For general readers.

Smith, David G., ed. *The Cambridge Encyclopedia of Earth Sciences.* Cambridge, England: Cambridge University Press, 1981. Organized as a compilation of high-quality and authoritative scientific articles rather than a typical encyclopedia. A chapter in this well-written textbook covers mass spectrometry, radioactive decay schemes, and several dating methods. Both rubidium-strontium and samarium-neodymium techniques are discussed. Includes an example of an isochron diagram and a table of trace element abundances. For the reader desiring a clear yet not simplistic source.

Walther, John Victor. *Essentials of Geochemistry.* 2d ed. Jones & Bartlett Publishers, 2008. Contains chapters on radioisotope and stable isotope dating and radioactive decay. Geared more toward geology and geophysics than toward chemistry, this text provides content on thermodynamics, soil formation, and chemical kinetics.

Zalasiewicz, Jan. *The Planet in a Pebble: A Journey into Earth's Deep History.* New York: Oxford University Press, 2010. An easily accessible account of Earth's formation and history, written for the layperson. Summarizes many studies in geology, explaining basic physics and chemistry, and even delving in to radiometric dating. This text is indexed and also provides further readings and bibliographies for each chapter.

See also: Earth's Age; Earth's Oldest Rocks; Environmental Chemistry; Fission Track Dating; Mass Spectrometry; Potassium-Argon Dating; Radioactive Decay; Radiocarbon Dating; Rubidium-Strontium Dating; Uranium-Thorium-Lead Dating.

SAN ANDREAS FAULT

The San Andreas fault has been recognized as a major geologic feature of California and of North America for nearly a century. It is hoped that research into this seismically active fault will help geologists to develop a simple, characteristic model to explain the behavior of the fault and possibly forecast potentially destructive earthquakes in California.

PRINCIPAL TERMS

- **epicenter:** the point on the earth's surface directly above the focus of an earthquake
- **geodetic surveying:** surveying in which account is taken of the figure and size of the earth and corrections are made for the earth's curvature
- **geodimeter:** an electronic optical device that measures ground distances precisely by electronic timing and phase comparison of modulated light waves that travel from a master unit to a reflector and return to a light-sensitive tube; its precision is about three times as great as that of a tellurometer
- **right-slip (right-lateral strike-slip):** sideways motion along a steep fault in which the block of the earth's crust across the fault from the observer appears to be displaced to the right; left-slip faults are exactly the opposite
- **seismic:** pertaining to an earthquake
- **tectonics:** a branch of geology that deals with the regional study of large-scale structural or deformational features, their origins, mutual relations, and evolution
- **tellurometer:** a portable electronic device that measures ground distances precisely by determining the velocity of a phase-modulated, continuous microwave radio signal transmitted between two instruments operating alternately as a master station and a remote station; it has a range up to 65 kilometers
- **triple junction:** a point on the earth's surface where three different global plate boundaries join

CALIFORNIA'S EARTHQUAKE HISTORY

The San Andreas fault is the longest fault in California and perhaps the longest in North America. Its total length is more than 1,600 kilometers. Because about three-quarters of California's population lives within 80 kilometers of the fault, its existence—and its potential for a major unannounced earthquake—is responsible for formulating many public policy decisions and for establishing the design criteria of many engineering projects in the state. In spite of much being known about the fault, geoscientists who have studied it readily admit that their knowledge remains far from complete.

California's earthquake history is short. The earliest recorded earthquake in the state was on July 28, 1769. It was experienced by the Spanish explorer Gaspar de Portolá while he camped along the Santa Ana River southeast of Los Angeles. One of the state's earliest recorded large earthquakes occurred on the Hayward fault, a branch of the San Andreas system in the East Bay on June 10, 1836, during which surface breakage occurred along the western base of the Berkeley Hills. Another large earthquake on October 21, 1868, also centered on the Hayward fault, caused 30 fatalities and ground breakage for about 32 kilometers, accompanied by as much as 1 meter of right-slip between San Leandro and Warm Springs. Other notable nineteenth-century earthquakes along the San Andreas system in the Bay Area occurred in June 1838, on October 8, 1865, and on April 24, 1890. Unfortunately, there are no seismographic records of these early earthquakes because it was not until 1887 that the first seismographs began being used in the United States.

Since 1850, California has experienced three great earthquakes of Richter magnitude 8 or greater. The Richter scale, which was introduced in 1935, defines earthquakes of magnitude 2 as "felt," those of magnitude 4-4.5 as "causing local damage," those of magnitude 6-6.9 as "moderate," those of magnitude 7-7.5 as "major," and those exceeding magnitude 7.5 as "great." Two of these three great earthquakes—the San Francisco earthquake of 1906 and the Fort Tejon earthquake of 1857—resulted from movement of the San Andreas fault.

RECOGNITION OF THE SAN ANDREAS FAULT

It was not until after the Great San Francisco earthquake of April 18, 1906, that the San Andreas was first recognized as a continuous regional geologic structure of California. This earthquake caused sudden right-slip of up to 5 meters, ground rupturing for 420-470 kilometers, an estimated 700 fatalities, and damage estimated at between $350 million and $1 billion. The earthquake is generally considered as having attained a magnitude of 8.3 on the Richter scale. Much of the property loss in the San Francisco disaster was caused by the extensive fires following the earthquake that resulted from ruptured gas lines and a lack of water from broken water mains. The strongest earthquake in the Bay Area associated with the San Andreas fault since 1906 was the magnitude 7.1 earthquake of October 17, 1989, which resulted in 63 casualties and caused more than $1 billion in damage.

The other important California earthquake of magnitude comparable to the 1906 San Francisco disaster caused by the San Andreas fault was the Fort Tejon earthquake of January 9, 1857. This earthquake had an estimated magnitude of 8.3 and caused ground breakage for 360-400 kilometers on the southern part of the San Andreas fault zone, from the Cholame Valley in the Coast Ranges to the Transverse Ranges as far south as the present site of Wrightwood. Ground motion was felt from north of Sacramento to Fort Yuma on the lower Colorado River. The ground accelerations during this earthquake were so great that mature oak trees in the Fort Tejon area were toppled, buildings collapsed, fish in a lake near the fort were tossed out of the water, and the Los Angeles River was thrown out of its banks. An informative account of the effects of the 1857 earthquake, taken from correspondence and newspaper accounts of the time, was published by D. Agnew and Kerry Sieh in 1978. Right-slip movement ranging from 4.5 to 4.8 meters occurred in the 1857 event. Although the 1857 earthquake had the same estimated magnitude as the 1906 earthquake, the 1857 earthquake was potentially more destructive than the 1906 disaster because the ground shaking is reported to have lasted from 1 to 3 minutes. Compare this time interval with the short 10- to 30-second shaking accompanying the moderate 1971 San Fernando earthquake, magnitude 6.3, in which 60 lives were lost, and the 1989 San Francisco earthquake of 7.1, which lasted for approximately 15 seconds and whose damage, although severe in certain local areas, was minor in comparison to that of earlier, larger-magnitude, longer-lasting earthquakes. (Evidence, although inconclusive, points to the possibility that the 1989 event was a strike-slip earthquake along the San Andreas fault.) Such long duration and low seismic

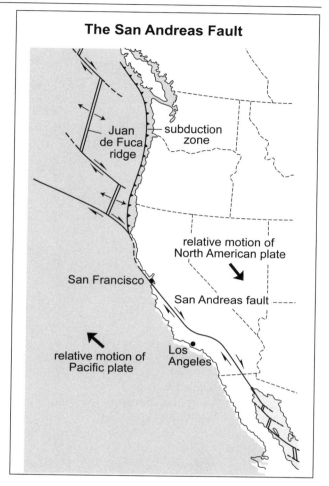

The San Andreas Fault

frequencies associated with events such as the 1857 earthquake occurring in California's highly populated regions are more likely to cause serious damage to large buildings and claim more lives than are moderate earthquakes.

DEFINITION OF THE SAN ANDREAS FAULT

The name "San Andreas" was first used in 1895 by Andrew C. Lawson, a geologist at the University of California, Berkeley. Later, it was Lawson who headed the California Earthquake Investigation Commission, whose monumental report on the San Andreas system was published in 1908. This report describes surface ruptures that developed in the 1906 earthquake that extended for more than 420 kilometers from Humboldt County to San Juan Bautista in San Benito County and continued to follow the fault southward through the Coast Ranges into San Bernardino County. The report used the term "San

Andreas rift" for the rift-valley surface expression in the San Andreas Lake area on the San Francisco Peninsula, and from this term have evolved the terms "San Andreas fault," "San Andreas fault zone," and "San Andreas system."

Curiously, in spite of the ample evidence of right-slip motion described for the segment that ruptured in the 1906 earthquake, the 1908 report maintained that the dominant characteristic movement on the San Andreas fault throughout its geologic history had been vertical (up and down) by as much as thousands of meters. This idea of vertical motion became the prevailing idea of motion for the San Andreas for almost fifty years. It has since been agreed, however, that activity along the fault has shown dominantly horizontal motion, with the North American plate moving southward relative to the Pacific plate. This kind of horizontal motion is termed right-lateral strike slip, or, more simply, right-slip, because the sideways displacement of the block across the fault from the observer appears to be to the right. Thus, if the historically observed motion is typical of the movement of the San Andreas fault in the geological past, rocks that once faced each other across the fault should now be separated. Because the amount of displacement of the faulted rock unit is known, it is possible to establish the ages of many rocks that were once a single unit. Scientists are therefore able to determine the movement rate (called slip rate) during the geologic history of the fault.

Current usage defines the San Andreas fault proper as the strand of the San Andreas zone that reveals surface rupturing produced by recent movements within the zone. The definition of the term "San Andreas zone" incorporates numerous parallel to subparallel related fractures that may be separated by 10-15 kilometers. Many of these subordinate fractures record that the locus of movement in the zone has shifted from one branch to another through time. Some examples of active or formerly active subordinate strands are the Pilarcitos, Calaveras, and Hayward faults in northern California and the Punchbowl, San Gabriel, San Jacinto, and Banning faults in southern California.

EXTENT OF THE SAN ANDREAS FAULT

On land, the San Andreas fault extends from Shelter Cove in Humboldt County southward, crossing the Golden Gate west of the Golden Gate Bridge, and under southwestern San Francisco, where the fault trace is obscured by housing tracts in Daly City. It appears that city planners, developers, and residents have ignored the hazards of living on this infamous fault. The fault continues southward along the length of the Santa Cruz Mountains of the San Francisco Peninsula, where it forms the rift valley occupied by the San Andreas and Crystal Springs reservoirs, and through the Coast Ranges past its intersection with the Garlock fault at Frazier Mountain near Tejon Pass. Beginning here, the fault turns more eastward—geologists call this segment of the fault the "big bend"—and marks the approximate northern boundary of the San Gabriel Mountains. It passes through Cajon Pass, which separates the San Gabriel and San Bernardino mountains, and continues southeastward along the southern margin of the San Bernardino Mountains to the vicinity of San Gorgonio Pass. Here, the San Andreas divides into the North and South branches, with the North Branch joining the Mission Creek fault and the South Branch joining the Banning fault. North of Indio, the branches rejoin, forming what some geologists call the "southern big bend," and the San Andreas fault resumes its more southerly trend to border the eastern edge of the Salton Sea trough and continue south into the Gulf of California.

Near the big bend at the Frazier Mountain intersection of the San Andreas and the Garlock faults are also the ends of the San Gabriel and Big Pine faults and the Frazier Mountain thrust. This area is known to geologists as the "knot," where both right-slip faults, the San Gabriel and San Andreas, are intertwined with left-slip faults, the Garlock and Big Pine. An important parallel branch of the San Andreas system is the San Jacinto fault. It splays off the San Andreas fault at the northeastern end of the San Gabriel Mountains and runs southeastward under San Bernardino, San Jacinto, and Hemet to El Centro.

The northern and southern ends of the San Andreas fault terminate in ways that are still not entirely clear to geologists. At its northern end, the San Andreas disappears into the Pacific Ocean at Shelter Cove near Point Mendocino, apparently to join the Mendocino fracture zone and the Juan de Fuca oceanic trench at a point known to geologists as a "transform-transform-trench triple junction." The southern end of the San Andreas disappears beneath the waters of the Gulf of California, where it appears to develop into a number of northwest-trending transform faults that offset the East Pacific Rise. The San Andreas system, therefore, links the East Pacific Rise to the Juan de Fuca Ridge as a transform fault and marks the approximate position of a very major structural feature of the earth's crust, the sliding boundary between the Pacific and North American crustal plates. Most of California rests on the North American plate, but much of southern California is on the Pacific plate.

SEISMOLOGY AND FIELD STUDY

Virtually every technique available to geoscientists has been used in studying the San Andreas fault. Most prominent, perhaps, is the work of seismologists and geophysicists who study the waves of energy generated in earthquakes. Seismologists in California study the activity of the state's fault systems by monitoring all seismic activity in the state with seismometers, instruments that record the passage of earthquake waves. The data from these instruments allow seismologists to determine the epicentral location, depths of foci (hypocenters), surface area, and movement directions of faults responsible for given earthquakes.

The geologic history of the San Andreas fault zone has been learned by fieldwork, which began in earnest following the San Francisco earthquake of 1906. Fieldwork is the process of working in the field by foot, car, or aircraft to locate and plot on aerial photographs and topographic base maps the locations of the rock units, faults, and folds in an area of interest. A primary goal of field mapping is correlation—that is, the determination of age relationships between rock units or geologic events in separate areas. For decades following the 1906 disaster, field mapping accumulated data on the structures and rock units exposed along the fault. For many years, the field data were interpreted to support a model supposing that the movement history of the San Andreas was primarily vertical. A tentative proposal of 38 kilometers of right-slip motion for the fault was advanced in 1926, but it was not widely accepted.

In 1953, in what may be the most important benchmark paper ever presented on California geology, two California geologists, Mason L. Hill and Thomas W. Dibblee, Jr., suggested that as much as 560 kilometers of movement had taken place along the San Andreas

fault in a period of 150 million years. Their proposal resulted from careful analysis of field observations that allowed correlation of rocks and fossils in separate exposures on opposite sides of the fault. They illustrated that not only had the oldest rocks studied been offset by 560 kilometers, but younger rocks—rocks that had not existed long enough to be offset as much—had also been displaced by progressively lesser distances. For example, they showed that the correlation of uniquely similar rocks in the Gabilan and San Emigdio ranges illustrated 282 kilometers of offset in about 25 million years, that offset beds of gravel in the Temblor Hills and San Emigdio Mountains revealed 22 kilometers of offset in less than 1 million years, and that the Big Pine fault had been displaced from its eastern extension, the Garlock fault, by 10 kilometers in the last 200,000 years.

Few geologists at the time could accept the idea of such great right-lateral displacement along the San Andreas fault. Nevertheless, so much interest was aroused by the tentative conclusions of Hill and Dibblee that many geologists began to conduct their own investigations. Consequently, subsequent field studies clearly confirmed the thesis. The work of Hill and Dibblee has proved to be the fundamental framework to which explanations of the geology of the Coast, Transverse, and San Bernardino ranges; the Salton Trough; and the Great Central Valley must conform. In retrospect, a large cumulative right-lateral offset by the San Andreas fault is a central component of the global tectonics theory. Work began in 2004 just north of Parkfield in central California on the San Andreas Fault Observatory at Depth (SAFOD), funded by the National Science Foundation (NSF), the U.S. Geological Survey and the National Aeronautics and Space Administration. Parkfield is a small town on the San Andreas fault that has experienced moderate earthquakes (magnitude 6) about every 22 years. By 2007, a borehole was angled into the fault at a depth of about 3 kilometers. Instruments have been placed in the hole at several points to monitor the behavior of the fault, with more holes planned. Work to predict future earthquakes continues at SAFOD.

FORECASTING THE NEXT "BIG ONE"

Estimates are that a repeat of the 1906 San Francisco earthquake would cause tens of thousands of casualties and tens of billions of dollars in damages. A repeat of the 1857 event would jolt southern

California with a similarly destructive earthquake, which could cause a similar amount of casualties and damages and leave hundreds of thousands of homes unfit for habitation.

Scientists are regularly asked to predict when the next "Big One" will occur. To arrive at answers to such questions, a different type of fieldwork was undertaken in the 1960's and 1970's. Studies by Tanya Atwater, a geologist who worked at Scripps Institute of Oceanography, showed that plate tectonics, which was at the time considered to be strictly an ocean-bound concept, was sometimes critical in understanding the deformation of crustal rocks within a continent. In terms of global tectonics, the San Andreas fault plays a very important role in understanding the geology of the western United States. Within the framework of plate tectonics, the San Andreas fault is a transform fault along which strain is occurring between the two moving plates.

To measure strain across the San Andreas fault, precision geodetic surveying began in the 1970's, using tellurometers, geodimeters, and long baseline interferometry (LBLI),a "space-geodetic" technique developed in the late 1970's. LBLI relies on extraterrestrial reference points, such as quasars, to measure distances to an incredible precision of 1 centimeter or less in 1,000 kilometers. Precision geodetic surveying and LBLI reveal that the relative motion between the Pacific and North American plates is about 5.5 centimeters per year, right-slip. About 4 centimeters of this motion is taken up each year by the San Andreas fault, with the remaining motion being accommodated by the slippage of other faults and in other ways within the San Andreas zone.

Thus, seismology, geologic field mapping during the first half of the twentieth century, large-scale tectonic and geodetic evidence of the 1970's and 1980's, and the recorded history of earthquakes along the San Andreas fault all suggest that Californians can expect the fault to move again—but questions remain about when, where, and with what frequency great earthquakes will occur along the San Andreas fault.

DETERMINING AVERAGE RECURRENCE INTERVAL

Researchers who estimate the frequency of earthquakes on an active fault use a still different kind of fieldwork: logging and mapping the walls of 20-foot-deep trenches dug across a fault in association with carbon-14 isotopic age dating and statistics. Through

these efforts, geologists establish an average recurrence interval (RI) of earthquakes for that particular fault. The reason for attempting to establish the history of a fault's activity is that much of geological interpretation is based on the premise that geologic events occurring today result from the same processes that caused them in the past, or uniformitarianism. Uniformitarianism holds that the activity of the San Andreas fault in the geological past is the best clue to forecasting the fault's behavior in the future.

To develop a regional overview of the San Andreas fault, the fault has been divided into four segments based on existing knowledge of seismic activity: northern, central, south-central, and southern. At least two recorded large earthquakes have occurred on the northern segment, the 1838 and the 1906 events. Although little detailed work has been done on the northern segment, it is believed that the frequency of earthquakes here is similar to that on the south-central segment. The central segment has a historic record of as much as 3 centimeters of offset per year. The strain along this segment of the fault, however, appears to be released slowly by aseismic creep (very slow, incremental movement along the fault that is unaccompanied by earthquake activity). The southern segment, from Indio south, is considered dormant, with no record of seismic activity since 1688.

Using this information as a starting point, Kerry Sieh and his associates at the California Institute of Technology have trenched three sites on the south-central and southern segments of the San Andreas fault to obtain the information necessary to determine the average RI. The first of these sites is at Pallett Creek near Palmdale, about 55 kilometers northeast of Los Angeles; another is just north of the big bend at Wallace Creek on the Carrizo Plain west of Bakersfield; and the third locality is on the southern segment near Indio. Ray Weldon, of the University of Oregon, has studied a fourth area astride the south-central segment at Lost Lake in Cajon Pass.

RESULTS OF RECURRENCE INTERVAL STUDY

The results of the Pallett Creek work reveal the RI for ten episodes of faulting in a period of about 2,000 years to be about 132 years. Because the Pallett Creek area was affected by the great southern California earthquake of 1857, Sieh's research seems to suggest that this segment of the San Andreas should break again soon. Sieh is quick to point out, however, that the site's average RI could have limited meaning because of newly perfected methods of carbon-14 dating, developed by Minze Stuiver of the University of Washington and statistician David Brillinger of the University of California at Berkeley, that allows much tighter precision in age dating than has been possible in the past. Consequently, age determinations of the ten events reveal that the earthquakes are clustered in bunches of two or three, with individual earthquakes within the clusters separated by tens of years. Five of the intervals between clusters are less than a century, and three of the remaining four intervals span from 200 to 330 years. From this field, laboratory, and statistical information, Sieh's interpretation of the Pallett Creek record is that it is probably near the middle of a two-hundred-year-long dormant period. Thus, if the pattern of earthquakes at Pallett Creek is truly representative of the fault's behavior of the south-central segment, it might be eighty years or more before Los Angeles could expect another great earthquake along this portion of the San Andreas fault.

Weldon's investigation at Lost Lake, about 40 kilometers southeast of the Pallett Creek site, reveals six earthquakes in the last thousand years with an average RI of 150-200 years. Thus, both the Pallett Creek and Lost Lake sites reveal relatively short RI values that appear to be at odds with what is known about the frequencies of earthquakes on the segments of the San Andreas to the north and south. Sieh's work at Wallace Creek, about 100 kilometers northwest of the Pallett Creek site and north of the big bend, results in a longer average RI of 250-450 years, and investigations on the southern segment show that this segment has not had a major, or great, event since about 1688, even though the RI for this segment is about 220 years. Comparing the results from the Indio, Lost Lake, Wallace, and Pallett creeks, Weldon and Sieh have developed several alternate models of behavior of the south-central and southern segments of the fault. Two of their models suggest that the southern segment is the most likely site for the next large earthquake. This is a sparsely populated region at present, about 120 kilometers east of Los Angeles, far enough that a great earthquake on the southern segment is currently not a serious threat to the residents of Los Angeles.

D. D. Trent

FURTHER READING

Abaimov, S. G., et al. "Earthquakes: Recurrence and Interoccurrence Times." *Pure and Applied Geophysics* 165 (2008): 777-795. Provides an analysis of the statistical probability of earthquake recurrence at the San Andreas fault. A background in statistics is needed to fully comprehend the equations.

Agnew, D., and K. Sieh. "A Documentary Study of the Felt Effects of the Great California Earthquake of 1857." *Bulletin of the Seismological Society of America* 68 (1978): 1717-1729. An informative collection of eyewitness accounts and newspaper articles about the great 1857 southern California earthquake.

Clarke, Thurston. *California Fault: Searching for the Spirit of State Along the San Andreas.* New York: Ballantine Books, 1996. Clarke traveled the length of the San Andreas fault collecting first-hand accounts from earthquake survivors and predictors. Along with the entertaining stories, Clarke provides historical and scientific information about the fault.

Collier, Michael. *A Land in Motion: California's San Andreas Fault.* San Francisco: Golden Gates National Parks Association, 1999. Filled with beautiful color photographs that accompany text intended for the nonscientist, *A Land in Motion* gives the reader excellent insight into earthquakes and their aftermaths. There are also many diagrams and graphs that explain subduction, faults, and orogeny.

Condie, Kent C. *Plate Tectonics and Crustal Evolution.* 4th ed. Oxford: Butterworth Heinemann, 1997. An excellent overview of modern plate tectonics theory that synthesizes data from geology, geochemistry, geophysics, and oceanography. Of special interest is Chapter 6, on seafloor spreading, and Chapter 9's treatment of the Cordilleran system, including a discussion of the evolution of the San Andreas fault. A very helpful tectonic map of the world is enclosed. The book is nontechnical and suitable for a college-level reader. Useful "suggestions for further reading" follow each chapter.

Davis, Craig A. "Los Angeles Water Supply Impact from a M7.8 San Andreas Fault Earthquake Scenario." *Journal of Water Supply: Research and Technology* 59.6-7 (2010): 408-417. Provides predictions of the possible effects of a severe earthquake in California.

Fradkin, Philip L. *Magnitude 8: Earthquakes and Life Along the San Andreas Fault.* Berkeley: University of California Press, 1999. Written for the layperson, this book can sometimes read overdramatic or un-scientific. However, *Magnitude 8* traces the seismic history, mythology, and literature associated with the San Andreas fault.

Hill, M. L. "San Andreas Fault: History of Concepts." *Bulletin of the Geological Society of America* 92 (1981): 112-131. A historical account of the recognition, mapping, and changing interpretations of the San Andreas fault, written by one of the key authorities on the geology of the fault.

Iacopi, Robert, ed. *Earthquake Country: California.* 4th ed. Menlo Park, Calif.: Sunset Books/Lane, 1996. Probably the best single source of information for the layperson on the San Andreas fault. Because of its photographs, maps, and diagrams, it can be well used by the traveler as a guide in finding the many fault-formed features along the fault's length.

Jordan, T. H., and J. Minster. "Measuring Crustal Deformation in the American West." *Scientific American* 259 (August 1988): 48-58. Describes the method of measuring the actively deforming continental crust in the western United States by the use of "space-geodetic" techniques that rely on extraterrestrial reference points, such as quasars, to a precision of 1 centimeter or less in 1,000 kilometers.

Kerr, R. A. "Take Your Choice: Ice Ages, Quakes, or Impacts." *Science* 243 (January 27, 1989): 479-480. A brief overview of the December 1988 meeting of the American Geophysical Union that includes an overview of Kerry Sieh's research and the state of knowledge in forecasting the next great earthquake on the San Andreas fault. Easily understandable by the layperson with minimal technical background.

Norris, R. M., and R. W. Webb. *Geology of California.* 2d ed. New York: John Wiley & Sons, 1990. The definitive textbook on the geology of California. Written for use by college students who have a year of freshman introductory geology courses as a background. A full chapter is devoted to the San Andreas fault and should be readily understandable to the interested layperson. Also contains a geological map of California. An excellent glossary.

Plummer, Charles C., and Diane Carlson. *Physical Geology.* 12th ed. Boston: McGraw-Hill, 2007. A college-level introductory geology textbook that is clearly written and wonderfully illustrated. An excellent sourcebook of basic information on geologic terminology and fundamentals of geologic processes. The chapters on structural geology and global plate tectonics are especially relevant to understanding the San Andreas fault in the context of large-scale geologic processes. Three full pages are devoted to the San Andreas fault. An excellent glossary.

Powell, Robert, et al., eds. *The San Andreas Fault System: Displacement, Palinspastic Reconstruction, and Geological Evolution.* Boulder, Colo.: Geological Society of America, 1993. This book provides a clear description of the power and size of the San Andreas fault. It details the history of the fault as well as its constant evolution. Illustrations and folded leaves; includes a bibliography and index.

Sieh, K. E., M. Stuiver, and D. Brillinger. "A More Precise Chronology of Earthquakes Produced by the San Andreas Fault in Southern California." *Journal of Geophysical Research* 94 (January 10, 1989): 603-623. Although this article is published in a technical journal, the section comparing the results of the investigations at the Pallett Creek site with other sites on the south-central and southern segments of the San Andreas fault and the conclusions are relatively nontechnical and worth reading.

Stoffer, Philip W. *Where's the San Andreas Fault?: A Guidebook to Tracing the Fault on Public Lands in the San Francisco Bay Region.* University of Michigan Library, 2006. An excellent field guide for the geologist, student, or adventurer. Contains pertinent background information on geology and the San Andreas fault. Also provides a history of the "Great San Francisco Earthquake," 100 years after its event.

Walker, Bryce. *Earthquake.* Alexandria, Va.: Time-Life Books, 1982. Features a nontechnical essay on the San Andreas fault that includes magnificent illustrations and a map.

See also: Creep; Cross-Borehole Seismology; Discontinuities; Earthquake Prediction; Elastic Waves; Experimental Rock Deformation; Faults: Normal; Faults: Strike-Slip; Faults: Thrust; Faults: Transform; Lithospheric Plates; Notable Earthquakes; Plate Motions; Plate Tectonics; Seismic Observatories; Seismic Tomography; Seismic Wave Studies; Seismometers; Stress and Strain.

SEISMIC OBSERVATORIES

A global network of seismic observatories detects and records seismic waves produced by earthquakes and other energy releases. Their data can be used to locate the sources and magnitudes of earthquakes, to interpret the earth's internal structure, to delineate seismically active zones, to study dynamic processes in the earth's crust, and to monitor nuclear tests.

PRINCIPAL TERMS

- **epicenter:** the spot on the earth's surface directly over the focus of an earthquake
- **focus:** also called the hypocenter, the region in the earth's crust or upper mantle where an earthquake begins; larger earthquakes have a focus several tens of kilometers in size
- **seismic waves:** oscillatory vibrations generated by an earthquake that travel outward in all directions through the earth (as body waves) or along and near the earth's surface (as surface waves)
- **seismogram:** a recording, by ink pen, film, or digital data, that measures the train of seismic waves arriving by the variation in ground motion as time advances
- **seismograph:** a sensitive instrument that mechanically or electromagnetically detects and records arriving seismic waves, usually by measuring the motion of the ground with respect to a relatively fixed mass
- **time of origin:** the time of an earthquake's occurrence in local time or—more conveniently for analysis of worldwide events on a standard time scale—in Coordinated Universal Time (CUT, or Greenwich Mean Time)

MONITORING SEISMIC WAVES

Earthquakes are one of nature's most sudden and terrifyingly unexpected phenomena. They are dangerous to human life and cause the destruction of buildings and property. However, they do not occur randomly in time and space. Seismology—the study of earthquakes and the outgoing seismic waves they produce—has yielded much understanding of the earth and its dynamic physical environment, including knowledge, since the 1960's, of plate tectonics. Large tectonic plates in the earth's lithosphere (crust and upper mantle) move at rates of several centimeters per year. It is at the margins of these plates that earthquakes have been most common in the past and are most likely to occur in the future. At their margins, tectonic plates spread apart, slip sideways, or collide in great processes of mountain-building or deep-ocean trench subduction (diving down into the earth's mantle).

The seismic wave vibrations generated by earthquakes travel through the entire earth and, by observing them, scientists can yield information about the earth's interior structure, physical properties, and likely composition. Study of seismic waves can also provide understanding of other sudden energy releases, such as volcanic eruptions, landslides, meteorite impacts, and large subsurface explosions such as nuclear weapons tests.

All this is possible because of the detection, recording, and analysis of seismic waves at seismic observatories. By sharing data with other observatories that have appropriate equipment, scientists can determine a seismic event's time of origin, size (energy release), geographic location, and depth. By considering many such events, scientists can investigate wave travel through the earth and interpret its structure and properties; map the distribution of earthquakes over the earth's surface and over time; delineate seismically active zones, major plate tectonic boundaries, and local faults (where crustal rock masses are slipping and crunching past one another); and assess seismic hazards.

Advances in seismology were made possible by the invention of seismographs in the 1870's; the establishment of progressively larger, more coordinated, and more sensitive global networks of seismic observatories beginning in the early twentieth century; and the introduction of computers in the 1960's. Computers have allowed researchers to deal with massive amounts of data and complicated mathematical analysis procedures, as well as subtle and difficult problems of interpretation.

EARTHQUAKES AND SEISMIC WAVES

An earthquake occurs when rock masses in the crust and upper mantle suddenly break and shift along a plane or zone called a fault. This can occur when stress has built up, as from plate tectonic movement in the lithosphere, and exceeds the breaking strength of the rock. The accumulated strain energy is released as frictional heating on the fault zone, plus seismic waves (vibrations) that are transmitted rapidly out from the focus in all directions through the earth. These waves can range from barely perceptible to catastrophically destructive.

There are different types of waves produced by an earthquake. "Body" waves are generated by the faulting, designated as either P (for "primary," or compressional, having forward-and-back motion like coils moving in a slinky spring) or S (for "secondary," or shear, having sideways motion like a wave in a rope snapped sideways at one end). P waves always travel faster than S waves and thus arrive at a distant point sooner. When body waves travel up to the earth's surface, they can produce surface waves that travel out across the earth's surface like large ripples from a stone dropped into a pond. Surface waves are termed L waves (for "large"), since they are typically larger in amplitude (or extent of oscillation) than body waves. Because of their amplitude and the nature of ground shaking, they can be quite destructive.

When earthquakes are detected and analyzed by seismic observatories, information on their time of origin, epicenter, magnitude (energy release), and depth of focus can be catalogued and mapped at a central data repository. Two prime sites for global data collection are the U.S. Geological Survey's National Earthquake Information Center (USGS/NEIC) in Golden, Colorado, and the International Seismological Center in Cambridge, England. The data are publicly available on the Internet. One example would be a map that shows seismicity around the Pacific Rim. This margin around the Pacific Ocean basin is defined by the major plate tectonic collision and subduction as the Pacific sea floor spreads apart on the East Pacific Rise and is forced to dive down upon collision with the other plates rimming the Pacific. The pattern of earthquake epicenters effectively outlines these active plate margins.

The amplitude (height, or deflection, of the measured vibration) of seismic waves can be used to calculate a "magnitude" for the earthquake. This is a measure of the energy released by the event. Several slightly differing versions of magnitude exist, depending on the wave type analyzed and the distance to the earthquake epicenter. Richter scale magnitudes vary numerically from about 1 (very minor earthquake) to about 9 (extremely large and rare)

in a logarithmically increasing scale. Each year there are several hundred thousand earthquakes that are instrumentally detectable, at least locally. Many of these are too small to be felt by people. A "major" earthquake has a Richter magnitude of at least 7. There are typically several of these events worldwide each year.

When incoming seismic waves arrive at a seismological observatory, they are detected and ultimately recorded for display on a seismogram—a line trace related to the variation of ground motion from the passing wave train as time advances. The seismogram, in conjunction with other seismograms from other observatories, can be analyzed by computers and by human interpreters to give the precise arrival times of various waves. The waves include those that have traveled directly from the focus, with travel paths shaped by the wave velocities and properties of the layers in the earth's interior, and those that have been multiply reflected inside the earth. Major earthquakes produce so many waves that they often continue to arrive with enough energy to be detected for several hours.

The travel velocity of seismic waves in the earth is several kilometers per second, depending on the type of wave, the composition of the material at depth, and the temperature and pressure at depth. For example, a seismic station at the antipode (the point on the earth's far side opposite the epicenter, 12,740 kilometers straight through the earth) would detect the first-arriving P wave about twenty minutes after the earthquake occurred. It would be followed by a wave train of many waves having different travel paths.

SEISMOGRAPH EQUIPMENT

Any collection of data at a seismological observatory is only as useful as the quality of its staff and equipment. A seismograph is a sensitive device designed to detect the waves of energy generated by earthquakes. It consists of a seismometer, which mechanically or electromagnetically detects arriving seismic waves and other ground vibrations by measuring the movement of the ground; an amplifier, which uses electrical or optical systems to magnify the small vibrations being sensed by a factor of several thousand; and a recording system, which makes a record, for storage or display (on a seismogram), of the arriving seismic waves as a function of time advancing.

Recording methods include an ink pen on a rotating paper drum, a light spot on photosensitive paper, or analog or digital data stored on magnetic tape or disk. Analog-to-digital converters are also used.

A common seismometer technique is to measure the movement of the ground—at the surface, in a buried vault, or down a borehole or mine—with respect to a suspended mass such as a pendulum. The mass, not being fixed rigidly to the ground, briefly "hangs back" as the ground suddenly moves, because of its inertia. This indicates the relative motion.

The world's first seismometer is attributed to Zhang Heng of China in the year 132 C.E. during the Han dynasty. Its design featured a bronze kettle with eight dragon heads holding balls over eight toads. Inside was probably an inertial pendulum with linkages to the dragons' heads. When earthquake ground motion moved the kettle sideways, the pendulum hung behind and opened one or more dragons' mouths so that those balls dropped into the open mouths of the toads below. This ingenious seismoscope measured the occurrence of a vibration but could not make a record of its duration and behavior. Instead, it indicated the relative intensity of the effects felt and possibly the direction from which the seismic waves came. The principle of earthquake duration was neglected until European geologists again addressed the problem of earthquake detection and measurement during the nineteenth century.

John Milne, an English professor of mining and geology, arrived in Japan to teach in Tokyo in 1875. Interested in the frequent earthquakes there, he began developing seismographs. He, along with John Gray and James Ewing, who were also visiting Tokyo, designed swinging pendulums that detected and recorded wave motion by scratching a trace on a moving smoked glass plate. The instruments measured three components of motion, which, when combined, could give the net ground motion. Milne left Japan for England in 1894. By 1900, his efforts had encouraged the installation of seismographs at a couple of dozen seismic observatories around the

Seismographs measure the ground movement from recording stations in several key areas of California. (Russell Curtis/Photo Researchers, Inc.)

world. Thus began the systematic collection of global seismic data.

Other work to develop or refine seismographs was done in Germany by Ernst von Rebeur Paschwitz and Emil Wiechert, who introduced a damped pendulum in 1898 so motions could be more quickly recovered, and in Russia by Boris Galitzin, who introduced moving-coil electromagnetic recording of motion in the early twentieth century.

In practice, seismographs must be designed to accurately sense a wide range of wave amplitudes and periods. The period of a wave is the time (in seconds) between successive oscillation, or vibration, peaks (from crest to crest). The frequency is the inverse—that is, the number of peaks passing per unit time (in cycles per second, or Hertz). A seismograph might be designed, or tuned, to emphasize quarry blasts with wave periods of less than one second (frequency greater than 1 Hertz), nearby earthquakes with body wave periods of one to ten seconds and surface wave periods of ten to sixty seconds. Seismographs might also specialize in measuring distant earthquakes (called teleseisms), with arriving waves having longer periods (the shorter-period energy is progressively absorbed during travel through the earth), or earth oscillations and even earth tides having periods of hours.

A typical seismological station would have six seismographs. Three would be short-period (about a one-second response for nearby earthquakes and vibrations) and would measure the three components (north-south, west-east, and vertical). Three would be long-period (about a twenty- to thirty-second response for teleseisms and surface waves). All would record continuously. Apart from the need to detect and record a wide range of periods, it may also be necessary to have both low- and high-sensitivity instruments for global data collection and interpretation. This is because a high-sensitivity instrument, designed to respond to small local events or moderate distant events, would be deflected or driven "off scale" (exceeding its reading limits) by a large seismic event.

A variety of filtering, damping, and data-analysis techniques can assist in yielding a useful data set. One innovation is to restrain the relative motion of the inertial mass with a "forced feedback" mechanism. The restraining force is measured as a signal. This allows a greater range of sensitivity, and, without large excursions of the inertial element, the device can be more compact. An example is the Wielandt-Streckeisen STS-1 leaf-spring seismometer, developed in 1986, in which the mass is on a flexible strip rather than on a pendulum.

GLOBAL NETWORK OF OBSERVATORIES

In the early twentieth century, a worldwide network of seismological stations that used the rudimentary seismograph technology available at the time was established for the purpose of studying earthquakes and the interior structure of the planet. The 1906 earthquake in San Francisco was recorded, for example, at dozens of seismological stations around the world, including in Japan, Italy, and Germany. By 1960, about seven hundred seismic observatories were operating worldwide using various types of seismographs and standards, which led to incomplete data exchange and analysis. Most observatories were operated by government agencies and universities.

A standardized global network of calibrated instruments was needed, with coordinated accurate timing and central data collection. This was not only a vexing organizational problem, but one that required significant funding. Nonetheless, scientists recognized that seismology and the mapping of earthquakes would contribute to the emerging study of global plate tectonics and the dynamic processes at the plate boundaries during the 1960's.

By convenient coincidence, the impetus and funding for these earth-centered interests emerged from the new military and international political need to monitor underground nuclear testing. The 1963 Test Ban Treaty prohibited atmospheric, oceanic, and space testing of nuclear technology—all of which could be monitored fairly directly. Nations with existing or developing nuclear programs now needed to test underground. Military competition between the United States and the Soviet Union necessitated a program of remote surveillance, and the techniques of seismology—particularly the ability to distinguish small, shallow earthquakes from buried nuclear explosions—were needed and quickly applied. In the early 1960's, the United States began a program that deployed a series of stations in the World Wide Standardized Seismograph Network (WWSSN). Over time, the network would grow to about 120 stations in total. The stations used Benioff short-period (a period of one second) seismographs and Sprengnether long-period (a period of fifteen to sixty seconds) seismographs with moving-coil electromagnetic seismometers and galvanometers to record on seismogram drums.

During the 1970's, seismic recording began a conversion from analog recording to digital recording onto magnetic tape. Digital technology samples the seismic signal at short time intervals and stores the data. It can retain greater dynamic range and is convenient for direct computer processing. The combination of new digital seismic observatories and some of the WWSSN stations that were upgraded to digital recording had a major impact on global seismology from the mid-1970's to the mid-1980's.

In 1986, seismographs based on the forced-feedback mechanism to restrain motion and increase range of response, along with digital recording of signals onto magnetic tape, began being used to form the Global Seismic Network (GSN). This is a joint effort of a consortium of universities (the Incorporated Research Institutions for Seismology [IRIS], the U.S. Geological Survey (USGS), and upgraded stations run by European countries, Canada, Australia, and Japan. Funding comes from the U.S. National Science Foundation and the USGS. The GSN's plan was to have 128 seismic stations when the network was complete. The one hundredth station was installed in 1997. The high-quality global network, designed to replace the WWSSN, exemplifies international scientific collaboration. Within hours of an earthquake, data are automatically collected and made available to government and university scientists. Data are also made available to the general public over the Internet.

REGIONAL NETWORKS

In addition to the coordinated global collection and analysis of seismic data, a variety of regional, or more localized, networks exist. These are set up for more detailed and rapid analysis of local earthquakes or other events, such as for detection of patterns of small earthquakes. They also aim to detect foreshocks that can occur prior to a larger event in the hopes of possibly predicting a larger forthcoming earthquake.

These networks are commonly set up by technologically advanced nations in regions at high risk of suffering seismic hazards, such as the United States (particularly California, the Pacific Northwest, and the New Madrid area of southeast Missouri), Japan, and Canada. The U.S. National Seismograph Network (USNSN), under development since 2000, uses sensitive forced-feedback seismometers with satellite telemetry of data back to the USGS/NEIC in Golden, Colorado. Stations are being developed in the continental United States, Alaska, Hawaii, Central America, and the Caribbean region. The network is designed to detect, locate, and analyze earthquakes of magnitudes as small as 2.5 but also has seismographs with high dynamic range to handle large earthquakes. The National Science Foundation's Earthscope project includes the fifteen-year USArray program, begun in 2004, to place a dense network of permanent and portable seismic stations across the United States at 2,000 locations over a ten-year period to better understand the structure and evolution of the crust and lithosphere under North America. The network is designed to detect, locate, and analyze earthquakes of magnitudes as small as 2.5 but will also have seismographs with high dynamic range to handle large earthquakes.

The occurrence of tsunamis (seismic sea waves) generated by some large earthquakes or subsea volcanic eruptions, particularly in the Pacific Ocean, prompted the United States set up the Pacific Tsunami Warning System in 1948. This was in response to a destructive tsunami that hit Hawaii in April 1946. It traveled south from an earthquake off the coast of the Alaskan peninsula, killing 179 people in Alaska and Hawaii with waves that rose to crests of up to 30 meters high on Unimak Island, Alaska, and up to 10 meters high in Hawaii. The system is administered by the National Oceanic and Atmospheric Administration (NOAA), with coordination, data processing, and alerts issued from a warning center located in Honolulu, Hawaii. The system quickly activates when any of its thirty or so participating seismographic stations, located around the Pacific Rim and on Pacific islands, detects an earthquake or other disturbance that could potentially generate a spreading tsunami wave. Another seventy-eight stations have tide gauges for monitoring unusual changes in sea level which can detect a tsunami as it passes by. If a tsunami is detected, an alert can be issued, with prediction of tsunami arrival times at Pacific-bordering nations or islands. Coastal people can be evacuated inland, and ships or fishing boats can be taken out to sea to ride out the much more subdued offshore waves.

Portable seismograph systems are used for rapid and temporary setup of local seismic networks. These are deployed to monitor precursory earthquakes before a suspected major earthquake or aftershocks after a large one has occurred. They can also be used around volcanoes before or after eruptions, or in anticipation of an underground nuclear test or some other energy-releasing disturbance. Local temporary arrays have also been used during the seismic probing of deep geological structures and to study local seismicity patterns.

Robert S. Carmichael

FURTHER READING

Frechet, Julien, Mustapha Meghraoui, and Massimiliano Stucchi, eds. *Historical Seismology: Interdisciplinary Studies of Past and Recent Earthquakes.* New York: Springer, 2010. A compilation of articles which provide new approaches and a historical review of seismic observation.

Havskov, Jens, and Lars Ottemoller. *Routine Data Processing in Earthquake Seismology.* New York: Springer, 2010. This text provides practical application of software and data analysis in the field of seismic observation. The text comes with software and exercises. It has use within an undergraduate course, in observatory processes, and in research.

Lane, N., and G. Eaton. "Seismographic Network Provides Blueprint for Scientific Cooperation." *EOS/Transactions American Geophysical Union* 78 (September 1997): 381. The authors discuss attempts to install a global network of modern seismological stations.

Lay, T., and T. C. Wallace. *Modern Global Seismology.* San Diego: Academic Press, 1995. Chapter 5 discusses the development of seismographs as well as their deployment in regional and global networks for interpreting the earth's internal structure, earthquake characteristics, and nuclear weapons testing. Written at a more technical level, suited for graduate students with a background in mathematics.

Natural Resources Canada/Canadian National Seismology Data Center Website: earthquakescanada. nrcan.gc.ca. Database of earthquakes as recorded in Canada.

Simon, R. B. *Earthquake Interpretations: A Manual for Reading Seismograms.* Los Altos, Calif.: W. Kaufmann, 1981. Instructions for reading earthquake seismograms, with examples.

U.S. Geological Survey Earthquake Website: earthquake.usgs.gov. Contains useful information about earthquakes and seismicity.

U.S. Geological Survey/National Earthquake Information Center Website: earthquake.usgs.gov/ regional/neic. Database of earthquakes in the United States and around the world, as located and assembled by NEIC in Golden, Colorado. Includes data and listings on worldwide seismological stations, earthquake listings by geographical region, and listings of significant historical earthquakes.

Walker, Bryce. *Earthquake.* Planet Earth Series. Alexandria, Va.: Time-Life, 1982. Chapter 3, "Secrets of the Seismic Waves," discusses seismology, recording instruments, and historical developments. Map included.

See also: Creep; Cross-Borehole Seismology; Discontinuities; Earthquake Distribution; Earthquake Locating; Earthquake Magnitudes and Intensities; Elastic Waves; Experimental Rock Deformation; Faults: Normal; Faults: Strike-Slip; Faults: Thrust; Faults: Transform; San Andreas Fault; Seismic Reflection Profiling; Seismic Tomography; Seismic Wave Studies; Seismometers; Stress and Strain.

SEISMIC REFLECTION PROFILING

Seismic reflection profiling is a method of applied exploration geophysics that allows scientists to determine the location of subsurface geological structures. It is accomplished by using one of various methods of generating seismic waves. These waves are reflected to the surface from the subsurface and are received and analyzed. This analysis enables geologists to locate oil and, less frequently, mineral-bearing formations.

PRINCIPAL TERMS

- **attenuated:** becoming less intense as distance from the source increases
- **discontinuity:** the sudden change in physical properties of rock with increased depth
- **electromagnetic radiation:** forms of energy, such as light and radio waves, that consist of electric and magnetic fields that move through space
- **focus:** the source of earthquake waves; the actual point of rock breakage
- **propagated:** conducted through a medium
- **strata:** rock layers produced by sediment deposition in layers or beds

BODY SEISMIC WAVES

Seismic reflection profiling enables earth scientists to determine what the earth's subsurface looks like without having to drill exploratory wells. This study of applied seismics is related to seismology, which is the study of earthquake waves. When an elastic body such as rock is stressed and suddenly breaks, the energy released is transferred through the material in the form of various types of waves. This is what happens during an earthquake. When stress builds up and the rock fractures, energy radiates out from the focus or zone of breakage in the form of an ever-enlarging sphere of wavefronts. As the sphere gets larger, the energy along any part of the wavefront is diminished or attenuated. This sphere will continue to enlarge and maintain its basic shape as long as the properties of the rock through which the waves are traveling remain the same.

The four types of seismic waves are classified into two major categories. Those waves that travel beneath the surface of the earth (that is, in a three-dimensional medium) are called body waves. Waves that travel at or near the surface of the earth (in a two-dimensional medium) are called surface waves. Body waves are caused by earthquakes that take place well beneath the surface of the earth. Surface waves are caused by near-surface earthquakes or by human-made explosions.

The two types of body waves are P, or primary, waves and S, or secondary, waves. The P waves are compressional waves and are the faster of the two types. P waves, for example, can be generated by pinching several coils of a "Slinky" toy together and releasing them. Sound waves are an example of this type of wave; sound will travel through any medium that is capable of being compressed. In a compressional wave, the particles of the conducting medium vibrate parallel to the direction of wave propagation, or travel. A compressional wave is set up by a vibration. In the case of sound waves, the vibration may be that of the human vocal chords; in the case of seismic compressional waves, it is the breaking of rock. The particles of the conducting material (air in the case of sound, rock in the case of seismic waves) nearest the point of vibration also begin to vibrate. These particles, as they move back and forth, strike additional particles and then return to their original position—the moving forward is known as compression, the moving back to the original position as rarefaction. This next group of particles moves forward and strikes another group of particles and then returns to their original positions. This repeated succession of compressions and rarefactions is how a compressional wave travels. The S waves, which are sometimes known as shear waves, are transverse waves: The particles in the conducting medium travel perpendicular to the direction in which the wave is traveling. S waves can be generated by tying a rope to a doorknob and rapidly moving the other end up and down with a flick of the wrist. Light and other forms of electromagnetic radiation are propagated by means of transverse waves. Unlike compressional waves, the secondary seismic waves will not travel through liquids.

SURFACE WAVES

The two types of surface waves are named for the scientists who demonstrated their existence, Lord Rayleigh (English physicist John William Strutt) and English mathematician A. E. Love. Both of these types of seismic wave are of the S or transverse variety but travel at a slower speed than the body type of S wave. The Rayleigh wave travels along the surface in the vertical plane not unlike the waves of the ocean. This type of wave differs from a transverse body wave in that instead of moving back and forth along a straight line perpendicular to the direction of wave motion, the particles in a Rayleigh wave move in an elliptical motion with the long axis of the ellipse usually vertical. Unlike other types of waves that may cause particles in their path to move back and forth a few times before coming to rest, the Rayleigh wave vibration lasts much longer. The particles will travel in their elliptical path many times before coming to rest.

The other type of seismic surface wave is the Love wave. It is a shear wave but, unlike the Rayleigh wave, the motion is parallel to the surface and at right angles to the direction of wave transmission. The vibration of a Love wave lasts much longer than that of a transverse body wave.

WAVE REFLECTION AND REFRACTION

By studying the paths of seismic waves generated by earthquakes, scientists have learned much about the interior of the earth. As stated earlier, waves expand in a spherically shaped shell from the focus, or center, as long as the rock properties remain the same. Should the rock properties change—for example, if the waves may encounter more or less dense material or perhaps a boundary or contact where rock properties change instantly rather than gradually—there is a significant change in the wave pattern. Some waves are reflected from that contact; others enter the new rock body and are refracted. For example, consider only the waves that a shallow earthquake will send straight down from the focus into the earth. Some of these waves will strike rock layers at various depths and will be reflected at various angles; some will return to the surface of the earth. Other waves will be refracted through the rock layers at different angles and will emerge elsewhere on the earth's surface.

The study of reflection and refraction of seismic waves has revealed that the earth is divided into three major zones: the crust, the mantle, and the core. It has been found that secondary or shear waves will not pass through the core but are reflected from it, while the primary waves pass through easily. Based on this evidence, scientists have concluded that a portion of the earth's core is a liquid. In addition to being

able to find discontinuities or rock boundaries by studying the paths of seismic waves, from the study of the velocities of these waves scientists can determine the density of the rock through which the waves pass.

During World War I, both the Allies and the Germans made some progress in using seismic detectors to locate the positions of large field artillery pieces. Some scientists who were involved with this study during the war became active in seismic prospecting development in the United States during the early 1920's. The method of seismic reflection was first used in an attempt to find oil-bearing rock in the area of the Texas Gulf Coast in the late 1920's. These early attempts were not very successful. The same techniques were being successfully employed in Oklahoma, however, and with this experience, seismic reflection prospecting became well established by 1931.

Later improvements in instrumentation and in field techniques enabled reflection profiling to be used under a wide range of geologic conditions. It is now used successfully in all oil-producing regions of the world.

SEISMIC SURVEYING

One of the methods that geologists and geophysicists use to study the earth's subsurface is seismic reflection profiling. The great advantage of this technique is that earth scientists can have a very accurate picture of subsurface geologic strata without going to the considerable expense of drilling numerous core samples or exploratory wells. The petroleum industry, more than any other concern, uses seismic reflection techniques. Seismic surveying is used to a lesser extent for construction site evaluation, groundwater exploration, and mineral exploration.

Since the early days of seismic surveying, explosives such as dynamite have played an important role. In modern times, however, dynamite has been replaced by safer types of explosives such as ammonium nitrate. The size of the charge can be varied depending on the nature of the survey. For most surveys, the explosive is placed in a hole called the shot hole. The holes may vary in depth from about a meter to a few hundred meters. These holes are drilled by a small drilling rig attached to a truck. There are always certain disadvantages to the use of explosives. First is the always-present potential for destructive side effects, and second is the inconvenience and cost of drilling shot holes. In addition, an explosion introduces an uncontrolled range of wave frequencies.

Surveying Techniques

In practice, seismic surveying consists of placing receivers, known as geophones, at various intervals (usually about 30-60 meters apart) and using them to detect vibrations that have been reflected back to the surface from subsurface geological features. Those vibrations may be caused by an explosive charge (impulsive technique), the explosion of a mixture of gases inside a closed steel chamber in contact with the ground (dinoseis technique), the repeated dropping of a 2-ton mass from a height of 2-3 meters (thumper technique), or a vibrator located on the surface (vibroseis technique).

When the dinoseis method is used, several canisters of gas (usually three to six) are fired simultaneously, which forces more of an energetic reaction from the surface. Like the conventional explosives technique, the dinoseis method can be used to reach and identify the location of deep structures. The thumper technique does not work well except for small engineering surveys because this method produces very little energy in the form of waves that penetrate into the surface. Much of the energy generated by dropping the weight is dissipated by surface waves. The vibroseis technique was developed in the 1950's as an alternative to the use of an impulse source. In the vibroseis system, energy is produced by a vibrating pad that is pressed firmly to the ground. The pad is attached to the underside of a truck by means of hydraulic jacks; when these jacks are employed, most of the weight of the truck forces the pad against the ground. Unlike the case of the impulsive source, the frequency of the vibrations can be controlled. Typically, the frequency is varied or swept from 15 cycles per second to 90 cycles per second over a period of several seconds. The reverse is also sometimes used, with frequencies being changed from higher to lower.

Geophones

The receiver used to detect vibrations from any one of these methods is known as a geophone. This device consists of a piece of magnetized metal attached to a container and surrounded by a suspended coil of wire. When the ground vibrates, the coil picks up these vibrations and oscillates up and down around the magnet. This action induces an electric current in the coil, which is detected: The greater the vibration of the ground, the greater the current generated. As the ground vibrates, the geophone produces a continuously varying signal.

The geophone signal is transmitted to the recording systems by means of the seismic cable. The recording device produces a record of the vibration of the ground, called a seismogram. In some seismic recorders, the signal coming in from the geophones is first amplified and then sent to galvanometers, which are devices that detect small currents. Each geophone sends a signal to a different galvanometer. This device contains a suspended coil that rotates in response to an electrical current. Attached to each coil is a tiny concave mirror that reflects light to a photosensitive paper. As the coil and mirror rotate back and forth and the paper slowly advances, an irregular line is projected on the paper. This line shows the vibrations of the ground. Marks are also placed on the chart by a timing device. It is common practice to record the signals from six, twelve, twenty-four, forty-eight, or even ninety-six different geophones simultaneously. Each geophone-amplifier-galvanometer of the system is known as a channel. With the growth of the electronics and computer industries, new types of recording devices have been developed. A modern digital seismic recording system records the incoming vibrations on magnetic tape, which can later generate the seismogram display on electrostatic paper. As geophones detect the arrival of seismic waves, the signals can be digitized for further signal processing by computers.

Seismic Crew

In seismic exploration studies, the equipment that has been described is operated by a unit known as a seismic crew. Before any actual seismic work is done, the necessary permits must be obtained. Next, the land surface must be surveyed and a decision made on where the actual tests will be conducted. If explosives are to be used, shot holes must be drilled. Two or more "shooters" handle the explosives. If vibroseis is used, one to five technicians are required to operate the large vibrator truck. A ground crew consisting of a foreman and several crew members place the geophones in their proper positions and lay out and connect the seismic cables. Depending on the size of the operation, the ground crew may consist of from about six to twenty-four people. Many more may be added to the crew when operating in a rugged terrain.

Until the 1960's, seismic crews were accompanied by two or more trained seismologists who interpreted the data. Now the interpreters are found primarily at data-processing centers. Digital tapes of the field data are delivered to the centers on a regular basis. Cornell University's Consortium for Continental Reflection Profiling (COCORP) pioneered the use of multichannel seismic reflection profiling for exploring the continental lithosphere. COCORP has collected more than 11,000 kilometers of profiling at thirty sites in the United States.

DATA INTERPRETATION

Although the actual method of interpretation of the data is quite technical, the basic principle is rather simple. A signal (vibrations) from an explosive or another source is transmitted through the ground. When this seismic energy encounters discontinuities of various types in the earth, part of that energy is reflected back to the surface, where it is detected by geophones. The signal is then recorded. The time at which the signal was sent is known, and the time at which the reflected signal was received at the geophone can be determined. If the speed of the waves through the various types of rock is known, the depth to the reflecting boundary is a matter of velocity of the wave multiplied by the time of travel.

USE IN OIL EXPLORATION INDUSTRY

The goal of seismic reflection is to reveal as clearly as possible the subsurface structure of the earth. Because of the great versatility of this method, some 90 percent of all seismic studies done in the world today are seismic reflection studies. The greatest use of seismic reflection continues to be in the oil exploration industry. An important advantage of this technique is that reflections are obtained from boundaries at several different depths. The depth of any reflection can be determined if the velocity of waves in that particular type of rock is known or can be determined by another method. The exact depth to oil- or mineral-bearing strata can therefore be determined without the costly drilling of test wells or drilling cores. The extent of the rock strata in question can be determined by moving the survey equipment and redoing the test as many times as necessary.

David W. Maguire

FURTHER READING

Dohr, Gerhard. *Applied Geophysics.* New York: Halsted Press, 1981. A well-illustrated volume dealing with both basic principles of exploration geophysics and how these principles are actually applied to such areas as seismic, gravitational, magnetic, and geoelectrical methods. Some topics involve the use of trigonometry and differential calculus. Suitable for college-level students of physics or geophysics.

Gochioco, Lawrence. "Advances in Seismic Reflection Profiling for US Coal Exploration." In *Geophysics: The Leading Edge of Exploration,* 1991. Explains the use of seismic reflection principles in determining subterranean profiles. The technology is outdated, but it contains a good description of the theory which is still relevant.

Howell, Benjamin F. *Introduction to Geophysics.* New York: McGraw-Hill, 1959. A technical volume dealing extensively with various areas in the study of geophysics. Topics such as seismology and seismic waves, gravity, isostasy, tectonics and continental drift, and geomagnetism are covered. The reader should have a working knowledge of differential and integral calculus. Suitable for college students of physics or geophysics.

Judson, Sheldon, and Marvin E. Kauffman. *Physical Geology.* 8th ed. Englewood Cliffs, N.J.: Prentice-Hall, 1990. An excellent introductory text on the principles of physical geology. Suitable for the high school or college introductory geology course.

Meissner, Rolf, et al., eds. *Continental Lithosphere: Deep Seismic Reflections.* Washington, D.C.: American Geophysical Union, 1991. Meissner and his coauthors provide a brief account of seismic reflections and imaging through an examination of the continental lithosphere. Illustrations and maps reinforce the ideas and concepts described.

Nettleton, Lewis L. *Geophysical Prospecting for Oil.* New York: McGraw-Hill, 1940. A complete volume on the methods of applied exploration geophysics by one of the pioneers in the field. There are several chapters each on gravity studies, geomagnetism, seismic methods, and electrical methods. Trigonometry and calculus are used extensively in the book. Suitable for college-level students of physics and geophysics.

Pugin, Andre J. M., Susan E. Pullman, and James A. Hunter. "Multicomponent High-Resolution Seismic Reflection Profiling." *The Leading Edge* 28 (2009): 1248-1261. Provides overview of seismic profiling using three-component and nine-component records. Requires a strong understanding of geophysics. Provides many figures of data.

Robinson, Edwin S., and Cahit Coruh. *Basic Exploration Geophysics*. New York: John Wiley & Sons, 1988. A well-illustrated volume dealing with the science of geophysics both in theory and in applications. Contains well-developed chapters on seismic, gravity, and magnetic exploration techniques. The reader should have a working knowledge of algebra and trigonometry. Suitable for college students of geology, geophysics, or physics.

Scales, John Alan. *Theory of Seismic Imaging*. Berlin: Springer-Verlag, 1995. For the reader with a strong interest in earth sciences, Scales describes the theories and applications of the study of seismology and seismic imaging. Illustrations, charts, and maps.

Spencer, Edgar W. *Dynamics of the Earth*. New York: Thomas Y. Crowell, 1972. An introduction to the principles of physical geology, the book covers all aspects of geology, from introductory mineralogy through a study of the agents that shape the planet's surface. Concludes with chapters on global tectonics and geophysics that tend to be somewhat technical, requiring the use of algebra. Suitable for college-level geology students.

Stewart, S. A. "Vertical Exaggeration of Reflection Seismic Data in Geosciences Publications." *Marine & Petroleum Geology* 28 (2011): 959-965. Author comments on methodologies used in papers published between 2006 and 2010, claiming seismic reflection values were vertically exaggerated. Suggests precautions taken by readers and writers when interpreting seismic reflection data.

Tucker, R. H., et al. *Global Geophysics*. New York: Elsevier, 1970. A technical volume covering such topics as geodesy, seismology, and geomagnetism. The reader should have an understanding of trigonometry and differential calculus to comprehend the material as presented. Suitable for college students of physics or geophysics.

See also: Deep-Earth Drilling Projects; Earthquakes; Earth's Lithosphere; Engineering Geophysics; Ocean Drilling Program; Ocean-Floor Drilling Programs; Seismic Observatories; Seismic Tomography; Seismic Wave Studies; Seismometers.

SEISMIC TOMOGRAPHY

Seismic tomography is a technique for constructing a cross-sectional image of a slice of the earth from seismic data. Measurements are made of seismic energy that propagates through or reflects from subsurface geological materials. The measured time of travel and amplitude of this energy are used to infer geometry and physical properties of the geological materials, from which an image of the inside of the earth is generated.

PRINCIPAL TERMS

- **amplitude:** the maximum departure (height) of a wave from its average value
- **attenuation:** a reduction in amplitude or energy caused by the physical characteristics of the transmitting medium
- **imaging:** a computer method for constructing a picture of subsurface geology from seismic data
- **inversion:** the process of deriving from measured data a geological model that describes the subsurface and that is consistent with the measured data
- **lithology:** the description of rocks, such as rock type, mineral makeup, and fluid in rock pores
- **resolution:** the ability to separate two features that are very close together
- **seismic reflection method:** measurements made of the travel times and amplitudes of events attributed to seismic waves that have been reflected from interfaces where changes in seismic properties occur
- **seismometer:** an instrument used to record seismic energy; also known as a geophone or a seismic detector
- **travel time:** the time needed for seismic energy to travel from the source into the subsurface geology and arrive back at a seismometer

METHOD PIONEERED

Seismic tomography is a means of making an image of a slice of the earth using seismic data. "Tomography" is derived from a Greek word meaning "section" or "slice." Since the 1970's, seismic techniques have been used to create subsurface pictures. Although some methods that have been used in exploration geophysics for a number of years can be classified as tomographic, it is only since the mid-1980's that seismic tomography has been specifically developed for geophysical exploration and exploitation. Increased interest in seismic tomography in geophysical exploration and global seismology is the product of many factors, including the interaction between different scientific disciplines, along with advances in seismic field-data acquisition, imaging and inverse-problem theory, and computing speed.

The basic idea of tomography is to use data measured outside an object to infer values of physical properties inside the object. This method was pioneered by J. Radon in 1917. Radon showed that if data are collected all the way around an object, then the properties of the object can be calculated. In fact, Radon derived an analytical formula that relates the object's internal properties to the collected data.

METHOD APPLIED

Since the mid-1970's, the ideas of tomography have been applied to a number of fields of study. Applications of tomographic techniques are found in fields as diverse as electron microscopy and astronomical imaging. In medicine, the process of computed tomographic scanning ("CT" or "CAT" scanning) has developed rapidly since its inception in the early 1960's and its use has been integral in diagnostic medicine.

In tomography, geophysicists apply techniques similar to those of medicine to geophysical problems. Beginning in the early 1970's, interest and applications began rapidly expanding. Since then, a number of papers on the applications of seismic tomography have been presented. These range from attempts to estimate the internal velocity structure of the subsurface to formulations that provide a complete image of the subsurface geology. Since the late 1980's, tomographic reconstruction has become a standard technique in analyzing data between drill holes (crosshole analysis). Thus, while tomography is relatively new to exploration geophysics, it is a broad, powerful concept that has made a significant impact. Seismic tomography has led to many useful new applications as well as insightful reinterpretations of some existing imaging methods.

Seismic tomography is a type of "inverse problem"—that is, measurements are first made of some energy that has propagated through and reflected from within a medium (in this case, the earth). The received travel times and amplitudes of this energy are then used to infer the values of the medium through which it has transmitted. The parameters that are extracted are velocities and depths; therefore, a gross model of the earth's structure can be derived. Initially, this was considered the ultimate goal of seismic tomography. However, accurate measurements can be used effectively for other purposes, such as constructing an accurate depth image of the subsurface.

In CT scanning, an X-ray source and a number of X-ray detectors are used to acquire data around the human body. The X-ray source sends out X-rays, and receivers record the transmitted X-ray intensity. This intensity is related to the attenuation of the X-rays along their ray paths inside the object. In turn, the amount of attenuation is related to the density of the object encountered by the X-rays. Thus, a CT scan is an actual estimate of the density distribution within a body. CT scans can be done over various parts of the body, and these scans can be put together to form a three-dimensional image. This kind of image can show with great clarity the internal structure of the body or damage inside the body. Interpreting three-dimensional images of the body's interior is similar in many ways to interpreting the interior of the earth from three-dimensional seismic data.

SEISMIC SURVEYS

Tomographic geophysical exploration attempts to determine from seismic data the velocities with which sound propagates through a section of the earth, as well as other properties of the earth, such as density and compressibility. Classical tomography is typically associated with transmitted energy and requires a distribution of sources and receivers around the object to be imaged. In medical X-ray tomography, the source and receiver rotate all the way around the object being imaged. In contrast, by far the most pervasive seismic measurement is the surface reflection survey. Its measurements are made on just the upper boundary of the medium of interest.

Since the first seismic detectors (seismometers or geophones) were placed on the surface of the earth near the end of the nineteenth century, seismic waves have been used to locate remote objects. The first applications involved the location of earthquake epicenters in faraway regions. Efforts to locate heavy

artillery by seismic means during World War I later evolved into the first exploration methods for oil and gas. Imaging techniques in exploration seismics have continued to be improved ever since. At first, the process involved the interpretation of travel times of observed seismic pulses in terms of the depth and slope of reflecting surfaces. Beginning in the 1970's, complete seismic records were used, and imaging methods were developed that were based on sophisticated mathematical techniques.

In a seismic survey, geophysicists typically arrange seismic detectors along a straight line and then generate sound waves by vibrating the earth. Earthquakes release the large amounts of energy needed to probe the deep layers (mantle and core) of the earth. Other methods can produce seismic waves that can be focused on the geologic features closer to the earth's surface. These waves can be generated by explosions, such as a charge of dynamite, or by dropping a weight or pounding the ground with a sledgehammer. To eliminate environmental risks associated with the use of explosives, a system called "vibroseis" is used. In this system, a huge vibrator mounted on a special truck repeatedly strikes the earth to produce sound waves. A seismograph records how long it takes the sound waves to travel to a rock layer, reflect, and return to the surface. The recorded data display the amplitudes of the reflected sound waves as a function of travel time. Such a graphic record is called a "seismogram." The equipment is then moved a short distance along the line, and the process is repeated. This procedure is known as the seismic reflection profiling method.

METHOD ADAPTED TO SEISMIC DATA

Since seismic waves traveling in the earth readily spread, refract, reflect, and diffract, classical tomographic methods must be adapted to produce realistic seismic pictures, and effective software and interactive graphics are required to process the seismic data into a relevant image. It has taken some time for tomographic concepts to spread to seismic imaging, for appropriate data to be acquired, and for effective processing and interpretation techniques to be developed. Using a variety of computer programs, seismograms are processed to yield seismic sections that represent the earth's reflectivity in time. Geologists, though, would really like to have a lithologic picture illustrating such features as rock velocity, seismic wave attenuation, and elastic constants of the rocks as a function of depth.

Reflectivity is a property associated with interfaces between rocks; a rock sample held in one's hand does not have an intrinsic reflectivity. Therefore, reflectivity is not an actual rock property, and it must be converted to some parameter that really describes the rock. In addition, seismic time data must be converted to depth measurements in the imaging process. Consequently, conventional reflection sections are being greatly improved by the use of tomographic techniques to produce subsurface images as a function of depth and to estimate rock properties from some of the images.

The basic procedure of seismic tomography is an extension of the notion of transmission tomography. This process can also be classified as a generalized linear inversion of travel times. The first part of the procedure is to locate reflected events on the raw seismograms and then associate these events with the structure of a proposed or guessed geological model. Next, the laws of physics are applied to trace ray paths of seismic waves through the proposed model from given seismic sources down to a particular reflector and back to the seismic detectors. The ray-traced travel times are then compared through the model with the travel times recorded on the seismogram, and the geological model is updated to make the ray tracing consistent with the observed data. Seismic tomography is distinct from classical tomography in that only reflected waves are used, and the source-detector coverage of the object or area of interest is far from complete. These aspects of the problem create difficulties, but the tomographic velocity determination is still very useful, especially in areas of significant lateral velocity variations. By including all available data in the tomographic process, as well as any other available geophysical data, the resolution and certainty of subsurface images can be greatly improved. In 2009 the Federal Lands Highway Program of the U.S. Department of Transportation used cross-borehole seismic tomography to investigate an active sinkhole causing damage to property in a residential neighborhood of central Florida. Personnel were able to obtain images that showed the "throat" of the sinkhole at a depth of 24 meters.

RESOURCE EXPLORATION

Various survey geometries and tomographic constructions are used to assist the solving of geophysical problems. Geophysicists can use the velocities of seismic waves recorded by a seismograph to determine the depth and structure of many rock formations, since the velocity varies according to the physical properties of the rock through which the wave travels. In addition, seismic waves change in amplitude when they are reflected from rocks that contain gas and other fluids. Sometimes the fine details of seismic records can be used to infer the type of rocks (lithology) in the subsurface. Some tomographic studies have used subsurface velocities determined from the inversion of seismic travel times to construct geological cross sections of the geology inside the earth, while other studies have used reflection amplitudes for the same purpose. Based on the characteristic geometries and amplitudes for oil and gas traps and for mineral ore deposits, these tomographic images are used to predict where oil, natural gas, coal, and other resources such as groundwater and mineral deposits are most likely to be found in

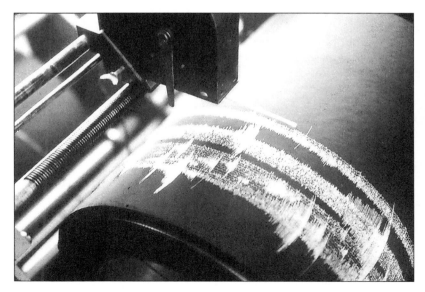

A smoke-drum seismograph record of seismic acitivity on Mount Kilauea's east rift, on the island of Hawaii. (U.S. Geological Survey)

the subsurface. The tomographic cross sections constructed from seismic data make the odds of finding such resources much greater than would be the case if exploration were based on mere random drilling.

In seismic tomography, various source and detector geometries are used, such as drill-hole-to-drill-hole, surface-to-drill-hole, and surface-to-surface. The greater the degree of angular coverage around the rock mass, the greater the reliability of the constructed tomographic image. By making numerous measurements from various source-detector positions and analyzing the travel times and amplitudes from a number of source-detector locations, the velocity and attenuation of the intervening rock can be calculated from the recorded energy—energy that is either reflected or transmitted. This technique has found applications not only in locating subsurface natural resources but also in areas such as determining the location of nuclear-waste dumps and the monitoring of stream floods, which are used to help produce hydrocarbons from a reservoir.

EXPLOITATION OF HYDROCARBON RESERVOIRS

In many hydrocarbon development areas, adjacent drill holes may be available. In these situations, it is desirable to have a very high-resolution description of the rock mass between the drill holes. For this purpose, it is often effective to use crosshole tomography. A seismic source, such as dynamite caps or downhole air guns, is placed in one drill hole, and appropriate detectors are placed in an adjacent drill hole. The source is fired, and the resulting seismic energy propagates through the rock and is detected in the other drill hole. The travel times and amplitudes of seismic waves that have been reflected or transmitted through the rock mass between the drill holes are recorded. The source and detectors are then moved to another position, and the process is repeated. This procedure is continued until the region of interest is adequately covered by the propagating energy. Seismic crosshole tomography has been used for a number of applications, including mineral exploration in mines, fault detection in coal seams, stress monitoring in coal mines, delineation of the sides of a salt dome, investigation of dams, and mapping of dinosaur-bone deposits. The resolution of crosshole tomography is typically better than for surface reflection tomography.

The broad objective of geophysics is to produce images that represent the subsurface geology as accurately as possible, and tomography-based imaging algorithms provide seismic depth sections that are consistent with drill-hole data in regions of resource exploration and exploitation. Integrating drill-hole and surface seismic reflection data in a tomographic approach can provide a better, less ambiguous subsurface picture. This correlation holds considerable promise to increase knowledge of the subsurface. The resulting seismic depth sections assist in interpreting the structure (geometry), stratigraphy (depositional environment), and lithology (rock and fluid types) of potential and established hydrocarbon reservoirs and mineral deposits.

The geologic detail needed to develop most hydrocarbon reservoirs substantially exceeds the detail required to find them. For effective planning and drilling, a complete understanding of the lateral extent, thickness, and depth of the reservoir is absolutely essential. This can be found only with detailed seismic interpretation of three-dimensional seismic reflection surveys integrated with drill-hole data. A common practice in three-dimensional seismic reflection surveying is to place the seismic detectors at equal intervals and collect data from a grid of lines covering the area of interest. Based upon integrated seismic tomographic imaging of the drill-hole and seismic reflection data, more wells are drilled in the area, and the three-dimensional data volume evolves into a continuously utilized and updated management tool that influences reservoir planning and evaluation for years after the seismic data were originally acquired and imaged.

GLOBAL SEISMOLOGY

Imaging in global seismology (whole-earth geophysics) has lagged behind the developments in exploration geophysics for several reasons. In contrast to artificial sources, earthquakes are uncontrolled, badly placed sources of seismic-wave energy, and the earth is only sparsely covered with seismometers. In addition, instrument responses were for a long time widely different, and recording was not done digitally. Thus, seismologists were faced with the paradox that the available data, despite the enormous volume, often contained crucial gaps.

In global seismology, the whole three-dimensional earth is considered as an object to be imaged. Seismic energy generated by earthquakes travels through the earth and is recorded by a distribution of seismic detectors, such as the World Wide Standardized Seismographic Network. By examining the travel times of the propagating energy for a number of earthquakes and stations, researchers can construct a model representing the velocity structure inside the earth. Likewise, by measuring the shapes and sizes of the amplitudes of the recorded energy, they can estimate a seismic attenuation model of the earth. Based upon these models, a three-dimensional tomographic image of the earth can be constructed. In 2004, seismic tomography revealed large pieces of subducted lithosphere "floating" in the upper mantle.

Global seismic tomography has been used to image convective flow within the mantle. Changes in seismic wave velocity have been used to identify sinking (cold) and rising (warm) mantle materials. Estimates can be made of the variation of seismic velocities inside the earth using seismic tomography, and these variations in turn depend upon the variations in composition, structure, and temperature of the materials inside. Mantle regions that are relatively hot have lower velocities compared to cooler regions at the same depth, because the higher temperatures reduce the values of the elastic constants of the mantle material. Seismic tomography supports a hybrid convection theory that postulates the existence of shallow, small-scale convection currents as well as of deep, large-scale convection currents in the mantle. Convection of the mantle is the primary driving force of plate tectonics. In general, tomographic results show a strong correlation at shallow depths with present plate boundaries, such as fast movement under cold, old shields and in subduction zones, and slow movement under hot, spreading ridges and other volcanically active areas. Three-dimensional images of the earth's interior reconstructed with seismic tomographic procedures have had a major effect on the understanding of the structure and dynamics of the earth.

Alvin K. Benson

FURTHER READING

Bording, R. P., et al. "Applications of Seismic Travel-Time Tomography." *Geophysics* 90 (1987): 285-303. Discusses the basic principles of tomography and how they can be applied to seismic data to create a velocity model of the earth from recorded travel times.

Iyer, H. M., et al., eds. *Seismic Tomography: Theory and Practice.* London: Chapman and Hall, 1993. Various essays explore all aspects of seismic tomography and the inversion of seismic data.

Lines, L. R. "Cross-Borehole Seismology." *Geotimes* 40 (January 1995): 11. Discusses applications of seismic tomography to the shallow subsurface.

Lo, Tien-When, and Philip L. Inderwiesen. *Fundamentals of Seismic Tomography.* Tulsa, Okla.: Society of Exploration Geophysicists, 1994. Provides a fine introduction to seismic tomography for students with little knowledge on the subject.

Nolet, Guust. *A Breviary of Seismic Tomography.* New York: Cambridge University Press, 2008. A text accessible to undergraduate students, this book provides an introduction to seismic tomography, wave propagation theory, travel-time tomography and more.

Nolet, Guust, ed. *Seismic Tomography.* Boston: D. Reidel, 1987. Describes the methods and reliability of seismic tomography. Contains many qualitative discussions that will be useful to the general reader as well as more technical discussions for those with the appropriate background. Primarily discusses applications of tomography to whole-earth geophysics, with some discussion of applications to exploration geophysics.

Poupinet, Georges. "Seismic Tomography." *Endeavour* 14, no. 2 (1990): 52. Good description of seismic tomography as it is applied to the study of the deep structure of the earth by integrated analysis of seismic wave patterns generated from earthquakes.

Russell, B. H. *Introduction to Seismic Inversion Methods.* Tulsa, Okla.: Society of Exploration Geophysicists, 1988. Discusses techniques used for the inversion of seismic data, including principles of seismic tomography. Good illustrations.

Stewart, R. R. *Exploration Seismic Tomography.* Tulsa, Okla.: Society of Exploration Geophysicists, 1991. Recounts the historical development of tomography. Reviews the fundamentals of seismic tomographic techniques and discusses applications of seismic tomography, mainly to exploration geophysics.

Tarbuck, Edward J., Frederick K. Lutgens, and Dennis Tasa. *Earth: An Introduction to Physical Geology.* 10th ed. Upper Saddle River, N.J.: Prentice

Hall, 2010. This college text provides a clear picture of the earth's systems and processes that is suitable for the high school or college reader. It has excellent illustrations and graphics. Bibliography and index.

Valentine, Andrew P., and John H. Woodhouse. "Reducing Errors in Seismic Tomography: Combined Inversion for Sources and Structure." *Geophysical Journal International* 180 (2009): 847-857. Presents new solutions to problems in seismic tomography. A background in mathematics is necessary. The paper focuses on full-waveform inversion, but the authors suggest solutions may be useful in other areas.

See also: Creep; Cross-Borehole Seismology; Discontinuities; Earth's Mantle; Elastic Waves; Experimental Rock Deformation; Environmental Chemistry; Faults: Normal; Faults: Strike-Slip; Faults: Thrust; Faults: Transform; San Andreas Fault; Seismic Observatories; Seismic Reflection Profiling; Seismic Wave Studies; Seismometers; Stress and Strain.

SEISMIC WAVE STUDIES

To understand earthquakes, scientists must study the seismic waves that are produced during such events. There are a number of different types of seismic waves, each of which acts differently during a seismic event. Body waves emanate through Earth's interior outward to the crust, while surface waves occur only on Earth's surface. Scientists rely on seismic waves to provide a glimpse into the planet's inner workings. Seismologists also use seismic waves as indicators of the epicenters and hypocenters of earthquakes.

PRINCIPAL TERMS

- **epicenter:** the surface-level geographic point located directly above an earthquake's hypocenter
- **hypocenter:** the point of origin of an earthquake
- **Love waves:** surface seismic waves that occur in a side-to-side motion
- **primary wave (P wave):** a longitudinal, compression-body seismic wave that can move through rock
- **Rayleigh waves:** surface seismic waves that occur in a circular, rolling fashion
- **secondary wave (S wave):** a slower-moving body wave that has a rippling, shear-particle motion
- **wave propagation path:** the directions in which seismic waves travel during an earthquake

BASIC PRINCIPLES

Seismologists study earthquakes and seismic activity, primarily seismic waves. Seismic waves are energy waves released when massive rock formations (plates) far under Earth's surface break away from one another. Using seismographs (devices that measure the intensity of seismic waves), seismologists attempt to record the severity of a quake and the distance seismic waves have traveled from the event's surface-level, the geographical point of origin (the epicenter), and the subterranean point of origin directly beneath it (the hypocenter).

A number of different types of seismic waves occur during an earthquake. Body waves travel to the surface from Earth's interior. The first of these waves is known as the primary wave (P wave), which radiates quickly through rock and liquid. The P wave is followed by the secondary wave (S wave), which moves more slowly to the surface. The time elapsed between the two types of body waves helps seismologists determine the distance traveled from the epicenter to the locations experiencing the quake. Meanwhile, surface waves are seismic waves that occur only at the crust level; they do not radiate from the interior.

Arriving after body waves are the Love and Rayleigh surface waves, which are the primary causes of an earthquake's destructiveness.

BACKGROUND AND HISTORY

Before the eighteenth century, earthquakes were not given a great deal of scientific attention, although a wide range of theories and myths were offered through the millennia. In the fourth century B.C.E., Aristotle speculated that seismic events were caused by Earth's interior winds, which whipped upward so strongly that they rattled the planet's surface. Five centuries later, Chinese philosopher Zhang Heng invented the first known seismoscope, a device that resembled a large jar with eight decorative dragon heads facing in different directions. Each head held a ball directly above a decorative toad at the base of the device. When an earthquake occurred, the shaking would cause a ball to drop, indicating the direction from which the quake came.

Zhang Heng's device was among only a few attempts to understand seismic waves in a scientific framework for more than one thousand years. This trend changed after 1755, when a massive earthquake struck Lisbon, Portugal, killing seventy thousand people. In addition to assessing the damages, the Marquis of Lisbon sought information from his people about the duration of the quake and aftershocks. It is believed that the Lisbon earthquake fostered the start of the field of seismology.

In the early nineteenth century, scientists offered theories about the results of earthquake, including elasticity (the point at which a surface-level wave passes and the surface returns to normal) and seismic waves. In 1830, French seismologist and mathematician Siméon Denis Poisson identified the P and S waves as the only types of seismic waves that could pass through rock. Half a century later, British physicist John William Strutt (who later became the third Baron Rayleigh) mathematically calculated the presence in an earthquake of what would be named

Rayleigh surface waves. In 1911, another British scientist and mathematician, Augustus Edward Hough Love, calculated the second type of surface wave, which would be known as Love waves.

PLATE TECTONICS

Below Earth's surface lies the lithosphere, a rigid outer shell consisting of massive rock plates known as tectonic plates. These plates are in constant motion, floating on the planet's upper mantle in a puzzle-like formation that spans the globe. Frequently, these plates come into contact with one another, locking along their jagged edges. However, the force moving these plates eventually causes them to separate. When they do separate, the plates release large quantities of energy known as seismic waves. These waves are primary causes of earthquakes.

SEISMIC WAVES

The first form of seismic wave is the body wave, known as such because it travels through Earth's body, including rock. The two general types of body waves are the P and S waves. P waves move in a longitudinal direction and are compressional (having an elastic, rather than rolling, effect on the surface). P waves are also the fastest of the seismic waves, traveling at seven times the speed of sound. S waves, meanwhile, move only through water and the air, at a much slower pace than P waves. These seismic waves move in a ripple motion, similar to the motion of a spring that has experienced a sudden sideward deflection. S waves are also known as shear waves: Rather than change the volume of the material through which they travel, they shear it, vibrating the ground perpendicular to the direction in which the wave is traveling. Body waves are important to seismologists because the difference between the arrival of the P and S waves helps scientists determine the distance to an earthquake's hypocenter.

The second form of seismic wave is the surface wave. These waves do not radiate from the interior but rather along the surface during an earthquake. The first of these types of seismic waves is the Rayleigh wave. Rayleigh waves are transverse, which means that the particle displacement that occurs with each wave is perpendicular to the direction in which the wave is traveling. As Rayleigh waves pass through a solid object, the particles move in a counter-clockwise, elliptical path.

Love waves, the second type of surface waves, are also transverse. Whereas Rayleigh waves move in a motion similar to rolling ocean waves, Love waves move in a side-to-side motion. Although Love and Rayleigh waves move more slowly than body waves, they are also major sources of the destruction that occurs in an earthquake, particularly in light of the contortions caused by the combined rolling and back-and-forth movements.

SEISMOGRAPHS AND SEISMIC WAVE DETECTION

Key to the study of seismic waves is the seismograph. Seismographs are sensitive pieces of equipment that detect seismic waves, operating in much the same way as Zhang Heng's seismoscope. The principle of the seismograph is simple: A motorized wheel of paper scrolls under an ink-tipped needle. The device is attached to bedrock to prevent vibrational pollution (such as passing trains or traffic). When seismic activity occurs, the tape on the seismograph shows the incoming waves as they are received. The overall shape or profile of each wave as it appears on the seismograph is known as the wave coda.

On a seismograph, seismic waves radiating from a rupture (the point at which the two rock plates separate, causing an earthquake) frequently appear scattered, showing trains of differently shaped codas. The prevailing view of the causes of this scattering effect is that the variable shapes seen in the wave train are caused by Earth's heterogeneities (materials of different density and composition through which the waves travel). Research indicates that there may be other factors contributing to the scattering of seismic waves, including the nonlinear movement in which some seismic waves travel, which demonstrates varying degrees of elasticity and causes differently formed seismic waves.

Seismographs provide a profile of an earthquake, detecting P and S waves that otherwise would not have been detected. In seismically active zones, such as the San Andreas fault that spans the western part of California, there are entire networks of seismograph stations. The small town of Parkfield, near the San Joaquin Valley in central California, has more than one dozen seismograph stations along the fault in that community alone. These seismographs cast a wide and complex net across the fault. Any activity along this line is quickly detected and then documented and shared with other members of the network and the scientific community. Seismographs do not predict earthquakes, but they help scientists understand their nature.

Seismographs are not limited to studying the wave trains of earthquakes. Scientists also use data collected from seismic waves to study Earth's subterranean systems and processes. In the greater Lake Superior region of Canada, for example, seismologists studied Earth's mantle and lithosphere beneath the province of Ontario. Researchers utilized the extensive network of seismographs throughout this region to gather data on the velocities of the region's S waves. Using the seismograph network's data on the area's shear waves, scientists were able to determine the relative thickness of the lithosphere in the region and glean a better understanding of the mantle below.

SEISMIC WAVE STUDIES AND EARTHQUAKES

The study of seismic waves is a critical facet of analyzing earthquakes and tracing seismic activity back to its source or sources. However, scientists also use seismic waves to study the many components and systems under Earth's crust. For example, scientists have used data obtained from seismic waves as part of their research into the planet's core. In this case, the speed and scattering of seismic waves help researchers glean more information about the heterogeneities of Earth's inner core. Such data are invaluable, considering that it is virtually impossible to directly study the planet's inner regions. To analyze these data, scientists use mathematical equations and computer models.

Although each wave in each region (and from each seismic event) is different, scientists are able to piece together such data using mathematical equations known as algorithms. These formulae establish a process or framework for analysis of a specific concept. With the framework in place, data are simply entered as one of the variables in the equation.

In 2009, for example, mathematicians introduced the Bayesian single event location (BSEL) algorithm, which is used to estimate the distance to the hypocenter of a seismic event. In most cases, the areas of hypocenter depth and the event origin time are addressed with great inconsistency. The BSEL takes into account the physical characteristics of the area

in and around a hypocenter and adds those characteristics to a nonlinear equation, giving the analysis of a seismic event more dimensions. When data from seismic events, such as tremors in Montana and in the Sea of Japan, were input into this mathematical formula, scientists reported a greater degree of accuracy in determining both the origin time and the depth of those seismic events.

Much in the same way they use mathematical equations and algorithms to collate seismic data, seismologists and other scientists use computer models to help them study Earth's seismic activity and inner workings. In some cases, the data on a given seismic event are so extensive that seismologists are able to create a complex, multidimensional profile of an earthquake. Using these computer models, scientists may test theories and examine certain aspects of seismic activity without the need to wait for another event.

In 2007, for example, a significant earthquake took place in the Ishikawa prefecture of Japan, about 300 miles west of Tokyo. The Noto-hant earthquake killed one person, injured nearly three hundred, and damaged more than six hundred homes in the area. Seismologists, attempting to study the origins of this earthquake, created a three-dimensional computer model of seismic wave velocities from that event. Using data from hundreds of other earthquakes in the area and in the Sea of Japan, these scientists determined that, based on the relatively low wave velocities in these events, the ruptures that caused these quakes (including the Noto-hant quake) were caused by fluid flow into the fault lines rather than by dehydrated ruptures. This example demonstrates that although direct study of deep seismic activity remains beyond scientific reach, computer modeling using the study of seismic wave activity is a viable alternative.

RELEVANT NETWORKS AND ORGANIZATIONS

Although earthquake prediction is impossible, governments have a stake in investing the monies needed to study seismic activity. The leading U.S. government agency involved in the study of seismic waves is the U.S. Geological Survey (USGS). This organization is highly concerned with studying earthquakes and with preventing significant earthquake-related damage. The USGS created an earthquake hazards program, which focuses on the seismic events as they happen.

For many years, the USGS has focused on the initial shock (called the moment magnitude) of an earthquake. However, the USGS and other related organizations have expanded their areas of concern to include the smaller (yet still destructive) seismic waves that precede and follow major events. The purpose of this broadening approach is so that governments can work with emergency officials to formulate updated public safety responses to major earthquakes.

Because seismic studies depend heavily on filtering out vibrational pollution, many seismographs and other sensitive equipment are located in universities. For example, the University of Utah has established an extensive network of seismograph stations that monitors waves radiating throughout the region, which includes Wyoming, Montana, Utah, Idaho, Nevada, and Arizona. The university's network also features the seismically active Grand Teton mountain range and Yellowstone National Park. Similar networks have been established through San Diego State University and through universities in Japan.

A growing number of energy companies are turning to seismic wave study to locate potential oil and gas deposits for extraction. Failure to accurately conduct exploratory drilling can result in, at best, the permanent closure of pores in which those deposits exist, should seismic waves trigger a collapse. There is also the potential for loss of lives and for the destruction of drilling equipment. In this regard, it is critical that energy industry businesses research and pay careful attention to the seismic profiles of potential drilling and extraction operations.

IMPLICATIONS AND FUTURE PROSPECTS

The field of seismology is relatively young. It has grown out of necessity, marking its first major steps forward in association with the Lisbon earthquake of 1755. The study of seismic waves in particular has seen a great deal of growth, as scientists have come to understand the important role they play not only in presenting a profile of an earthquake but also in providing insights into the nature of Earth's core and mantle.

The study of seismic wave is likely to continue to evolve, driven by improvements to sensory equipment, by the introduction of computer modeling software, and by the spread of Internet access. Students and scientists alike can observe in real time seismic waves as they are detected in stations around the world. Seismic wave studies today rely heavily on the quick global dissemination of information via the Internet. It is likely that scientists will continue to develop a better understanding of seismic waves and their implications.

Michael P. Auerbach

FURTHER READING

Ben-Menachem, Ari, and Sarva Jit Singh. *Seismic Waves and Sources.* Mineola, N.Y.: Dover, 1998. Presents data on seismic waves from nearly two centuries of seismic events. Discusses the basic elements of seismic wave studies and offers a detailed analysis of how seismic waves form and radiate throughout the globe.

Chapman, Chris. *Fundamentals of Seismic Wave Propagation.* New York: Cambridge University Press, 2010. This book presents an analysis of the elasticity of seismic waves as they radiate from a hypocenter. The author discusses a number of theories on wave elasticity and proposes new ideas based on the latest in three-dimensional computer models and mathematical calculations.

DiGiacomo, Domenico, et al. "Suitability of Rapid Energy Magnitude Determinations for Emergency Response Purposes." *Geophysical Journal International* 180, no. 1 (2010): 361-374. This article suggests that government agencies may benefit from the study not only of initial seismic waves during an earthquake but also of the event's subsequent waves. A comprehensive profile created in this arena, the authors argue, can help officials formulate a viable emergency management plan.

Pujol, Jose. *Elastic Wave Propagation and Generation in Seismology.* New York: Cambridge University Press, 2003. The author applies mathematical theory and approaches to understanding the generation and radiation of seismic waves. The book serves as a guide for students of seismology, teaching the fundamentals of seismology and the problems seismology presents for scientists.

Razin, A. "Excitation of Rayleigh and Stoneley Surface Acoustic Waves by Distributed Seismic Sources." *Radiophysics and Quantum Electronics* 53, no. 2 (2010): 82-99. The author of this study discusses surface seismic waves, including those detected on an acoustic level (such as Stoneley waves). The author describes some of the surface-level elements that can affect the power of surface seismic waves.

Wu, Jiedi, John A. Hole, and J. Arthur Snoke. "Fault Zone Structure at Depth from Differential Dispersion of Seismic Guide Waves: Evidence for a Deep Waveguide on the San Andreas Fault." *Geophysical Journal International* 182, no. 1 (2010): 343-354. In this article, the authors argue that seismic wave studies may be used to illustrate a fault's structure. The authors describe a method that includes the study of fault seismic waves traveling from the fault's depth and along its structure.

Yoon, Choonhan, et al. "Web-Based Simulating System for Modeling Earthquake Seismic Wavefields on the Grid." *Computers and Geosciences* 34, no. 12 (2008): 1936-1946. The authors discuss a web-based modeling system they developed that creates theoretical seismic waveforms. The examples this Internet model creates help students understand how seismic waves form and radiate.

See also: Continental Drift; Cross-Borehole Seismology; Deep-Focus Earthquakes; Earthquake Distribution; Earthquake Engineering; Earthquake Hazards; Earthquake Locating; Earthquake Magnitudes and Intensities; Earthquake Prediction; Earthquakes; Earth's Interior Structure; Earth's Mantle; Elastic Waves; Faults: Normal; Faults: Strike-Slip; Faults: Thrust; Faults: Transform; Geodynamics; Lithospheric Plates; Mantle Dynamics and Convection; Mountain Building; Notable Earthquakes; Plate Motions; Plate Tectonics; San Andreas Fault; Seismic Observatories; Seismic Reflection Profiling; Seismic Tomography; Seismometers; Slow Earthquakes; Soil Liquefaction; Stress and Strain; Subduction and Orogeny; Tectonic Plate Margins; Tsunamis and Earthquakes; Volcanism.

SEISMOMETERS

A seismograph is a device that detects, measures, and records the ground motion at a point. The sensor that detects and, in part, measures the motion is called a seismometer. The recorded ground motion, a seismogram, is the output of a seismograph.

PRINCIPAL TERMS

- **earthquake:** a sudden release of strain energy in a fault zone as a result of violent motion of a part of the earth along the fault
- **natural period of vibration:** the period at which structures undergo oscillation if they are set in motion by an impulse
- **particle motion:** the motion of a particle in a material volume when it experiences the passage of seismic waves
- **seismic waves:** the propagation of a disturbance in the form of energy release in a solid medium; the released energy propagates in the solid from one region to another by setting individual particles in motion in a particular direction
- **seismogram:** the recorded output of a seismograph
- **seismograph:** a device that detects, measures, and records the ground motion
- **wave velocity:** the velocity at which a particular seismic wave travels through a medium

HISTORY AND DEVELOPMENT

According to the historical records, the first seismometer was invented by Chinese astronomer and mathematician Zhang Heng (78-139 C.E.), during the Han Dynasty. The records indicate that it was a bronze vessel containing a suspended pendulum. The pendulum was connected to an eight-spoked wheel, and each spoke terminated in the mouth of one of eight externally mounted dragon heads with movable jaws. Each mouth contained a bronze ball, and eight open-mouthed frogs were located around the base. During an earthquake, the ground motion would displace the pendulum laterally, ejecting one of the balls into a frog's mouth. The ejected balls would give some idea of the direction of the traveling waves and the source of the earthquake.

The first truly precise seismometer was developed by an English mining engineer, John Milne, in the late 1800's. This instrument was improved by J. J. Shaw in the early 1900's, when the Milne-Shaw seismograph was introduced. In the United States,

the first seismographs were installed in 1887 at the University of California, Berkeley. The 1906 San Francisco earthquake was the first large event in the United States to be recorded.

In the 1960's, largely as a result of the development of the atomic bomb and concern over the Cold War, the improvement and development of seismographs took a large leap forward. There was a national security need to be able to discriminate between an underground nuclear explosion and natural events. Such effort led to the construction and the deployment of 120 seismic stations in sixty countries, called the World Wide Standardized Seismograph Network (WWSSN). This period also marks a turning point in bringing the science of observational seismology to the forefront of the physical sciences. At the same time, countries such as the Soviet Union, France, and Canada modernized their earthquake observation systems.

MECHANICS

Currently, many types of seismographs are available. Most of them incorporate similar physical principles, utilizing some sort of spring or pendulum to detect and measure the ground motion. The exception is a strain seismograph, invented by Hugo Benioff, which uses a long horizontal bar (20-30 meters long) to measure the ground deformation between two points.

A simple seismometer can be considered as a pendulum attached to a rigid frame, which is anchored to a horizontal ground surface. When an earthquake occurs, seismic waves are radiated from the earthquake source in all directions through the earth. For a sufficiently strong earthquake, the seismometer site experiences ground vibration. Assuming that there is no slippage between the ground and the rigid frame of the seismometer, the frame experiences the same motion as does the ground. If the pendulum could stay motionless, which is the ideal case, the relative motion of the frame with respect to the pendulum would be the true ground motion. Yet, that is not the case; the pendulum undergoes motion, and because of its resistance to motion (inertia), it tends to lag the motion

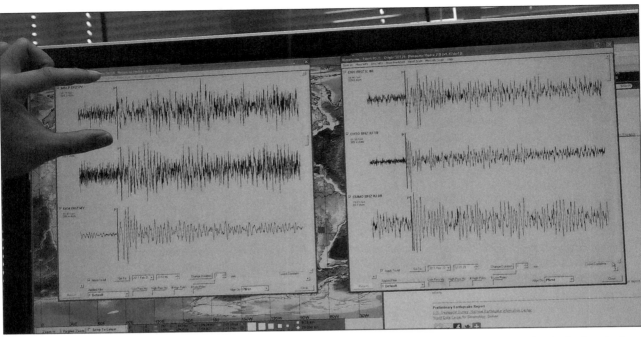

Philippine seismologist Winchelle Sevilla points to a seismograph of an earthquake which struck New Zealand earlier in the day, during a press briefing at the Philippine volcanology office in Manila on February 22, 2011. At least 65 people died in the 6.3 magnitude earthquake that hit the New Zealand town of Christchurch. (AFP/Getty Images)

of the frame (ground). This lag results in a complex differential motion between the pendulum and the ground.

Pendulums and springs have the characteristic that, if they are set in motion, they oscillate with period independent of the amplitude of motion as long as the amplitude does not become extreme. This characteristic period or frequency (frequency is the inverse of period) is called the free-period or the natural frequency of a pendulum or spring, respectively. A seismometer pendulum or spring can be designed to have a high- or a low-period sensitivity. The relative frequency content of the seismic waves with respect to the natural frequency of the pendulum or spring-mass determines in part the nature of the recorded seismograms. The deflection of a pendulum that experiences ground motion with frequencies much higher than its natural frequency is proportional to the ground displacement. The deflection, however, is proportional to the ground acceleration if the frequency content of the ground motion is much lower than the natural frequency of the pendulum, and the

seismograph is called an accelograph. Finally, if the pendulum has a natural frequency close to the frequency content of the ground motion, the deflection is proportional to the ground velocity. Nevertheless, displacement, velocity, and acceleration are mathematically related. For example, ground velocity and displacement could both be determined from an accelogram. Early seismometers amplified small motions with mechanical linkages or optical levers with a mirror to reflect light onto recording paper. Modern instruments use electronic amplification.

The relative motion of the pendulum with respect to the ground must be magnified, allowing very small wave amplitudes to be distinguished. The most sensitive seismometers, at a quiet observatory, can detect ground displacement of a thousandth of a millionth of a meter, which is on the order of atomic size. The record of such ground motion might show an amplitude of 1 centimeter, a magnification of 10 million. The ratio of the largest to the smallest amplitudes which are undistorted is called the dynamic range of the seismograph. Conventional visible seismographs (those that produce records directly on paper or film) have a dynamic range of approximately one hundred to three hundred.

COMPLICATIONS AND LIMITATIONS

Unfortunately, there is an unwanted complication in detecting useful seismic signals. The ground is always in motion. This motion, microseismicity, is caused by natural sources such as wind, storms, and human activities. Seismographs should be designed so that the recorded seismograms are least affected by the unwanted microseismic noise—to get an optimum signal-to-noise ratio. Most microseismic noise has a period in the range of 5-8 seconds. These noises are effectively avoided in the commonly used short-period (1 hertz) seismographs. The moon is the ideal place for readings from advanced seismographs because of the absence of wind, ocean waves, and human-made noise. The lunar seismographs left behind by the Apollo mission astronauts can detect the seismic waves generated by a 1-kilogram meteor striking anywhere on the moon's surface.

Borehole seismographs have been developed which can operate within deep boreholes. The advantage of a downhole recording is that the noise level at depth is much lower than at the surface, and, with certain geometrical arrangements, the recording of seismic waves that travel desired paths can be made. Seismologists have employed such data-acquisition techniques in conjunction with mathematical models similar to those used in medical tomography to construct a three-dimensional picture of the surveyed volume of the earth.

Using mechanical vibration as the only means of measuring and recording motion, although relatively simple, has a few drawbacks. Such instruments have limited magnification and sensitivity, and the recording is mechanical, with inherent friction. In older seismographs, very large pendulum masses were used to reduce the friction—for example, 1,000 kilograms in many Wiechert seismographs. Some Wiechert seismographs built with a pendulum mass of roughly 20 tons were installed in the beginning of the century in several places in central Europe. The older seismographs all use mechanical methods for transferring the motion of the pendulum to the recording pen. The recording pen can have several variations; the most common are a stylus on a rotating smoked paper drum and a reflected light beam onto a photographic paper. The highest measurable frequency of mechanical recording depends upon the speed of the recording medium relative to the stylus pen or light spot. The typical recording speed ranges from 0.1 millimeter per second for frequencies at roughly 0.05 hertz, up to 10 millimeters per second for frequencies as high as 20-50 hertz. The dynamic range of these systems is approximately one hundred to three hundred.

INTRODUCTION OF ELECTROMAGNETICS

In 1906, Russian physicist Boris Golitsyn elegantly utilized the principles of electromagnetics to translate mechanical motion into electrical voltage. It is known that if a conducting loop (a coil) crosses lines of magnetic flux, there will be an induced electromotive force that will generate an electric current proportional to the velocity at which the coil crosses the magnetic field. This idea has had a significant effect on the later development of seismographs. A system of this kind consists of a coil of wires and a magnet: One of the components is fixed to the rigid frame and undergoes the same motion as the ground, while the other element is suspended by a spring from the frame. The relative motion of the coil with respect to the magnet will produce electromotive force between them, with voltage proportional to the velocity of motion. The current generated in the coil is sent through a sensitive galvanometer (an apparatus to measure electrical current) that makes a continuous record on photographic paper, a mirror, or a hot stylus. In most cases, the coil is fixed to the frame and the magnet is suspended, but there is a wide variety of possible arrangements. The sensitivity of an electromagnetic seismometer depends upon the magnet's strength, the number of turns of wire in the coil, and the geometry of the crossing of the magnetic flux by the coil. Modern seismometers have become quite small as magnetic materials with greater strengths have become available.

The introduction of the electromagnetic principles in seismic instrumentation seismographs has been a great step forward in the development of seismographs. The virtual elimination of friction also eliminates the need for the large pendulum mass used in the mechanical systems to overcome friction. Also, a much higher recording magnification with respect to the mechanical models is possible.

DAMPING

When the motion of a pendulum has started, the oscillation will continue at its natural frequency for a time, depending upon the rate of energy dissipation. To measure and detect the arrival of various waveforms separately, the motion of the pendulum must be damped to prevent its free-period swinging. If the damping is small, any seismic impulse will set the suspended spring into motion with an oscillatory or "ringing" output at a period of roughly the pendulum free-period. This problem can be pronounced for long-period seismometers. The damping of a pendulum or a spring-mass can be visualized by pulling a horizontal mass-spring, resting on a smooth oily surface, to one side by 1 centimeter and then releasing it. The distance that the spring overshoots the original rest position determines the damping ratio; for example, an overshoot of 0.1 centimeter corresponds to a damping ratio of 0.1, or 10:1. In older seismographs the damping was achieved in various ways, such as air damping in Wiechert seismographs or oil damping in the early version of the Benioff seismograph. Damping in modern-day seismographs takes advantage of the same electromagnetic principles as discussed previously. As the coil moves with respect to the suspended magnet, an electric current is induced in the coil, which in turn generates a magnetic force opposing the motion. This resisting force acts as the damping mechanism.

SEISMOGRAPH DESIGN

In designing a seismograph (a seismometer and the recording device), one must address issues such as the frequency content of the seismic waves of interest, the direction of the ground motion (vertical or horizontal), the smallest and the largest amplitudes of motion to be measured accurately, the plausible magnification of the pendulum motion relative to the ground motion, the method to be used for relative or absolute timekeeping, the recording of the measured ground motion (a visual recording on paper or recording on a magnetic tape), and the means of recording (an on-site or a telemetry recording). The nature of a particular application will determine the type of seismometer required, the frequency range of interest, the accuracy, and the resolution of the produced record.

Seismic waves generated by earthquake or explosive sources can have a broad frequency band of about 100-0.00033 hertz. Most of the primary earthquake-generated seismic waves have periods of between 0.05 and 20 seconds (frequencies of 20-0.05 hertz). Large earthquakes, such as the 1960 Chilean earthquake with a magnitude of 8.3, can put the whole earth into an oscillatory motion with a period of roughly one hour. Such an unusually broad frequency band creates a need for seismographs to record waveforms of different frequency bandwidth. A good seismological station such as WWSSN often houses six seismometers for obtaining a complete description of ground motion: one vertical and two horizontal (north-south and east-west) short-period seismographs, which are sensitive to the arrival of waveforms in the 0.05- to 2.0-second period range, and three long-period instruments with wave periods in the 15- to 100-second range. Seismographs are usually designed so that the output has a relatively constant magnification for displacement, velocity, or acceleration over some design frequency range. For example, a Wood-Anderson seismograph—the instrument whose seismograms are used to define the Richter magnitude scale—is a horizontal displacement seismograph with a flat response to ground displacement over the frequency range of greater than approximately 1.25 hertz with a magnification factor of 2,800. An earthquake with such a ground displacement at the distance of 100 kilometers is defined to have magnitude of 3.0 on the Richter scale.

TIMING SYSTEMS

Without exact time data, a seismic record is not very useful. Seismologists use the Greenwich Mean Time (Universal Time) for timing seismograms. The timing system can be based on either internal or external clocks, or on a combination of the two. The internal timing system is a precise crystal oscillator, with temperature compensation, for keeping the time base. This low-cost and low-power-drain clock has a drift rate as low as 0.1 second per month. The timing system of this type of oscillator can be as simple as a series of minute and hour pulses for visible records, or as complex as a digital time code with day, hour, minute, and second information repeated once per second. The external timing system is used to keep the internal clock correct to 1 millisecond by continuous reception of a radio time signal. The most common radio signal comprises the time signals transmitted by a standard world time service (such as WWV in the United States). These short wave signals can usually be received, with adequate equipment, anywhere in the world. The reception of this signal, however, is not of good quality. Low-frequency time broadcasts in the 15-100 kilohertz range, such as WWVB, can be received reliably enough for continuous recording over a large area.

RECORDING AND CALIBRATION SYSTEMS

In the 1970's, the development of analog and digital circuits, coupled with increasing access to computers, resulted in the development and use of magnetic tapes for seismic recording. The recording on magnetic tapes can be either analog or digital. In analog, the motion is directly converted into proportional magnetization on the tape, although the analog data cannot be directly used by computers and have to be digitized. In digital recording, the analog ground motion is magnified and digitized at some selected time intervals, and the measured signals are recorded onto a magnetic tape.

A more recent development in seismic recording systems has been integrated circuit modules, allowing a huge amount of information to be held in memory on a single printed circuit board. That enables the recording system to screen the data regarding their significance and, thus, to make the decision if the data should be recorded. Such "self-editing" systems with the use of low-cost cassette tapes have made digital recording very attractive.

Seismologists are interested in the true amplitude of the ground motion. Therefore, seismographs must be calibrated to construct the true ground motion from the recorded seismograms. It is impossible to do a theoretical calibration because of the difficulty in modeling the physical behavior of each component. A direct method of calibration is to use shaking tables, where a seismometer is placed on a table, which is set into harmonic motion with a known amplitude and period. The record of such motion gives direct information on the magnification at that period, assuming that the period is not contaminated by the free-period of the shaking table. The experiment is repeated for different periods.

EARTHQUAKE HYPOCENTER AND MOTION
DETECTION

Earthquake prediction and the design of earthquake-resistant structures have important social

and economic value. A successful program would save lives and billions of dollars; the key to success is understanding the physical mechanisms of earthquakes, if there is to be hope of predicting the time, the location, and the magnitude of future earthquakes. Considering that the majority of earthquakes are caused by the jerky motion of a volume of earth along a fault plane, their spatial distribution could be used to delineate the fault zone. Seismologists use the relative travel times of the primary seismic waves to determine the location (hypocenter) and the time of occurrence of earthquakes. To accomplish that, seismographs are distributed around the expected seismogenic region to record the earthquake-generated seismic waves. The accuracy of pinpointing the hypocentral locations depends upon the accuracy in the relative timekeeping and the location of seismographs. For example, an error of 0.1 second in the travel time of a P wave (pressure wave) could translate to some 600 meters in error in the hypocentral location (P-wave velocity in upper crust is roughly 5,000-6,000 meters per second).

Seismologists determine the direction of motion along the fault surface by studying the relative direction, upward or downward, of the initial motion of the P wave at sites surrounding an earthquake. The mechanism is similar to moving two blocks in opposite directions in a sandbox. The upward or downward motion of the sand particles on the surface can be directly related to the direction of the motion of the blocks.

Perhaps the most important single factor driving the development of seismographs has been the need to detect and differentiate underground nuclear tests from earthquakes. At the present time, seismographs are the main source of information for monitoring the time, the location, and the size of such tests.

EARTHQUAKE-RESISTANT STRUCTURAL DESIGN

For the earthquake-resistant design of structures, it is necessary to determine the expected level of ground shaking from future earthquakes. Seismologists use the recording of past earthquakes to obtain relationships between the maximum ground motion, the magnitude of earthquakes, and the source-site distance for earthquakes in various regions (attenuation equations). Such relationships are used to predict the maximum ground acceleration, velocity, or displacement of future earthquakes of given magnitudes and distances for a desired site.

Another aspect of the seismic design of structures is the determination of the structural response to strong ground shaking. For this purpose, accelographs are mounted in various levels of structures to record the vibration. From the study of such records, earthquake engineers determine the frequency-dependent magnification factors of typical structures. Such information, coupled with the knowledge of the natural period of the building and the site (the site period is the period at which ground motion is magnified), is critical for the design of structures safe against future earthquakes.

Seismic waves from large earthquakes travel thousands of kilometers through the center of the earth. These waves carry much information about the physical properties of the earth and have been used to construct a picture of the interior of the earth. Any discontinuity in the earth's material properties that crosses the traveling path of seismic waves reflects part of the seismic energy and transmits the rest into the adjacent region. Examples of major discontinuities are the earth's crust-mantle and mantle-core discontinuities. The reflected waves reach the recording stations as distinct waveforms; seismologists identify these waveforms on seismograms and determine the location of the discontinuity in the earth by modeling the travel times of such waveforms.

Mehrdad Mahdyiar

FURTHER READING

Bath, Markus. *Introduction to Seismology.* New York: John Wiley & Sons, 1981. Chapter 2, "Seismometry," gives a very good and comprehensive description of seismometers and seismic recording systems. Well written and includes much interesting information.

Berlin, G. Lennis. *Earthquakes and Urban Environment.* 3 vols. Boca Raton, Fla.: CRC Press, 1980. Chapter 2, "Earthquake Descriptors," gives very easily read information on the history and the fundamentals of seismic recording systems. A well-written book worth reading.

Bormann, P., ed. *New Manual of Seismological Observatory Practice.* 2d ed. Potsdam, GFZ. 2002. An excellent resource for anyone studying applied seismology. Chapters 5 and 6 reference seismometer technology and other equipment used in seismological observation. This text also includes useful datasheets, information sheets, and exercises.

Bullen, K. E., and B. A. Bolt. *An Introduction to the Theory of Seismology.* Cambridge, England: Cambridge University Press, 1985. Chapter 9, "Seismometry," gives relatively technical details of the principles of seismometers and the recording systems. One may ignore the equations and still get some good information from this chapter.

Carlson, Shawn. "The New Backyard Seismology." *Scientific American* 274 (April 1996). This article provides easy-to-follow instructions for building a rudimentary seismograph.

Dobrin, M. B. *Introduction to Geophysical Prospecting.* 4th ed. New York: McGraw-Hill, 1988. Gives a general description of the principles of the electromagnetic seismometers and the analog and digital recording systems.

Garland, G. D. *Introduction to Geophysics: Mantle, Core, and Crust.* 2d ed. Philadelphia, Pa.: W. B. Saunders, 1979. Chapter 5, "Seismometry," gives a general background on seismometry. The text is semitechnical. Also discusses seismic arrays. The whole work is a good introductory book on the subject of geophysics.

Hutt, C. Robert, et al. *Albuquerque Seismological Laboratory: 50 Years of Global Seismology.* U.S. Geological Survey: FS 2011-3065. 2011. This fact sheet provides a comparison of seismology practices from the 1960's to 2011. There are many excellent images reflecting the change in technology over that time period.

Manukin, A. B., et al. "Compact High-Sensitivity Accelerometer-Seismometer." *Cosmic Research* 48 (2010): 346-351. This article presents a design for an accelerometer-seismometer, a description of the instrument's characteristics, and techniques for use. The article is very technical; a background in seismology is recommended.

Plummer, Charles C., and Diane Carlson. *Physical Geology.* 12th ed. Boston: McGraw-Hill, 2007. A college-level introductory geology textbook that is clearly written and wonderfully illustrated. An excellent sourcebook of basic information on geologic terminology and fundamentals of geologic processes. An excellent glossary.

Reynolds, John M. *An Introduction to Applied and Environmental Geophysics.* 2d ed. New York: John Wiley, 2011. An excellent introduction to seismology, geophysics, tectonics, and the lithosphere. Appropriate for those with minimal scientific background. Includes maps, illustrations, and a bibliography.

Wiegel, R. L., ed. *Earthquake Engineering*. Englewood Cliffs, N.J.: Prentice-Hall, 1970. Chapter 6, "Ground Motion Measurements," was written by D. E. Hudson and concentrates more on the development of accelographs. The whole book is highly recommended, as it contains valuable information on the history of earthquake engineering and seismic hazards.

See also: Creep; Cross-Borehole Seismology; Discontinuities; Earthquake Engineering; Earthquake Hazards; Earthquake Magnitudes and Intensities; Earthquake Prediction; Earthquakes; Earth's Lithosphere; Elastic Waves; Experimental Rock Deformation; Faults: Normal; Faults: Strike-Slip; Faults: Thrust; Faults: Transform; Notable Earthquakes; San Andreas Fault; Seismic Observatories; Seismic Tomography; Seismic Wave Studies; Slow Earthquakes; Stress and Strain; Tsunamis and Earthquakes.

SLOW EARTHQUAKES

The earth's outer shell, or crust, is composed of huge blocks called plates that regularly move small distances, usually several inches each year, causing them to collide with other plates. Slow earthquakes—the movements of the earth that result from small plate movements—are barely felt but occur regularly in some areas.

PRINCIPAL TERMS

- **asthenosphere:** flexible rock in layers beneath the earth's brittle crust
- **fault:** a deep fissure in the earth's surface along which rock moves
- **Richter scale:** one of several scales used to measure an earthquake's magnitude
- **seismic wave:** a wave of energy released during an earthquake
- **seismologist:** a scientist, often a geophysicist, who specializes in studying earthquakes
- **tectonic plate:** any one of about ten enormous pieces that form the earth's outer layer

KINDS OF EARTHQUAKES

Most people who think of earthquakes immediately visualize huge, destructive, earth-shattering movements of the earth's surface that last only seconds but that bring down buildings, rupture gas and water mains, crush people under tons of rubble, and often are followed by fires. When such movements occur, newspaper headlines are filled with statistics about the amount of damage they have done and about the numbers feared dead.

Among the most severe earthquakes in modern history are the one in the Kansu Province of China in 1920, which killed 180,000 people; the Japanese earthquake of 1923, which killed some 143,000 people in Tokyo and Yokohama; the 1935 earthquake in Quetta, India, which killed more than 60,000 people; the 1970 earthquake in Peru, which killed more than 60,000 people; and the colossal 1976 great Tangshan earthquake in the Hebei Province of China, which killed 240,000 people. Severe earthquakes that occur in heavily populated areas result in heavy casualties. This is particularly true of those that strike less affluent countries in which buildings are often badly constructed, resulting in their collapse when the earth shakes violently.

Another type of earthquake, often as severe as those felt on land, are deep earthquakes beneath the ocean's surface. The ocean floor has been drastically changed by such earthquakes. If they result in casualties on land, it is usually from the tsunamis, or enormous waves, that they generate. These waves can attain heights of more than 30 meters. When they hit developed and heavily populated areas of a shoreline, they can crush everything in their paths and leave behind incredible destruction and thousands of casualties.

Slow earthquakes, also called silent or quiet earthquakes, are undramatic and receive no headlines in the press. A slow earthquake is a discontinuous event that releases energy over a period of hours to months, rather than the seconds to minutes of a typical earthquake. They occur with considerable frequency, although most people are unaware of their existence even if they are in an area where considerable seismic activity is taking place. Seismographs may record their occurrence, but few people feel threatened by slow earthquakes because they do not cause the earth to tremble, books to fall from library shelves, or walls of canned goods to fall on the floors of supermarkets. Their destructive force is cumulative; it takes place over substantial periods of time, causing such minute changes on a day-to-day basis that these changes are not apparent to the naked eye.

STRUCTURE OF THE EARTH

Regardless of which kind of earthquake one is considering, all earthquakes have similar underlying causes. To understand these causes, one must consider how the earth is constructed. The planet is composed of three basic parts. The one with which people are most familiar is the crust, the outer layer that, below the surface, consists of solid rock. The crust has an average thickness of about 32 kilometers beneath the earth's seven continents, although it is considerably thinner beneath the sea, where its thickness averages about 5 kilometers.

Underlying the crust is the mantle. It, like the crust, is composed of solid rock, but this rock is extremely hot. The mantle is thick, extending in many places more than 3,000 kilometers below the crust. Its rigid upper portion is called the lithosphere.

Beneath it is a weaker area of the mantle, the asthenosphere, which, being closer to the earth's molten core, is much hotter than the lithosphere. The lithosphere may be nearly 100 kilometers thick beneath the continents and some oceanic areas, but it shrinks to just 8 to 10 kilometers in thickness beneath submerged ridges in mid-ocean.

Inward toward the earth's center from the lithosphere and asthenosphere are the two major parts of the earth's core, the liquid core and the solid core. The liquid core has a radius of about 2,300 kilometers and consists of molten iron and nickel whose temperature averages about 5,000 degrees Celsius. The solid core, which is at the earth's very center, has a radius of just under 1,300 kilometers and is composed of solid iron and nickel.

Ancient people had various quaint explanations about what caused earthquakes. The ancient Greeks thought that the titan Atlas carried the world on his shoulders and that every time he shifted the weight of this great burden, the earth moved, causing earthquakes. In India, people conceived of the earth as an object balanced on the head of an elephant riding on the back of a huge tortoise. Whenever either animal moved, an earthquake resulted. Other theories viewed the world as being carried by giant catfish, whales, or oversized gods who rode in sleds pulled by dogs. In the past, many religions viewed earthquakes as expressions of God's anger and punishment for humankind's transgressions.

Modern science has been slow to offer rational physical explanations for earthquakes. English geologist John Michell, in 1760, suggested that they are caused by the movement of subterranean rocks. Few accepted this explanation, and those who did thought that such movement was caused by gigantic explosions deep inside the earth. It took another hundred years before Robert Mallet, an engineer from Ireland, contended, in 1859, that the causes of earthquakes were strains in the earth's crust. After 1960, most seismologists accepted the theory of plate tectonics as the cause of earthquakes.

PLATE TECTONICS

According to the theory of plate tectonics, which has gained wide acceptance, the earth was once a solid landmass surrounded by a great sea. The planet began to cool after its fiery formation some 5 billion years ago. As it cooled, its surface cracked.

Over hundreds of millions of years, parts of the once-solid landmass drifted away, forming seven large and twelve small islands, all with ragged edges. The large islands are the earth's seven continents. These islands, or tectonic plates, float and are in constant but often quite limited motion. For example, the two plates that exist on the western part of the North American continent, the Pacific plate and the North American plate, hardly move at all. The Pacific plate drifts north at the barely perceptible rate of about 5 centimeters per year. The North American plate drifts southwest at a similar rate.

Despite the slow movement of the North American plate, it has been estimated that over hundreds of millions of years, the North American continent could, through continental drift, collide with Australia. Between the Pacific and North American plates lies the San Andreas fault, a gash in the earth's surface that runs more than one-half the length of California. The movement of these plates results in slow or silent earthquakes.

When the edges of the two plates collide, however, as they did in the San Francisco earthquake of 1906 and the Northridge earthquake of 1994, the result is a major earthquake. The Northridge quake, which registered 6.8 on the Richter scale, revealed the existence of a hidden thrust fault and a horizontal fault that had previously gone undetected. The Northridge earthquake was followed by more than one thousand aftershocks as the earth beneath the area resettled. Many slow earthquakes preceded the Northridge disaster as foreshocks and followed it as aftershocks. Slow earthquakes often presage the coming of major earthquakes. As increased knowledge about slow earthquakes evolves, seismologists are beginning to understand more fully how to interpret the often subtle signals they send. The interpretation of these signals can help predict future deep earthquakes.

Signs of impending earthquakes exist—probably quite often in the form of slow earthquakes—that cause animals to react in anticipation of severe, deep earthquakes. The behavior of animals in zoos in the hours preceding a severe earthquake shows clear signs that they are disquieted and sense something, perhaps minute subterranean vibrations, that humans are not able to perceive. The fields of plate tectonics and seismology are becoming more and more sophisticated as technology produces ever more sensitive instruments for detecting seismic activity.

TYPES OF SLOW EARTHQUAKES

Most people think of earthquakes as cataclysms in which a crack breaks through the earth's crust at a speed of several kilometers per second, causing a violent shaking of the ground, severe damage to structures, and injury to living things in the quake zone. In many parts of the world, however, the development of a crack along a fault line occurs at a speed of less than 1 meter per second, with some slips even measured in millimeters per year. Three faults in California—the Hayward, San Andreas, and Calaveras faults—demonstrate the great variety of seismic activity, ranging from the ordinary earthquakes that occur from rapidly developing breaks in the earth's crust that suddenly release waves of stored elastic-strain energy, to a variety of smaller tremors.

Among the types of earthquakes that geologists and seismologists have discovered and named are slow earthquakes, defined as having speeds of hundreds of feet per second; silent earthquakes, defined as having speeds of tens of feet per second; strain migration events, measured at speeds of centimeters per second; and creeping earthquakes, with speeds measured in millimeters per second. These varieties are not always measurable on typical seismographs. Few of them attract attention as they are taking place.

Slow earthquakes can, at times, cause rapid ruptures that produce high-frequency sound waves, but more often they take a much longer time to rupture through the earth's crust than ordinary earthquakes of comparable magnitude. Some slow earthquakes occur in oceanic transform faults, as happened on June 6, 1960, in the Chilean transform fault, which ruptured for about one hour as a series of small, barely detectable breaches.

Silent earthquakes have been so named because they are never accompanied by the high-frequency sound waves that most seismographs need to register seismic events. Some researchers have employed delicate instruments that measure tectonic strain to detect silent earthquakes. Such instruments also have revealed creeping movement of about 10 millimeters per second in parts of California's San Andreas fault. The low-frequency waves of a silent earthquake moving about 0.3 meters per minute were recorded shortly before the 1976 earthquake in Fruili, Italy, and again in 1983 before a severe earthquake hit the Japan Sea. Seismologists think that the occurrence of some slow and some silent earthquakes may be warning signs that, if heeded, could prevent substantial loss of life when an ensuing ordinary earthquake is on the brink of shattering a region. The stick-slip earthquake, with its jerky, sliding motion at a fault, usually comes after a slip with propagation speeds of 20 to 200 meters per second. Not all such slips can be detected on the typical seismographs that most geophysicists and seismologists use. Close to the earthquake, silent earthquakes can be recorded geodetically and by using strainmeters. Only digital, broadband seismographs are able to record seismic waves of such low frequencies.

Pacific Rim nations are shaken yearly by thousands of tiny earthquakes resulting from the collision of oceanic and continental plates. Some of these are slow or silent earthquakes that relieve the pressure that is built up in subterranean rocks when one tectonic plate rams into another. Severe earthquakes are the result of several years of pressure buildup, but slow and silent quakes relieve the pressure more gradually and possibly act as the safety valves that prevent the earth from experiencing more numerous major earthquakes than it does.

SLOW CREEP AT WORK

Hollister, California, about 160 kilometers southeast of San Francisco, lies close to the San Andreas fault. Studies of major fault lines near Hollister have revealed a gradual slippage beneath the earth's surface, although it has not been possible to measure this slippage with total accuracy. A winery constructed in Hollister in 1939 is located almost on top of the San Andreas fault. In 1956, the winery began to experience damage that could not be easily explained but that could no longer be ignored. Strong reinforced concrete walls and floors in one of the warehouses were gradually crumbling. None of the local people remembered any overt seismic action that could explain the phenomenon.

Finally, because of the winery's location close to the known fault line, the owners engaged seismologists to assess the situation. They found that an active branch of the fault zone ran directly below the building. They discovered that the two sides of this fault line were moving past each other at an estimated rate of 1.3 centimeters every year. Although such movement does not attract immediate attention, over fifty years the distance involved is more than 0.6 meter, which causes damage readily observable by anyone who looks at the building.

As such movement continues, structures are weakened and are finally felled by it. Because the San Andreas is a right-lateral fault, the winery's west side was steadily moving north of its east side. When cracks appeared in the floor and walls, they were patched up. Sagging walls were reinforced. Since 1956, the situation has been monitored carefully. It has been determined that the creep continues, as it surely will do in the foreseeable future.

This sort of creep is related to slow earthquakes in that it does not involve a dramatic underground upheaval that happens in a matter of minutes, although creep is thought not to be entirely gradual. It often occurs in a matter of seven to ten days at a time, after which there is a period of quiescence for weeks or months. It can, however, continue for decades and be barely detected in areas that are neither built up nor heavily populated. In 1960, recorders clocked an earthquake in Hollister in which instant creep of about 0.3 centimeter occurred. The frequency with which earthquakes occur may have an influence on periods in which seismic activity takes place. An earthquake in the area near Hollister in 1939 separated the winery's adobe walls from side walls and pulled girders from their brick moorings. Rents appeared in the ground around the winery. A major jolt in 1960 severely shook the winery, causing damage to it.

The slow creep beneath the winery and in other areas in the Hollister area is under the constant scrutiny of seismologists, who are trying to determine why slow creep occurs along some areas of the fault but is not observed in other nearby areas. A tunnel for the Los Angeles Aqueduct that was constructed in 1911 and crosses the San Andreas fault has remained intact for its entire existence, with no signs of seismic activity.

R. Baird Shuman

FURTHER READING

Beroza, G. C., and T. H. Jordan. "Searching for Slow and Silent Earthquakes Using Free Oscillations." *Journal of Geophysical Research* 95 (1990): 2485-2510. The authors relate how free oscillations, which ring like a bell, were recorded over a decade, most of them caused by large, ordinary earthquakes. In some instances, the earthquake involved was not big enough to cause free oscillation, suggesting that they were slow earthquakes. Of the 1,500 free-oscillation earthquakes, 164 were not accompanied by a recorded earthquake.

Bolt, Bruce A. *Earthquakes.* 5th ed. New York: W. H. Freeman, 2005. This comprehensive overview of earthquakes is easy to understand and highly informative. Its material on seismic waves and seismography is of great significance to those interested in various types of earthquakes and in where and how they occur. Bolt writes clearly and with authority in this field.

Ebert, Charles H. V. *Disasters: Violence in Nature and Threats by Man.* Dubuque, Iowa: Kendall/Hunt, 1988. Chapter 1, "Earthquakes," and Chapter 4, "Tsunami Waves and Storm Surges," should prove of particular interest to readers interested in earthquakes, although parts of other chapters also contain relevant information. Chapter 3, for example, relates how earthquakes can trigger landslides and avalanches.

Koyhama, Junji. *The Complex Faulting Process of Earthquakes.* New York: Springer, 2010. This well-documented, carefully researched study, though quite technical, is excellent in the scope of its coverage. Koyhama explains how various faults have developed and what courses they have taken.

Levy, Matthys, and Mario Salvadori. *Why the Earth Quakes: The Story of Earthquakes and Volcanoes.* New York: W. W. Norton, 1995. Written with general readers in mind, this volume is exceptionally clear. Its illustrations, both verbal and graphic, add to the accessibility of what the authors are saying. Chapter 9, which focuses on California's San Andreas fault, should be of particular interest to American readers, illustrating as it does how various forms of seismic activity can occur simultaneously along the fault.

Melbourne, Timothy I., and Frank H. Webb. "Slow but Not Quite Silent." *Science* 300 (2003): 1886-1887. Provides a short overview of slow earthquakes and the discovery that such events are a result of deep tremors.

Rundle, John B., Donald L. Turcotte, and William Klein, eds. *Reduction and Predictability of Natural Disasters.* Reading, Mass.: Addison-Wesley, 1996. Among the nine contributions to this book on earthquakes, "Thoughts on Modeling and Prediction of Earthquakes," by S. G. Eubanks, and "A Hierarchical Model for Precursory Seismic Activation," by W. I. Newman, D. L. Turcotte, and

A. Gabrielov, are the most relevant to the topic of slow earthquakes. The contributions to this volume, while remarkably significant, are highly specialized and may be difficult for beginners.

Schenk, Vladimir, ed. *Earthquake Hazard and Risk.* Dordrecht, Netherlands: Kluwer Academic Press, 1996. The twenty contributions to this volume, all of which were written by acknowledged specialists in the field, focus on the prediction and management of earthquakes. The volume covers the field thoroughly but would be more useful if it contained an index. The essays seem more specialized than most beginners can easily handle.

Walker, Sally M. *Earthquakes.* Minneapolis: Carolrhoda, 1996. Written with the young audience in mind, Walker's account is lively, interesting, and accurate. The illustrations are colorful and cogent. Various easily comprehended charts and tables add considerably to the substance of the book's engaging text. The glossary is of special value to readers, as is the index.

Wright, Karen. "The Silent Type." *Discover* 23 (2002): 26-27. This article describes the mechanics of a slow earthquake and how it relates to the regular surface earthquakes. It also addresses the methodology used to detect slow earthquakes. Easily understood by the nonscientific reader.

Yeats, Robert, Kerry Sieh, and Clarence R. Allen. *The Geology of Earthquakes.* New York: Oxford University Press, 1997. This comprehensive study of the geology of earthquakes, written essentially as a textbook, is well presented and thorough. Its material on slow, silent, creeping, and strain migration occurrences, though brief, is as solid as any in the field. The writing style, even in the presentation of the more technical material, is extremely appealing.

Zebrowski, Ernest, Jr. *Perils of a Restless Planet: Scientific Perspectives on Natural Disasters.* New York: Cambridge University Press, 1999. Chapter 1, "Life on Earth's Crust," and Chapter 6, "Earth in Upheaval," are the most useful to those seeking more information about types of earthquakes. Zebrowski obviously has a strong background in ancient Greek and Roman mythology, as well as a scientist's grasp of the mechanics of earthquakes. The illustrations are well chosen, and the appendices include a great deal of technical information in charts that make it easily understandable.

See also: Continental Drift; Creep; Deep-Focus Earthquakes; Earthquake Distribution; Earthquake Engineering; Earthquake Hazards; Earthquake Locating; Earthquake Magnitudes and Intensities; Earthquake Prediction; Earthquakes; Faults: Strike-Slip; Faults: Thrust; Faults: Transform; Notable Earthquakes; Plate Motions; San Andreas Fault; Seismic Observatories; Seismic Reflection Profiling; Seismic Tomography; Seismometers; Soil Liquefaction; Tsunamis and Earthquakes.

SOIL LIQUEFACTION

Soil liquefaction is the group of processes by which otherwise solid soil particles are shaken apart by earthquakes or collapse away from edge-to-edge contact with one another and become temporarily supported by the pore water contained within them. The resulting fluid mush can allow buildings to sink; this phenomenon is responsible for considerable loss of property and life.

PRINCIPAL TERMS

- **bearing capacity:** the ability of granular soil materials to support the weight of building structures
- **clay minerals:** the diverse group of very finely crystalline mineral structures that are predominately composed of silicon, aluminum, and oxygen and that have very different water retention capabilities
- **cohesion:** molecular attraction by which the particles of a body are united throughout the mass, whether alike or unlike
- **electrostatic charge:** the fundamental atomic force in which objects that have a similar electric charge repel each other whereas those with unlike charges attract each other
- **flocculation:** the sedimentation process by which a number of individual minute suspended particles are held together in clotlike masses by electrostatic forces
- **remolding:** the property of some sensitive clays upon disturbance to reorient their particles, which softens them, and to flow in a liquid form
- **sensitive (quick):** describes fine-grained deposits that are characterized by considerable strength in the undisturbed condition, but upon disturbance their ability to support themselves declines dramatically
- **van der Waals force:** a weak electrostatic attraction that arises because certain atoms and molecules are distorted from a spherical shape so that one side of the structure carries more of the charge than does the other

TYPES OF LIQUEFACTION FAILURE

Soil liquefaction is the abrupt and temporary change of seemingly solid soil materials into liquid mush. The soil can lose all of its cohesion and bearing capacity in a matter of moments. Once the process begins, most commonly when an earthquake strikes, ground that had supported a hospital or a high-rise apartment can suddenly become a fluid into which the buildings sink like rocks into quicksand. Anything built upon such materials can slip or sink into the new liquid. Buried gasoline or septic tanks can suddenly become buoyant and float to the surface.

A number of different sorts of liquefaction failure are recognized; several occur on land and a few are subaqueous. Quick condition failure (not to be confused with sensitive or quick-clay landslips) is the complete loss of bearing capacity caused by liquefaction of sands and silts so that structures sink or rise in material that appears otherwise solid. Flow landslips, or flowslides, can occur on moderate slopes on land or beneath water and involve either sands or clays. These types tend to retrogress, or work their way backward, slice by slice as the material beneath them liquefies. Lateral spread landslips can occur on gentle, or nearly flat, slopes. Many lateral spreads are in quick clays. In extreme cases, the liquefaction slides result from spontaneous subaqueous liquefaction that propagates, or spreads, in all directions.

WATER-SATURATED SAND LIQUEFACTION

Two classes of soil materials can liquefy: water-saturated sands and silts and the unusual sensitive, or quick, clays. In the first example, layers of loosely packed, well-sorted, fine- to medium-grained sands and coarse silts are subject to liquefaction where groundwater tables are within 10-15 meters of the ground surface. When this water-saturated sediment starts to shake apart in an earthquake or other vibration, the grains temporarily lose rigid contact with one another and collapse inward. Much of the pore water (water in the gaps between soil grains) is then superfluous but does not escape at once, so that the rearranged grains cannot fit close to one another. The particle weight is thereby transferred to the pore water, its pressure increases (which reduces friction between the grains), and the soil becomes liquid for a short time.

When water-saturated sands liquefy during the course of an earthquake, sand boils erupt muddy water and sand from ground fractures and turn the surface into a quagmire. Sand boils tend to be roughly circular in plan and can have a depressed center like a volcano. After a very large earthquake, sand boils may form as much as several hundred meters across and several meters high. Their presence in an area is a reliable sign of past earthquake activity.

Water-saturated earth-fill dams are also subject to soil liquefaction, with obvious disastrous consequences. For example, during the 1971 San Fernando Valley earthquake in California, the Lower Van Norman dam collapsed. The 40-meter-high dam had been built with a core of clay surrounded by a fill of water-saturated sand. After about twelve seconds of strong shaking, a large, wedge-shaped segment of this water-saturated sand fill liquefied. Eight large blocks on the upstream side of the dam slid into the reservoir as parts of the sediment fill liquefied. Before the quake, the crest of the dam was a safe 10 meters above the water level of the reservoir. Afterward, however, only a thin barrier of 1.5 meters was left above the water level. Had this minor amount of remaining dam surface moved down but a fraction, the dam would have quickly failed, because water would have rushed down and eroded the downstream side. The 80,000 people who lived directly below the reservoir thus escaped disaster by a slim margin.

QUICK-CLAY LIQUEFACTION

In the quick-clay type of soil liquefaction, because of a rather special geologic history, certain clays develop the ability to liquefy, or to become quick, as with quicksand. Two main theories have been advanced to account for the unusual distribution of these quick clays. The saltwater theory holds that sensitive clays are first formed where glaciers erode very fine-grained clay mineral platelets from the soils and bedrock over which they ride. Where the weight of such glaciers depresses the surface of the land close to the sea, the small platelets can be deposited directly into the salty waters.

The small clay mineral platelets of a sensitive clay carry a negative charge, and as they settle in ocean water they tend to pick up positively charged sodium ions from the salt in solution. These oppositely charged, mutually attractive forces act as a glue to cause the

particles to clump together into a sort of coagulated honeycomb structure. The platelets thus develop an edge-to-edge, or "house of cards," structure held together by the electrostatic forces of the salt ions attached to the platelet edges.

Glacial marine clays may remain quite stable as long as the salt water remains in the pore spaces of the open card-house structure. These deposits can later be raised above sea level by rebound of the land following removal of the glacier ice load. Then the salty pore waters may be flushed out of the clays by the fresh waters that normally occur above sea level. Removal of the mutually attractive electrochemical charges then sets the stage for eventual liquefaction.

A different theory explains the freshwater type of quick clays. In this case, glacial erosion also provides extremely fine particles of other than clay minerals. Instead, the fine particles (mostly quartz minerals) are so small that the normally weak van der Waals attraction is sufficient to hold them together. When the particles are small enough, the ratio of these weak attractions to the weight of the particles is greater, and the material will be cohesive, strong but brittle, and sensitive. When the cohesive bonds are broken through shaking, the short-range forces are ineffective, and a total loss of strength results. If there is

Anchorage, Alaska, was hit by one of the most damaging earthquakes in history, on Good Friday, March 27, 1964; the earthquake had a magnitude of 8.3 (higher, by some estimates). Much of the damage to Anchorage was the result of liquefaction of an unstable layer of clay a few meters below the surface. (U.S. Geological Survey)

sufficient pore water, the material will liquefy and flow on gentle slopes.

An example of a well-documented failure in quick clay is the St. Jean-Vianney failure of May 4, 1971, in Quebec, Canada. This landslip was located in the middle of a much larger prior failure that was at least five hundred years old. At about 7:00 P.M., instability first began with no warning on the steep west bank of the small Petit Bras tributary to the Saguenay River. The first actual failure, however, did not occur until 10:15 P.M., when, in fifteen minutes, the failure surface retrogressed 150 meters west into the slope. The resulting debris moved into the river and blocked both the river valley and the opening to the failure. The dam blocked the outflow of the liquefied material from the crater until the pressure became too high. Finally, the dam burst and allowed the remolded fluid to flow down the river valley with a wavefront about 18 meters high and traveling about 26 kilometers per hour. The mass carried with it thirty-four houses, one bus, and an undetermined number of cars. Thirty-one lives were lost as well.

SOIL STRENGTH

The strength, or load-carrying capacity, of all soils varies considerably. In addition, the strength of any specific soil can vary under different conditions of moisture and density. Sensitive soils have natural water contents already above the liquid limit, which is the moisture content at which a soil passes from a plastic to a liquid state. Plasticity is a characteristic of clayey soils that allows them to be squeezed and easily deformed without disintegration, whereas being above the liquid limit, as the name implies, allows the soil to flow easily and remold itself.

Sensitive clays are generally considered to be thixotropic in the sense that strength diminishes upon disturbance and is regained when disturbance ceases. Once collapse of the card-house matrix is initiated and flow begins, the clays seem to be remolded by lining up of the platelets and particles in a parallel or linear fashion. Remolded clays slip easily over one another in the watery mixture. Once these instabilities develop, they can spread rapidly throughout an area of retrogressive failure. The regaining of strength is not well understood but is thought to result from the gradual rearrangement of particles into positions of increasing mechanical stability under the action of new bonding forces. Perhaps the

compaction and gradual expelling of excess water from the moving mass also are responsible for some restoration of strength. The artificial addition of new salts to such materials can be used in certain cases to regain strength and thereby stabilize hazardous areas.

REMOLDING

The typical lateral spread and flow landslips in sensitive clays have certain characteristics that make them distinctive. The processes are essentially a gravitational remolding that transforms the clay into a viscous slurry, or mixture of liquid and solid. Overlying sediments can break into strips or blocks that then become separated. The cracks between the blocks fill with either soft material squeezed up from below or detritus (loose material resulting from disintegration) from above.

Such failures commonly start in the lower part of a slope as a result of local oversteepening through stream or other erosion. After the initial slip, the failure spreads retrogressively, slice by slice, farther and farther into the bank. Movement generally begins quickly, without appreciable warning, and proceeds with rapid to very rapid velocity. A large, bowl-shaped crater commonly results.

SOIL LIQUEFACTION STUDY

Analysis of soil liquefaction is not extensive, because the phenomenon has been known as an important hazard for a relatively short period of time and is an unusual event in any case. The rarity of the phenomenon means that scientists have had trouble installing instruments and making measurements of the ground in places that would eventually liquefy and provide a well-documented record. In most cases, studies had to be done on ground that had liquefied in the past but had since become stable or in the rather artificial situations of the laboratory.

After soil liquefaction was first discovered following catastrophic ground failure, interviews with survivors provided most of the early information. Borehole cores and measurements of displaced ground showed types of sediments and characteristic depths and amounts of water involved. In many engineering applications, the undisturbed cores are subjected to various stresses to determine their behavior when shaken or loaded. Stirring of undisturbed quick clays to a liquid state in the lab, followed by resolidification upon addition of salt, is a means of analyzing the condition. Such studies allow the hazardous condition of certain areas to be represented on maps of liquefaction susceptibility that indicate places in which the phenomenon is likely to occur.

Success in observing soil liquefaction in action has occurred where instruments were previously installed in the ground to measure pore-water pressures at various depths during carefully measured earthquakes. Surprisingly, and in contrast to laboratory experiments in which liquefaction occurs at the same time as does strong shaking, pore-water pressures rose only slowly as the shaking intensified, and sandy layers completely liquefied only well after strong earthquake motions had ceased. The delay appears to occur as uneven pore-water pressures are redistributed in the ground.

EVER-PRESENT EARTHQUAKE HAZARD

The significance of soil-liquefaction potential is enormous. Many of the world's major cities are partly built upon weakly consolidated, water-saturated sediments. Understanding the mechanisms of liquefaction is also important in analyzing earthquakes long past; old sand boils have been used to date and to estimate magnitudes of prehistoric earthquakes. Major earthquakes in Niigata, Japan, and Anchorage, Alaska, brought about a heightened appreciation of the importance of soil liquefaction as a general geologic process. Following these events, geologists, engineers, and planners have come to recognize soil liquefaction as an ever-present earthquake hazard and have associated it with almost all major earthquakes since.

In June 1964, for example, a magnitude 7.3 earthquake occurred 55 kilometers from the city of Niigata on Japan's west coast. In 50 seconds of shaking, the city of 300,000 was subjected to dramatic soil liquefaction that affected thousands of dwellings and industrial structures. Much of the city was built originally upon sand deposits about 100 meters thick along the Shinano River and upon younger lowland sediments and reclaimed riverfront land. During the earthquake, subsurface sand and water flowed up and out of cracks in the ground. The liquefaction caused major destruction of highways, bridges, railroads, utilities, oil refineries, and harbor facilities. There were 3,018 houses destroyed outright and 9,750 damaged moderately or severely because of cracking

and unequal settlement of the ground; much of this damage occurred on the newly reclaimed areas. In a most spectacular occurrence, a number of large apartment buildings, which had been designed to be earthquake-resistant, tipped over on their sides to settle at angles of as much as 80 degrees, though the structures themselves remained intact. People were able to escape by walking down the sides of the buildings. Several of these apartment houses were later jacked up, reinforced, and opened for reoccupation.

The Good Friday (March 27) earthquake of 1964 in Alaska was, at 8.4-8.6 magnitude, the largest ever recorded in North America. The Turnagain Heights landslide in Anchorage, Alaska, took place in flat terrain along the steep coastal bluffs that border the Knick Arm of the Cook Inlet there. Before the earthquake, the bluffs rose steeply some 30-35 meters above sea level. Marine clays and silts with layers and lenses of sand of the Bootlegger Cove Clay were exposed in the bluffs. A 6-meter thickness of sand and gravel on the flat terrace above had provided an apparently fine place to build homes. During the 1964 earthquake, giant blocks of Bootlegger Cove Clay and the overlying sands and gravels were set in motion toward the sea as the sand layers in the clay formation were liquefied. The first movements of the blocks began about two minutes after the onset of intense earthquake shaking. In the next five minutes, the previously flat terrain was transformed into a jumble of blocks capped by tilted trees and broken buildings. Seventy-five homes and three lives were lost in the breakup of the ground. More than seven huge blocks became widely separated as they moved toward the sea.

One positive result of soil liquefaction is the tendency for earthquake shaking to be significantly reduced, as liquids do not support shear stress. Once the soil liquefies due to shaking, shear waves are not transferred to buildings at the ground surface.

INCIDENCES OF SUBAQUEOUS FAILURE

In a subaqueous example of failure, the sand beds along the coast of Zeeland in the Netherlands periodically liquefy. The coast is located on a thick layer of fine quartz sand that consists of rounded quartz grains. The slope of the beach is only about 15 degrees. Once every few decades, especially after exceptionally high spring tides, the structure of the sand breaks down beneath a short section of the coastal belt. The sand flows out and spreads with great speed in a fan-shaped sheet over the bottom of the adjacent body of water. The tongue of such a flowslide is always much broader than is the source; the flowslides themselves commonly have surface slope angles of as small as 3-4 degrees. Such a failure occurred at Borssele in 1874 and involved nearly 2 million cubic meters of sand.

A liquefaction process also seems to be significant in the fine silts and clays of the Mississippi Delta region, particularly in the formation of collapse depressions and elongate flows. In these sediments, the pore-water pressures are extremely large, and the pore spaces also contain large amounts of methane gas from decaying organic matter. Following initial failure of these materials, softening of the highly pressured clay/water/gas system causes remolding and strength loss similar to a type of liquefaction or quick behavior. Collapse of offshore structures and sinking of pipelines and seafloor monitoring equipment vertically into the sediment may result. Increased pore pressures during storms also cause bottom movement, collapse, and indications that the sediment can become active as a fluid. On very low-angle slopes (0.1-0.2 degree), distinct collapse depressions are formed, whereas on slightly steeper slopes (0.3-0.4 degree), more elongate flows can result.

John F. Shroder, Jr.

FURTHER READING

Das, Braja M., and G. V. Ramana. *Principles of Soil Dynamics.* Stamford: Cengage Learning, 2011. Provides background information on physics of vibrations, elastic waves, stress waves, and earthquakes. Chapter 10 covers aspects of soil liquefaction procedures and tests.

Dennen, W. H., and B. R. Moore. *Geology and Engineering.* Dubuque, Iowa: Wm. C. Brown, 1986. This volume is one of the most nontechnical but accurate books available on general geotechnical subjects. The sections on soil liquefaction and related phenomena are not long but are easy to understand.

Doyle, Hugh A. *Seismology.* New York: John Wiley, 1995. A good introduction to the study of earthquakes and the earth's lithosphere. Written for the layperson, the book contains many useful illustrations.

Holzer, T. L., T. L. Youd, and T. C. Hanks. "Dynamics of Liquefaction During the Superstition Hills, California, Earthquake." *Science* 244 (April 7, 1989): 56-59. This article reports the details of the first record of a natural liquefaction event after an earthquake. An array of instruments had been installed up to 12 meters deep that recorded excess pore pressures generated once horizontal ground acceleration from the earthquake exceeded a certain threshold value.

Idriss, I. M., and R. W. Boulanger. *Soil Liquefaction During Earthquakes.* Earthquake Engineering Research Institute, 2008. This text covers the basic principles of liquefaction behavior of soils and analysis methods.

Lade, Poul V., and Jerry A. Yamamuro. "Evaluation of Static Liquefaction Potential of Silty Sand Slopes." *Canadian Geotechnical Journal* 48 (2011): 247-264. The article compares sand types and their liquefaction potential.

Lundgren, Lawrence. *Environmental Geology.* 2d ed. Englewood Cliffs, N.J.: Prentice-Hall, 1998. A general book, with several excellent sections and photographs of soil liquefaction phenomena. Not too difficult or technical for the average interested reader.

Penick, J. L., Jr. *The New Madrid Earthquakes.* Rev. ed. Columbia: University of Missouri Press, 1981. This well-written account covers the most intense earthquakes ever to strike the North American continent and details their effects upon people, animals, waterways, and land. Some of the largest sand boils from soil liquefaction ever recorded are described here. The vivid description of the devastation wrought upon the face of the land gives a picture of the dramatic changes caused by the upheaval of natural forces.

Pipkin, Bernard W., and Richard J. Proctor. *Engineering Geology Practice in Southern California.* Belmont, Calif.: Star Publications, 1992. This book provides a detailed description of geological engineering in areas prone to earthquakes, including attempts to prepare for the destructive effects of soil liquefaction. Illustrations and bibliographic references.

Plescan, Costel, and Ancuta Rotaru. "Aspects Concerning the Improvement of Soils Against Liquefaction." *Bulletin of the Polytechnic Institute of Iasi* (2010): 39-45. Presents options for improving foundation soils to protect against soil liquefaction. Discusses jet grouting as an effective preventative method.

Plummer, Charles C., and Diane Carlson. *Physical Geology*. 12th ed. Boston: McGraw-Hill, 2007. A college-level introductory geology textbook that is clearly written and wonderfully illustrated. An excellent sourcebook of basic information on geologic terminology and fundamentals of geologic processes. An excellent glossary.

Spangler, M. G., and R. L. Handy. *Soil Engineering*. New York: Harper & Row, 1982. This book is a detailed engineering text with many equations. Nevertheless, readers will find the discussions on soil liquefaction useful even if they disregard the numerical equations.

Terzaghi, K., and R. B. Peck. *Soil Mechanics in Engineering Practice*. New York: John Wiley & Sons, 1948. This text is the classic work on soil liquefaction. Although it includes many equations, the authors have also provided plentiful written descriptions that are easy to understand.

See also: Creep; Deep-Focus Earthquakes; Earthquake Distribution; Earthquake Engineering; Earthquake Hazards; Earthquake Locating; Earthquake Magnitudes and Intensities; Earthquake Prediction; Earthquakes; Notable Earthquakes; Slow Earthquakes; Tsunamis and Earthquakes.

SOLAR WIND INTERACTIONS

Scientists are able to study solar winds as they reach Earth's orbit, but to know how those winds interact with planets and other objects in the solar system requires scientists to understand the objects themselves and the nature of solar wind. The study of solar winds is important in terms of understanding the manner by which the sun's heat is carried to Earth, especially in an era of increasing reliance on telecommunications, satellite, and other technologies.

PRINCIPAL TERMS

- **aurora:** phenomenon in which highly charged particles in Earth's magnetic field generate a multicolored light show in the skies above the planet's polar regions (and sometimes farther south)
- **corona:** the sun's outer atmosphere
- **geomagnetic storm:** short-term disturbance in a planet's magnetosphere, frequently causing disruptions within the magnetic field
- **heliosphere:** area within a solar system affected by the radiation emitted by the sun
- **magnetosphere:** field consisting of magnetically charged particles from both the sun and Earth
- **radio telescope:** instruments used to detect and track radio waves emanating from other planets and celestial bodies
- **solar wind:** the emission of superheated particles from the sun's corona
- **Van Allen belt:** one of Earth's two radiation belts located outside of Earth's atmosphere

BASIC PRINCIPLES

Solar wind is the emission of superheated particles from the sun's corona (outer atmosphere). Although scientists have not conclusively determined the cause of this emission, what is known is that these particles (mainly electrons and protons) become so charged that the sun's gravity can no longer contain them. In some cases, this emission is known as a coronal mass ejection (CME), a sudden and violent release of the charged particles. Once released through coronal holes (dark regions within the corona that have an open magnetic field and low density), the winds travel as plasma away from the sun at varying speeds, sometimes as fast as 900 kilometers (560 miles) per second and with a temperature of 1.8 million degrees Fahrenheit.

Solar winds carry with them magnetic fields that, when interacting with Earth, can become trapped within Earth's magnetic field, causing a collision of these charged particles. These particles are visible as a natural light show known as the aurora. In some cases, solar winds cause geomagnetic storms, disturbances in Earth's magnetosphere (the field consisting of magnetically charged particles from both the sun and Earth) surrounding Earth. These storms can be severe enough to cause power outages and disruptions to cellular and satellite systems.

BACKGROUND AND HISTORY

In the early seventeenth century, the renowned German astronomer and mathematician Johannes Kepler noticed that the visible tails of comets, which consisted of dust, always seemed to point away from the sun regardless of their position. Kepler speculated that the pressure of sunlight that radiated outward caused such phenomena. Comets also have ion trails, visible in other light spectra (not just sunlight). These trails also point away from the sun, but with varying appearances. Some appear bent or appear to track in a different direction, suggesting that a force other than sunlight is at work.

In the 1940's, Germans astronomers Cuno Hoffmeister and Ludwig Biermann theorized that this force was solar corpuscular radiation, which worked at varying speeds and flow velocities. The Hoffmeister and Biermann theories worked to explain what this radiation was but failed to explain what caused it.

In 1958, solar astrophysicist Eugene Parker offered a different take on this phenomenon. He speculated that the sun, like Earth, had an atmosphere (the corona), but that the corona operated much more differently than that of Earth. As the intense heat generated by the sun reached the corona, Parker's theory posited, the outermost layers of the corona would be pushed away into space. The speed of this radiation would be determined by the temperature and by the distance traveled. Parker's model for solar wind was an ambitious approach but relied on a number of assumptions. In particular, the model determined that the corona's equilibrium (balance) was not in flux. Later theories asserted that Parker's model did not account for the variable speeds and intensities of solar winds.

SOLAR WIND AND THE MAGNETOSPHERE

Studying how solar winds form and interact with the magnetic fields of planets is a developing science. Solar winds are an even mix of highly charged protons and electrons that are emitted through coronal holes. These holes are found primarily at the sun's poles; on X-ray images of the sun's corona, they appear as dark, loose structures. These holes have low-level magnetic fields, which allow the hot gas to escape at high speed and radiate throughout the solar system in the form of solar wind. The area impacted by solar winds within the solar system is known as the heliosphere.

As they travel through space, solar winds (which possess magnetic fields) come into contact with the magnetic fields of other celestial bodies, including that of Earth. These winds ultimately affect Earth's magnetosphere (a broad area surrounding Earth's magnetic field that blocks highly charged particles from directly impacting the planet). Studies show that solar winds contribute to the shape of the magnetosphere, exerting pressure on it and reacting to the outward pressure it exerts as well. In 2005, three separate solar wind monitors in orbit captured a solar wind impact on the magnetosphere, demonstrating such an encounter. The event also was highlighted by a brief increase in strength of the ground-based magnetic field when the magnetosphere was pushed inward. When the magnetosphere stabilized, the ground-level magnetic field returned to normal levels.

Scientists also believe that solar winds that approach Earth's magnetosphere on a northward or southward track can cause a drag effect, tugging the magnetosphere away as particles within the wind connect with particles within the magnetosphere (a process known as reconnection). Experts continue to seek to understand the nature of reconnection and whether this process, as it occurs between solar winds and the magnetosphere, is a large-scale event or a patchier, small-scale phenomenon.

SOLAR WEATHER

The interaction between solar winds and objects within the heliosphere varies based on the strength of the magnetosphere, the speed and distance of the wind, and other factors. Earth has one of the strongest magnetospheres among the terrestrial (rocky) planets. Mars, however, does not have as strong a magnetosphere and, as a result, is subject to occasional geomagnetic storms (short-term solar weather disturbances in the magnetosphere) that are caused by solar winds. Data from the Mars Global Surveyor showed a number of such storms (periods of significant changes in the magnetic field at the surface level) taking place, some of which lasted nearly two full days and were severe in nature.

Earth's magnetosphere is one of the most durable regions in the solar system, but the planet is not immune to magnetic storms caused by the solar winds. Indeed, when the activity of sunspots (dark, cooler areas on the sun's surface that facilitate intense magnetic fields) is at a peak, solar winds are more intense. Magnetic storms with such severity can charge the particles within the magnetosphere, particularly within the two radiation belts (one of which is called

Aurora borealis or northern lights display. Aurorae are caused by the interaction between energetic charged particles from the sun and gas molecules in the upper atmosphere of Earth, about 100 kilometers up. A stream of charged particles, called the solar wind, flows out into space continuously from the sun at speeds of 400-500 kilometers per second. On reaching Earth, the charged particles are drawn by Earth's magnetic field to the poles, where they collide with gas molecules in the upper atmosphere, causing them to emit light. (Pekka Parviainen/Photo Researchers, Inc.)

the Van Allen belt, named for its discoverer, James Van Allen, in 1958). The protons and electrons that are trapped in these belts, energized by the solar winds, interact with the oxygen and nitrogen in Earth's atmosphere, causing a bright, multicolored hue in the polar regions (and sometimes regions closer to the equator) called the aurora. In more severe cases, however, solar storms have been known to cause electrical surges, blackouts, and major disruptions to telecommunications, satellite, and navigational systems. Adding to the risks is the fact that Earth's magnetic field flips every few hundred thousand years. It is unclear how solar winds will affect Earth should such a major shift occur.

SATELLITES

One of the most useful approaches to studying the solar winds and how they interact with the rest of the solar system is satellite technology. Such orbiting probes can be positioned outside the magnetospheres of their target of study and, as a result, provide data that are not susceptible to radio and magnetic interference. Some satellites are placed in a wide orbit of the sun, gathering data that have never before been seen. For example, in 2006, National Aeronautics and Space Administration (NASA) launched the Solar Terrestrial Relations Observatory (Stereo) probes. These two probes were placed in a wide orbit of the sun, one ahead of Earth's orbit and one behind. From this vantage point, Stereo has provided views of the far side of the sun, creating a three-dimensional image of the sun for the first time. The Stereo satellites have also been useful in studying CMEs and for recording the different speeds at which solar winds travel. Such research can lead to an early warning system for Earth, should a particular wind move with the intensity needed to create a geomagnetic storm.

Satellite technologies also are invaluable in observing the interactions between solar winds and Earth's magnetic field. In 2009, scientists used the German small satellite mission, Challenging Minisatellite Payload, to monitor the magnetic field above Earth's northern polar region from the lower atmosphere. This study created a new and previously unseen illustration of how the solar winds can influence Earth's upper atmosphere.

TELESCOPES

In addition to employing space-based systems and technologies, astronomers and astrophysicists frequently employ the use of ground-based telescopes (such as radio telescopes) to monitor coronal activity and to track solar winds. For example, scientists anticipate using the Advanced Technology Solar Telescope in Hawaii (Maui) and the Frequency-Agile Solar Radiotelescope in California to monitor magnetic activity on the sun and its transit through solar winds to Earth. Experts expect to use the two systems to create models of global magnetic fields as these fields interact with the solar winds.

Ground-based telescopes also are used in the observation of how solar winds interact with the magnetospheres of other planets. For example, astronomers have long shown interest in studying the causes of radio emissions of the planets in the solar system. Their hope is that by understanding the emissions of planets such as Jupiter, they may find a way to detect new planets. However, a 2007 radio-telescope study of the emissions of Jupiter and its moons revealed that many of their most pronounced radio emissions came not from within the magnetic fields of Jupiter and its moons but rather from intense solar winds and their effects on those objects. Therefore, tracking solar winds in the solar system and in other systems could also help scientists locate new planets.

COMPUTER MODELS

In the study of solar winds, it is useful to study more than just the phenomena they foster (such as geomagnetic storms and aurora). Scientists recognize the need to also place these characteristics and occurrences into a broader and more intensive framework. It is in this capacity that computer models are deemed essential. For example, in 2011, scientists analyzed eight years (1998-2006) of solar wind occurrences, including CMEs and solar winds of varying intensity and velocity. Using a statistical program to compile data from these events (which included wind speed and similar characteristics), the resulting models can accurately predict the severity and scope of incoming solar winds.

The study of the interaction between solar winds and the planets and other objects of the solar system is not limited to just magnetic fields. Solar winds are believed to have played a major role in how much of the solar system appears today. Using advanced supercomputer models, NASA scientists recently completed a simulation of how dust particles carried on solar winds outward into the heliosphere formed the Kuiper belt (a large area near Neptune that contains millions of icy rocks and debris, including what was formerly considered a planet, Pluto). This simulation demonstrates how these particles were carried on solar winds to their current location and then collided with one another, forming larger, detectable objects.

Computer modeling systems, used in this arena, can help scientists consider how the solar system formed. Such models also can help scientists discover the existence of other objects within and beyond the solar system. Observing how the Kuiper belt was formed and how Neptune appeared during this period, astronomers can use these models to locate similar planetary systems.

RELEVANT NETWORKS AND ORGANIZATIONS

The study of the interaction between solar winds and objects in the solar system is critical to a wide range of public and private organizations and entities. For many in the scientific arena, an understanding of solar winds provides clues about how the solar system was formed, how it currently works, and how it will continue to develop. Others outside the scientific world look upon this field as useful in avoiding disruptions of the technologies on which twenty-first century civilization depends. Among the different groups that have a stake in understanding solar winds are governments, universities, and the global positioning systems industry.

Governments remain among the largest stakeholders in space exploration and astronomy. NASA, the European Space Agency, and other organizations are heavily reliant on the budget dollars that come from their respective national governments. From deep space probes to satellites to the ground-based Very Large Array radio-astronomy observatory (comprising twenty-seven massive radio antennae in New Mexico), government agencies such as NASA and the National Science Foundation are among the organizations exploring solar winds and their relationships with Earth and other planets.

Universities play an integral role not just in the study of existing data surrounding solar winds but also in providing the tools useful for such analyses. At the University of California, San Diego, for example,

scientists have developed a three-dimensional modeling program that reconstructs the effects of solar winds on Earth's magnetic fields. This Solar Mass Ejection Imager creates a clear and comprehensive illustration of the behavior of charged particles that occurs when solar winds and the magnetosphere come into contact with one another. Such university-based technologies and, more important, the theories that arise from analyses of the data they produce, are essential to understanding solar winds.

Satellite-based global positioning devices have greatly benefited modern technology, but widespread reliance on this type of navigation means that a larger percentage of the population is at a loss when this worldwide network fails. A number of geomagnetic storms have cast a light on the vulnerability of the global positioning system (GPS). A 2006 storm, for example, affected receivers across half of the world. Such events have fostered a call within the GPS industry to locate ways to safeguard against large-scale outages. Few options exist, however, that are both proven and cost-effective. These ideas include altering the satellite antennae to filter solar signals or simply replacing all GPS satellites with ships with more broadcasting power.

Some experts are examining whether a lower orbit may provide answers. That is, by moving GPS satellites from the fringes of the magnetosphere, it may be possible to avoid significant disruptions caused by solar winds and some geomagnetic storms. However, research in this arena is ongoing, particularly as scientists examine the differences between orbits and how those differences affect service.

IMPLICATIONS AND FUTURE PROSPECTS

The study of solar winds and how they interact with Earth and other planets continues to develop, particularly in light of the advances in astronomical technology, which have unveiled new information about these phenomena. In less than one-half century, humans have developed the ability to monitor solar activity, chart CMEs, and predict the arrival of the solar winds that travel through the heliosphere to their destination. Scientists have sent a number of probes throughout the solar system to monitor solar winds and their interactions with Mercury, Jupiter, Saturn, Earth's moon, and other celestial objects and planets.

The potentially destructive effects of solar winds on the technologies on which humanity is so dependent are also a driving force behind this field of study. Scientists are continually seeking ways to safeguard Earth's satellite systems and power grids from the geomagnetic storms that solar winds produce. As more and more segments of the economy become reliant on GPS, advanced telecommunications, satellite, and other technologies, the demand for an understanding of how space weather occurs and ways to avoid widespread disruptions and damage will likely continue.

Michael P. Auerbach

FURTHER READING

Burch, James L. "The Fury of Solar Storms." *Scientific American* 14, no. 4 (2004): 42-49. This article discusses how solar winds affect satellites and human space missions. The author cites a major solar flare and its effect on satellites in orbit of Earth.

Den Hond, Bas. "Scientists Predict GPS Failures." *Astronomy* 34, no. 4 (2006). In this article, the author discusses the effects of solar storms on GPS satellites. Also outlines how charged particles in the ionosphere can disrupt the systems on which pilots and drivers alike rely for navigation.

Hanslmeier, Arnold. *The Sun and Space Weather.* 2d ed. New York: Springer, 2010. This book provides a review of the various forms of space weather, including magnetic storms and solar winds. The author also offers a synopsis of efforts to study such phenomena, including satellite and deep space missions and projects.

Miralles, Mari Paz, and Jorge Sanchez Almeida, eds. *The Sun, the Solar Wind, and the Heliosphere.* New York: Springer, 2011. This book features the findings of the International Association of Geomagnetism and Aeronomy's solar wind and interplanetary division. The volume discusses the sun's interior, the processes that exist throughout the heliosphere, the corona, and other aspects of the relationship between Earth and the sun.

Velli, Marco, Roberto Bruno, and Francesco Malara, eds. *Solar Wind Ten: Proceedings of the 10th International Solar Wind Conference.* College Park, Md.: American Institute of Physics, 2003. This conference discussed topics such as the physics of the sun's corona, the solar winds, and how the sun's emissions affect the objects of the heliosphere. Among areas examined were the sun's magnetic field, CMEs, and the geomagnetic effects of solar activity on planets and other celestial bodies.

Yakolev, O. I., J. Wickert, and V. A. Anufrief. "Effect of the Solar-Wind Shock Wave on the Polar Ionosphere According to the Radio Occultation Data on Satellite-to-Satellite Paths." *Doklady Physics* 54, no. 8 (2009): 363-366. This article analyzes the effects of solar winds and solar storms have on the satellite systems in Earth's ionosphere.

Zubinaite, Vilma, and George Preiss. "Investigation of the Effects of Specific Solar Storming Events on GNSS Navigation Systems." *Aviation* 15, no. 2 (2011): 44-48. Using Norway as a point of reference, the authors discuss the disruptive influences of solar winds on telecommunications and satellite systems in that region. The article also describes the reliance in this area on such twenty-first century technologies.

See also: Earth's Magnetic Field; Remote-Sensing Satellites.

STRESS AND STRAIN

Stress and strain have to do with why and how a solid body deforms. Each point within a body under a load will have a set of stresses associated with it, varying in direction, magnitude, and the planes on which they act, according to the intensity of the forces acting within the body at that point. Each point within a deformed body will have a set of strains associated with it that indicate the translation, rotation, dilatation, and distortion experienced by the material at that point during the deformation.

PRINCIPAL TERMS

- **dilatation:** the change in the area or volume of a body; also known as dilation
- **distortion:** the change in shape of a body
- **normal stress:** that component of the stress on a plane that is acting in a direction perpendicular to the plane
- **rotation:** the change in orientation of a body
- **shear stress:** that component of the stress on a plane that is acting in a direction parallel to the plane
- **translation:** the movement of a body from one point to another

PHYSICS OF RIGID BODIES

Stress and strain are concepts that help to explain how and why rocks deform. Strain describes the deformation, and stress pertains to the system of forces that produce it. In considering strain, first it will be helpful to review some aspects of the physics of rigid bodies to appreciate the significance of these concepts.

When dealing with many problems in mechanics, it is common to assume that the bodies involved are perfectly rigid; that is, they do not deform. Such problems usually involve the balancing of forces, or, if forces do not balance, determining the resulting accelerations. Movement of a rigid body can involve translation, rotation, or both, but the individual points within the body do not move relative to one another. In a translation, all points within the body move the same linear distance and in the same direction. In a rotation, all points within a body rotate through the same angle around the same center of the rotation. Any rigid body motion can be described in terms of a translation plus a rotation. Deformation introduces further complications. A volume-conserving, shape-changing deformation is called distortion. A change in volume, without a change in shape, is called dilatation (or dilation). Strain combines all four of these possibilities: translation, rotation, distortion, and dilatation.

NET STRAINS

If the beginning and ending locations and orientations of a rigid object are known, it is easy and straightforward to determine the net translation and the net rotation of the object and of every point within the object. For example, if an airplane begins in New York facing north as it is loaded and ends up in Madrid facing northwest as it is unloaded, then it can be said to have moved 5,781 kilometers to the east and rotated 45 degrees counterclockwise. Similarly, every piece of luggage on that airplane had a net translation of 5,781 kilometers to the east and a net rotation of 45 degrees counterclockwise. Very little information is needed to determine such net displacements and rotations, but the path that the airplane took is not well represented by them. The plane probably flew along a great circular route, changing its bearings constantly, and it very likely circled a bit after taking off and again before landing. To describe the path of the plane, one would need much more data. These data might consist of a series of translations and rotations taken at one-minute intervals. Each item of luggage, rigidly fixed within the hold of the aircraft, would move through an exactly identical series of translations and rotations. Furthermore, by applying the basic laws of mechanics, one could attribute each acceleration (linear or angular) to the forces resulting from the interplay of the thrust of the engines, the force of gravity, air resistance, prevailing winds, and other relevant factors.

In much the same way that net translations and rotations can be determined by knowing the original and final locations and orientations, net distortions and dilatations can often be determined relatively easily when initial and resultant shapes and volumes are known. Analysis is simplified if the area of study can be divided into subareas such that straight, parallel lines within each area remain straight and parallel after deformation. Such deformation, called homogeneous strain, is often assumed in the study of strain. Under these conditions, initially circular objects deform into ellipses.

STRAIN PATHS

Determining the strain path requires a series of known translations, rotations, distortions, and dilatations; in turn, to tie the strain to the series of forces and stresses that produced it, one needs to know the strain path. Just as there are an infinite number of ways to fly from New York to Madrid, so there are an infinite number of strain paths that could result in identical net strains.

As an indication of the problem, consider a circle 1 centimeter in radius that deforms into an ellipse with a semimajor axis of 2 centimeters and a semiminor axis of 0.5 centimeter. Although there is no net dilatation, the deformation may have consisted of stretching in one direction and shrinking in the direction perpendicular to it. Alternatively, this deformation could have been produced entirely by distortion, as can be seen by drawing a circle on the edge of a deck of cards and then moving each card slightly to the right of the one below it. Each card will have two spots on it, one from each side of the circle. Since the distance between the spots on an individual card does not change, and the number of cards does not change, the area inside the resulting ellipse will remain constant. Continuing to deform the deck in this manner (a process called shearing) will result in the ellipse getting longer and thinner.

STRESS MEASUREMENT

Strains are produced by stresses similar to the way movements of rigid bodies are produced by forces. More specifically, unbalanced forces acting on a rigid object cause it to accelerate. Within the elastic limit, the applied uniaxial stress is proportional to the resulting uniaxial strain (Hooke's law). In brittle rocks, faulting occurs at approximately 30 degrees to the greatest compressional stress (Byerlee's law). The amount of acceleration can be calculated if the net force and the mass of the object are known. The intensity of the forces acting within a body causes it to deform, and this force intensity is called stress. The units used to measure stress are the same as those used to measure pressure and are given in terms of force per unit area. Data may be presented in terms of atmospheres, pounds per square inch, bars, or similar units. The appropriate unit (based on the International System of Units) is the pascal, defined as 1 newton per square meter, or 1 kilogram per meter-second-squared. It is important to note that stress measurements contain

an area term, and therefore they cannot be added, subtracted, or resolved as if they were forces. By multiplying a stress by the area over which it is applied it can be converted to a force, which can then be treated like any other force. It is customary to resolve it into forces parallel and perpendicular to a plane of interest. Finally, by dividing by the area of this plane, stresses can be obtained once again, yielding the shear stress and normal stress, respectively.

These factors can be demonstrated with a simple case in which a cube, 1 square meter on a side, has two forces acting on it in the vertical direction: One force of 10 newtons is pushing down on the top, another force of 10 newtons is pushing up on the bottom. The forces balance, so there will be no acceleration. Any horizontal plane within this cube will have an area of 1 square meter. It is subject to stresses of 10 pascals, perpendicular to the plane, acting on each side of it. A diagonal plane through this cube, cutting the cube in half from one edge to the other, has an area of 1.414 square meters. The component of the vertical downward force acting perpendicular to this plane (the normal force) will be 7.071 newtons, and another component of the vertical downward force acting parallel to this plane (the sheer force) will be 7.071 newtons. When these forces are divided by the area over which they act, it is apparent that there will be a normal stress of 5 pascals and a shear stress of 5 pascals acting on the upper surface of this plane.

Similar stresses can be shown to exist on the lower surface of the plane. Planes with different orientations will have other combinations of normal stresses and shear stresses, even though the forces responsible for those stresses remain the same. There are equations to manipulate the general situation, which give the normal and shear stresses acting on any plane as functions of the size and directions of the boundary forces. These result in what is called the stress ellipse, in two dimensions, or the stress ellipsoid in three dimensions. A graphical way of representing these equations (and the equations for strain) was developed by Otto Mohr in 1882 and is now called Mohr's circle.

MODES OF DEFORMATION

The results obtained previously may be compared with those for a hydrostatic condition, where stresses are the same in every direction. If stresses of 10 pascals are acting on all six sides of the cube, no matter which plane one considers inside the cube, there will be normal stresses of 10 pascals acting on each side of it. The stress ellipses and ellipsoids one might construct will be circles and spheres, and there will be no shear stresses anywhere.

Different modes of deformation are favored by different combinations of stresses. Movement on a fault plane, for example, is favored by low normal stresses and high shear stresses on that plane. Through the simple analysis described previously, it becomes clear that faulting is much more likely to occur along diagonal planes than along horizontal ones.

STRESS AND STRAIN FIELDS

The study of stresses often involves determining the stress field in a particular area, either at present or at some time in the past. After some simplifying assumptions are made concerning the geometry, mechanical properties, and boundary stresses of the area, a model is constructed that will indicate certain aspects of the stress field. Sometimes the model can be a physical one, produced from photoelastic plastic, for example. Such models can display the magnitudes of shear stresses when viewed appropriately with polarized light. More often, though, the model is constructed on a computer, and the stress ellipses are calculated for points of interest throughout the area. If geological stress indicators exist, such as the igneous sheet intrusions called dikes, the results of the model can be compared with the observed indicators, and the model can be adjusted until it fits the observations as closely as possible.

Determining at least parts of the strain field is in some ways more direct. Objects are sought in those rocks of the area that have net strains that can be determined. Frequently the distortion experienced by such an object can be easily observed and measured. A fossil that is elliptical but is known from its appearance in other areas to have been circular when it was alive provides a simple example. Strain ellipses showing the distortion of such objects can be constructed by measuring the shapes of these objects in the field.

These ellipses can then be plotted on a map. (A map with ellipses on it is a way of representing a tensor field—for example, a stress field or a strain field.) Most of the time, however, the initial size, location, and orientation of the objects are not known. One cannot tell whether a particular fossil is small because it never grew very large or whether it was once large and became smaller by deformation. If all the deformation occurred within a limited period of time, this map will represent the distortion part of the net strain field for that deformation. When similar data obtained from rocks deformed at different times are combined, a partial strain path can be obtained. The effects of more recent deformations are removed from the effects of earlier ones to isolate the earlier distortions.

The next step is to determine the stresses responsible for each increment of strain observed. To do so, it is necessary to know how each of the rocks responded to stress. Such data on mechanical behavior come from studies of experimental rock deformation. With these data, estimates can be made of the stress field present at different times in the history of the area. Finally, all this work can be applied to the known geological history of the area, permitting quantitative assessments of the various forces thought to have been active in the past.

APPLICATIONS FOR STRUCTURAL GEOLOGY

The stresses in a body are a function of the geometry of the body and the distribution of loads acting on it and within it. Determining the distribution of stresses is usually considered to be an exercise in statics, a branch of mechanical engineering, but it also plays a significant role in the earth sciences. Earthquakes occur when rock fails suddenly. Mine collapses, landslides, and dam failures are other catastrophes that occur when stress exceeds the strength of the material involved.

When a load is placed on a solid, the distribution of stresses within that solid is usually uneven. If the solid deforms, the deformation will usually also be uneven. To measure such deformation, one examines strain, which includes movements and changes in size and shape. As with stress, strain usually varies throughout the region being deformed.

A structural geologist is often concerned with determining how a region of the crust of the earth became deformed, and then why it deformed that way. Rocks often contain objects that are presently deformed but whose original shapes are known; such strain indicators include fossils, raindrop impressions, and bubbles. Using these indicators, a geologist seeks to reconstruct the strain field that existed at some time in the past. With enough indicators, along with dates for each, it may be possible to construct a strain history for the area in question. The next step is to guess, using the known mechanical behavior of the rocks involved, what the stresses were that produced the reconstructed strains. A final goal might be to seek causes for those stresses in terms of a larger picture of earth history, perhaps involving plate tectonics.

Otto H. Muller

Further Reading

Davis, George H. *Structural Geology of Rocks and Regions.* 2d ed. New York: John Wiley & Sons, 1996. Chapter 2, "Kinematic Analysis," deals with strain, and Chapter 3, "Dynamic Analysis," deals with stress. The treatment is the most descriptive and least technical of the references listed here, although some knowledge of stereonets is assumed. The double subscripts often used in the field are not employed here, and calculus is carefully avoided. Suitable for the general reader.

Fossen, Haakon. *Structural Geology.* New York: Cambridge University Press, 2010. This text is well written and easy to understand. An excellent text for geology students or resource for geologists. Provides many links between structural geology theory and application. Photos and illustrations add great value to the text. Contains a glossary, references, an appendix of photo captions, and an index.

Hatcher, Robert D., Jr. *Structural Geology: Principles, Concepts, and Problems.* 2d ed. Englewood Cliffs, N.J.: Prentice Hall, 1995. This undergraduate textbook covers folds and stress in three chapters. Intended for the more advanced reader, it is accompanied by charts and illustrations.

Hobbs, Bruce E., Winthrop D. Means, and Paul F. Williams. *An Outline of Structural Geology.* New York: John Wiley & Sons, 1976. The first section, "Mechanical Aspects" (71 pages), covers stress, strain, and the response of rocks to stress. The approach is largely descriptive, with no calculus, but double subscripts and some linear algebra are used. Although presumably written by Means, this section is sufficiently different from his book to complement it. Suitable for college students.

Johnson, Arvid M. *Physical Processes in Geology.* San Francisco: Freeman, Cooper, 1984. Chapter 5, "Theoretical Interlude: Stress, Strain, and Elastic Constants" (43 pages), develops a fairly rigorous treatment of stress and strain. Includes partial differential equations and other elements of calculus. Suitable for technically oriented college students.

Marshak, Stephen, and Gautam Mitra, eds. *Basic Methods of Structural Geology.* Englewood Cliffs, N.J.: Prentice-Hall, 1988. Chapter 15, "Analysis of Two-Dimensional Finite Strain," by Carol Simpson (26 pages), presents an overview of strain, with easily understood analogies, followed by an excellent survey of the techniques that have been used to measure strain in the field. The approach is definitely "how-to," with each step of each method clearly spelled out. Suitable for college students.

McEvily, A. J., Jr., ed. *Atlas of Stress-Corrosion and Corrosion Fatigue Curves.* Materials Park, Ohio: ASM International, 2000. An excellent reference guide filled with illustrations and graphs. Includes sections on stress-strain curves, corrosion fatigue, and stress corrosion. Bibliographical references.

Means, W. D. *Stress and Strain.* New York: Springer-Verlag, 1976. This 339-page text is intended to be used for self-study by college undergraduates. Each of the twenty-seven chapters is followed by a number of problems, and solutions to the problems are provided. Although several of the chapters go into partial differential equations and other technical subjects, most of the book is easily read by nonspecialists. Suitable for college students.

Plummer, Charles C., and Diane Carlson. *Physical Geology.* 12th ed. Boston: McGraw-Hill, 2007. A college-level introductory geology textbook that is clearly written and wonderfully illustrated. An excellent sourcebook of basic information on geologic terminology and fundamentals of geologic processes. An excellent glossary.

Ragan, Donal M. *Structural Geology: An Introduction to Geometrical Techniques.* 4th ed. New York: John Wiley & Sons, 2009. Chapter 9, "Stress"; Chapter 11, "Concepts of Deformation"; and Chapter 12, "Strain in Rocks," introduce the concepts in a mathematically straightforward way. The use of a card deck as a means of demonstrating strain was pioneered by Ragan and is included in this text. Suitable for college students.

Ramsay, John G. *Folding and Fracturing of Rocks.* New York: Blackburn Press, 2004. Originally published in 1967, a classic book on strain and strain analysis. Chapter 3, "Strain in Two Dimensions"; Chapter 4, "Strain in Three Dimensions"; and Chapter 5,

"Determination of Finite Strain in Rocks," have provided the background in these subjects to generations of geologists. Although much of what is presented is fairly technical, the abundant photographs and line drawings and some of the descriptions in the text are useful to the general reader. A quick browse through this book furnishes valuable insight into what is involved in strain analysis. Suitable for college students.

Reynolds, John M. *An Introduction to Applied and Environmental Geophysics.* 2d ed. New York: John Wiley, 2011. An excellent introduction to seismology, geophysics, tectonics, and the lithosphere. Appropriate for those with minimal scientific background. Includes maps, illustrations, and bibliography.

Stacey, Frank D., and Paul M. Davis. *Physics of the Earth.* 4th ed. New York: Cambridge University Press, 2008. Chapter 17 covers seismological methods for determining the earth's structure. Chapter 11 has good information on rock mechanics and stress forces on the earth's crust. Content is organized into accessible units, with additional mathematics and physics concepts geared toward graduate students. This text also includes many appendices and student exercises.

Suppe, John. *Principles of Structural Geology.* Englewood Cliffs, N.J.: Prentice-Hall, 1985. Chapter 3, "Strain and Stress" (33 pages), presents an elegant development of the subject. It is the only reference listed here that uses the Einstein summation convention and relies heavily on matrix manipulations. Thus, mathematically, this work may be above the level of most nontechnical college students. The eight photographs and the discussion of strain versus displacement are excellent, however, and useful to the general reader.

See also: Creep; Cross-Borehole Seismology; Discontinuities; Earthquake Magnitudes and Intensities; Earthquake Prediction; Earthquakes; Elastic Waves; Experimental Rock Deformation; Faults: Normal; Faults: Strike-Slip; Faults: Thrust; Faults: Transform; San Andreas Fault; Seismic Observatories; Seismic Tomography; Seismometers; Subduction and Orogeny; Volcanism.

SUBDUCTION AND OROGENY

Subduction and orogeny are fundamental consequences of plate tectonics and are the two processes that build mountains on the edges of continents. Through the recognition of subduction, scientists have been better able to determine regions where risks of earthquakes and volcanic explosions are significant.

PRINCIPAL TERMS

- **continental margin:** the edge of a continent that is both exposed on land and submerged below the water that marks the transition to the ocean basin
- **crust:** the outermost layer of the earth, which consists of materials that are relatively light; the continental crust is lighter than oceanic crust, which allows it to float while oceanic crust sinks
- **faulting:** the process of fracturing the earth such that rocks on opposite sides of the fracture move relative to one another; faults are the structures produced during the process
- **folding:** the process of bending initially horizontal layers of rock so that they dip; folds are the features produced by folding and can be as small as millimeters and as big as kilometers long
- **geosynclines:** major depressions in the surface of the earth where sediments accumulate; geosynclines lie parallel to the edges of continents and are long and narrow
- **intrusion:** the process of forcing a body of molten rock generally derived from depths of tens of kilometers in the earth into solidified rock at the surface
- **magma:** molten rock that is the source for volcanic eruptions
- **orogeny:** mountain building by tectonic forces through the folding and faulting of rock layers

THEORY OF GEOSYNCLINES

Subduction and orogeny are two processes that are fundamental to the evolution of continents. All continents contain long, narrow mountain chains near their edges that are composed of folded and faulted rocks that are younger than the rocks in the continental interiors. The event that formed the mountains is termed an orogeny, and the mountain chain itself is called an orogenic belt. Because of the proximity of mountain chains to the edges of continents, scientists have believed for centuries that orogenies reflected movements localized along continental margins. It is only recently, however, that orogeny has been coupled with subduction, the process in which sea floor descends below a continent or another piece of sea floor. Earlier views of orogeny were part of the theory of geosynclines. Geosynclines are linear basins that form on subsiding regions of the earth's surface adjacent to continental margins, fill with sediment, and evolve into mountains composed of folded and faulted sedimentary strata. The origin of the compressive forces responsible for the creation of the mountains was not known. Erosion of the newly created mountains provides sediment for new geosynclines that develop seaward of the mountain belt, thereby completing one geosynclinal cycle, which typically lasts on the order of a few hundred million years. The advent of plate tectonics theory in the 1960's led the majority of the scientific community to abandon the geosynclinal cycle in favor of the subduction process as an explanation for orogenies. Subduction was attractive because it readily provided a mechanism by which the large compressive forces needed to form mountains could be produced.

Orogenic belts are characterized by the folding and faulting of layers of rock, by the intrusion of magma, and by volcanism. Folds and faults form parallel to the continental margin and extend hundreds of kilometers toward the continental interior. Folding bends layers of rocks, whereas faulting takes rocks that were side by side and stacks them on top of each other in sheets up to 20 kilometers thick. Both processes significantly shorten the horizontal and thicken the vertical dimensions of the continents. At the same time as they are folded and faulted, the rocks are intruded by magmas derived from tens of kilometers below the surface. Some of the magmas eventually erupt, building volcanoes on the deformed rocks. An additional feature of orogenic belts is the juxtaposition of sequences of rock that have nothing in common with each other. The rocks in the two sequences may be different in age, composition, or style of folding. The origin of this juxtaposition was unreconciled by the theory of geosynclines, which holds that all rocks in a mountain belt were originally deposited near one another and were derived from the same source. The theory of plate tectonics, however, easily explains the juxtaposition.

THEORY OF PLATE TECTONICS

To understand subduction and orogeny, one must have a clear grasp of the theory of plate tectonics. This theory states that the surface of the earth is composed of about twelve rigid plates which are less than 100 kilometers thick. Plates are either oceanic or continental. Below the plates is a partially molten layer that allows the rigid plates to float and move relative to each other at speeds between 2 and 10 centimeters per year. The motions are defined primarily by the oceanic plates; the continental plates drift passively.

The relative motions of the plates define three types of boundaries: convergent, where plates move toward one another; divergent, where plates spread apart; and transcurrent, or conservative, where plates slide smoothly past each other. Convergent boundaries are frequently along the margins of continents, and divergent boundaries are commonly in the ocean basins. For example, the west coast of South America is a convergent boundary, and the Mid-Atlantic Ridge— the mountain range that runs down the middle of the Atlantic Ocean—is a divergent boundary. Divergent boundaries are zones along which two plates separate. This type of plate boundary is typically demarcated by a linear ridge system in an ocean basin where magma rises from deep in the earth to fill the gap created by the diverging plates. When the hot magma contacts the cold seawater, it solidifies into new oceanic crust. As the plates continue to separate, additional magma wells up from the earth's interior, allowing the continuous creation of oceanic crust at the ridge. This process is known as seafloor spreading and is responsible for the drifting of the continents on the surface of the earth.

Convergent boundaries are where two plates move toward each other and one plate subducts, or descends below, the other. The subducting plate is always oceanic, but the overriding plate may be either oceanic or continental. This reflects the greater density of oceanic crust relative to continental crust, which allows the oceanic plates to sink readily into the earth's interior, whereas the continental plates remain afloat. When two continental plates collide, neither plate subducts—they are too light—but the plates push against each other with tremendous force such that their edges buckle and huge mountain ranges grow.

This process built the world's tallest mountains, the Himalaya, which are the result of the collision between the subcontinent of India and the continent of Asia.

SUBDUCTION ZONES

Subduction zones are characterized by a progression from the subducting to the overriding plate of deep trenches, high mountains, and many volcanoes that occupy an area hundreds of kilometers wide and thousands of kilometers long. The deep trench, frequently filled with sediments eroded from the adjacent mountains, marks the point in the ocean floor where the subducting plate bends to descend below the overriding plate. As the oceanic plate descends, these sediments are scraped onto the overriding plate. Slivers of oceanic crust may also scrape off and mix with the sediments. The off-scraped rocks form an intricately folded and faulted region tens of kilometers wide and several kilometers high at the edge of the overriding plate. These complexly deformed mixtures of sediments and slivers of oceanic crust are called mélanges and are characteristic of most ancient subduction zones now exposed on land.

Another important feature of subduction zones is the linear belt of volcanoes on the overriding plate that parallels the plate boundary. The volcanoes grow from the eruption of magma that is generated at the interface between the subducting and overriding plates at depths between 100 and 200 kilometers. At these depths, the temperature of the earth is high enough to melt small areas of either the subducting or the overriding plate. The magma rises, intruding the rocks at the surface and eventually erupting to build the volcanic belt. Some of the magma, however, may solidify between the top of the oceanic plate and the surface.

The similarity of features in orogenic belts and subduction zones is striking and forces the obvious conclusion that subduction leads directly to orogeny. An orogeny can occur either during subduction of an oceanic plate below a continental plate, such as on the west coast of South America, or during the collision of two continental plates, such as in the Himalaya. Because continents do not subduct, the compressive forces are much greater in a continent-continent collision than in seafloor subduction. The mountains produced during collision (Himalaya), therefore, are much taller than those generated during subduction (Andes).

Consequences of Plate Tectonics

The theory of plate tectonics elucidates important differences between the oceans and the continents and provides a mechanism by which different rock sequences can be juxtaposed in orogenic belts. The ocean basins are transient features that are constantly modified by the growth and destruction of new sea floor at divergent and convergent boundaries, respectively. In contrast, the continents are too light to be subducted and are permanent features of the earth's surface. This consequence of plate tectonics is supported by the 200 million year age of the oldest sea floor and the 4 billion year age of the most ancient rocks on the continents. Continents, therefore, drift, fragment, and collide as relative plate motions change through geologic time. The collision of continents that were once widely separated allows the bringing together of rocks that have had very different histories. As the collision leads to orogeny, these different sequences of rock may be juxtaposed in the same mountain belt.

The difference between the age of orogenic belts and the interiors of continents implies that the continents have evolved through time by the addition of material at their edges during orogenies. Orogenic belts are also of different ages, ranging from a billion years to zero (actively forming). Two or three belts whose ages decrease away from the continental interior may define one edge of a continent. This suggests that orogenies have occurred repeatedly through geologic time and that continents have added material continuously to their margins since the formation of their interiors. Because the ocean floor is so young, orogenic belts are the only record of subduction and collision events prior to 200 million years ago. If subduction is the only mechanism responsible for orogeny, plate tectonics must have been active since early in the history of the earth.

Analysis of Earthquakes

Subduction and orogeny are studied by hundreds of scientists, each of whom looks at only a small part of the picture. One may determine the composition of volcanic rocks that are characteristic of subduction zones; another may examine the styles of folds and faults in orogenic belts. Three techniques, however, are dominant in the study of subduction and orogeny: the analysis of the locations and sizes of earthquakes, the discrimination of relationships between different

types of rocks in the field, and the investigation of features in deep-sea trenches and in the submerged region of folded and faulted rocks. The first defines where subduction and orogeny occur today, whereas the second determines what the physiographic expressions of these processes are, how they are preserved in the rocks, and where they were active in the past. The third technique provides a direct link between subduction and orogeny and illustrates the early stages of development of a mountain belt.

One of the most important discoveries of plate tectonics was that earthquake zones define plate boundaries. Earthquakes occur when a fracture, or fault, forms in the earth's crust, and the two pieces on either side of the fault move, or slip, past each other. For large earthquakes, the slip is on the order of 10-20 meters. The forces responsible for faulting are simply the result of the relative motions of the plates at the plate boundaries. The motion can accumulate in the rocks for hundreds of years prior to causing a rupture. When the crust finally breaks, the energy stored by the rocks is released suddenly as waves that travel through the earth and generate the intense vibrations associated with an earthquake. The rupture continues for as much as 1,000 kilometers and moves at speeds in excess of 10,000 kilometers per hour.

The energy carried by the waves is recorded on seismographs, which are instruments that monitor ground motion. Seismographs are composed of a mass attached to a pendulum. The mass remains still during an earthquake, measuring the amount the earth moves around it. The motion is recorded on a chart as a series of sharp peaks and valleys that deviate from the background value measured during times of no earthquake activity. The arrival of the waves at different times at different places allows the geophysicist to calculate the location, or epicenter, of the earthquake. The amount of the deviation of the peaks and valleys from the background noise is an estimate of the magnitude of the earthquake.

Earthquakes near mountain belts define zones that extend at an angle from the surface of the earth at the deep-sea trench to depths of hundreds of kilometers below the continents. This zone corresponds to the subducting plate at a convergent boundary. As a result, the locations of subduction zones that are currently active are very well known. The descent of a subducting plate below an overriding continent has triggered some of the deepest and largest earthquakes ever recorded. Continued motion of the plate and rupturing of the earth's crust in response translate into mountain ranges on the earth's surface.

STUDY OF ROCKS AND DEEP-SEA TRENCHES

Analysis of earthquakes is essential to evaluate the modern plate tectonic setting of the earth, but it reveals nothing about the geologic past. Information about plate tectonics of the past must be obtained from looking at ancient mountain belts. Recognition of relationships among rocks in the field involves determining the ages, compositions, and histories of the rocks. This process led to the discovery that mountain belts on different continents contained rock sequences that were very similar. For example, rocks in the Appalachian Mountains of the east coast of North America were found to match closely those in the Atlas Mountains of the west coast of Africa. Conversely, recognition of relationships determined that dissimilar rock sequences frequently are adjacent to each other in the same orogenic belt. Both phenomena are most readily explained by continental drift, seafloor spreading, and subduction.

The critical link between ancient orogenic belts and modern subduction zones identified by earthquake activity was provided by deep-sea trenches. Using highly sophisticated techniques to "see" the ocean floor, scientists discovered the region of off-scraped rocks that lies on the overriding plate in a subduction zone. These regions sometimes continue to the continental margin, where they are exposed on land as mountains. Thus, subduction was observed to cause folding and faulting in rocks and to build mountains—both important processes in orogenic belts.

EARTHQUAKE AND VOLCANO HAZARD ASSESSMENT

The theory of plate tectonics provides scientists with a process that can be observed—subduction—to explain the origin of mountains. Because young mountain chains are the locus of most of the large earthquakes that occur today, understanding subduction yields insight into the potential for destructive earthquakes in any given area. This is extremely important because most of the global population lives along convergent plate boundaries. The identification of subduction zones at the margins of the Pacific Ocean has explained the "Ring of Fire," a region of abundant earthquakes and volcanoes that had long

Drilling into the San Andreas fault at Parkfield, California. An international team of scientists are embarking on a project to drill an angled hole through a seismically active portion of the San Andreas fault zone, creating a San Andreas Fault Observatory at Depth (SAFOD). In the summer of 2004, SAFOD began drilling directly through the fault to a depth of 3.2 kilometers, in order to obtain samples and make geophysical measurements within and adjacent to the fault zone, and to install instruments to continuously monitor variations in rock deformation and other parameters during the earthquake cycle. (Science Source)

puzzled the scientific community. Restriction of most earthquakes to plate boundaries allows the assessment of earthquake hazards anywhere in the world if the locations of plate boundaries are known. For example, the city of Santiago in Chile, which is above a subduction zone, has a high risk, whereas the city of Chicago in the United States, which is in the continental interior, has a low risk.

Additional information can also be gathered about the type of earthquakes that may occur. In subduction

Subduction and Orogeny

zones, the piece of the crust that is above the rupture typically moves upward relative to the piece below, which generates waves that shake the ground in certain directions. At divergent plate boundaries, however, the piece of crust that is above the rupture moves downward relative to the piece below. This motion produces waves that vibrate the ground in directions different from those generated by earthquakes in subduction zones. Additional differences between convergent and divergent boundaries that may affect ground motion include the depth and size of the earthquakes. Subduction zones generate the deepest and largest earthquakes; earthquakes at divergent plate boundaries are more frequent, smaller, and shallower. Knowledge of the way the ground may move helps civil engineers to design and construct buildings able to withstand large earthquakes.

Eruptions of volcanoes that lie above subduction zones can be devastating. These volcanoes typically erupt violently and explosively in contrast to volcanoes near mid-ocean ridges, which erupt quietly and smoothly. This reflects the greater viscosity (resistance to flow) of magmas at convergent boundaries relative to those at divergent boundaries. Because of their greater viscosity, the magmas above subduction zones tend to plug the volcanoes at the surface, preventing any eruptions. Finally, when the pressure below the plug is great enough, the volcano erupts with such force that cities nearby are damaged considerably. For example, in 79 C.E., the entire city of Pompeii, Italy, was destroyed, and hundreds of people were killed by the volcano Vesuvius. Clearly, the investigation of subduction and orogeny is beneficial to understanding the forces of nature that are harmful to humankind. Perhaps someday in the future, large earthquakes and violent volcanic eruptions may be predicted far enough in advance that precautions can be taken to prevent the loss of human life.

Pamela Jansma

FURTHER READING
Bebout, Gray E. *Subduction Top to Bottom*. Washington, D.C.: American Geophysical Union, 1996. Bebout's book gives clear definitions and explanations of subduction, folding, faults, and orogeny. Illustrations and maps help to clarify some difficult concepts.

Kearey, Philip, Keith A. Klepeis, and Frederick J. Vine. *Global Tectonics*. 3rd ed. Cambridge, Mass.: Wiley-Blackwell, 2009. This college text gives the reader a solid understanding of the history of global tectonics, along with current processes and activities. The book is filled with colorful illustrations and maps.

Lowrie, William. *Fundamentals of Geophysics*. 2d ed. New York: Cambridge University Press, 2007. Excellent overview of geophysics topics written for a student with strong physics background. Lowrie presents the mathematics at a level understood by mid-level university students.

Oncken, Onno, et al., eds. *The Andes: Active Subduction Orogeny (Frontiers in Earth Sciences)*. Berlin: Springer-Verlag, 2006. The book is a comprehensive overview of subduction orogeny. Complete with chapter abstracts, high-resolution images, and a DVD.

Osterihanskay, Lubor. *The Causes of Lithospheric Plate Movements*. Prague: Charles University, 1997. This college-level text examines geography and geoecology in relation to plate tectonics and the earth's lithosphere. Many theories are illustrated with graphics and maps.

Press, F., and R. Siever. *The Earth*. 4th ed. New York: W. H. Freeman, 1986. An excellently illustrated introductory text on geology, this book has five chapters that deal with folding, faulting, plate tectonics, earthquakes, and orogeny. A map of the major plates appears on the inside back cover. The glossary is huge and indispensable. Recommended for senior high school and college students.

Roeder, D. H. "Subduction and Orogeny." *Journal of Geophysical Research* 78 (1973): 5005-5024. A classic article on the subject. A great overview of the topics of subduction and orogeny written in a nontechnical manner.

Shelton, J. S. *Geology Illustrated*. San Francisco: W. H. Freeman, 1966. This book has a superb collection of photographs and sketches drawn from the photographs that illustrate specific geologic features such as folds, faults, and volcanic landforms. Although it does not talk about subduction and orogeny directly, this source does help the reader in visualizing the features representative of convergent plate boundaries.

Short, Nicholas M., and Robert W. Blair. *Geomorphology from Space: A Global Overview of Regional Landforms*. Washington, D.C.: National Aeronautics and Space Administration, 1986. This book contains beautiful pictures taken by various satellites that orbit the earth. Text accompanies each picture and explains the tectonic setting. Many mountain belts and volcanic chains are shown. Although the text is fairly technical, the photographs are worth examining by anyone. Recommended for college-level students.

Tatsumi, Y. "The Subduction Factory: How It Operates on Earth." *GSA Today* 15 (2005): 4-10. This article is technically written and requires the reader to have a science background. A good explanation of the theories behind subduction and orogeny, with a well-defined image (Figure 1) modeling major concepts.

Uyeda, Seiya. *The New View of the Earth: Moving Continents and Moving Oceans.* Translated by Masako Ohnuki. San Francisco: W. H. Freeman, 1978. Although slightly dated, this book discusses the historical context of the theory of plate tectonics in addition to explaining the theory very well. Nontechnical and designed for the nonscientist. Interesting stories about the people responsible for the theory abound. Suitable for anyone interested in plate tectonics.

Wilson, J. T., ed. *Continents Adrift.* San Francisco: W. H. Freeman, 1972.

_____. *Continents Adrift and Continents Aground.* San Francisco: W. H. Freeman, 1977. These two volumes are collections of articles originally printed in *Scientific American* magazine. The amount of overlap is very small, which amply illustrates the rapid advances in plate tectonics in the late 1960's and early 1970's. The articles are very well illustrated, and the introductions in each volume are very helpful. Provides a historical perspective of plate tectonics. Suitable for anyone interested in any aspect of plate tectonics.

See also: Earthquake Engineering; Earthquake Prediction; Earthquakes; Earth's Core; Earth's Differentiation; Earth's Mantle; Heat Sources and Heat Flow; Lithospheric Plates; Mantle Dynamics and Convection; Mountain Building; Plate Motions; Plate Tectonics; Plumes and Megaplumes; Seismometers; Volcanism.

T

TECTONIC PLATE MARGINS

The outer 70 to 200 kilometers of the earth comprises a number of rigid plates that move about independently of one another. Each plate interacts with the adjacent plate along one of three types of plate margin: convergent, divergent, or conservative. The interactions of the moving plates along these margins cause most earthquakes as well as much of the volcanic activity on the surface of the earth.

PRINCIPAL TERMS

- **accretionary prism:** a complex structure composed of fault-bounded sequences of deep-sea sediments mechanically transferred from subducting oceanic lithosphere to the overriding plate; it forms the wall on the landward side of a trench
- **asthenosphere:** the soft, partially molten layer below the lithosphere
- **convection cell:** a single circular path of rising warm material and sinking cold material
- **lithosphere:** the outer, rigid shell of the earth that contains the oceanic and continental crust and the upper part of the mantle
- **rifting:** the process whereby lithospheric plates break apart by tensional forces
- **seafloor spreading:** the concept that new ocean floor is created at the ocean ridges and moves toward the volcanic island arcs, where it descends into the mantle
- **subduction zone:** a region where a plate, generally oceanic lithosphere, sinks beneath another plate into the mantle
- **transform fault:** a fault connecting offset segments of an ocean ridge along which two plates slide past each other
- **volcanic island arc:** a curving or linear group of volcanic islands associated with a subduction zone
- **Wadati-Benioff zone:** the inclined band of earthquake focus points interpreted to delineate the subducting oceanic lithosphere; better known as the Benioff zone

PLATE TECTONICS

The surface of the earth is a mosaic of several large and more numerous smaller plates that move about laterally. The boundaries, or margins, of these plates are the sites of most of the volcanic and earthquake activity on the earth. The geological concept of the plate and plate margins ultimately has its origins in the theory of continental drift. The first comprehensive theories of continental drift were independently proposed around 1910 by the American geologists F. B. Taylor and H. H. Baker and the German meteorologist Alfred Wegener. Wegener's work was particularly thorough, and he is generally considered to be the person who first made the theory of continental drift an important scientific issue. Wegener devoted much time and effort to matching geological features and fossil types on both sides of the Atlantic Ocean. He argued, based on this work, that approximately 230 million years ago, all the continents were joined as one supercontinent that he named Pangaea. Furthermore, Wegener suggested that Pangaea broke apart approximately 170 million years ago. Since then, the various parts of the supercontinent—the modern continents—have moved to their present positions. Wegener also postulated that as the continents move through the oceans, their leading edges become crumpled and often collide with other continents, thereby forming the mountain belts.

In the period after World War II, several important discoveries made by oceanographers studying the ocean basins lent support to Wegener's scoffed-at theory. This postwar research ultimately led to the formulation of the theory of seafloor spreading proposed by Harry Hammond Hess of Princeton University. According to Hess's theory, new ocean floor is continuously forming at the ocean ridges, or the large subsea mountain ranges that traverse the earth, in a conveyor belt fashion. As this process continues, the newly formed sea floor moves laterally away from the ocean ridge on both sides of the ridge. The opening in the earth's surface created by the spreading of the sea floor at the ridge is filled with magma, or molten rock, from the mantle, which cools to form new sea floor. Hess and other scientists suggested that if the sea floor is continuously moving, the continents must also be moving with it. Thus, the concepts of continental drift and seafloor spreading were combined into the more comprehensive theory of plate tectonics. This revolutionary theory in the earth sciences describes the movement of rock in the earth's 70 to 200 kilometer-thick outer brittle shell, the lithosphere, as it moves over the deeper, more ductile, partially molten asthenosphere. The lithosphere, which includes continental and oceanic crust and the upper part of the mantle, comprises a number of large and small, rigid lithospheric plates that move independently of one another. As these plates move, they interact along one of three types of plate margin: divergent or accreting margins, convergent or destructive margins, and conservative or neutral margins.

DIVERGENT PLATE MARGINS

Divergent or accreting plate margins are tensional plate boundaries that correspond to the ocean ridges. According to plate tectonics theory, ocean ridges, also referred to as spreading centers, are sites on the earth's surface where new oceanic lithosphere is formed by the process of rifting, or the tensional separation of plates. Rifting can occur in the oceans as well as on the continents, as in the case of the East African rift zone. If rifting begins on a continent and continues for an extended period of time, a new ocean basin will ultimately form—as is now occurring in the Middle East, where the Red Sea rift is creating a new ocean between Saudi Arabia and northeastern Africa.

The crests of ocean ridges are characterized by deep valleys believed to have been caused by tensional faulting in response to oppositely directed lateral movement of the plates bound by the ocean ridge. This valley is referred to as the rift valley and is marked by a high degree of earthquake activity. The rift valley is also the site of voluminous outpourings of basaltic magma, or molten rock enriched in iron and magnesium.

Some of the most dramatic evidence of the processes occurring at divergent plate margins has come from a series of observations made from submersibles on the Galápagos spreading ridge near the equator, just west of South America and that part of

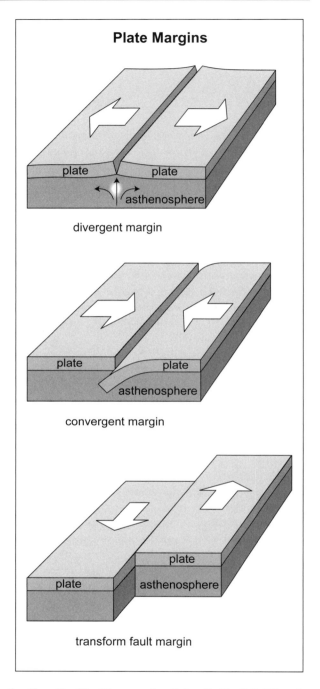

Plate Margins

divergent margin

convergent margin

transform fault margin

the East Pacific Rise south of the Gulf of California. Researchers observed undersea hot springs and mounds of iron-rich clay minerals and manganese dioxide precipitated from the hot ore-carrying springs. Similar submersible dives to the Mid-Atlantic Ridge southwest of the Azores permitted observation of submarine volcanism and yielded abundant evidence of tensional faulting within the narrow rift valley.

CONVERGENT PLATE MARGINS

Convergent or destructive plate margins are those areas of the earth's surface where the lithospheric plates grind together head-on and are then recycled back into the asthenosphere. Thus, these margins are characterized by compressive tectonic forces. Submarine features typical of convergent plate margins are long, narrow troughs on the sea floor referred to as trenches. Hess postulated that the trenches mark the positions where ocean lithosphere created at ocean ridges is drawn down or sinks into the mantle. The trenches, which are the deepest points of the oceans, are closely associated with volcanic island arcs, or linear or arcuate groups of volcanic islands such as Japan or the Aleutian Islands.

Convergent margins are characterized by a high incidence of earthquakes, many of which originate at depths greater than 600 kilometers within the earth. Scientists have demonstrated that the focus points of these earthquakes (those points within the earth from where the seismic energy is first generated) are generally found within a band inclined from the trench toward the volcanic island arc or continent called the Wadati-Benioff zone (better known as the Benioff zone). Realization of this distribution of earthquakes at convergent margins led scientists to speculate that the Wadati-Benioff zone delineates a slab of dense lithosphere formed at an ocean ridge that, according to plate tectonics theory, is sinking at the trench into the mantle. This process of lithospheric sinking at convergent margins is referred to as subduction. In general, the most powerful earthquakes at convergent margins are the shallow-focus earthquakes generated close to the trenches. These earthquakes occur when a sinking lithospheric plate moves beneath an island arc or a continent and drags the overriding plate down a bit. Eventually, this process reaches a critical point, and sudden slip occurs along the boundary of the sinking plate and the plate beneath which it is moving, thereby creating the earthquake.

SUBDUCTION

The volcanic island arcs, like the distribution of earthquake focus points, can be considered in terms of the process of subduction. The basic question concerns how the magma that spewed from the volcanic islands forms beneath the island arcs. It is generally agreed that island arc volcanism is caused by melting of the subducting plate as it descends into the hot mantle. Generation of magma may also be assisted by frictional melting along the upper surface of the subducting plate as it moves beneath the island arc.

A final point regarding convergent margin processes concerns the fate of marine sediments on subducting oceanic lithosphere. Initially, it was postulated that all the marine sediments on descending oceanic lithosphere should be piled up and folded on the bottom of the trench. Studies of modern trenches, however, indicate that trenches generally contain only minor amounts of sediment. A more complete understanding of this problem was gained when more sophisticated geophysical techniques were applied to the study of convergent margins. Results of these investigations suggest that marine sediments carried on a subducting plate are stripped off the plate at the trench. These sediments are attached to the leading edge of the overriding plate to form a complexly deformed sequence of sedimentary rocks called an accretionary prism that builds the landward wall of the trench.

In summary, convergent plate margins display a series of features that have been interpreted in terms of subduction, the dominant plate tectonic process occurring at these margins. Trenches mark the locations on the earth where oceanic lithosphere, formed at a divergent plate margin, sinks into the mantle. At least a portion of the deep marine sediment carried atop this plate is mechanically transferred to the leading edge of the overriding plate to form an accretionary prism. As the plate moves deeper into the mantle, it fractures and generates earthquakes along its length. Additionally, it begins to melt, thereby producing magma that rises to the surface of the earth to form a volcanic island arc. If the oceanic plate is attached to a continent, that continent eventually reaches the trench. Because continental lithosphere is less dense and therefore more buoyant than oceanic lithosphere, however, the continent cannot be subducted, and instead it collides with the volcanic island arc or the overriding continent. This collision results in the formation of a mountain belt, as in the case of the collision of the Indian subcontinent and the Tibetan plateau to the north, which is still forming the Himalaya belt.

CONSERVATIVE PLATE MARGINS

The third type of plate margin is the conservative, or neutral, margin. These margins, along which lithosphere is neither created nor destroyed, are characterized by oppositely directed horizontal movement of adjacent plates. The actual boundaries of the moving plates are marked by transform faults, or faults along which plates slide horizontally past one another. The San Andreas fault of California, perhaps the best-known example of a transform fault, marks the boundary between the northwest-moving Pacific plate and the North American plate. This plate boundary is characterized by contrasting geology on both sides of the fault, by little if any volcanic activity, and by powerful shallow-focus earthquakes such as the kind that devastated San Francisco in 1906.

Transform faults separate offset segments of ocean ridges. Although ocean ridges extend continuously for thousands of kilometers, they are actually broken into much smaller segments separated by transform faults that are oriented at nearly right angles to the ridge segments. The relative movement of the two plates along a transform fault is caused by creation of new sea floor at the two offset ridge segments. As ocean lithosphere forms at and moves away from one ridge segment, it slides in the opposite direction past lithosphere forming at the other ridge segment. The transform fault, therefore, marks the contact of the oppositely moving plates and is situated between the ridge segments.

TRIPLE JUNCTIONS

Plate tectonics theory requires that there be single points, called triple junctions, at which three lithospheric plates meet. In the Middle East, for example, three divergent plate margins—the Gulf of Aden, the East African rift, and the Red Sea rift—meet at what is referred to as a ridge-ridge-ridge triple junction. Almost any combination of the three plate margins—ridge, trench, and transform fault—can form triple junctions. Some types of triple junctions move with the plates, and they may even be subducted.

ECHO SOUNDING AND SEISMIC REFLECTION PROFILING

Much of what is known about plate margins, particularly convergent and divergent margins, has come about through detailed study of the surface and interior structure of the ocean floor. One of the

most important techniques developed in this regard is echo sounding. In echo sounding, a sound pulse generator-receiver system mounted on the hull of a ship emits sound pulses at regular intervals. Each pulse travels to the ocean floor at a known velocity and echoes back to the ship, where its return is detected by the pressure-sensitive receiver. The recording apparatus, a precision depth recorder, indicates the travel time of the sound pulse to and from the ocean bottom on an advancing paper chart. As the ship moves across the ocean, travel time marks for a succession of pulses detected by the receiver are displayed on the chart profile. The depth to the sea bottom is then calculated by multiplying the velocity of the sound pulse by one-half its travel time. This method was most instrumental in defining the ocean ridges and trenches on the ocean floor.

A somewhat more sophisticated approach using seismic waves allows scientists to study the internal structure of the upper part of oceanic lithosphere. In this approach, referred to as seismic reflection profiling, a ship emits sound waves powerful enough to penetrate the bottom of the ocean and then to reflect back to the ship. More specifically, these sound waves, which may be generated either by undersea explosions or by compressed air, reflect from the surface of the ocean floor and from internal sediment layers and faults back to the ship, where they are picked up by a receiver, or hydrophone, towed behind the ship. The travel times of the waves reflected off and from within the sea floor are recorded on charts by a sparker-profiler. Seismic reflection profiling has helped scientists to understand better the interior structure of the ocean bottom. For example, the faults bounding the slivers of marine sediments in accretionary prisms at convergent margins were recognized through the use of this technique.

MULTINATIONAL RESEARCH PROJECTS

The Deep Sea Drilling Project (DSDP), a multinational research program initiated in 1968, attempted to understand better the evolution and geologic history of the modern oceans by drilling through the deep-sea sediments into the underlying igneous floor of the ocean. A specially designed ship, the *Glomar Challenger,* was used in the drilling. Among other things, results of the DSDP indicated that the age of the igneous ocean floor increases away from ocean ridges, thereby substantiating the major tenet of seafloor spreading: that oceanic lithosphere is produced at, and moves away from, the ridges. The Deep Sea Drilling Project was superseded by a new program of research with many of the same goals, the Ocean Drilling Program (ODP).

Many details of plate margins, particularly divergent ocean ridges, have been revealed through direct observation of the sea floor. The French-American Mid-Ocean Undersea Study (project FAMOUS) of 1973 and 1974, for example, concentrated on a small area of the Mid-Atlantic Ridge southwest of the Azores. Several deep-sea submersible submarines were used to dive to the ridge to map the shape of the rift valley and to collect samples of the ocean floor. This project permitted observation of the tensional faults that formed the rift valley and extrusion of basaltic magma. In several submersible dives to transform faults, scientists recovered igneous rock samples that showed evidence of the shearing associated with horizontal plate movement along the transform faults. A number of submersible dives to the East Pacific rise in the eastern Pacific Ocean have allowed marine geologists to observe submarine hot springs associated with lava extrusion at a divergent margin.

Heat Flow and Rock Sequence Studies

Measurement of terrestrial heat flow (the amount of heat that escapes from the earth's interior through the sea floor) yielded particularly valuable information regarding the nature of plate margins. Results of heat-flow studies indicated that the ocean ridges, once thought to be dormant submarine mountain ranges, are actually sites where large amounts of heat from the interior of the earth reach the earth's surface. This finding fit in well with Hess's convection-cell interpretation of ocean ridges. In addition, heat-flow studies demonstrated extremely low heat-flow values in the trenches, an observation consistent with Hess's proposal that convection cells sink into the mantle at convergent margins.

Finally, studies of modern plate margins have been supplemented by investigations of rock sequences exposed on land and interpreted to have formed at ancient plate margins. This approach is particularly useful to the study of convergent plate margins. For example, highly deformed or chaotic rock units, referred to as mélanges and exposed in the Appalachian belt, along coastal California, and elsewhere, have

been interpreted as marine sediments that were incorporated into ancient accretionary prisms. By studying these exposed sedimentary rocks, geologists can understand better the processes occurring at modern convergent plate margins.

Dangers and Benefits

Plate margins are generally not a major concern of most people unless they happen to live or work near one. Nevertheless, one has only to pick up a newspaper to see the effects of plate margins and their attendant processes on human life. Convergent margins, for example, are characterized by frequent earthquakes and are generally prone to volcanic activity, some of which may be violent. The powerful earthquakes and deadly volcanoes of the Aleutian Islands, Central and South America, Japan, and Indonesia attest the potentially dangerous conditions of convergent plate margins. Conservative plate margins, like the San Andreas fault, are susceptible to powerful earthquakes, although volcanic activity is not likely. The instability of plate margins must be kept in mind by community planners so that proper building codes can be created and followed to reduce the potential for catastrophe in these areas.

Despite the obvious dangers of living close to or along plate margins, there can be some benefits. In Iceland, for example, the heat emanating from the Mid-Atlantic Ridge is used as geothermal energy. Indeed, Reykjavík, the capital of Iceland, is heated entirely by geothermal energy.

Understanding the relation of plate tectonics and metal deposits is of paramount importance given the growing global need for various metals. At ocean ridges, for example, marine geologists and oceanographers have observed the formation of metallic sulfide ores. These deposits, which form in association with the basalt magma extruded at the ridge, precipitate out of the hot water that circulates through the newly erupted basalt. Convergent plate margins are characterized by various types of metal deposits formed in association with magma generated during subduction. In the Andes belt of South America, iron, copper, and gold ores accumulated in response to subduction of the Pacific Ocean floor beneath the western coast of South America.

Gary G. Lash

FURTHER READING

Bonatti, E., and K. Crane. "Ocean Fracture Zones." *Scientific American* 250 (May 1984): 40. Excellent discussion of transform faults and associated oceanic fractures. Suitable for the college-level reader.

Condie, Kent C. *Plate Tectonics and Crustal Evolution.* 4th ed. Oxford: Butterworth Heinemann, 1997. An excellent overview of modern plate tectonics theory that synthesizes data from geology, geochemistry, geophysics, and oceanography. A very helpful tectonic map of the world is enclosed. The book is nontechnical and suitable for a college-level reader. Useful "suggestions for further reading" follow each chapter.

Dewey, J. F. "Plate Tectonics." *Scientific American* 226 (May 1972): 56. A good overview of the theory of plate tectonics and plate margins. Can be read by the high school or college student.

Heirtzler, J. R., "Seafloor Spreading." *Scientific American* 219 (December 1968): 60. This article provides an excellent discussion of the theory of seafloor spreading. Suitable for high school students.

Heirtzler, J. R., and W. B. Bryan. "The Floor and the Mid-Atlantic Rift." *Scientific American* 233 (August 1975): 78. Excellent discussion of divergent plate margins, with the Mid-Atlantic Ridge as the example. Can be read by high school and college students.

Kearey, Philip, Keith A. Klepeis, and Frederick J. Vine. *Global Tectonics.* 3rd ed. Cambridge, Mass.: Wiley-Blackwell, 2009. This college text gives the reader a solid understanding of the history of global tectonics, along with current processes and activities. The book is filled with colorful illustrations and maps.

Kious, Jacquelyne W. *This Dynamic Earth: The Story of Plate Tectonics.* Washington, D.C.: U.S. Department of the Interior, United States Geological Survey, 1996. Kious is able to explain plate tectonics in a way suitable for the layperson. The book deals with both historic and current theory. Illustrations and maps are plentiful.

Ladd, J. W., et al. "Caribbean Marine Geology: Active Margins of the Plate Boundary." In *The Caribbean Region.* Geological Society of America, edited by G. Dengo and J. E. Case. *The Geology of North America,* Vol. H (1990): 261-290.

Marsh, B. D. "Island-Arc Volcanism." *American Scientist* 67 (March/April 1979): 161. A detailed discussion of volcanic activity at convergent margins. Suitable for college students.

Ogawa, Yujiro, Ryo Anma, and Yildirim Dilek. *Accretionary Prisms and Convergent Margin Tectonics in the Northwest Pacific Basin.* New York: Springer Science+Business Media, 2011. Discusses new techniques in plate tectonics studies. One volume of the series *Modern Approaches in Solid Earth Sciences. Accretionary Prisms, Tectonics, and Pacific Ocean Events.*

Sutherland, Lin. *The Volcanic Earth: Volcanoes and Plate Tectonics, Past, Present, and Future.* Sydney, Australia: University of New South Wales Press, 1995. Although Sutherland focuses on volcanic activity in Australia, the book provides an easily understood overview of volcanic and tectonic processes, including the role of igneous rocks. Includes color maps and illustrations, as well as a bibliography.

Tokosoz, M. N. "The Subduction of the Lithosphere." *Scientific American* 233 (November 1975): 88. This article describes the process of subduction at convergent margins and can be read by high school and college students.

Uyeda, Seiya. *The New View of the Earth: Moving Continents and Moving Oceans.* San Francisco: W. H. Freeman, 1971. An excellent presentation of the evolution of the theory of plate tectonics from continental drift. Convergent and divergent plate margins are particularly well discussed, with numerous examples from the Pacific Ocean. Probably most suitable for college-level readers.

See also: Continental Drift; Creep; Earth's Core; Earth's Differentiation; Earth's Mantle; Heat Sources and Heat Flow; Lithospheric Plates; Mantle Dynamics and Convection; Plate Motions; Plate Tectonics; Plumes and Megaplumes; Subduction and Orogeny; Volcanism.

TSUNAMIS AND EARTHQUAKES

A tsunami is a series of traveling ocean waves of extremely long length and depth generated by violent submarine disturbances associated primarily with earthquakes. The mass movement of water represented by a tsunami poses a significant danger to low-lying coastal regions, particularly along the Pacific Rim.

PRINCIPAL TERMS

- **continental shelf:** the gently sloping surface that extends between the shoreline and the top of the continental slope, which slopes steeply to the deep ocean bed
- **earthic crust:** the thin, outermost layer of the earth that varies in thickness from about 30 to 50 kilometers
- **ocean wave:** a disturbance on the ocean's surface, viewed as an alternate rise and fall of the surface
- **Richter scale:** the measurement of the magnitude of seismic disturbance (an earthquake) using the amplitude of seismic waves, named after American seismologist and physicist Charles Richter
- **seismic sea wave:** an enormous wave in the ocean generated by an earthquake under the floor of the ocean or along the seacoast
- **subduction zone:** the zone, at an angle to the earth's surface, down which an upper layer of oceanic or continental plate descends
- **wave refraction:** the process by which the direction of waves moving through shallow water is altered by local submarine conditions
- **wavelength:** the distance between the crest of one water wave to the crest of the next

UNDERSEA ENVIRONMENT

While other catastrophic events can trigger a tsunami, undersea earthquakes are responsible for the majority of them. Many tsunamis occur around the Pacific Rim in regions designated by geologists as subduction zones, where the dense earthic crust of the ocean floor slips below the lighter continental shelf crust and into the earth's mantle. Subduction zones are most common along the west coasts of North and South America and the coasts of Japan, eastern Asia, and the Pacific island chains. Another subduction zone is located in the Caribbean Sea but is not considered as active as those in the Pacific.

Scientists believe the sudden movement in the sea floor during an earthquake disrupts the equilibrium of the overlying expanse of water, raising or lowering enormous amounts of it all the way from the sea floor to the surface. Once the sea floor settles into its new position, it has nowhere to go until the next disturbance. However, the water mass above it is still subject to the downward pull of gravity. Consequently, as the swell of water returns to its original position, the water around it is pushed up, creating a rippling effect or series of waves called a tsunami. The primary factor in determining the size of the tsunami is the amount of sea floor uplifted during the undersea disturbance. A side-to-side movement of the sea floor is unlikely to cause a severe tsunami. For example, the San Andreas fault does not generate tsunamis, since its primary movements are horizontal.

Tsunami waves are not to be confused with tidal waves, which are simply the movements of water associated with the rise and fall of tides generated by the gravitational pull of the sun and moon. Since they often are the result of a sudden movement in the earth's crust, tsunamis often are referred to as seismic sea waves. This can be somewhat misleading because the term "seismic" indicates an earthquake-related event, when other natural phenomena, such as landslides, volcanic eruptions, or even meteor strikes, can also generate tsunamis, albeit on a much less frequent basis.

WAVE MOVEMENT

The waves that form a tsunami are different from the surface waves observable from a beach. The latter are produced by winds blowing over the surface of the sea, and their size is directly dependent on the strength of the wind that creates them. The distance between these waves can range from a few centimeters to nearly 300 meters, though the normal separation is about 9 to 18 meters. The speed of the common surface wave can range from a few kilometers per hour to more than 90 kilometers per hour, with the lapse of time between two successive waves running from about five to twenty seconds.

Unlike the common surface wave, the tsunami wave is categorized as a shallow-water wave because of its extensive wavelength, which can stretch up to 500

Fishing boats and vehicles are carried by a tsunami wave at Onahama port in Iwaki city, in Fukushima prefecture, northern Japan, on March 11, 2011. A massive 8.9-magnitude earthquake shook Japan, unleashing a powerful tsunami that sent ships crashing into the shore and carried cars through the streets of coastal towns. (AFP/Getty Images)

kilometers and run for a period of ten minutes to two hours at speeds up to 800 kilometers per hour, depending on the depth of water in which it is traveling. A wave is classified as a shallow-water wave when the ratio between the water depth and the wavelength becomes very small. The formula for making the determination is as follows. The speed of a shallow-water wave is equal to the square root of the product of the acceleration of gravity (9.8 centimeters per second per second) and the depth of the water in meters. For a depth of 100 meters, this gives a speed of 99 m/s, or 360 km/hr. This formula enables seismologists to alert coastal communities about the potential of a tsunami following an earthquake. The formula also points out another significant distinction between a tsunami wave and common surface wave: The rate at which a wave runs out of energy is inversely related to its wavelength. Because of its long wavelength, a tsunami can travel thousands of kilometers across ocean waters without dissipating; in contrast, the average surface wave begins to lose its energy after a distance of a few kilometers. In 1960, an undersea earthquake

off the coast of Chile generated a tsunami that had enough lasting power to kill 150 people in Japan, following an earlier strike on the Hawaiian Islands, where it killed 51 people.

Tsunamis not only can travel great distances but also can reverberate through an ocean for extended periods, as they bounce back and forth between continents. At speeds of up to 800 kilometers per hour, a tsunami can travel across an ocean in about the same time it takes a jet aircraft to cross. The swell of the waves is of such magnitude that it takes only a few surges and collapses for it to span the sea. The tsunami is barely discernible to the naked eye, since the crest of one of its waves represents only the tip of the vast amount of water that extends deep into the ocean. The depth of the tsunami is the reason that its course can be altered by undersea mountain ranges, valleys, or other landforms that stand in its way.

WAVE TRANSFORMATION

As a tsunami approaches land, it undergoes a major transformation. While its total energy level remains constant, it begins to slow from the friction it encounters in the shallow waters above the continental shelf. The process is called "wave refraction," and it occurs when the tsunami's wave train travels in shallow water and begins to move at a slower pace than the portion still advancing in deeper water. As its speed decreases, the wave's height grows because of a "shoaling" effect in which the trailing waves pile onto the waves in front of them. At this stage, the tsunami takes on a more visible appearance, as its waves often reach up to 9 meters or more in height. The incoming waves approach much like an incoming tide, except at a much faster pace. The maximum vertical height a wave reaches in relation to the sea is called a "run-up." The maximum horizontal distance attained by a wave is termed an "inundation." The contours of local reefs, bays, mouths of rivers, and undersea features, as well as the angles of beaches, can have a significant effect on the shape and impact of the tsunami as it nears land.

Tsunami waves normally do not curve and break like common surface waves. Survivors of tsunamis often describe them as walls or plateaus of water being driven by what appears to be the whole weight of the ocean behind them. Although greatly drained of their energy, the waves retain sufficient momentum to wash away nearly everything in their path, including buildings, houses, and trees. Tsunamis can cause enormous erosion, stripping beaches of sand and vegetation that has taken years to accumulate. The fact that a tsunami consists of a series of waves poses a hidden danger to coastal residents. In some cases, curious onlookers and homeowners have returned to an exposed area following the initial wave, only to be overwhelmed by a succeeding one.

Damage from tsunamis generally falls into three categories: inundation of coastal structures resulting from rapid flooding, destruction of buildings and beaches caused by water velocities, and a combination of the two in which velocity and flooding result in a complete tidal-like inundation. The most destructive form of tsunami is one that transforms itself into a bore, a concentrated wave of great force created when the tsunami moves from deep water into a well-defined shallow bay or river. This was the case during the 1960 Chilean tsunami when one of its waves struck Hilo Harbor in Hawaii as a high-velocity bore.

Destructive tsunamis strike somewhere in the world an average of once or twice each year, with most occurring in the Pacific basin. During the four-year period from 1992 to 1996, seventeen recorded tsunamis in the Pacific claimed 1,700 lives. Thirteen major tsunamis hit the Hawaiian Islands during the twentieth century; all were generated by earthquakes along the Pacific basin. The largest recorded wave heights were nearly 16 meters on the islands of Hawaii and Molokai in 1946, as a result of an earthquake off the Aleutian Islands that registered 7.1 on the Richter scale. The waves from this tsunami crested at about fifteen-minute intervals.

HISTORICAL RECORD

Since scientists are unable to predict exactly when earthquakes will occur, they cannot determine the precise moment when a tsunami will be generated. However, with the aid of historical records and numerical models, they can predict where they are most likely to occur. In addition, scientists have deployed sensors on the floor of the Pacific Ocean along the Aleutian Islands and the Pacific Northwest to enhance advance warning systems. The Pacific Tsunami Warning System (PTWS) in Hawaii, established after the 1946 tsunami devastated many coastal areas of the islands, monitors seismological and tidal stations throughout the Pacific basin to evaluate potential tsunami-causing earthquakes for the purpose of determining their direction and issuing adequate advance warnings.

For confirmation of past tsunamis, scientists have turned to geological evidence. Based on a technique called dendrochronology (the dating of trees by counting the ring patterns in their trunks), researchers discovered in the 1980's that a cataclysmic earthquake that struck the Pacific Northwest was the triggering mechanism for a giant tsunami that inundated the coastal Japanese island of Honshu in 1700. For some time scientists had suspected that an earthquake, centered somewhere along the Cascadia subduction zone—a fault that stretches from British Columbia, Canada, to Northern California—was to blame. They found their evidence in the marshlands along the Washington and Oregon coasts. Traces of a thin, unbroken sheet of sand were detected nearly 1 kilometer inland, indicating that a wave at least 9 meters high was likely to have hit the local coast. Calculations indicate that it would have taken an earthquake up to a magnitude 8, striking at high tide, to have produced such a wave. However, the energy of an earthquake of at least magnitude 9 would have been necessary to send waves of sufficient size across the Pacific to inflict substantial damage in Honshu. The link between the two events was established more firmly when researchers discovered that annual growth rings of drowned cedar trees and damaged spruce revealed they had died or were damaged around the same time of the Pacific Northwest waves and Honshu floods. The trees were located in regions where scientists have discovered geological evidence of previous earthquakes.

In 1998, an earthquake off the north coast of Papua New Guinea, registering a magnitude of 7.1, produced a tsunami that killed close to three thousand people at Sissano Lagoon with waves reaching 12 meters in height. Earthquakes of this size normally do not generate significant tsunamis, but researchers concluded that the earthquake probably occurred in relatively shallow water near the ocean floor and thus was able to induce a tsunami larger than normal. A later hypothesis was that the earthquake likely produced an undersea landslide that helped generate the giant waves. The lagoon is separated from the ocean by a narrow strip of land and represents an especially vulnerable section of the coast. Since the lagoon blocked the route inland, the families of the fishermen living on the sand bar had no way of escaping the waves. The entire coastal strip was swept clean by the tsunami, except for some stilts that the residents used to raise their houses off the sand.

In 2004, a 9.1-magnitude subduction earthquake off the west coast of Sumatra triggered a tsunami in the Indian Ocean. It produced wave heights reaching 30 meters, inundating coastlines of 14 countries and killing 230,000 people. The 2011 Tohuku earthquake in the Pacific Ocean caused a tsunami that caused widespread destruction on Japan's northeast coast. More than 15,600 people were killed.

EARTHQUAKE MAGNITUDE

Many questions remain to be answered regarding the relationship between earthquake activity and the formation of tsunamis. For a long time it was assumed that the magnitude of the earthquake, as registered on the Richter scale, was the major factor in determining the size of the tsunami. It was also believed that the shock would have to register somewhere near the 7.4 range to have a major impact on wave generation. However, scientists have learned that the

largest earthquakes do not always create the greatest seismic waves. Some earthquakes can release energy in subterranean settings where the earth's crust reacts slowly and with less violent convulsions than might be the case elsewhere.

Another assumption long held by scientists was that the highest waves of a tsunami were generated immediately following the disturbance. In 1992, however, the highest tsunamis measured along the Northern California coast arrived nearly six hours after a nearby earthquake. During the same year, an earthquake with a magnitude of 7 produced a series of giant waves that devastated a 320-kilometer stretch of the Nicaraguan coast, killing nearly 170 people and injuring 500 others. The earthquake generated only moderate shaking but shifted an estimated 193-kilometer stretch of sea floor nearly 1 meter in about two minutes. A slow-paced event of this type is very efficient in producing great amounts of water to supply the tsunami. The unusually large waves in the Nicaraguan incident resulted from the relatively shallow depth of the disturbance and an accompanying subterranean landslide. The waves arrived at some coastal regions only twenty minutes after the earthquake and struck in the evening, when most of the fishing boats were in dock, destroying or damaging many of them. Numerous homes and two schools also were destroyed. As in the Papua New Guinea disaster, the initial waves were relatively weak, misleading local residents about the potential danger.

An example of a landslide-induced tsunami occurred in Skagway, Alaska, in 1994, when a large accumulation of sediment along the eastern edge of the harbor was loosened by a drop in the tide. An estimated one-third of the landmass involved in the landslide was situated above the water. The collapse produced a wave nearly 12 meters in height at the shore, killing a construction worker and causing extensive damage to a railroad dock.

Perhaps the most destructive of the volcanic-induced tsunamis was the one that occurred in 1883, when the volcano Krakatau, located between the islands of Java and Sumatra in Indonesia, erupted in a series of violent undersea explosions. The upheaval destroyed Krakatau and created gigantic waves as high as 35 meters that inundated towns and villages on nearby islands, killing more than 36,000 people. In a desperate maneuver, the crew of the ship *Loudon* aimed the boat's bow into the oncoming tsunami and managed to ride out the waves. Scientists have debated whether it was the submarine explosions, the slumping of the cone into the crater, or the occasional surges of matter falling into the water that caused the event. Waves from the Krakatau tsunami were recorded as far away as South America and Hawaii. Nearly nine hours after the event, the waves smashed boats resting in the harbor at Calcutta, India. When the tsunami reached Port Alfred in South Africa, it was still nearly 0.5 meter high. The English Channel even recorded a minor surge.

SIGNIFICANCE

By exploring the relationship between earthquakes and tsunamis, scientists continue to learn more about the conditions that lead to a tsunami. The bottoms of the oceans, where most tsunamis originate, are largely unexplored. Only in the last few decades of the twentieth century did researchers begin to accumulate data in sufficient amounts to assess the destructive potential of the phenomenon, particularly in regard to the near-shore tsunamis similar to the one that struck Papua New Guinea.

Scientists already have determined that the hazard to American shorelines, especially along the heavily populated West Coast, is greater than previously thought. The likelihood that the earthquake that triggered the Papua New Guinea tsunami also generated an underwater landslide carries great significance for coastal settlements located close to the offshore seismic faults stretching from Northern California to the Aleutian Islands and near Hawaii. These areas feature canyons that slope deep into the ocean, making them vulnerable to the kinds of underwater landslides that occur in conjunction with moderate earthquakes. How such relatively small disturbances generate such large waves remains a key area for research. The fact that they take place only minutes from shorelines underscores the importance of the public's being made aware of the tsunami's potentially destructive force, if damages are to be limited.

Researchers have concluded that the technology utilized in early warning systems, from monitoring seismographs to measuring water-level changes at tide-gauging stations, cannot significantly alter the advance warning time for near-coastal tsunamis as it can for the disturbances that originate in deeper seas.

Instead, with the aid of computer-generated models and historical records, researchers seek to provide a basis for educational programs designed to alert the general public about the regions where tsunamis are most likely to occur and about the impact that can be expected.

William Hoffman

FURTHER READING

Associated Press. *The Associated Press Library of Disasters.* Danbury, Conn.: Grolier, 1998. Volume 1 of this work covers earthquakes and tsunamis and provides a compilation of news reports on the major events. Entries are arranged chronologically and are accompanied by supporting texts that place each event in context and explain any inaccuracies in the original reports. A note at the end of each entry indicates the outcome of the event.

Clague, John, Chris Yorath, and Richard Franklin. *At Risk: Earthquakes and Tsunamis on the West Coast.* Tricouni Press, 2006. A book geared to the layperson, with a good amount of technical depth. Presents an overview of seismology and historical perspective of geology research in Cascadia.

Dudley, Walter, and Min Lee. *Tsunami!* 2d ed. Honolulu: University of Hawaii Press, 1998. This work offers firsthand accounts and photographs of major tsunamis, including the 1946 event in the Aleutian Islands and Hawaii and the 1998 occurrence in Papua New Guinea. Other sections cover the development of the early warning system and its effectiveness in tracking actual tsunamis.

Folger, Tim. "Waves of Destruction." *Discover* 15 (May 1994): 66-73. Folger offers an account of the surprisingly large tsunami that hit the coast of Nicaragua in 1992 following an offshore earthquake of moderate size. The article reveals how the relatively low intensity of an earthquake's vibrations can mislead coastal residents into thinking the potential danger of a tsunami is minimal.

Montastersky, Richard. "Waves of Death." *Science News* 154 (October 3, 1998): 221-223. Provides a close look at the tsunami that struck Papua New Guinea in 1998 as a result of a moderate undersea earthquake. The data retrieved from studies of this event highlight the differences between near-shore and deep-sea earthquakes and their potential effects on similar coastal formations elsewhere along the Pacific Rim.

O'Loughlin, K. F., and James F. Lander. *Caribbean Tsunamis: A 500-Year History from 1498-1998.* Norwell, Mass.: Kluwer Academic Publishers, 2010. Presents characteristics of tsunamis and the history of their occurrences in the Caribbean Sea. Chapter 4 categorizes the tsunamis by type and describes the effects of the tsunamis. A great deal of data is fit nicely into this nontechnical book, accessible to the general public and natural disaster specialists. Extensive bibliography and appendices as well as an index.

Prothero, Donald R. *Catastrophes!: Earthquakes, Tsunamis, Tornadoes, and Other Earth-Shattering Disasters.* Baltimore: Johns Hopkins University Press, 2011. This text provides a detailed and clear explanation of the many natural and anthropogenic disasters facing our planet. Each chapter is devoted to a different catastrophe, including earthquakes, volcanoes, hurricanes, ice ages, and current climate changes.

Satake, Kenji, and Fumihiko Imamura, eds. *Tsunamis 1992-1994: Their Generation, Dynamics, and Hazards.* Boston: Birkhauser, 1995. Reviews the unusually active period between 1992 and 1994, when six destructive tsunamis occurred with waves up to 12 meters high and casualties totaling close to 1,500 hundred people. The book reports on the findings of geologists, seismologists, oceanographers, and specialists from other disciplines who were sent to investigate the circumstances surrounding the events.

Sutton, Gerard K., and Joseph A. Cassalli, eds. *Catastrophe in Japan: the Earthquake and Tsunami of 2011.* Nova Science Publishers, 2011. Editors compiled a number of reports on the effects of the earthquake and resulting tsunami, including the impact on agriculture and economics. This book focuses on the nuclear crisis following the earthquake and tsunami. One chapter describes the events within the nuclear power plant resulting from the natural disaster.

Tsuchlya, Yoshito, and Nobuo Shuto, eds. *Tsunami: Progress in Prediction, Disaster Prevention, and Warning.* Advances in Natural and Technological Hazards Research Series 4. Norwell, Mass.: Kluwer Academic Publishers, 1995. This work contains a series of studies conducted by scientists, engineers, and specialists on all aspects of the tsunami phenomenon. The articles are divided into three major categories: tsunami generation, propagation and inundation (including prediction and simulation), and observation and warning systems. Written for an advanced audience of scientists, engineers, and graduate-level students.

See also: Deep-Focus Earthquakes; Earthquake Distribution; Earthquake Engineering; Earthquake Hazards; Earthquake Locating; Earthquake Magnitudes and Intensities; Earthquake Prediction; Earthquakes; Notable Earthquakes; Plate Tectonics; Seismometers; Slow Earthquakes; Soil Liquefaction.

U

URANIUM-THORIUM-LEAD DATING

Radioactive decay of uranium and thorium into lead can be used as a natural clock to determine the ages of rock samples. Rocks from different locations on the earth have been dated in a wide range of ages, from 100 million to several billion years. The earth, the earth's moon, and meteorites all have a common age of about 4.6 billion years.

PRINCIPAL TERMS

- **common lead:** ordinary lead as it was formed at the time when all the elements in nature were created; also called primordial lead
- **concordant age:** a situation in which several naturally radioactive elements, such as uranium, thorium, strontium, and potassium, all give the same age for a rock sample
- **discordant age:** a situation in which several radioactive elements do not give the same age because of gain or loss of decay products from a rock sample
- **half-life:** the time for half the atoms in a radioactive sample to decay, having a different value for each radioactive material
- **isotope:** atoms of the same element but with different masses as a result of extra neutrons in the nucleus, such as the two uranium isotopes uranium-235 and uranium-238
- **mass spectrometer:** an apparatus that is used to separate the isotopes of an element and to measure their relative abundance
- **radiogenic lead:** lead formed from uranium or thorium by radioactive decay

RADIOACTIVE HALF-LIFE

Radioactivity was discovered by French physicist Antoine-Henri Becquerel, a professor of physics in Paris, in 1896. He found that a rock containing uranium was emitting radiation that caused photographic film to become exposed. Shortly afterward, his graduate student, French scientist Marie Curie, discovered two radioactive elements, polonium and radium, which are decay products of uranium. Over the next ten years, British chemist Ernest Rutherford and other scientists were able to unravel a whole series of radioactive processes in which uranium

(element 92) decayed into its stable end product, lead (element 82).

All radioactive materials have a particular half-life, which is the time required for half the atoms to decay. In the early days of radioactivity, scientists wanted to see if the half-life of an element could be changed by various processes. For example, they heated or cooled the radioactive material, made different chemical compounds, or converted it to a gas at high pressure. In all cases, the half-life did not change. This is because radioactivity comes directly from the nucleus of an atom, while heat and pressure affect only the outer electron cloud. Therefore, the radioactive half-life is like a built-in clock, keeping time at a fixed rate. No geological process, no matter how violent, can change the half-life.

Radioactivity can be applied to dating the age of rocks under certain conditions. The isotope uranium-238, for example, has a half-life of 4.47 billion years and eventually decays into lead-206. To make the situation as simple as possible, assume a rock sample contained some uranium but no lead at all when it first solidified. This phenomenon marks the starting time of the radioactive clock (time = 0). Suppose that for a very long time period, the rock remained a closed system; that is, no uranium or any of its decay products leaked out or were added from the surroundings. Because of radioactive decay, the uranium content will decrease and the lead will gradually increase. The ratio of lead-206 to uranium-238 in the rock, which will change with time, can be used to calculate the age.

For a numerical example, suppose the rock originally contained 10 grams of uranium-238 and zero lead. After 1 billion years, one can calculate from the half-life that the 10 grams of uranium-238 must have decreased to 8.56 grams because of radioactive decay. The lead-206 will have increased from zero up

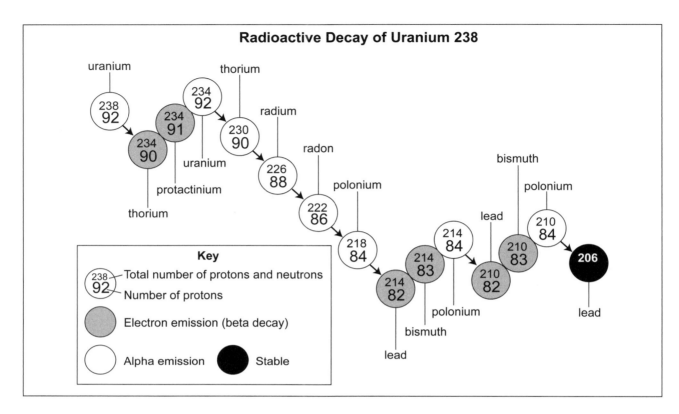

to 1.24 grams. (Note that 8.56 + 1.24 = 9.80 grams. The "missing" mass of 0.20 gram has gone into the creation of alpha, beta, and gamma rays.) The ratio of lead-206 to uranium-238 after 1 billion years would equal 1.24/8.56, or 0.145. In a similar way, the ratio can be calculated for any other elapsed time. The older the rock, the more lead it will contain, so the ratio of lead to uranium gradually increases.

ISOTOPES

In the early 1900's, large errors were made in age calculations because one vital item of information was missing. The idea of isotopes had not yet been discovered. The mass spectrometer, whose invention may be credited to the English physicist Sir Joseph John Thomson about 1914, was later refined by Canadian-American physicist Arthur Jeffrey Dempster and others. The mass spectrometer uses a magnetic field to separate atoms of slightly different mass. Many elements were shown to be a mixture of several isotopes. Lead, for example, has four stable isotopes, with masses of 204, 206, 207 and 208. (Atom masses are expressed relative to carbon = 12.) Gradually, it

became clear that uranium-238 decaying into lead-206 is only one of several radioactive decay processes. The situation is more complex. There are two other decay chains that produce lead as an end product: Thorium-232 decays into lead-208 with a half-life of 14 billion years, and another uranium isotope, uranium-235, decays into lead-207 with a half-life of 700 million years.

Radioactive dating requires that the individual isotopes are measured separately. Typically, three experimental ratios are measured with the mass spectrometer: lead-206/uranium-238, lead-208/thorium-232, and lead-207/uranium-235. Each of these three ratios is combined with the half-life for that decay process to calculate an age. If all three calculations give the same result, the age is probably quite reliable and is called concordant. If the calculations disagree, the ages are discordant, and further investigation is necessary.

CORRECTING FOR DISCORDANT AGES

What can cause discordant ages when several radioactive clocks are compared? In the uranium-thorium-lead (U-Th-Pb) method (described above), one probable source of error is the possibility that the rock, when it first solidified, already contained some natural lead. In other words, the lead content of the rock is the sum of radiogenic lead (from radioactive decay of uranium and thorium) plus the primordial lead. Only the radiogenic lead should be used to calculate an age. A method is needed to subtract out the primordial lead, which did not come from radioactive decay.

Fortunately, a good method to correct for primordial lead is available. Three of the lead isotopes come from the decay of uranium and thorium; there is a fourth lead isotope, lead-204, which is not formed by a radioactive decay process. If a rock contains any lead-204, it means that it already must have contained some lead at the time it was formed. Ordinary lead contains a mixture of isotopes whose normal relative abundance has been measured. The normal ratio of lead-206 to lead-204 is about 17 to 1. Suppose a rock contains 50 milligrams of lead-206 and 1 milligram of lead-204. The investigator would subtract 17 milligrams of lead-206 from the total, which leaves a net excess of 50 – 17, or 33 milligrams of lead-206 that must have come from the decay of uranium-238. A similar correction can be made for lead-207 in normal lead.

There is an uncertainty in the correction factor. One cannot be sure that the natural lead, when it was incorporated into the rock, had the same relative abundance of lead isotopes as lead does today. It is very likely that the normal ratio of lead-206 to lead-204 was smaller than 17 to 1 at an early time in the history of the earth because lead-206 has gradually been added as a result of decay. The most accurate age measurements are obtained if the natural lead correction is small—that is, if most of the lead in a rock sample under study came from radioactive decay.

Sometimes discordant ages can be corrected in a systematic way to calculate a consistent result. For example, some samples of the mineral zircon from the Montevideo area of southern Minnesota gave uranium-lead ages that vary from 2.6 to 3.3 billion years. Suppose a loss of lead occurred, perhaps because of temperature or weathering. It is reasonable to assume that all the lead isotopes decreased by the same percentage because they are chemically identical. One can extrapolate backward in time to show that the zircon samples must have been formed about 3.55 billion years ago and probably had a lead

loss resulting from regional heating at a later time. Fortunately, zircon crystallization strongly rejects lead, so any lead content is probably radiogenic unless the crystalline structure has been damaged so that lead can escape.

LOSS OF INTERMEDIATE ELEMENTS

Another reason that discordant ages sometimes are measured may be that intermediate elements escape in the decay chain between the starting element, uranium or thorium, and the end product, lead. For example, it may happen that radium forms a chemical compound that is relatively soluble in water and is leached out from a rock sample. Also, radon gas may escape. (Scientists know this happens at least some of the time because of the radon buildup in the basements of houses in various parts of the country.)

Any loss of intermediate elements means that too little of the lead end product accumulates. Therefore, the apparent age of a rock sample would be calculated to be too short. An experienced investigator will try to select those minerals for analysis whose crystal structures are known to be relatively impervious to losses. Also, it is desirable to analyze many samples from an area. If most of the results are consistent, one may be able to reject those ages which are discordant for some reason.

DETERMINING EARTH'S AGE

Radioactive age determinations made before 1930 are considered to be unreliable because the mixture of isotopes with different half-lives was not well understood. With improvements in the mass spectrometer in the 1930's, it was shown that uranium consists of two main isotopes: about 99 percent uranium-238 and less than 1 percent uranium-235. Rutherford suggested how such data could be used to estimate at least a rough upper limit for the age of the earth. His reasoning was as follows: The isotope uranium-235 has such a very low abundance presently because it has a relatively short half-life; most of it has decayed away. The amount of uranium-235 that existed on Earth 700 million years ago (one half-life) would have been twice as much as currently; 1.4 billion years ago, the amount of uranium-235 would have been four times as much. By comparison, the amount of uranium-238 would have been only a little greater than at present because of its long half-life. If one calculates back far enough, the amounts of uranium-235 and uranium-238 would have been equal about 6 billion years ago; that is, uranium would have been a fifty-fifty mixture of these two isotopes. It is very unlikely that uranium-235 ever was more abundant than uranium-238 because odd isotopes in general are less abundant in nature than are even ones. Therefore, 6 billion years sets an upper limit for the age of the earth.

A much-improved procedure to determine the age of the earth was developed in the 1950's. As described by American geochemist Harrison Brown, the lead isotopes in the Canyon Diablo meteorite (which created the famous Meteor Crater in Arizona) were analyzed. The ratio of lead-206 to lead-204 was only about 9.4, much lower than any samples on Earth. The argument is made that this ratio represented primordial lead, uncontaminated by any radiogenic lead from uranium decay. Over the history of the earth, this ratio should gradually increase for terrestrial samples because additional lead-206 is produced from uranium, but lead-204 remains constant. Samples that are representative of modern lead on Earth contain about twice as much lead-206 as the meteorite. This amount of extra lead-206 would have required about 4.5 billion years to accumulate. The data about primordial lead in meteors, when combined with the accumulated radiogenic lead from terrestrial samples, give the most reliable result for the age of the earth—about 4.5 billion years.

Major improvements in the sensitivity of mass spectrometers have made it possible to measure the abundance of both parent and daughter isotopes. For example, fine surface material and small rocks brought back from the moon by the Apollo 11 astronauts in 1971 were analyzed for uranium and lead isotopes. The results were in good agreement (concordant) for the uranium-238/lead-206 and the uranium-235/lead-207 decay chains. The so-called moon-dust was dated to be between 4.6 and 4.7 billion years old. It appears that the age of the earth, the moon, and meteorites all cluster around 4.6 billion years. This value would be representative for the age of the solar system.

DETERMINING EVOLUTION OF GEOGRAPHIC REGIONS

In general, rocks that solidified much later in the evolution of the earth contain considerable uranium and thorium. Their lead content is largely radiogenic. The ratio of lead isotopes to uranium and thorium will vary greatly, depending on the time of solidification.

Uraninite is a radioactive mineral containing uranium (in the form of UO_2) and thorium. It is similar to the "pitchblende" that was used by Curie in her famous experiment to isolate the new element radium. In a typical age analysis, samples of uraninite from the Black Hills of South Dakota were dated using three different isotopes, with the following results: uranium-238/lead-206 gave an age of 1.58 billion years, uranium-235/lead-207 gave 1.6 billion years, and thorium-232/lead-208 gave 1.44 billion years. The three measurements agree fairly closely and therefore are said to be concordant. The overall goal of such age measurements is to understand the stages of geological evolution for a whole geographical region on the earth's surface.

LOCATING URANIUM DEPOSITS

Another application of U-Th-Pb dating has been to study the worldwide distribution of uranium resources. All over the surface of the earth, the crust contains about one part per million of uranium. Because of the combined action of high temperature, chemical reactions, and water flow, concentrated uranium mineral deposits were formed when suitable geological conditions existed. Some high-grade uranium ore from Gabon, on the west coast of Africa, contains more than 20 percent uranium. This deposit took place about 2 billion years ago, according to uranium-lead dating. Much larger deposits, but with a much lower percentage yield, are located in northern Canada at Elliot Lake and at Witwatersrand, South Africa. These deposits occurred considerably earlier, about 2.5 billion years ago. In the United States, the major deposits are located in the Colorado plateau extending from Wyoming to Texas, with a relatively recent age of less than 200 million years. Such information about the age of uranium deposits is useful to understand the process of mineralization and possibly to locate new deposits.

Nuclear power plants in 1989 contributed about 16 percent of the world's electricity. Coal-burning plants generate acid rain and carbon dioxide in the environment, and the oil supply is limited, so it is likely that nuclear power will continue to be used, especially in Europe, Japan, Russia, Canada, and the United States. The location and size of the world's major uranium deposits are of great importance to supply the necessary fuel. The mining industry needs to know as much as possible about uranium ore deposits so that the present resources can be estimated accurately and exploration for new deposits can receive helpful guidance.

INVESTIGATING RADON RELEASE

Another area where the uranium-lead decay chain plays an important role is in regard to the radon hazard. Uranium in the soil decays in several steps into radium, which in turns decays into the radioactive gas radon. Because it is a gas, radon mixes with the air in small quantities and is ingested into the lungs. In the open, this natural radioactivity in the air is very dilute, so it is not a hazard. The problem comes when radon seeps into the basement of a house through cracks in the floor or through a sump hole. If the house is located in a geographic region where the soil contains considerable uranium, the radon level may be hazardous to the occupants.

The radon problem came to national attention in 1984 when an engineer at the Limerick nuclear power plant in Pennsylvania set off a radiation alarm when he entered the plant, not when he was leaving. The radioactivity was traced to the engineer's home. The radiation level in that house was found to be about one hundred times greater than exposures permitted for workers in uranium mines. Other homes in the area were also found to have relatively high levels. Surveys of radon levels have been made recently in other areas of the United States to investigate the extent of the problem. The Environmental Protection Agency (EPA) has estimated that radon in homes may be responsible for between 5,000 and 20,000 cancer fatalities per year—a cause for great concern.

Another application of radon release may be in earthquake prediction. In Uzbekistan, the city of Tashkent is in a major earthquake zone. The radon content of well water was monitored in the area. A graph of the data starting in 1956 showed a low level of radon at first, increasing slowly for several years. After 1964, the rate of increase became very steep, until the earthquake came in 1966. Immediately after the quake, the radon decreased rapidly. The explanation for this phenomenon is based on the idea that stresses in the ground cause microfracturing of rocks with release of radon from the pores. This method of study is very promising, but more work needs to be done to see if the radon signal can predict the magnitude and epicenter of a quake with any quantitative accuracy.

Hans G. Graetzer

FURTHER READING

Allison, Ira S., and Donald F. Palmer. *Geology.* 7th ed. New York: McGraw-Hill, 1980. A college-level introductory textbook in geology that has gone through many revisions since the first edition was published in the 1930's. Chapter 5 gives a clearly written and up-to-date overview of how the ages of rocks and geologic time can be measured.

Dosseto, Anthony, Simon P. Turner, and James A. Van-Orman, eds. *Timescales of Magmatic Processes: From Core to Atmosphere.* Hoboken, N.J.: Wiley-Blackwell, 2010. Covers many aspects of the earth's history from the formation and differentiation of the earth, to magma ascent, cooling, and degassing. Uranium series Isotopes are referenced multiple times in evaluating the timescales of multiple concepts.

Durrance, E. M. *Radioactivity in Geology.* New York: Halsted Press, 1986. The author shows the wide scope of radioactivity measurements in geological investigations. Up-to-date information is presented on environmental radioactivity (including the radon hazard), heat generation, and various isotope-dating procedures. A bibliography of articles published in professional as well as popular journals follows each chapter.

Eicher, Don L. *Geological Time.* 2d ed. Englewood Cliffs, N.J.: Prentice-Hall, 1976. This thin volume of six chapters gives a historical overview of various methods to estimate age and time sequence in the evolution of the earth. The evidence from heat loss, rock strata, fossils, and eventually radioactivity is described in a nontechnical narrative style.

Faure, Gunter. *Isotopes: Principles and Applications.* 3rd ed. New York: John Wiley & Sons, 2004. An intermediate-level book, originally titled *Principles of Isotope Geology,* addressed to students of geology as well as to practicing geologists who may not be trained in this area of investigation. Both radioactive and stable isotope analyses are described. After each chapter, some numerical problems with actual experimental data are given. Numerous references to published scientific articles are listed.

_____. *Origin of Igneous Rocks: The Isotopic Evidence*. New York: Springer, 2010. Descriptions of multiple radioactive isotope dating methods are contained within this book. Principles of isotope geochemistry are explained early, making this book accessible to undergraduates. Data are presented in diagrams, there are more than 400 original drawings, and a long list of references is included at the end.

Russell, R. D., and R. M. Farquhar. *Lead Isotopes in Geology*. New York: Interscience Publishers, 1960. A compact discussion of methodology, followed by a 120-page appendix giving specific data on many samples taken worldwide. The authors are particularly concerned about discordant age measurements and how to interpret occasional large variations in lead-isotope ratios.

Skinner, Brian J., and S. C. Porter. *Physical Geology*. New York: John Wiley & Sons, 1987. A widely used college-level textbook for an introductory course in geology. One chapter deals with geological time and its determination, using radioactivity and other physical methods. The uranium/lead and thorium/lead techniques are described in a readable way.

Wagner, Gunther A., and S. Schiegl. *Age Determination of Young Rocks and Artifacts: Physical and Chemical Clocks in Quaternary Geology and Archaeology*. New York: Springer, 2010. The authors cover various materials and dating methods. Well organized, accessible to advanced undergraduates and graduate students.

Walker, Mike. *Quaternary Dating Methods*. New York: Wiley, 2005. This text provides a detailed description of current dating methods, followed by content on the instrumentation, limitations, and applications of geological dating. Written for readers with some science background, but clear enough for those with no prior knowledge of dating methods.

Walther, John Victor. *Essentials of Geochemistry*. 2d ed. Jones & Bartlett Publishers, 2008. Contains chapters on radioisotope and stable isotope dating and radioactive decay. Geared more toward geology and geophysics than toward chemistry; this text provides content on thermodynamics, soil formation, and chemical kinetics.

See also: Earth's Age; Earth's Oldest Rocks; Fission Track Dating; Mass Spectrometry; Nucleosynthesis; Potassium-Argon Dating; Radioactive Decay; Radiocarbon Dating; Rubidium-Strontium Dating; Samarium-Neodymium Dating.

V

VOLCANISM

Volcanism is the investigation of Earth's volcanic activity in a number of environments. Volcanoes comprise not only the well-known cone-shaped mountains, but can also be found surrounded by dense vegetation, embedded in glaciers, and at the bottom of oceans. Some volcanic activity is sudden and violent, while other volcanic activity is more incremental and subtle.

PRINCIPAL TERMS

- **caldera:** a steep, bowl-shaped depression formed after an eruption, when a volcano collapses into a depleted magma chamber
- **lithosphere:** a layer of large plates believed to be floating on molten rock beneath Earth's outer crust
- **magma:** molten rock pushed outward from Earth's core
- **magma chamber:** a reservoir of molten rock that builds under Earth's crust
- **mantle:** the superheated layer of molten rock located between Earth's core and outer crust
- **plate tectonics:** the theory that beneath the outer crust there is a series of plates in constant motion and through which magma flows
- **plinian eruption:** a powerful eruption that forces a column of ash and other material high into the sky
- **pyroclastic flow:** the fluidized and superheated mixture of gases and materials that slide down the side of a volcano at hurricane-like speeds
- **Ring of Fire:** a seismically and volcanically active region along the perimeter of the Pacific Ocean

BASIC PRINCIPLES

Volcanism is the eruption of materials from Earth's inner layers. The planet features three general layers: the crust (the outermost layer, on which life exists), the mantle, and the core. High pressures and heat located in the core and mantle push molten rock (magma) outward into the cooler rock layers under the outer crust. Magma, according to the theory of plate tectonics (a model that argues that the various plates of the mantle, known collectively as the lithosphere, drift across a sea of superheated rock), is pushed through the boundaries of those plates and outward through openings in the crust. These openings are called volcanoes.

A volcano releases material in many ways. A key factor in the magnitude of a volcanic eruption is gas pressure. High degrees of gas pressure can eject magma, steam, ash, and other materials at an explosive rate, while low degrees of pressures can lead to a more subdued eruption or no eruption at all. Explosive eruptions, however, send magma, ash, rock, and smoke high into Earth's atmosphere. Effusive eruptions, in contrast, are extremely slow and sometimes move no faster than human walking speed. In some cases, volcanism involves no eruption, as is the case when magma pushes slowly through the crust, cooling as it moves upward to form a volcano.

BACKGROUND AND HISTORY

Volcanoes have played a role throughout Earth's history. Although they are noted for their destructiveness, volcanoes have also been instrumental in the creation of much of the planet's landscape and composition. Mountains, oceans, and even the atmosphere were formed as volcanoes pushed gas and material through the crust. As a result, volcanoes are largely responsible for the conditions that support life on Earth.

While this creative element has piqued scientific interest, it may be said that most research on volcanism has been inspired by the destructive history of volcanoes. Some scientists believe that volcanoes may be responsible for ending the reign of the dinosaurs, linking their extinction to a series of tremendous volcanic explosions 65 million years ago. This show of volcanism would have sent enormous amounts of soot and ash into the atmosphere, choking the air and preventing plants from growing. In 1650 B.C.E., what is believed to be the largest volcanic eruption in the last 10,000 years took place under what is now the Greek island of Santorini. A tremendous eruption killed millions of people in the Mediterranean region. In 79 C.E., Mount Vesuvius erupted, sending a cloud of hot gas and ash into the sky that instantly killed everyone in the Italian cities of Pompeii and Herculaneum.

In the nineteenth century, the massive eruptions of two Indonesian volcanoes, Mount Tambora and another under the island of Krakatoa, affected weather around the world for years. In 1980, Washington State's Mount St. Helens exploded, sending ash around the world within two weeks. In 2010, the Icelandic volcano Eyjafjallajökull sent an ash cloud into the sky that disrupted airline travel in Europe for weeks. These high-profile events are but a sample of the volcanism that has affected life on Earth.

LOCATIONS AND TYPES OF VOLCANISM

Volcanic eruptions are some of the most violent natural occurrences on Earth. They endanger lives, threaten property, and even cause major changes in the planet's topography. Although scientists are unable to predict major eruptions, they have an extensive knowledge of how volcanism occurs and can interpret indicators of the likelihood of eruption.

Volcanoes are typically found along the edges of the earth's tectonic plates. The majority of the world's volcanoes (about 90 percent) are found along the Ring of Fire, a seismically and volcanically active region along the "edges" of the Pacific Ocean. Iceland, in the northern Atlantic region, also has a significant number of active volcanoes.

Volcanoes erupt in three general ways. The first of these eruptions centers on magma buoyancy. The mass of rock remains the same as it melts, even when its volume increases. When the rock melts, however, its density becomes lighter than that of the surrounding rock, causing the lighter magma to flow up through the earth's surface. The rate of magma flow increases as the magma's density decreases. The buoyancy element relies on the notion that magma is pure (that is, not polluted by compounds that can affect its structure).

In the second scenario, magma flow is aided by the presence of volatile substances (or volatiles). Some examples of volatiles are water, sulfur dioxide, and carbon dioxide. Volatiles also include certain forms of crystallized rock, which increase the density of the magma, causing greater pressure as it moves outward. When rock melts into magma, volatiles start to bubble, causing the magma to flow upward. As the

magma approaches the surface, the bubbles intensify, releasing gas as the flow accelerates and causing a violent eruption. Scientists believe that the presence of these volatiles leads to explosive volcanism, whereas buoyancy alone leads to magma pooling and, possibly, cooling at the crust level unless the magma assumes an increased degree of buoyancy.

The third scenario involves the injection of new magma into a highly pressurized magma chamber (an underground reservoir of molten rock located beneath Earth's crust). As magma flows outward toward the crust, it may pool in a magma chamber. As more magma is pushed into the chamber, however, pressure increases steadily. If a conduit presents itself under these conditions, magma will flow upward using the opening. However, if there is no conduit, the magma will continue to push until the crust fractures and allows the magma to escape. Volcanoes above magma chambers are known to erupt repeatedly over time as new magma is injected. This ongoing volcanism can lead to significant changes in a volcano's structure.

DIFFERENT DEGREES OF INTENSITY

Volcanic eruptions range from simple magma emissions to violent explosions. Some eruptions, as was the case with Mount St. Helens and Mount Etna, take place within the conical shape of mountain volcanoes. Others, like the Kilauea volcano in Hawaii, simply burst from a fissure in the ground. The Eyjafjallajökull event started out as an effusive eruption but later sent a large plume of steam and ash several kilometers into the sky.

The presence of volatiles within the magma is a major determinant in an eruption's degree of intensity. Volatiles may also determine the shape of the eruption. For example, the Vesuvius eruption was considered a plinian eruption, forcing a column of ash and other material high into the sky. Plinian eruptions are some of the most dangerous types, capable of producing pyroclastic flows of fluidized and superheated gas that can slide down the side of a volcano at hurricane-like speeds.

Sometimes an eruption completely depletes the magma chamber beneath a volcano. The volcano then collapses into the empty chamber, creating a bowl-shaped indentation called a caldera. A volcano's collapses into a caldera after an eruption does not meant that that volcano will become extinct or dormant. The Mount Aso volcano in Japan, for example, collapsed into a tremendous caldera more than 100,000 years ago, but its caldera remains the site of frequent volcanic activity.

METHODS OF STUDY

In the study of volcanism, the data that are collected at a given eruption site are both voluminous and complex. Adding to this challenge is the pursuit of general theories based on more than one eruption and, therefore, comparing the data sets from each event. It is here that mathematics proves an invaluable tool.

Scientists seeking to understand the eruption process at Mount Etna, for example, sifted through data from sixty-one earthquakes that took place before a 2001 eruption of the volcano. Researchers used a tensor (a mathematical concept that assigns nonlinear vectors to various geographic sites—in this case, points in and around the volcano at which the seismic activity was recorded) and compared the data input to other vector data. The tensor approach helped provide scientists with a more comprehensive profile of the processes and mechanisms that powered the volcano's eruption.

To analyze the conditions of a volcano, scientists deploy a large array of sensor equipment at a given site. For example, seismographs are placed in key locations to detect any tremors and seismic waves radiating from deep beneath the site. Tiltmeters are used to measure any changes in the horizontal level of the ground. Furthermore, pressure sensors and water detectors are deployed at and near water sources; a change in pressure or the appearance of new streams can be a useful indicator of volcanic activity. Scientists also take gas samples using specialized bottles; as magma approaches the surface, the increased presence of sulfur dioxide and carbon dioxide (among other

Aerial view of the caldera of Mt. Ruapehu, an active volcano in Tongariro National Park, New Zealand. (William D. Bachman/Photo Researchers, Inc.)

gases) emanating from that substance is a strong indicator of volcanic activity.

In addition to ground-based sensor equipment, aerial sensors are deployed near the volcanic site. Aircraft fly over the volcano using infrared scanners, thermal imaging cameras, and radar to detect any changes in the volcano's environment. Satellite technology is utilized as well, focusing on heat sources, cloud plumes, and any other characteristics of the site. The use of such technology helps scientists maintain vigilance over an active volcanic site, generating scientific data and enabling greater awareness of the risks of potential eruption.

One of the most challenging aspects of studying volcanism is the impossibility of directly studying the mantle and the superheated rock beneath the lithosphere. In some cases, even sensors are rendered unusable because of the risk of eruption. Scientists are still able to study volcanism through chemical analysis.

Studying lava that flowed from a fissure, for example, can reveal the type of magma that was released and the gases that were contained in the magma (based on the types of crystals found in the composition). Scientists also can study the chemical properties of the ash and gas released after an eruption. These airborne elements, upon returning to the earth, can reveal a great deal about the volcanic activity.

An example of the value of chemical analysis may be found in a study of the Piton de la Fournaise volcano eruption. This volcano, located on an island near Mauritius in the Indian Ocean, experienced its largest eruption in April 2007. The remoteness of the island prevented scientists from installing sensor equipment before the eruption. However, volcanologists were able to study the chemical composition of steam, ash, and gas ejected above the volcano. Using the Ozone Monitoring Instrument's special aerosol-sensing technologies (this device is onboard a National Aeronautics and Space Administration satellite), scientists gathered data about the volume and a mixture of sulfur dioxide that was part of the plume. The data collected during helped scientists gain vital information about the processes that created Piton de la Fournaise's eruption.

RESEARCH

There are several U.S. government agencies that have departments dedicated to the study of volcanic activity. Within the Department of the Interior, for example, is the U.S. Geological Survey (USGS). The USGS operates the Volcano Hazards Program, which monitors volcanism within the United States. The USGS also focuses on active volcanoes whose activity

has the potential to affect U.S. interests. In addition, the agency maintains several volcano observatories. Meanwhile, the National Oceanic and Atmospheric Administration (through the U.S. Department of Commerce) also works in volcanism, particularly concerning eruption plumes and how they affect air quality and transportation routes.

Universities and their faculties and research staff play a key role in studying volcanism. The University of Hawaii, for example, operates the Hawaii Center for Volcanology, an institution comprising leading researchers. The University of Utah, Oregon State University, and the University of California, Berkeley, all have leading volcano researchers on their faculties.

Because information can be shared so quickly via the Internet and Web, the need for global volcano research networks remains high. The Smithsonian Institution, for example, operates a global network of volcano researchers, whose data and theories are shared within the group through the Smithsonian's central repository. Meanwhile, the European Union formed the Network for Observation of Volcanic and Atmospheric Change, which allows experts from all over the world to collaborate on research on such topics as gas emission monitoring and volcanic risk assessment.

IMPLICATIONS AND FUTURE PROSPECTS

There is no evidence to suggest that there is either less or more volcanic activity in the twenty-first century than in previous eras. However, human civilization has advanced significantly in terms of its transportation and telecommunications abilities. When a massive eruption occurs in these areas, the potential for regional and even global economic disruption goes hand in hand with risks to public safety. For example, the massive eruption of Eyjafjallajökull in Iceland caused a major disruption in world travel patterns. For several days, planes to and from Europe were either diverted or grounded, affecting travel around the world.

The development of computer modeling, cutting-edge chemical analysis, and photographic technologies (including satellite technologies) has helped volcanologists study these phenomena with greater clarity than ever before. One of the most important technological developments, the Internet, has brought the study of volcanic activity to a whole new level. Now, scientists from around the world can almost instantly share data from active volcanic sites and laboratory experiments. In addition to the ability to transfer large data sets and documents, scientists can share photographic images and videos. Furthermore, scientists can use the Internet to monitor active volcanic sites from around the world first-hand, simply by logging into an observatory's network. Although the need for field research in the arena of volcanism is still high, a great deal of time can be saved through the application of modern technologies.

Michael P. Auerbach

FURTHER READING

Castro, Jonathan M., and Donald B. Dingwell. "Rapid Ascent of Rhyolitic Magma at Chaiten Volcano, Chile." *Nature* 461 (2009): 780-783. This article discusses the inclusion of the mineral rhyolite in magma at the Chaiten volcano, which erupted in 2008. The authors argue that the mineral's presence in the magma played a significant role in the explosiveness of that eruption.

Gottsman, Joachim, and Joan Marti, eds. *Analysis, Modeling and Response.* Vol. 10 in *Caldera Volcanism.* Atlanta: Elsevier Science, 2008. The editors present a selection of articles on calderas, how they form, and ways to use caldera monitoring to forecast volcanic activity.

Lockwood, John P., and Richard W. Hazlett. *Volcanoes: Global Perspectives.* Hoboken, N.J.: Wiley-Blackwell, 2010. This book provides an overview of the various types of volcanic eruptions, based on observer accounts. The examples provided, from eruption sites around the world, range from effusive to explosive eruptions and include analyses of the processes that create such volcanism.

Paone, Angelo. "The Geochemical Evolution of the Mt. Somma-Vesuvius Volcano." *Mineralogy and Petrology* 87, nos. 1/2 (2006): 53-80. This article discusses the presence of a number of minerals found in the rocks at Mount Somma-Vesuvius in Italy. The presence of these minerals may contribute to an eventual plinian eruption similar to the famous first-century eruption of the same volcano.

Schminke, Hans-Ulrich. *Volcanism.* New York: Springer, 2005. A comprehensive review of the forces that cause volcanic eruptions and other similar geologic activity. Presents a detailed analysis of how volcanoes form and the processes by which volcanic eruptions occur.

Walter, Thomas R. "Structural Architecture of the 1980 Mount St. Helens Collapse: An Analysis of the Rosenquist Photo Sequence Using Digital Image Correlation." *Geology* 39, no. 8 (2011): 767-770. The author reviews the eruption and subsequent collapse of Mount St. Helens, using photographic evidence and sensor data that were collected from each stage of the eruption and enhancing them using the latest in photographic analytical equipment.

See also: Climate Change: Causes; Earthquakes; Earth's Core; Earth's Interior Structure; Earth's Mantle; Geodynamics; Heat Sources and Heat Flow; Lithospheric Plates; Mantle Dynamics and Convection; Mass Extinction Theories; Metamorphism and Crustal Thickening; Mountain Building; Plate Motions; Plate Tectonics; Plume and Megaplumes; Seismic Wave Studies; Stress and Strain; Subduction and Orogeny; Tectonic Plate Margins.

WATER-ROCK INTERACTIONS

Water-rock interactions occur as fluids circulate through rocks of the earth's crust. Isotopes of common elements are exchanged between a fluid and its host rock. As a result of these reactions, a rock may preserve a record of the fluids that have passed through its pore spaces. Studies of water-rock interactions provide information on the nature of fluid movement through rocks and on the origin of economic ore deposits.

PRINCIPAL TERMS

- **connate fluids:** fluids that have been trapped in sedimentary pore spaces
- **dehydration:** the release of water from pore spaces or from hydrous minerals as a result of increasing temperature
- **exchange reaction:** the exchange of isotopes of the same element between a rock and a liquid
- **fractionation:** a physical or chemical process by which a particular isotope is concentrated in a solid or liquid
- **isotopes:** atoms of the same element with identical numbers of protons but different numbers of neutrons in their nuclei
- **juvenile water:** water that originated in the upper mantle, also called magmatic water
- **mass spectrometer:** a laboratory instrument that separates isotopes of a particular element according to their mass difference
- **meteoric water:** water that takes part in the surface hydrologic cycle
- **volatiles:** dissolved elements and compounds that remain in solution under high-pressure conditions but would form a gas at lower pressures

STABLE ISOTOPES

Fluids that circulate within the crust of the earth take part in chemical reactions involving an exchange of elements between the fluid and the host rock. Such reactions, referred to as fluid-rock interactions, are an important mechanism in the concentration of minerals in economically valuable ore deposits. A useful means of studying fluid-rock interactions is by measurement of the stable isotopic composition of rocks that have undergone such an exchange history.

Rocks are exposed to fluids that are diverse in their origin and composition. The most important volatile constituents of natural fluids are water, carbon dioxide, carbon monoxide, hydrogen fluoride, sulfur compounds, and light hydrocarbons such as methane. In addition, fluids contain dissolved solids derived from the crustal rocks through which they pass. As the list of volatiles makes clear, the dominant elements present in fluids are oxygen, hydrogen, carbon, and sulfur.

An element may be characterized by its atomic number and atomic weight. Atomic number refers to the number of protons present in the nucleus of an atom and is a constant value for each element. Atomic weight is determined by adding together the number of protons and neutrons contained in an atom. Because the number of neutrons present often varies within a limited range, atoms of a particular element may have several different atomic weights. These atoms are referred to as isotopes. Oxygen, with an atomic number of 8, occurs most commonly with eight neutrons but may have nine or ten. Therefore, three isotopes of oxygen occur in nature: oxygen-16, oxygen-17, and oxygen-18. Similarly, carbon occurs as carbon-12 or carbon-13. (Carbon-14 is a radioactive isotope that forms in the earth's upper atmosphere and will not be considered here.) Hydrogen contains only one proton; however, a small fraction of hydrogen atoms also contain a neutron, which doubles the mass of the atom. This heavy hydrogen isotope is called deuterium. Sulfur has four stable isotopes: sulfur-32, -33, -34, and -36.

STABLE ISOTOPE RATIOS

For any given mineral or fluid, the relative concentrations of the isotopes of a particular element may be expressed as a stable isotopic ratio, or the ratio of the second most abundant isotopic species over the most abundant isotope. Any physical process that results in the enrichment or depletion of the concentration of a heavy isotope is referred to as fractionation. A common fractionation process is evaporation of water. Water molecules containing the lighter isotope of oxygen (oxygen-16) will preferentially evaporate so that the remaining liquid will be enriched in the heavier oxygen isotope (oxygen-18). Fractionation also occurs during the growth of minerals in either a magma or a water-rich solution. Some minerals, because of the nature of their chemical bonds and their crystal structure, tend to contain a greater number of heavy isotopes than do other minerals. Quartz, dolomite, and calcite are common examples of minerals that contain high concentrations of heavy oxygen, while oxides such as ilmenite and magnetite have very little of the heavy isotope. The effectiveness of fractionation during mineral growth is dependent upon temperature. Low temperatures permit minerals to be more selective in choosing atoms for growing crystal sites, resulting in large differences in isotopic ratios between different minerals. At high temperatures, the selection of atoms is a more random process, and differences between isotopic ratios become progressively smaller.

Because the heavier isotopes of elements naturally occur in such small concentrations, isotopic ratios of oxygen and carbon, for example, have numerical values that are very small and difficult to measure accurately. For this reason, isotopic ratios for a particular sample are presented as a relative enrichment or depletion of the heavy isotope of an element as compared with a defined standard. The difference between the sample and the standard is measured in parts per thousand or per million and is expressed by the Greek letter delta (δ). For carbon isotopic values, for example, the accepted standard is called PDB and is obtained from a belemnite fossil of the Cretaceous-age Pee Dee formation of North Carolina.

EXCHANGE OF STABLE ISOTOPES

As fluids migrate through rocks, reactions occur that involve the exchange of stable isotopes between the fluid and the solid. The exchange process may be pervasive, where fluid movement is diffusive and affects the entire rock mass, or localized along specific fluid channelways, such as fractures, where only the wall of the rock is altered along the route of water

movement. The degree to which the isotopic composition of the rock is altered depends on the initial composition of the rock and the fluid, the temperature at which isotopic exchange is occurring, and the amount of fluid present. Typically, the oxygen and hydrogen isotopic values for crustal water are light compared to those for most rocks. Therefore, as isotopic exchange proceeds, the isotopic composition of the rock becomes progressively lighter, while the water becomes increasingly enriched in the heavier isotopes. If the fluid-rock interaction has occurred under constant temperature conditions, the final isotopic value of the rock is proportional to the volume of water that has passed through the rock.

RESERVOIRS OF CRUSTAL FLUIDS

There are four principal sources, or reservoirs, of crustal fluids. Each of these fluid reservoirs contains stable isotopic ratios that reflect the fractionation mechanisms at work and the result of chemical reactions between the fluids and their host rocks.

"Meteoric water" is a term applied to fluids that take part in the surface hydrologic cycle. Water that undergoes evaporation, precipitation, and runoff to lakes and to the ocean is capable of penetrating the earth's crust to a depth of several kilometers. This penetration is usually accomplished by fluid migration along weaknesses in the crust, such as faults or fracture systems. Natural hot springs are an example of meteoric water that has been heated deep in the crust and then reemerges at the surface. The oxygen and hydrogen isotopic ratios associated with meteoric water are controlled principally by the distillation effect of evaporation and precipitation. As a result of the general transport of air masses from the equator toward the poles, meteoric water isotopic values vary in a systematic manner, with heavier ratios found near the earth's equator and progressively lighter isotopic values occurring toward the poles.

Meteoric water that is trapped in the pore spaces of accumulating sediments is referred to as connate water. When loose sediments have lithified to form hard sedimentary rock, the enclosed pore fluids may become isolated for very long periods. Connate water reveals an isotopic trend similar to that of meteoric water, except that the oxygen values tend to be heavier because of the capacity of the lighter oxygen (oxygen-16) contained in the trapped pore fluids to be exchanged for some of the heavier oxygen

(oxygen-18) of the host sedimentary rock. Isotopic exchange between connate water and the host sediments continues until equilibrium is achieved.

As rocks undergo increases in temperature and pressure associated with metamorphism, water is frequently released in a process called dehydration. The escaping water may be from pore spaces in the rock or from hydrous minerals such as micas or amphiboles. Because of their origin, these dehydration fluids are also referred to as metamorphic fluids. Dehydration water has a very wide range of isotopic compositions, which reflects the diversity of the original sediments.

Juvenile water, which originates in the upper mantle, escapes from ascending magma and represents the fourth important fluid source. Not all water derived from magma should be considered juvenile, as meteoric or connate water will frequently be present in sediments that undergo melting deep in the crust. True juvenile water has a very narrow range of isotopic compositions. Because of mixing of crustal fluids and exchange reactions with igneous rocks, samples of unaltered juvenile water are found very rarely.

ISOTOPIC COMPOSITIONS OF ROCKS

The isotopic compositions of rocks reflect the formation history of the particular rock type. Igneous rocks are controlled by the composition of their magmatic source area, which is usually in the lower crust or upper mantle. Other factors include the temperature of crystallization, the type of minerals, and the degree to which the magma remains isolated or mixes with other constituents. Unaltered igneous rocks typically have a narrow range of isotopic values as compared with natural fluids. Deep mantle rocks contain minerals that are low in oxygen-18 and therefore fall in a narrow range of isotopically light compositions. Crustal rocks, with a higher proportion of silicate minerals, which concentrate oxygen-18, are isotopically heavy. Sedimentary rocks have two distinct modes of formation. Clastic sedimentary rocks are made of transported particles of weathered material and therefore have isotopic ratios that reflect the individual components. Chemical sedimentary rocks precipitate directly and often involve biological activity. The oxygen isotopic values of these rocks are usually much heavier than those of fluids. Because of fractionation factors associated with organisms, carbon isotope ratios are highly variable. Rocks that

have undergone metamorphism contain the widest range of isotopic compositions, as a result of chemical reactions in the presence of fluids that may be derived from any of the reservoirs previously described.

MASS SPECTROMETRY

The most important analytical tool in the study of fluid-rock interactions is the stable isotope mass spectrometer. This instrument separates isotopes according to their mass differences, as determined by the deflection of charged ions within a magnetic field. Elements of interest are extracted from minerals through appropriate chemical reactions and then converted to a gas, which is entered into the mass spectrometer. The sample gas is bombarded by a stream of electrons, converting the gas molecules to positively charged ions. The ions are accelerated along a tube, where a powerful magnet deflects the charged molecules into curved pathways. Lighter particles are deflected more than heavier ones, so several streams of ions result, each with a particular mass. The relative proportions of each isotope are measured by comparing the induced current produced by each ion stream at a collector.

RESEARCH INVOLVING IGNEOUS ROCKS

Studies of oxygen and hydrogen isotopes have been useful in research on water-rock interactions involving igneous rocks. Hydrous minerals, such as biotite and hornblende, are separated out of granitic rocks in order to extract hydrogen isotopic values. Oxygen isotopic compositions are usually measured from feldspars and quartz. Two areas of water-rock study have been of particular interest. The first concerns the origin and quantity of water responsible for the isotopic alteration of large igneous intrusions within the continental crust. The second area of investigation addresses the interaction of ocean water with ocean-crust basalt and the formation of associated sulfide ore deposits.

The initial isotopic composition of igneous rocks is predictable according to their mineral content and temperature of crystallization. Therefore, it is possible to recognize when igneous rocks have been influenced by exchange with a fluid. Many examples have been found of shallow igneous intrusions that have been depleted in oxygen-18 through interaction with large volumes of isotopically light meteoric water. Hydrogen isotope exchange is even more sensitive than with oxygen, because igneous rocks have so little hydrogen relative to the amount in water. A very small quantity of water may produce a large isotopic shift in a rock's hydrogen value, while the oxygen is largely unaffected.

Isotopes associated with the ocean crust have been

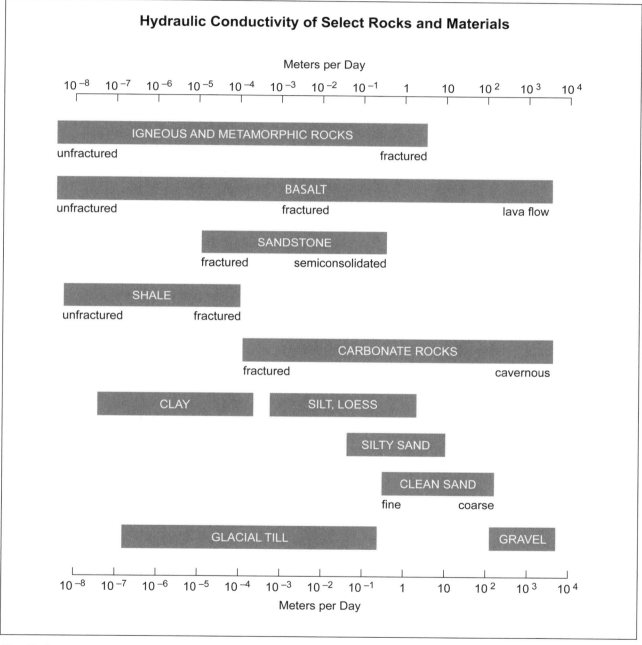

Note: Rocks vary tremendously in their ability to conduct water. The meters-per-day scale is logarithmic: Each increment to the right and left of 1 indicates a change by a power of 10. To the right, 10 meters, 1,000 meters, and 100,000 meters; to the left, 0.1 meter, 0.01 meter, 0.001 meter, and so on.

Source: Ralph C. Heath, Basic Ground-Water Hydrology, U.S. Geological Survey Water-Supply Paper 2220, 1983.

studied through cores obtained from seabed drilling and through the analysis of ophiolites, which represent portions of ocean crust exposed on land. In the vicinity of the mid-ocean ridge, where temperatures exceed 300 degrees Celsius, ocean water circulates to a depth of 3 or 4 kilometers and is responsible for depleting oxygen-18 by one or two parts per thousand.

STUDY OF METAMORPHISM AND PRECIPITATED ROCK

Fluids associated with metamorphism have also been intensively studied by stable isotopic methods. Increasing metamorphic grade is associated with progressively lighter isotopic values for oxygen and hydrogen. Areas of regional metamorphism may show two end-member types of fluid behavior. Consistent depletion of oxygen and hydrogen isotopic values throughout a large terrain points to exchange with an external source of water that flows pervasively through the region. Alternatively, only the trapped connate water may be involved in exchange reactions, leading to higher isotopic composition values and to enrichment of deuterium in hydrous minerals. Because of the large difference in isotopic composition between magma and sedimentary rocks, contact metamorphism is particularly appropriate for study. Samples from intrusions indicate that the margin of the igneous rock is enriched in oxygen-18 through exchange with the country rock, while the interior of the body remains unaltered. The early stages of fluid-rock interaction are dominated by magmatic water, while meteoric water becomes increasingly important as cooling proceeds.

Sedimentary rocks that form by precipitation, such as limestone and chert, have heavy isotopic compositions as a result of the large fractionation between water and either calcite or quartz at low temperature. Clastic sandstones, which contain transported quartz grains, are characterized by the lighter isotopic values of the component particles. Sedimentary rocks selected from a large sample area frequently show a progressive change in the amount of isotopic depletion that has resulted from water-rock exchange. As water continues the exchange process, the isotopic composition of the fluid also shifts. During the next increment of water-rock exchange, the potential amount of depletion of the rock will not be as great. By plotting isotopic values of sedimentary rocks, it is possible to determine directions of fluid motion.

SEARCH FOR ECONOMIC RESOURCES

Water-rock interactions are important primarily for their role in the formation of economically vital ore deposits. These are localized regions where metals such as gold, silver, lead, copper, zinc, and tungsten occur in unusually high concentrations and can be extracted. Water, circulating within the crust of the earth, plays an important role in the formation of most ore deposits by leaching elements from rocks and concentrating them in zones of new mineral growth. During this process, the isotopic compositions of the host rock and fluid are progressively changed. Analysis of the resulting stable isotope ratios of the ore rocks allows geologists to understand the sources and quantity of the mineralizing fluids. Better understanding of the formation of ore deposits has led to greater success in the discovery of new mineral resources.

Stable isotope research into water-rock interaction is not limited to studying water-rich fluids associated with precious mineral deposits. Petroleum and natural gas flow from source rocks rich in organic material to porous reservoir rocks, where they may become trapped. Rocks through which hydrocarbons have migrated frequently show well-depleted carbon isotope values and thus preserve a record of fluid movement. Oil companies have used carbon data to track the migration history of hydrocarbons and to locate regions where petroleum leaks to the surface. The use of carbon isotopes has also proved successful in identifying source rocks associated with producing oil fields. This technology will become even more important as resources become increasingly scarce.

The source of fresh drinking water for almost half the population in the United States is subsurface groundwater. This crucial resource is jeopardized by contamination with common pollutants, such as pesticides, and by depletion through the withdrawing of water faster than it is replenished at recharge zones. Research involving stable isotope studies has become important in hydrology to track fluid movement within aquifers and identify sources of recharge water.

HAZARDOUS WASTE STORAGE

Another area of concern associated with water-rock interaction is the safe, long-term storage of toxic and nuclear waste. One of the most important criteria for the isolation of dangerous wastes is that the enclosing rocks be relatively dry and impermeable to water movement so that hazardous material is not transported into water supply aquifers. The record of water flow recorded by stable isotopes is sensitive to even very small fluid volumes and provides one of the means of assessing the risks associated with a toxic disposal site.

Grant R. Woodwell

FURTHER READING

Albarede, Francis. *Geochemistry: An Introduction.* 2d ed. Boston: Cambridge University Press, 2009. A good introduction for students looking to gain some knowledge in geochemistry. Covers basic topics in physics and chemistry; isotopes, fractionation, geochemical cycles, and the geochemistry of select elements. Also includes water-rock reactions and ratios.

Blatt, Harvey, and Robert J. Tracy. *Petrology: Igneous, Sedimentary, and Metamorphic.* 3rd ed. New York: W. H. Freeman, 2005. Undergraduate text in elementary petrology for readers with some familiarity with minerals and chemistry. Thorough, readable discussion of most aspects of water-rock interactions. Abundant illustrations and diagrams, good bibliography, and thorough indices.

Bowen, Robert. *Isotopes and Climates.* London: Elsevier, 1991. Bowen examines the role of isotopes in geochemical phases and processes. This text does require some background in chemistry or the earth sciences but will provide some useful information about isotopes and geochemistry for someone without prior knowledge in those fields. Charts and diagrams help clarify difficult concepts.

Brantley, Susan, James Kubicki, and Art White, eds. *Kinetics of Rock-Water Interactions.* New York: Springer, 2007. Written by experts in the field of rock-water interactions. This text covers rates of reactions, transition state theory, the mineral water interface, mineral dissolution, and much more. Chapter 12 focuses on water-rock interactions.

Faure, Gunter. *Isotopes: Principles and Applications,* 3rd ed. New York: John Wiley & Sons, 2004. Originally titled *Principles of Isotope Geology.* A college-level text that covers both radioactive and stable isotopes. The first five chapters are introductory in nature and include a good historical review of the development of isotope geology and mass spectrometry. The last unit covers stable isotopes and includes figures reproduced from class research papers. Each chapter includes a detailed reference list.

Gregory, Snyder A., Clive R. Neal, and W. Gary Ernst, eds. *Planetary Petrology and Geochemistry.* Columbia, Md.: Geological Society of North America, 1999. A compilation of essays written by scientific experts, this book provides an excellent overview of the field of geochemistry and its principles and applications. The essays can get technical at times and are intended for college students.

Hoefs, Jochen. *Stable Isotope Geochemistry.* 6th ed. New York: Springer-Verlag, 2009. Suitable for an advanced college student who seeks a detailed discussion of isotope fractionation, sample preparation, and laboratory standards. The material is introduced in three sections. The first chapter provides theoretical principles; the second chapter is a systematic description of the most common stable isotopes; and the third summarizes the occurrence of stable isotopes in nature. An extensive list of references is included at the end of the book.

Krauskopf, Konrad B. *Introduction to Geochemistry.* 3rd ed. New York: McGraw-Hill, 2003. A comprehensive advanced text that covers most aspects of the chemistry of natural fluids. Radioactive and stable isotopes are briefly treated, along with a discussion of ore-forming solutions. This resource is particularly useful for students who seek detailed information on the chemistry and interaction of crustal water. Suggestions for further reading are provided at the end of each chapter.

Oelkers, Eric H., ed. *Thermodynamics and Kinetics of Water-Rock Interaction: Reviews in Mineralogy and Geochemistry.* Mineralogical Society of America, 2009. Contains multiple chapters on water-rock interactions. Each chapter complete with references.

O'Neil, J. R. "Stable Isotope Geochemistry of Rocks and Minerals." In *Lectures in Isotope Geology,* edited by Emilie Jäger and Johannes C. Hunziker. New York: Springer-Verlag, 1979. This source provides a brief and clear introductory section on stable isotope nomenclature. The remainder of the chapter outlines major conclusions drawn from isotope analysis of igneous, metamorphic, sedimentary, and ore deposit rocks. Examples are provided from pioneering research studies. Although the text is oriented toward the college level, high school students interested in the results of isotope studies will find this chapter useful.

Smith, David G., ed. *The Cambridge Encyclopedia of Earth Sciences.* New York: Crown Publishers, 1981. Organized as a compilation of high-quality and authoritative scientific articles rather than a typical encyclopedia. Chapter 8, "Trace Element and Isotope Geochemistry," is a brief, well-illustrated summary of the occurrence of trace elements, stable isotopes, and radiogenic elements. The chapter emphasizes how trace element and isotope studies have enhanced understanding of processes such as the generation of magma and the occurrence of ore deposits. The discussion of water-rock interaction associated with the ocean crust would be accessible to advanced high school students. Few additional references are offered.

See also: Elemental Distribution; Fluid Inclusions; Freshwater Chemistry; Geochemical Cycle; Geothermometry and Geobarometry; Isotope Geochemistry; Mass Spectrometry; Nucleosynthesis; Oxygen, Hydrogen, and Carbon Ratios; Phase Changes; Phase Equilibria.

X-RAY FLUORESCENCE

When a sample is placed in a beam of X-rays, some of the X-rays are absorbed and the absorbing atoms are excited, or raised to a higher energy state. X-rays with energies characteristic of the particular element are emitted when these atoms decay back to their normal energy states. The intensity of these emitted, or fluorescence, X-rays indicates the abundances of each element in the sample.

PRINCIPAL TERMS

- **atomic number:** the number of protons, or units of positive charge, in the nucleus of an atom
- **Bohr model:** a model of the atom in which electrons move in circular orbits around a positively charged nucleus, with orbits of only certain discrete energies being permitted
- **energy level:** the energy of an electron in one of the permitted orbits of the Bohr model of the atom
- **fluorescence:** light emitted as the result of the decay of an atom from an excited state back to its ground state
- **ground state:** the configuration of an atom such that all of its electrons are in the lowest energy levels that are permitted
- **X-ray:** light in the wavelength range from 10^{18} meter to about 10^{-10} meter, spanning the range from the ultraviolet to the gamma rays

BOHR MODEL

In the atomic model developed in 1913 by Danish physicist Niels Bohr, an atom consists of a positively charged nucleus surrounded by a number of negatively charged electrons that orbit around the nucleus in circles of fixed radii. Only orbits of specific radii are permitted in the Bohr model. The energy required to remove an electron completely from the atom is larger for orbits of smaller radii. Thus, the electrons in orbits of smaller radii are said to have less energy, or to be in a lower energy level. Only two electrons are permitted in each orbit, or energy level. Thus, elements heavier than helium, with an atomic number of two, must have electrons in several different energy levels. The specific energies of these levels depend on the atomic number of the atom involved.

When an atom is placed in a beam of X-rays, a collision between an X-ray from the beam and an electron in a lower electron energy level can result in the ejection of that electron from the atom. This action leaves a vacancy in a low energy level, which is filled by an electron from a higher energy level. The energy given up by the electron when it moves from the higher energy to the lower energy level goes into the emission of an X-ray with a characteristic energy equal to the energy difference between the two electron energy levels. This process, however, leaves a vacancy at the higher energy level, which is filled by an electron from an even higher energy level. Again, an X-ray with an energy characteristic of the energy difference between the two levels is emitted. This process continues until the atom that was disturbed or excited by the X-ray from the incident beam has returned to its ground state, or lowest energy state. The series of X-rays emitted as the atom returns to the ground state have energies characteristic of the atomic number of the atom that was excited. These emitted X-rays, called fluorescence X-rays, form the basis for the X-ray fluorescence method of chemical analysis.

The energy level structure of an atom depends on the charge of its atomic nucleus. Thus, each chemical element has a unique pattern of X-ray fluorescence emission. Chemical bonding into molecules generally disturbs the outer electron shell energy levels, because these electrons participate in the bonding. Because the fluorescence X-rays are associated with the loss of an electron from an inner electron shell, the energies of the emitted X-rays are virtually independent of the molecule in which the element is present in the sample. Thus, the X-ray fluorescence technique of chemical abundance determination is applicable to samples in their natural state and, generally, does not require significant sample preparation.

DEVELOPMENT OF X-RAY SPECTROSCOPY

The first demonstration of X-ray spectroscopy dates to around 1910, when British physicist C. G. Barkla obtained positive evidence of X-ray emission at characteristic energies by each element. By 1912, H. G. J. Moseley had established the relationship between the energies of the fluorescence X-rays and the atomic number of the atom responsible for the emission, laying the foundation for the identification of elements by X-ray emission analysis. This principle was used to validate the existence of the element hafnium, element 72, from its X-ray fluorescence.

Although the potential of this new technique was understood, practical difficulties limited its applicability. The early experiments used an incoming beam of electrons (not X-rays) to eject an inner electron from the sample. This process required the sample to be electrically conductive, and the electron beam caused considerable sample heating. In the mid-1920's, it was recognized that the use of an incoming beam of X-rays would eliminate many of the problems associated with the electron beam, but X-ray sources of high intensity and very sensitive detectors were required. By the mid-1950's, commercial instruments for X-ray fluorescence became available.

BENEFITS OF THE TECHNIQUE

A qualitative measure of the chemical composition of a sample can be obtained by measuring the energies of the fluorescence X-rays that are emitted. These energies can be compared to tables indicating the energies expected for each element. The presence of an element in the sample at a detectable concentration is indicated if X-rays of the energies corresponding to that element are seen in the emitted spectrum. Quantitative chemical analysis can be carried out on the sample by measuring the intensity, or number of X-rays detected, at the wavelengths characteristic of the elements(s) of interest. This intensity is proportional to the number of atoms of that element or elements present in the sample.

The nondestructive nature of X-ray fluorescence analysis makes it the preferred technique for small or rare samples that must be preserved for other types of experimental measurements. The X-ray fluorescence technique is easily adaptable to almost

complete automation; thus it is advantageous when a large number of samples needs to be analyzed, such as in mining operations to monitor ore quality.

COMPLICATIONS

In practice, corrections must be made for the efficiency of the X-ray fluorescence process, which depends on the element being detected and the energy distribution of the incoming beam of X-rays. If the sample is thick enough that fluorescence X-rays have a substantial probability of interaction with the sample before escaping and being detected, then corrections for this absorption process must also be made.

These complications delayed the routine application of X-ray fluorescence to the analysis of geological specimens until the early 1960's. By then, appropriate correction techniques had been developed. The most successful of these correction techniques is the preparation of a synthetic control sample of very similar composition to the rock to be analyzed and the comparison of the fluorescence intensities observed from the control to those from the rock under identical analysis conditions. Using this technique, by the mid-1960's most of the major rock-forming elements, particularly aluminum, phosphorus, potassium, calcium, titanium, manganese, and iron, as well as some minor and trace elements, could be measured as accurately using X-ray fluorescence as by the traditional wet chemical or optical spectrograph techniques. For most commonly analyzed rock specimens, the corrections are well determined, and X-ray fluorescence analyses can now be performed down to a sensitivity of about 10 parts per million with better than percent level precision.

TOOLS FOR ANALYSIS

The first requirement for X-ray fluorescence analysis is a beam of X-rays to shine on the sample, usually provided by an X-ray tube or a radioactive source. A typical X-ray tube consists of a high-energy electron beam striking a heavy element target. This target then emits fluorescence X-rays, which escape through a window in the X-ray tube and strike the sample. A smaller and more portable X-ray fluorescence apparatus frequently employs a radioactive source that emits X-rays in its decay sequence. Some commonly used sources are iron-55, cobalt-57, cadmium-109, and curium-242. Intense X-ray beams have become available at particle accelerators, such as the National Synchrotron Light Source at Brookhaven National Laboratory, permitting X-ray fluorescence analysis of smaller samples and at lower elemental concentrations than with conventional laboratory instruments.

Two types of detectors are commonly employed to determine the number and energy of the fluorescence X-rays. The first is a wavelength dispersive spectrometer, which uses a single crystal to diffract X-rays of a particular energy into an electronic counter. The wavelength dispersive detector provides very high energy resolution, allowing nearby peaks from two different elements to be separated. The disadvantage of the wavelength dispersive spectrometer is that only a few X-ray energies can be measured at one time, depending on the number of counters that are placed around the diffracting crystal.

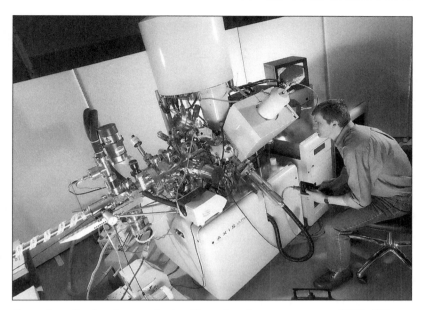

Researcher adjusting the controls of an X-ray photoelectron spectrometer (XPS). This machine is used to analyze the surface chemistry of a material. It measures the composition, empirical formula, chemical state, and electronic state, of elements found within the material. (Andrew Brookes, National Physical Laboratory/Photo Researchers, Inc.)

In the early 1970's, the energy dispersive X-ray detector was developed. It consists of a silicon semiconductor doped with lithium that produces an electronic pulse proportional to the energy of the X-ray absorbed by the semiconductor. Thus, this device is responsive to all energies simultaneously, allowing the entire elemental composition of the sample to be determined at one time. The energy dispersive detector has two disadvantages. First, it must be operated at very low temperature, requiring liquid nitrogen to cool it; second, its energy resolution is inferior to the wavelength dispersive detector, causing the energy peaks to broaden and frequently overlap. Mathematical modeling of the peaks' shapes is then required to recover information on the number of X-rays in each of the two overlapping energy ranges.

APPLICATIONS IN SCIENCE AND INDUSTRY

In mining, automated X-ray fluorescence systems are used for the continuous analysis of the zinc abundance in flowing slurries of zinc concentrates. Throughout the mining industry, the X-ray fluorescence technique is employed to analyze ores, tailings, concentrates, and drilled cores. In geology, the X-ray

fluorescence method of chemical analysis has been applied to all rock types. Because of its sensitivity to elements present in low abundances and its ease of application to a large number of samples, the X-ray fluorescence technique has been employed in a wide variety of geological investigations.

Recently, X-ray fluorescence has been applied to the analysis of particles collected from the air in order to determine the concentrations of toxic elements. In the analysis of airborne particles, the samples are usually collected by passing measured volumes of air through filter paper and then performing an analysis on the bulk material trapped on the filter paper. With the development of more sensitive X-ray fluorescence apparatus, it is sometimes possible to perform elemental analyses on individual dust particles. In addition, the elemental makeup of the particles is frequently useful in determining the source of the air pollution. In some cases, determination of the chemical composition of the particles has allowed identification of the source of the pollution. In agriculture and food science, the X-ray fluorescence method is used in determining the trace element content of plants and foods. This application has been used to monitor the concentrations of insecticides on leaves and fruits.

X-ray fluorescence has also been applied to problems in medicine. The sulfur content of each of the different proteins in human blood, determined by X-ray fluorescence, has proved useful in medical diagnosis. X-ray fluorescence has also been used to determine the strontium content of blood serum and bone tissue. This use was particularly important during the era of aboveground nuclear testing, when radioactive strontium absorption, particularly by children, was a major problem.

A simple X-ray fluorescence spectrometer flew on each of the two Viking spacecraft that landed on the planet Mars in 1976. These spacecraft provided the first determinations of the major element abundances in the soil of Mars. These measurements confirmed a basaltic composition for the soil, indicating that Mars had experienced planetary differentiation—that is, separation into a metallic core and a stony mantle.

ADDITIONAL APPLICATIONS

The development of the X-ray fluorescence technique for medical, chemical, geological, and industrial uses has led to a variety of additional applications. For example, X-ray fluorescence is used to observe automotive and aircraft engine wear by determining the concentrations of the metallic iron, curium, and zinc particles suspended in lubricating oils. By identifying the specific element, it is often possible to identify the actual engine component that is wearing.

The nondestructive nature of X-ray fluorescence has made it an ideal technique for the analysis and authentication of art objects and ancient coins. The elemental compositions of inks, paints, and alloys in the object are compared with the compositions in use at the alleged time of production of the object in question. The widespread availability of X-ray fluorescence apparatuses has given rise to a variety of applications for this technique that were not anticipated at the time of its initial development.

George J. Flynn

FURTHER READING

Adler, I. *X-ray Emission Spectrography in Geology.* New York: Elsevier, 1966. This classic 258-page text describes all aspects of the theory of X-ray fluorescence analysis as well as the practical aspects of sample preparation, sensitivity, and methods of interpreting the resulting data. Well illustrated and intended for college-level geology students, the book provides a clear introduction to the method of X-ray fluorescence analysis and a thorough bibliography.

Bush, Laura. "The Dynamic World of X-ray Fluorescence." *Spectroscopy* 26 (2011): 40-44. Provides current information on X-ray fluorescence technology and applications. Written in a nontechnical manner accessible to the layperson, but contains enough detail to be relevant to professionals in the field of X-ray fluorescence.

Dzubay, T. G. *X-ray Fluorescence Analysis of Environmental Samples.* Ann Arbor, Mich.: Ann Arbor Science, 1981. This 310-page book describes the applications of X-ray fluorescence analysis to problems of atmospheric science, particularly the chemical characterization of airborne particulate matter. Each chapter contains a reference list of scientific journal articles describing particular applications of the technique. While intended for professionals, most sections should be understandable by college-level science students.

Goldstein, Joseph, et al. *Scanning Electron Microscopy and X-ray Microanalysis.* 3rd ed. New York: Springer, 2003. An excellent resource for anyone working in a SEM-EMPA lab.

Jenkins, Ron. *X-ray Fluorescence Spectrometry.* 2d ed. New York: Wiley, 1999. This college-level textbook clearly describes the entire process of X-ray fluorescence analysis, beginning with a historical account of the development of the technique. The sources of X-rays, the X-ray fluorescence emission process, and the various types of detectors are described in detail.

Klockenkeamper, R. *Total-Reflection X-ray Fluorescence Analysis.* New York: Wiley, 1997. A clear description of the procedures and protocols associated with X-ray and fluorescence spectroscopy. Appropriate for the college student without much background with the field. Illustrations, index, and bibliographical references.

Liebhafsky, H. A., and H. G. Pfeiffer. "X-ray Techniques." In *Modern Methods of Geochemical Analysis,* edited by R. E. Wainerdi and E. A. Uken. New York: Plenum Press, 1971. The process by which X-rays are produced and the interaction of these

X-rays with matter are thoroughly discussed. A schematic electron-shell diagram clearly illustrates the various X-ray energies emitted in the X-ray fluorescence process. Although intended for college-level readers, this well-illustrated, 25-page chapter should be suitable for students who have completed a high school chemistry course.

Maxwell, J. A. *Rock and Mineral Analysis.* New York: John Wiley & Sons, 1968. This comprehensive textbook emphasizes the analysis of rock composition by wet chemical techniques but devotes Chapter 11 to the X-ray fluorescence technique. The emphasis in this 22-page chapter is on sample preparation, precision of the analyses, and experimental complications. Suitable for college-level readers.

Pella, P. A. "X-ray Spectrometry." In *Instrumental Analysis,* edited by G. D. Christian and J. E. O'Reilly. 2d ed. Boston: Allyn and Bacon, 1986. This well-illustrated chapter describes the process of X-ray fluorescence and the instrumentation normally employed. Some of the complications in inferring chemical compositions from the X-ray spectra are also described. The textbook, intended for undergraduate science students, includes an extensive bibliography.

Pinta, Maurice. *Modern Methods for Trace Element Analysis.* Ann Arbor, Mich.: Ann Arbor Science, 1978. Chapter 8 of this well-illustrated book describes all aspects of the X-ray fluorescence method and its application to geological, biological, and industrial samples. An extensive reference directs the reader to original sources for a variety of applications of X-ray fluorescence analysis. This 53-page chapter is suitable for advanced high school students.

Potts, Philip J., and Margaret West, eds. *Portable X-ray Fluorescence Spectrometry: Capabilities for In Situ Analysis.* Royal Society of Chemistry, 2008. This text provides an overview of the limitations and capabilities of the new instruments available in X-ray fluorescence spectrometry. Written for the undergraduate student.

Robinson, J. W., Eileen M. Skelly Frame, and George M. Frame III. *Undergraduate Instrumental Analysis.* 6th ed. New York: Marcel Dekker, 2004. This textbook, intended for undergraduate science students, includes a chapter on X-ray spectroscopy, which discusses the applications of the technique to chemical abundance determinations. The medical, industrial, and scientific applications of X-ray fluorescence are described.

Tertian, R., and F. Claisse. *Principles of Quantitative X-ray Fluorescence Analysis.* New York: Wiley, 1982. This comprehensive 385-page text describes all aspects of X-ray fluorescence analysis. Individual chapters describe the process by which fluorescence X-rays are emitted, the instrumentation employed, sample preparation, and the procedure for interpreting the observed spectra. Each chapter contains a comprehensive reference list. Suitable for college-level science students.

See also: Electron Microprobes; Electron Microscopy; Experimental Petrology; Geologic and Topographic Maps; Infrared Spectra; Mass Spectrometry; Neutron Activation Analysis; Petrographic Microscopes; X-ray Powder Diffraction.

X-RAY POWDER DIFFRACTION

X-ray powder diffraction is a technique applied to finely powdered crystals or mixtures of crystals to identify and determine the relative amounts of the crystal phase or phases present.

PRINCIPAL TERMS

- **Bragg's law:** the fundamental equation that relates X-ray wavelength, interatomic distances, and the angle between the X-ray beam and the lattice plane of crystals
- **ceramic:** a human-made mineral, crystal, or aggregate thereof, excluding metals
- **crystal:** a solid consisting of a regular periodic arrangement of atoms; its external form and physical properties express the repeated units of the structure
- **diffractometer:** an instrument used for X-ray powder diffraction analysis
- **d-spacing:** the distance between successive parallel layers of atoms in a crystal
- **glass:** a solid with no regular periodic arrangement of atoms; an amorphous solid
- **goniometer:** the mechanism that maintains the correct arrangement among the sample powder, the X-ray beam, and the X-ray detector in a diffractometer
- **mineral:** a natural substance of fixed or narrowly limited chemical and physical properties; most minerals are also crystals
- **phase:** a homogeneous, physically distinct, mechanically separable portion of matter present in a nonhomogeneous chemical system
- **X-ray:** a photon with a much higher energy and shorter wavelength than those of visible light; its wavelength is of the same order of magnitude as the spaces between atoms in a crystal

STRUCTURE OF CRYSTALS

X-ray powder diffraction is a technique used in the analysis of fine powders. It can be used to distinguish glasses from crystals, to identify crystal and mineral phases, and to determine the relative amounts of crystal phases in mixtures. It can also be used to determine the composition of crystals that have a range of ionic substitution, and the size and shape of the unit cell of a crystal substance.

The technique is based on the structure of crystals—that is, on their orderly, periodically repeating system of atoms and molecules. For example, if the minerals quartz (silicon dioxide) and calcite (calcium carbonate) are present in a rock, then X-ray powder techniques can be used to determine their presence and relative amounts. It is not a chemical technique for determining the presence and amount of particular elements, except for those minerals with limited ionic substitution. It does not specifically determine the presence of the silicon or calcium in quartz or calcite. The procedure depends on the fact that the wavelength of X-rays and the spacing between layers of atoms that make up the periodic structure of crystalline substances (d-spacing) are similar: 0.5 to 2.5 angstroms. As a consequence, when X-rays are swept over a crystal lattice and geometric conditions are correct, an energy peak will be emitted that represents each lattice plane of the crystal. Every mineral has a unique set of peaks whose position and size are characteristic of its crystal structure and chemical composition.

A crystal is a homogeneous solid with an orderly, periodically repeating atomic structure. This structure is responsible for the flat faces on large crystals; the orientation of these faces relative to one another is a consequence of the internal structure of the crystal. A two-dimensional analogy to the periodic structure in a crystal lattice is the repeating pattern in wallpaper. Each design unit in the wallpaper can be envisioned as representing an atom or molecule (a cluster of atoms). The pattern in a wallpaper design obeys the same mathematical laws that pervade nature, including the structure of crystals. For simplicity, the structure of crystals is commonly envisioned as a series of points periodically repeating in a three-dimensional space. The atoms form sets of parallel planes called lattice planes, and the distance between each lattice plane is symbolized by the letter *d*. Even in a simple rectangular array of two dimensions, there are many possible lattice d-spacings (d1, d2, d3, d4, and so on). Intersecting sets of lattice planes delimit a minimal group of atoms that forms the unit cell, the fundamental building block of each crystal. There are strict mathematical laws that govern the way atoms can repeat in space, and there are only six possible crystal systems: cubic, tetragonal, orthorhombic, hexagonal, monoclinic, and triclinic. These six systems together have 230 non-identical space groups, or arrangements of points in space. When combined with the variation supplied by the ninety-two natural elements, this diversity means that no two crystals have identical structures.

GENERATION OF X-RAYS

One of the ways X-rays can be generated is when high-energy photons bombard a metal. Copper is the metal most used in X-ray powder diffraction procedures, but many other metals can be used, including molybdenum, nickel, cobalt, iron, and chromium. An X-ray tube consists of a tungsten filament and a copper (or other metal) target in a vacuum. The tungsten filament supplies electrons that are accelerated into the copper target by voltages of 30 to 50 kilovolts and currents of 10 to 50 milliamperes. The photons interact with the copper target in two ways. First, many photons are absorbed by a variety of processes that give rise to a broad and continuous spectrum of X-ray energy. Second in powder diffraction, X-rays of a very precise wavelength will be emitted by electrons changing orbits within the copper atoms when these electrons absorb and then release a fixed amount of energy. Thus, the wavelength of X-rays emitted by a copper tube consists of a broad "hump" with "spikes" at highly specific wavelengths. A typical commercial X-ray diffractometer with a copper tube is designed to allow only the X-rays of wavelength 1.5418 angstroms to hit the target powder.

BRAGG'S LAW

When X-rays impinge on a crystal lattice, many interactions take place. The one of importance here is diffraction. The diffraction relation between X-ray wavelength and the d-spacing between lattice planes of a crystal is expressed by Bragg's law. Bragg's law is so named because it was developed by British chemist

English physicist Sir William Henry Bragg (1862-1942) (right) and his son Australian-born British physicist Sir (William) Lawrence Bragg (1890-1971). (Getty Images)

William Henry Bragg and his son. Bragg's law is $n\lambda = 2d \sin\theta$. In this equation, n is a whole number (usually taken as 1), λ is the wavelength of the X-rays (a known quantity), d is the distance between successive parallel planes in a crystal, and θ is the angle between the direction of incoming X-rays and the lattice plane of interest. (The angle is measured by a goniometer.) When the angle θ is such that the difference in the X-ray paths from adjacent crystal planes is not a multiple of a whole wavelength, the diffracted energy will be low, because the emitted wave will be out of phase. When the angle ABC is equal to whole multiples of the wavelength, however, then the emitted waves will be in phase; they will reinforce each other, and an energy peak will be emitted. Bragg's equation can be readily solved, because the X-ray wavelength is known, the angle θ can be read on the goniometer when a peak appears, and one can assume that n is 1. Thus, the only unknown in the equation is the lattice d-spacing (d).

SINGLE-CRYSTAL DIFFRACTION

There are two X-ray diffraction techniques: single crystal and powder. Single-crystal diffraction requires one crystal of the substance of interest. The crystal must be oriented in such a way that the relation between the crystal lattice, the X-ray beam, and the detector is precisely defined. The technique is difficult, requiring precise orientation, refined analysis, and considerable mathematical skill. Single-crystal techniques give a "Laue spot pattern" of energy peaks and are used for determination of the details of crystal structure.

X-ray powder diffraction is a comparatively simple and routine analytical procedure, producing a series of rings of peak energy. It depends on statistics, with the crystallites having uniformly random orientations. The fine powder is packed in its holder so that the millions of crystallites, or minute crystals, are randomly oriented relative to the X-ray beam. Usually, a powder with particles measuring 45 micrometers or less is required. The effect is as though there were one average-sized crystal present. If not already fine grained, the sample must be ground into a powder.

POWDER DIFFRACTION

There are two powder techniques: camera and diffractometer. In the camera method, a special cylindrical Debye-Scherrer camera is used. The sample powder is placed in a thin, glass tube, and the tube is placed on the axis of rotation on the centerline of the cylinder. Photographic film is placed along the inner circumference of the cylinder. X-rays enter along a hole in the side of the camera, hit the rotating powder, are diffracted, and then hit the film. Peaks are recorded as a series of circles (rings) on the film. Careful measurement of the line position relative to the hole where the X-rays entered and knowledge of the geometry involved yield the location of the peaks, which in turn leads to the solution of Bragg's equation. The camera technique was the first to be developed. It is still useful when the amount of sample is very small, but it is tedious and time-consuming. A typical run may take four or five hours.

The diffractometer method is the fastest, easiest, most quantitative, and most widely used powder technique. This technique uses a goniometer to correctly position the X-ray beam, the surface of the sample powder, and the X-ray detector so that proper geometry for the solution of Bragg's equation

is maintained. First, the proper operating conditions for the X-ray generation are set and the goniometer positioned at the desired start angle. Then, the powder is pressed into a holder, so that it has a smooth surface, and placed in the sample chamber. The goniometer scans from the starting angle to the ending angle. In modern machines, that procedure is controlled by a computer program. The output signal is received by a scintillation counter, electronically enhanced, and sent to an output device, typically a strip chart recorder. The output data are called the X-ray pattern, presented as a graph of peak position and size versus angle degrees. A typical run takes thirty minutes.

Each diffraction peak corresponds to an interplaner d-spacing. The size of the peak is a function of the electron density along the "surface" of the lattice. In general, the heavier the element, the larger the peak. For identification of a single-phase powder, the three largest peaks in the low-angle range are normally adequate. Data on their position and relative size are compared to the X-ray powder diffraction card index file published by the American Society for Testing and Materials. When a match is found for the three main peaks, the identification is verified using the balance of the peaks.

If there are two or more phases in the powder, identification becomes a matter of experience, and guesses are used until a phase is identified. Elimination of the peaks of an identified phase is followed by repetitions of the procedure until all peaks are accounted for. Computers then search for likely combinations of peaks. Where precise measurement of the d-spacings is desired, such as in the determination of the unit cell, peaks in the high-angle region are used. The d-spacings are partially controlled by composition variation. For example, when small ions substitute for larger ions, the d-spacing decreases. This decrease results in a shift of the peak toward higher angles.

LIMITATIONS AND BENEFITS

X-ray powder diffraction has several limitations. It is difficult to detect components that form less than 1 percent of a mixture. Furthermore, estimates of the amounts of specific minerals present in a mineral mixture are seldom more accurate than plus or minus 1 percent. Nevertheless, the technique does not alter the character of the sample, and after X-ray analysis, the powder can be used for further analyses.

In fact, X-ray diffraction is widely used in the study of rocks, sediments, meteorites, and any crystalline solid in which the particles are too fine-grained for analysis by standard optical techniques. Because of its simplicity, it is also used to identify large crystals after they have been ground into a suitable powder.

One area in which X-ray diffraction is used is the classification of silicate minerals. Attempts to classify this large and diverse family of minerals on chemical grounds resulted in contradictions and confusion. X-ray diffraction is used to divide silicates into structural groups—such as orthosilicates, phyllosilicates (which include clay minerals), and isosilicates—giving rise to a logical and meaningful classification scheme.

Role in Technological Developments

X-ray diffraction procedures are vital to the technological developments that one tends to take for granted in geology, oceanography, meteoritics, ceramics, electronics, and cements. The search for a diminishing body of finite earth resources means that geologists must look more and more closely at rocks that contain the needed materials. Many of these rocks are very fine grained, and X-ray diffraction is the only suitable analytical tool. In oceanography, the X-ray analysis of the very fine sediments and rocks on the sea floor is essential to understanding the origin and history of the ocean basins. Meteorites—rocky and metallic fragments from beyond the atmosphere—are typically fine grained. Nondestructive analysis by X-ray diffraction is essential to their classification. In the ceramics industry, the development of tougher and more durable pottery, grinding compounds, and insulators requires this analytical technique; in electronics, the development of new transistors, thermisters, and superconductors requires X-ray-based, single-crystal analysis. Cements form a complex paste of reactants that, upon curing, are best studied by X-ray powder diffraction.

In summary, whenever the samples are in the form of very fine crystallites, X-ray powder diffraction is the most powerful and easy-to-use system for mineral or crystal analysis.

David N. Lumsden

Further Reading

Azaroff, L. V., and M. J. Buerger. *The Powder Method in X-ray Crystallography.* New York: McGraw-Hill, 1958. This text is best suited to a second college course in X-ray diffraction analysis. Its focus is the use of powder cameras. As with all books written on this topic, the authors expect the reader to have a background in elementary physics and crystallography. They discuss the design and alignment of cameras, how to take photographs, how to interpret powder photographs in terms of unit cell size and geometry, the causes of errors, and how to overcome them.

Bowen, David Keith, and Brian K. Tanner. *High-Resolution X-ray Diffractometry and Topography.* London: Taylor and Francis, 1998. This book examines the procedures involved with and the equipment required within the field of crystallography. Bowen and Tanner lay the foundation for a thorough look at the processes and applications of X-ray diffraction and X-ray crystallography. A somewhat technical book intended for the specialist.

Buhrke, Victor E., Ron Jenkins, and Deane K. Smith, eds. *A Practical Guide for the Preparation of Specimens for X-ray Fluorescence and X-ray Diffraction Analysis.* New York: John Wiley & Sons, 1998. Provides the best techniques for issues with XRF and XRD analysis. Covers material usually left for manuals along with theoretical discussion.

Bunn, C. W. *Chemical Crystallography.* Oxford, England: Clarendon Press, 1958. A readable text for a graduate-level course in X-ray diffraction procedures. Its emphasis is on basic principles of crystallography, and it provides comparatively little information on the source and interaction of X-rays. It is a valuable resource for camera techniques, both powder and single crystal. There is a minimum of math and chemistry; the author relies instead on photographs and diagrams. Includes a chapter with examples of successful solutions of crystallographic structures.

Bush, Laura. "The Dynamic World of X-ray Fluorescence." *Spectroscopy* 26 (2011): 40-44. Provides current information on X-ray fluorescence technology and applications. Written in a nontechnical manner accessible to the layperson, but contains enough detail to be relevant to professionals in the field of X-ray fluorescence.

Cullity, B. D., and S. R. Stock. *Elements of X-ray Diffraction.* 3rd ed. Addison-Wesley, 2001. Some more recent information in addition to that from the classic second edition. This text uses Bragg's law, so the reader does not need knowledge of reciprocal lattice. Covers fundamentals, experimentation, and applications of XRD.

Hammond, Christopher. *The Basics of Crystallography and Diffraction.* 3rd ed. London: Oxford University Press, 2009. Hammond offers a clear understanding of the principles and practices of crystallography and X-ray crystallography. Index and bibliography.

Jenkins, Ron. *Introduction to X-ray Powder Diffractometry.* 2d ed. New York: John Wiley, 1996. This classic text is intended for an introductory college course in X-ray crystallography. It is a basic source for information on the principles and practice of X-ray powder diffraction as applied to inorganic materials. Jenkins discusses crystallography, X-ray production, the interaction of X-rays and crystals, and the details of X-ray diffractometer design.

Jones, Christopher, et al., eds. *Crsytallographic Methods and Protocol.* Totowa, N.J.: Humana Press, 1996. Part of the *Methods in Molecular Biology* series, this volume examines the use of X-ray diffraction to determine the structure of compounds such as nucleic acids and proteins. A large portion of the book is dedicated to discussing the practices and protocols surrounding X-ray diffraction and X-ray crystallography.

Klein, Cornelis, and Barbara Dutrow. *Manual of Mineral Science.* 23rd ed. New York: John Wiley & Sons, 2008. A classic college-level introduction to mineralogy, updated numerous times since its original publication in 1912. Contains a thorough discussion of crystal systems and concise descriptions of all common minerals, including essential optical data. Chapters 13 and 14 contain a summary of optical microscopy, X-ray and electron imaging methods, and mass spectrometry. Well illustrated and indexed, with key references after each chapter.

Klug, H. P., and L. E. Alexander. *X-ray Diffraction Procedures for Polycrystalline and Amorphous Materials.* New York: John Wiley & Sons, 1954. This classic text is intended for an introductory college course in X-ray crystallography. It is a basic source for information on the principles and practice of X-ray powder diffraction as applied to inorganic materials. It discusses crystallography, X-ray production, the interaction of X-rays and crystals, and the details of X-ray diffractometer design. The specific diffractometers discussed are dated, but the principles remain the same.

Nuffield, E. W. *X-ray Diffraction Methods.* New York: John Wiley & Sons, 1966. This relatively brief book combines information on powder and single-crystal techniques. It is intended as a laboratory aid for students with limited mathematical backgrounds. Discussions of elementary crystallography and X-ray generation are followed by chapters devoted to specific methods, techniques, and concepts. A good introduction to how single-crystal and powder techniques are related.

Potts, Philip J., and Margaret West, eds. *Portable X-ray Fluorescence Spectrometry: Capabilities for In Situ Analysis.* Royal Society of Chemistry, 2008. This text provides an overview of the limitations and capabilities of the new instruments available in X-ray fluorescence spectrometry. Written for the undergraduate student.

See also: Electron Microprobes; Electron Microscopy; Experimental Petrology; Geologic and Topographic Maps; Infrared Spectra; Mass Spectrometry; Neutron Activation Analysis; Petrographic Microscopes; Phase Changes; X-ray Fluorescence.

APPENDIXES

GLOSSARY

absolute date or age: the numerical timing of a geologic event, as contrasted with relative, or stratigraphic, timing.

absorption: the capture of light energy of a specific wavelength by electrons in an atom or molecule.

abyssal plains: flat areas that make up large areas of the ocean floor, analogous to terrestrial plains and covering more than half the total surface area of the earth.

accretion: a process by which celestial bodies grow as mutual gravitation draws gases, dust, and other matter together; the process by which Earth and other large planetary bodies are thought to have formed.

accretionary prism: a complex structure composed of fault-bounded sequences of deep-sea sediments mechanically transferred from subducting oceanic lithosphere to the overriding plate; it forms the wall on the landward side of a subduction trench.

acid: commonly, a substance that increases the concentration of free hydrogen ions in a solution.

acidification: the increased presence of hydrogen ions in water or soil, thus lowering the acid/alkaline balance, or pH.

acidity: the degree to which a solution contains excess free hydrogen ions as determined by the quantity of a basic material required to achieve the neutral pH, or neutralize the solution.

activity: the number of transmutations that occur in a specific process in a specified period of time, such as counts per minute.

aerial photography: observation and imaging of the terrestrial surface from a vertical altitude; in some respects, a limited form of remote sensing.

aftershocks: temblors that follow a primary or major earthquake, originating at or near the same focus and generally decreasing in frequency and magnitude with time; caused by residual stresses not released by the main seismic event.

alkalinity: the degree to which a solution is depleted of free hydrogen ions as determined by the quantity of an acidic material required to achieve the neutral pH, or neutralize the solution.

allochthonous: of rock or sediment that was not originally formed in its present location, but some distance away.

alpha (α) particle: a subatomic particle consisting of two protons and two neutrons, bearing two positive electrical charges, and emitted from an atomic nucleus through fission; the nucleus of a helium atom.

altimetry: determination of the topography of geographic features by measurement of their respective altitudes; a satellite-based method involves sending a signal and determining its return time.

aluminosilicate: rock and mineral molecular compositions of which aluminum and silicon are central atoms, primarily as various oxides.

amplitude: the measure of the maximum positive or negative displacement of a wave relative to its neutral or baseline value, indicated on a seismogram by the tracings of the recording pen (or light beam).

andesite: a volcanic rock that occurs in abundance only along subduction zones; lighter in color than basalt, it contains plagioclase feldspar and often hornblende or biotite.

angstrom: a unit of electromagnetic wavelength equal to one ten-billionth (10^{-10}) of a meter.

anion group: a combination of ions that carries a positive charge and behaves as a single anion.

anisotropic crystal: a crystal whose index of refraction varies according to angle of incidence with respect to the spatial axes of the crystal structure.

anode: the positive terminus of an electrical circuit; in cathode ray tubes, a positively charged plate that directs negatively charged electrons through a tiny aperture in the plate to form an electron beam.

anthropogenic: caused by or resulting from human activities; generally refers to carbon emissions from human-made sources.

aphelion: the greatest distance of a planet from the central star in its elliptical orbit .

Apollo: the National Aeronautics and Space Administration (NASA) lunar expedition program active from 1963 to 1972, which included the 1969 Apollo 11 mission that placed humans on the moon for the first time.

Archimedes' principle: the notion that a floating body displaces a mass of fluid equal to its own mass.

arenaceous: of rocks or sediments having a sandy composition, composed of grains of sand.

argillaceous: of rocks or sediments formed principally from clay or clay-mineral particles.

asteroid: a small, rocky body in orbit around a star; the majority of solar asteroids exist in a belt between Mars and Jupiter.

asthenosphere: a zone of low seismic velocity between the lithosphere and the mantle in which temperature and pressure are such that rocks have very little strength and yield to viscous flow; it is located 100 to 200 kilometers (62-125 miles) below the earth's surface.

atmosphere: the gaseous mass or envelope surrounding a celestial body.

atom: the smallest integral particle of an element that has all the properties of that element.

atomic number: the number of protons in the nucleus of an atom.

atomic spectroscopy: a method to identify elements in a material by measurement of the unique series of light waves that each one emits or absorbs.

attenuation: a reduction in amplitude or energy as a result of the physical characteristics of the transmitting medium as distance from the source increases.

aurora: a multicolored aerial display of light in the skies above the planet's polar regions, caused by highly charged particles in Earth's magnetic field; called *aurora borealis* in the north and *aurora australis* in the south.

axis: the locus of a line through a planet, about which the planet rotates.

Azolla: species of prehistoric fern that could reproduce rapidly; believed by some to be the initial agent of climate change that contributed to the occurrence of an Ice Age.

background extinction rate: the rate at which species become extinct given the absence of any extraordinary environmental influences or phenomena.

basalt: a dark-colored, fine-grained volcanic rock containing the minerals plagioclase feldspar, pyroxene, and olivine; typically the basement rock underlying sediments in the abyssal plains.

bathymetry: determination of ocean depths, previously carried out by depth sounding but now through satellite altimetry.

bearing capacity: the ability of granular soils to support the weight of building structures.

beta (β) particle: an electron or positron emitted from the nucleus of an unstable atom through the spontaneous decomposition of a neutron into a proton.

"big bang" theory: the theory that the universe was created by an initial explosion that resulted in the formation of hydrogen and helium.

bioaccumulation: the process whereby a pollutant material becomes concentrated in the body tissues of organisms as they consume other organisms in their food chain.

bioactive: of those materials that are capable of biological or biochemical activity when present in living systems.

biodiversity: the diversity of life forms within a certain area, which may be limited to a certain biome or may include the entire biosphere of the planet.

biogeochemical cycle: the cycle in which nitrogen, carbon, and other inorganic elements of the soil, atmosphere, and other parts of a region are converted into the organic substances of animals or plants and released back into the environment.

biomagnetism: the magnetic fields generated by living organisms.

biosphere: the zone of the earth where all living organisms, including plants and animals, naturally occur.

biotic: related to biological life or living things, being produced, caused or required by living organisms.

birefringence: the difference between the maximum and minimum indices of refraction of a crystal.

blocking temperature: the temperature at which a magnetic mineral becomes able to permanently record the presence and direction of a magnetic field.

body wave: a seismic wave, designated as S or P, that propagates through the interior of a planet, reflected and refracted by the various layered boundaries within its interior.

Bohr model: a structural model of the atom in which negatively charged electrons move in circular orbits around a positively charged nucleus, with orbits corresponding only to certain discrete energies being permitted.

bolometer: a device that measures radiant energy by the change in resistance of an electrical conductor as temperature changes.

Bouguer gravity: a residual value for the gravity at a point, corrected for latitude and elevation effects and for the average density of the rocks above sea level.

Bragg's law: the mathematical relationship in X-ray crystallography between X-ray wavelength, interatomic distances, and the angle of incidence of the X-ray beam with the lattice plane of the crystal structure.

breccia: broken fragments of rock or mineral that have been fused in a matrix of sand or clay; often produced by impact events.

brine: water with a higher content of dissolved salt than ordinary seawater.

brittle behavior: the sudden failure of a material by the catastrophic loss of cohesion.

brittle fracture: fracturing that occurs at less than 3 to 5 percent of the applied compressional or tensional strain.

caldera: a steep-sided, bowl-shaped depression formed after an eruption when a volcanic mountain collapses into its depleted magma chamber.

capture theory: a lunar origin theory suggesting that the moon was created elsewhere in the solar system and was subsequently drawn into its stable orbit around Earth by gravitation.

carbonate: a chemical salt or mineral whose principal anionic component is the carbonate ion, which is composed of one carbon atom and three oxygen atoms and bears two negative charges.

carbonate rocks: sedimentary rocks composed mainly of carbonate minerals.

carbon capture and storage (ccs): the capture of carbon dioxide from industrial processes for storage in geological formations, biological organisms, or bodies of water.

carbon sequestration: active removal of carbon dioxide from the atmosphere, typically through the processes of photosynthesis and carbonate formation, so that the carbon content is retained in a nonvolatile chemical form.

carbon sink: a reservoir of carbon that has been sequestered for a period of time.

cathode-ray tube: a tubular electronic device that has been evacuated and sealed, in which an electron beam is generated at one end to produce a dot-like image on a fluorescent screen at the other end.

celestial equator: the locus defined by the intersection of Earth's equatorial plane with the celestial sphere.

Centaur asteroids: dynamically unstable icy minor planets formed beyond orbit of Jupiter and orbiting in region of outer planets.

ceramic: a human-made material produced by heating of earthy raw materials, composed primarily of silicon oxides and complex compounds.

Chandrayaan-1: a lunar probe launched by India that landed on the moon in 2008.

Chang'e 1: a lunar orbiter launched by China in 2007 to map the moon's surface in preparation for a manned lunar mission in 2012.

charged-particle reaction: a nuclear reaction involving the addition of a proton, an electron, or other charged subatomic particle to a nucleus.

chemical bond: the intimate interaction of electrons and atomic orbitals that binds two or more atoms together in a molecule.

chemical stratification: differentiation of layers in a planetary body based on chemical affinities and characteristics rather than physical properties.

chert: a hard, well-cemented sedimentary rock that is produced by recrystallization of siliceous marine sediments buried in the seafloor.

chlorofluorocarbons: carbon-based compounds to which various numbers of fluorine and chlorine atoms are bonded instead of the hydrogen atoms of the parent hydrocarbon compounds.

chondrite meteorite: a rocky meteorite, the most common type of meteorite to fall to Earth, comprising as much as 90% of all meteorites.

clathrate hydrates or gas hydrates: crystal structures of gas molecules entrapped by or co-crystallized with water under conditions of high pressure and low temperature.

clay minerals: the diverse group of minerals having very fine crystalline structures and predominantly composed of hydrated aluminum silicates.

cohesion: molecular attraction by which the various particles in a body, whether alike or unlike, are united throughout the mass and maintain an integral body form.

colorimeter: an instrument that measures the intensity of a specific color in a solution; the intensity of color is used to quantify the amount of the substance in solution according to Beer's law.

colorimetric: relating to the measurement of the intensity of absorption of a specific color, or the light of a specific visible wavelength.

column: a cylindrical segment of the earth oriented on a line from the center of the earth to any point on its surface, beginning somewhere in the asthenosphere and ending somewhere within the atmosphere.

common lead: ordinary lead as it was formed at the time when all the elements in nature were created; also called primordial lead.

complex life: multicellular organisms with integrated organ systems comprising animals and plants.

component: a chemical entity used to describe the compositional variation within a phase.

concordant age: a situation in which radiometric dating analyses based on different natural radioactive decay processes all yield the same age for a rock sample.

condensation theory: a lunar origin theory suggesting that the earth and its moon were created from the same nebular materials that formed the rest of the solar system.

condenser lens: the first stage of electromagnetic focusing, which "bends" an electron beam into a tightly concentrated focal point before it passes through a specimen.

conduction: the transfer of heat energy in a material from one region to another of lower relative temperature.

confining pressure: pressure acting in a direction perpendicular to the major applied stress in a rock deformation experiment.

connate fluids: fluids that have been trapped in sedimentary pore spaces.

continental margin: the edge of a continent marking the transition to the ocean basin, typically exposed on land and submerged for a distance into the surrounding seas.

continental rift: a divergent plate boundary from which continental masses are being separated and pulled in opposite directions by the movement of underlying magma in the mantle layer of the planet.

continental shelf: the gently sloping undersea surface extending from the shoreline to the top of the continental slope, which then slopes steeply to the deep ocean bed.

contour lines: on a topographic map, lines of equal elevation that portray the shape and elevation of the terrain.

convection: the transfer of heat by the movement or circulation of matter because of density changes produced by heat variations.

convection cell: a cyclic pattern of movement of mantle material in which the central area of hotter, less dense material flows upward and the outer area of cooler, denser material flows downward.

convergent boundary: the locus at which the movement of two tectonic plates causes them to directly collide.

convergent plate margin: a compressional plate boundary at which an oceanic plate is subducted or two continental plates collide.

core: the center portion of Earth some 2,900 kilometers in diameter, consisting of a liquid outer portion and a solid inner section; the cylinder of material recovered from a core drilling operation.

core drilling: a method of extracting samples of the materials being drilled through in a drilling project.

core-mantle boundary: the seismic discontinuity 2,890 kilometers below the earth's surface that separates the mantle from the outer core.

Coriolis effect: the apparent deflection of an object or mass in motion due to Earth's rotation.

corona: the outer atmosphere of the sun consisting of streams of charged particles.

coronal mass ejection: a sudden, large burst of solar wind that can cause geomagnetic storms and aurorae in Earth's upper atmosphere.

correlation: matching the sequence of events (distinctive layers, fossils, magnetic polarity intervals) between two stratigraphic sections, relating them to the same time in the geologic past.

covalent bonding: a type of chemical bonding produced by the sharing of electrons in overlapping orbitals of adjacent atoms; covalently bonded solids usually have low solubility in water.

creep: the very slow downhill movement of soil and rock under the influence of gravity.

creep tests: experiments that are conducted to assess the effects of time on rock properties, in which environmental conditions (surrounding pressure, temperature) and the deforming stress are held constant.

cross section: the effective area that a nucleus presents to an oncoming nuclear particle, which determines the chance that the particle will strike the nucleus and cause a nuclear reaction to occur.

crust: the outer layer of the earth, ranging in thickness from 5 to 60 kilometers; it consists of rocky material such as silicon-rich igneous rocks, metamorphic rocks, and sedimentary rocks, and is less dense than the mantle.

crystal: a solid consisting of a regular periodic arrangement of atoms; its external form and physical properties express the repeated units of the structure.

crystal axes: directions in a crystal structure with respect to which its molecular units are organized.

Curie point: the temperature at which the imprint of a magnetic field becomes permanent in a magnetic mineral.

Curie temperature: the temperature above which a permanently magnetized material loses its magnetization and below which minerals retain ferromagnetism.

daughter product: an isotope generally produced by the radioactive decay of a parent isotope; it may or may not also be radioactive.

declination: the angle in the horizontal plane between true north and the direction of the paleomagnetism of a rock.

deep-focus earthquakes: earthquakes whose focus is greater than 300 kilometers below the surface.

deformation: the alteration of an object from its normal shape by a force; in geology, a change in the shape of a rock formation.

degree of freedom: the variance of a system; the least number of variables that must be fixed to define the state of a system in equilibrium.

dehydration: the release of water from pore spaces or from hydrous minerals as a result of increasing temperature.

dendrochronology: the study of the annual growth rings of trees as a means of the absolute age of events, also used to calibrate radiocarbon dates with absolute chronology.

density: a property of matter expressed in units of mass per unit volume.

depletion: the process by which the concentration of an isotope in a material is reduced from its normal or native concentration.

deposition: the process by which mobile sediment particles suspended in water are drawn out of suspension by gravity to accumulate as layers of sediment.

desertification: the conversion of biologically productive semiarid grasslands to unproductive arid (dry) lands or deserts; it may include such effects as soil erosion, loss of natural vegetation, and deterioration of soil quality.

detrital remanent magnetization (DRM): magnetization in sedimentary rock acquired by the alignment of magnetic sediment grains with the magnetic field.

deuterium: an atom whose nucleus consists of just one proton and one neutron; an essential stepping-stone in the proton-proton cycle in solar-type stars.

diagenesis: change that can occur in sedimentary rocks at temperatures and pressures below 200 degrees Celsius and 300 megapascals; a distinct process from metamorphism.

differentiation: the separation of the matter in a planetary body into distinct layers according to differences in chemical and physical characteristics.

diffraction: the apparent bending of waves passing the edges of obstacles, a process that allows photons of a specific wavelength to be analyzed by passing them through a diffraction grating.

diffractometer: an instrument used to measure the diffraction pattern produced by a material, typically in X-ray powder diffraction analysis.

dilatation: the change in the area or volume of a body; also known as dilation.

dip: a measure of slope; the angle between the plane of an inclined stratigraphic layer and the horizontal, measured downward from the horizon in the vertical plane perpendicular to the strike of the layer.

discontinuity: a zone in which seismic wave velocity abruptly changes due to variation in physical properties of rock with increased depth, typically corresponding to a boundary between two of the planet's interior layers .

discordant age: a situation in which radiometric dating analyses based on different natural radioactive decay processes all yield different ages for a rock sample because of gain or loss of decay products in the rock sample.

dislocation: a defect in a crystal structure caused by misalignment of the crystal lattice; the presence of dislocations greatly reduces the stress necessary to produce permanent deformation.

distortion: the change in shape of a body.

divergent boundary, divergent plate margin: the boundary between two tectonic plates that exists where the two plates are moving apart from each other, as is the case along mid-oceanic ridges.

d-spacing: the distance between successive parallel layers of atoms in a crystal.

ductile behavior: permanent, gradual, nonrecoverable deformation of a solid; sometimes called plastic deformation.

ductile fracture: fracturing in rock that is able to sustain, under a given set of conditions, 5 to 10 percent deformation before fracturing or faulting occurs.

ductility: the physical property of solid matter that expresses total percent deformation prior to rupture; in geology; the maximum strain a specific type of rock can endure before it finally fails by fracturing or faulting.

dynamo: the actions of a moving, liquid core of a planetary body resulting in the generation of a magnetic field.

dynamo theory: a set of three conditions that allow for a body of fluid, such as Earth's outer core, to generate a magnetic field that does not collapse from ohmic decay over time.

earthic crust: see "crust."

earthquake: violent motion of a part of the earth along a fault due to a sudden release of strain energy in the fault either by lateral slippage or subductive movement.

earthquake focus: the specific location in a seismic fault below the surface of the earth where active movement occurs to produce an earthquake.

earthquake waves: vibrations that emanate from an earthquake, measurable with a seismograph.

eccentricity: the degree of elongation of an elliptical orbit relative to a circular orbit in which the eccentricity is zero by definition.

ecliptic plane: the plane of Earth's orbit around the sun.

ecosystem: a system composed of the interrelated communities of animals, plants, and bacteria, together with their physical and chemical environment.

ejecta: the material that is ejected from an impact or volcanic crater during its formation or eruption.

elastic behavior: recoverable deformation in which the strain is proportional to the stress.

elastic deformation: a nonpermanent deformation that disappears when the deforming stress is removed.

elastic material: a substance that, when deformed by an applied force in any way, undergoes a degree of deformation that is proportional to the applied force and returns back to its original shape when the deforming force is removed.

elastic rebound theory: the theory that rocks across a fault remain attached while accumulating energy and deforming until the energy is released in a sudden slip, which produces an earthquake, allowing the deformation to recover and disappear.

electromagnetic radiation: forms of energy, such as light and radio waves, that consist of both electric and magnetic field components moving through space.

electromagnetism: the relationship between electric fields and magnetic fields; one of four fundamental interactions in nature.

electron: one of the fundamental particles of which all atoms are composed, located in the outer portion of the atom and whose negative charge is equal in magnitude to the positive charge of the proton.

electron capture: retention of the electron emitted from the nucleus as a particle to balance the positive charge of the newly formed proton and maintain electrical neutrality of the nuclide atom.

electronic lens: tunable electromagnetic fields that interact with an electron beam to bend and focus the trajectory of its particles, typically used in electron microscopes and similar devices.

electron shell: a region about the nucleus of an atom, containing electrons of a specific energy.

electrostatic charge: the net negative or positive electrical charge produced by an accumulated excess or deficiency of electrons, respectively.

ellipsoid of revolution: a three-dimensional shape produced by rotating an ellipse around one of its axes.

El Niño Southern Oscillation (ENSO): a fluctuation in the surface temperature and pressure of the Indian and Pacific Oceans that has an extensive influence on North American weather patterns as its associated relatively warm water current encroaches on the west coast of South America.

energy level: the specific allowed energy of an electron in one of the permitted orbits of the Bohr model of the atom.

enhanced oil recovery (EOR) method: the method based on the use of carbon dioxide to enhance oil recovery from depleted oil fields.

enrichment: the process by which an isotope is concentrated in either the reactants or the products of a chemical reaction or physical process.

epicenter: the point on the surface of the earth directly above the focus of an earthquake

equilibrium: the condition of a system at its lowest stable energy state in which any change in one system variable results in a change in all other system variables so as to achieve a new stable state of lowest energy.

equinox: the points on the celestial sphere where the sun appears to cross the celestial equator, moving northward at the vernal equinox and southward at the autumnal equinox; it corresponds to equal hours of night and day.

equipotential surface: a surface on which every point is at the same potential, used here to include gravitational and rotational effects; no work is done when moving along an equipotential surface.

escape velocity: the minimum speed required for an object to escape from a particular gravitational field, such as the gravity of the earth or other celestial object.

Euler pole: the point on the surface of the earth where an axis, about which rotation occurs, penetrates that surface.

eutrophication: the depletion of dissolved oxygen in water by the growth of algae and aquatic plants, and the subsequent decomposition of vegetable matter

exchange reaction: the exchange of atoms of the same element between two different sources (phases, molecules, and so on) or between two different locations within the same source.

exponential decay: a process of decomposition or reaction, particularly in regard to nuclear fission, whose kinetics are described by the time-related function $A_t = A_o e^{-kt}$.

extinction: the total disappearance of a species, as occurs when the last representative of a species dies without leaving offspring.

extremophiles: microbial life forms that live in extreme environments such as above or below the temperature range for liquid water, or in acidic or alkaline media.

facies: the characteristic lithology and paleontology of a sedimentary rock structure from which the environmental conditions at the time the sediment was deposited can usually be inferred.

failure: in engineering terms, the fracturing or giving way of an object under load.

fault: a fracture in the earth's crust at which there has been measurable displacement vertically and/or horizontally relative to each other.

fault drag: the bending of rock layers adjacent to a fault.

faulting: the process of fracturing the crust such that rock layers on opposite sides of the fracture move relative to each other; faults are the structures produced during the process.

feedback: a mechanism by which the output or result of a process becomes a positive or negative influence for the continuation of the process, increasing or decreasing the rate at which the process occurs.

felsic: descriptive of magma having both high silica content and high concentration of light-colored minerals like feldspar, the term being devised from *feldspar* and *silica*.

ferromagnetic: relating to substances with high magnetic permeability, definite saturation point, and measurable residual magnetism.

ferromagnetic material: any type of material that retains a magnetic field, such as iron or magnetite; also called permanent magnet.

fission: splitting of a nucleus through the emission of nuclear particles to form atoms of different elements.

fission fragment: one of the lighter nuclei or nuclear particles resulting from the fission of the nucleus of a heavier element.

fission theory: a lunar origin theory suggesting that the moon and the spinning Earth were separated from each other as the solar system formed.

fission track: the line of detectable damage along the trajectory of a fission fragment traveling through an insulating solid material.

flocculation: the sedimentation process by which a number of individual minute suspended particles are held together in clot-like masses by electrostatic or intermolecular attractive forces.

fluid: any material capable of flowing and hence taking on the shape of its container, characteristic of gases and liquids.

fluorescence: light emitted as an atom in an excited electronic state reverts back to its ground state.

fluvial: having to do with or being the result of flowing water or other liquids.

flux: the rate at which an element of a system transfers from one location to another.

focus: the hypocenter of an earthquake; the actual point of rock breakage in the earth's crust or upper mantle at a fault line under stress and from which the first P waves arrive.

folding: the effect on stratigraphic rock layers due to the pressure exerted by the collision of two tectonic plates, bending initially horizontal layers of rock so that they dip.

foliated: having distinct layers that are formed by sequential mineral deposits (such as mica or chlorite) as a rock undergoes stress or strain on one side.

footwall: the block of crustal rock that lies directly below the plane of a fault.

forcing: an event or phenomenon that drives an initial shift in climate, such as an increase in solar output.

forensic: the examination of material clues to determine the cause and progression of an event that occurred in the past.

fossil fuels: carbon-based fuels that are derived from the remains of once-living organisms, including coal, oil, and natural gas.

fossil record: the history of life on Earth as interpreted from the fossilized remains of extinct life forms.

fractional crystallization: the process of minerals precipitating in a sequential manner from a molten mixture.

fractionation: a physical or chemical process by which individual components are isolated from a common blend, as in the fractional distillation process and the separation of isotopes.

fracture zone: the entire length of the shear zone that cuts a generally perpendicular trend across a mid-oceanic ridge.

frame of reference: a reference point for the planet, with respect to which all velocities are quoted.

free-air gravity: a residual value for the gravity at a point, corrected for latitude and elevation effects, used to determine differences in the densities of subsurface rocks; see also "Bouguer gravity."

fusion: the combining of two separate nuclear entities to form a single entity having an identity different from either of the originals, as when a proton and an electron fuse to form a neutron.

gamma decay: the emission of high-energy electromagnetic radiation as a nucleus loses excess energy.

gamma radiation: high-energy electromagnetic radiation emitted when a nucleus emits excess energy.

gamma spectrum: the unique identifying pattern of discrete gamma energies emitted by each specific type of nucleus as it decays.

gas: a substance in the gas or vapor phase, characterized by the ability to spontaneously assume both the shape and entire volume of the space in which it is contained.

geochronology: the study of the absolute ages of geologic samples and events.

geodesy: study of the earth's shape, topography, and physical features and the forces acting on and in it.

geodetic surveying: surveying in which the figure and size of Earth are taken into account and corrections are made for the curvature of the planet.

geodimeter: an electronic optical device that precisely measures ground distances by electronic timing and phase comparison of modulated light waves traveling from a master unit to a reflector and returning to a light-sensitive tube.

geographic information system (GIS): a computer application designed to store and handle the large quantity of data generated by geographic research.

geographic poles: the north and south termini of Earth, located on either end of Earth's rotational axis where the longitude lines meet.

geologic map: a representation of the distribution of mappable units (formations).

geomagnetic poles: hypothetical magnetic poles located at the points where the axis of the simplified dipole-like model of Earth's magnetic field intersects with Earth's surface; not to be confused with magnetic poles.

geomagnetic storm: a short-term disturbance that occurs when solar wind particles penetrate Earth's magnetosphere, frequently causing disruptions within the magnetic field.

geomagnetism: the planetary magnetic field generated by the motion of Earth's metallic core and mantle.

geophysical survey array: a designated orientation and spacing of sensors and energy sources relative to one another for a geophysical survey.

geophysical target: the object or surface that one wishes to detect by means of a geophysical survey; knowledge of the target is essential to selection of a survey type.

geospatial: pertaining to the location of objects on the surface of Earth and their relationship to each other.

geosyncline: a depression formed by downward folding of a stratigraphic layer in which rock structures of a younger age have formed, usually long and narrow, parallel to the edges of continents.

geothermometers: minerals whose composition can be used to determine the temperatures at which the mineral formation was produced.

giant impact hypothesis: a widely accepted lunar origin theory stating that the impact of a Mars-sized planetesimal with Earth ejected parts of the planetary mantle and crust into orbit, where they formed the moon by accretion.

gigatonnes carbon (GtC): one billon tonnes of carbon.

glass: a solid with no periodic ordered arrangement of atoms, generally formed when molten material is rapidly cooled to a solid state before a regular crystalline structure can be achieved.

global positioning system (GPS): an array of geosynchronous satellites that emit signals used to locate positions on Earth and in near-Earth orbits within a fixed coordinate system.

global warming/cooling: also called "climate change", the increase or decrease in the average temperature of the earth, as caused by either naturally occurring terrestrial and solar processes, or by the activity of life forms on Earth, or both; also called climate change.

goniometer: the mechanism that maintains the correct arrangement among the sample powder, the X-ray beam, and the X-ray detector in a diffractometer.

graben: a long, narrow sunken crustal block bounded by normal faults that forms the floor of a rift valley.

grade: an indicator of the relative pressure and temperature conditions under which a metamorphic rock forms, described as low-grade or high-grade.

granite: a light-colored, coarse-grained igneous rock containing feldspar, quartz, and small amounts of darker minerals.

granitic rock: light-colored, intrusive igneous rock containing large grains of quartz, plagioclase feldspar, and alkali feldspar.

gravimeter: a device that measures the local magnitude of the gravitational force.

gravitational compression: the process by which an object becomes smaller and denser because of the force of gravity acting upon it locally, or within it in the case of a celestial body; heat is generated as the mass is compressed.

greenhouse effect: environmental phenomenon in which carbon dioxide and other gases act to absorb thermal energy and retain it in the atmosphere rather than allowing it to radiate from the planet out into space.

greenhouse gas: any atmospheric gas that contributes to the greenhouse effect by absorbing infrared radiation and re-emitting the energy as longer wavelengths that cannot escape from the atmosphere, tending to increase its average temperature; examples include water vapor, carbon dioxide, methane, and chlorofluorocarbons.

ground state: the configuration of an atom such that all of its electrons are in the lowest allowed energy levels.

guyot: an underwater flat-topped formation made by mantle plume activity below the sea floor.

habitable zone: the region around a star where planetary conditions permit the existence of liquid water.

half-life: the length of time required for one-half of any given amount of material to decompose or be consumed through a process of exponential decay.

half-width: the distance over which the amplitude of an anomaly falls from its maximum value to half the maximum amplitude.

hanging wall: a crustal block that overlies a fault.

head wall: the block of rock that lies directly above the plane of a fault; also known as a hanging wall.

heat island: a region marked by higher temperatures than its surroundings, usually encompassing a city in which the effect is created as solar energy is absorbed by urban building materials like asphalt and concrete.

heliosphere: area within a solar system affected by the radiation emitted from the local star.

high-frequency seismic waves: seismic waves produced during earthquakes that produce the most rapid vibration of the rock structures through which they pass; also called short-period waves.

homogeneous: having uniform composition and properties throughout.

horst: a long, narrow elevated crustal block bounded by normal faults that may form a fault-block mountain.

hot spot: a local area of Earth's crust, unrelated to plate boundaries, heated by a mantle plume of molten rock and subsequently having high volcanic activity.

hydrosphere: the planetary environment consisting of all liquid water, including rivers, lakes, and oceans.

hydrothermal vents: areas of the ocean floor, typically along fault lines or in the vicinity of undersea volcanoes, where water that has percolated into the rock reemerges as superheated, mineral-laden water.

hypocenter: the initial point of rupture along a fault that causes an earthquake; also known as the focus.

ice ages: periods in Earth's past when large areas of the planet, and particularly the continents, were glaciated.

ideal solid: a theoretical solid that is isotropic, is homogeneous, and responds elastically to applied forces such as stress, compression, tension, or shear.

igneous rock: any rock that forms by the solidification of molten material, either lava or magma.

imaging: a computer method for constructing a picture of subsurface geology from empirical seismic data.

immiscible fluids: two fluids incapable of mixing to form a single homogeneous blend, such as oil and water.

impactor: any free-flying object, such as a meteorite, that collides with another body, with the collision itself being known as an "impact event."

inclination: the angle in the vertical plane between horizontal and the direction of magnetization of a rock.

index of refraction: the ratio of the speed of light in vacuum to its speed in a particular transparent medium.

inner core: the innermost component of the internal structure of Earth, believed to be a solid iron-nickel sphere with a radius of about 1,200 kilometers (900 miles).

insolation: the amount of incident solar radiation on a unit area of the earth's surface at any given latitude.

intensity: an arbitrary measure of the strength of shaking that an earthquake produces at a given point, generally strongest near the epicenter of an earthquake, based on the modified Mercalli scale.

interface: the surface between two materials at their point of contact with each other, such as, for example, air and water, liquid and solid.

interference: the combination of waves or vibrations from different sources such that they coincide and reinforce each other or are discordant and detract from each other.

interference color: a color in a crystal image viewed under cross-polarizing filters, caused by subtraction (cancellation) of other colors from white light by interference.

interference figure: a shadow shape caused by the blocking of polarized light from certain areas of a crystal image.

interplanetary magnetic field (IMF): the magnetic field produced by the movement of charged particles in the solar wind.

intrusion: the process of forcing molten rock or magma from within the mantle into fractures and fissures in solidified rock structures of the crust.

intrusive rocks: igneous rock seams formed from magmas that have cooled and crystallized within other crustal rock structures.

inversion (inverse problem): using measured data to construct a geological model that describes the subsurface and is consistent with the measured data.

ion: an atom that has either lost or gained electrons and, therefore, is electrically charged.

ionic bonding: a type of chemical bonding that holds the constituents of a crystal together primarily by electrostatic attraction between oppositely charged ions.

iron catastrophe: an early Earth event leading up to the formation of Earth's iron core, in which iron and other dense elements precipitated out of the early molten silicate mixture and sank to the center of the planet.

irradiation: the exposure of some material to radioactive or electromagnetic emissions.

isochemical processes: processes that leave the chemical compositions of rocks unchanged.

isochron: a line connecting points representing samples of equal age on a radioactive isotope (parent) versus radiogenic isotope (daughter) diagram.

isostasy: the condition of equilibrium position of the continental plates atop the asthenosphere, equivalent to floating.

isotope: atoms of the same element having the same number of protons but differing in the number of neutrons, thus having the same chemical identity and properties but different atomic weights.

isotopic fractionation: changes in the isotopic composition of natural substances, which result from small differences in the physical, chemical, and biological properties of isotopes.

isotropic: having identical properties in all directions.

Jovian: a gas giant planet having structural characteristics like those of Jupiter.

Jupiter family comets: a group of comets that orbit in the outer reaches of the solar system beyond the orbit of Jupiter.

juvenile water: water that originated in the upper mantle, also called magmatic water.

Kepler's third law of planetary motion: a theory introduced by Johannes Kepler that states that the ratio of the cube of a planet's average distance from the sun to the square of its orbital period is the same for all planets.

K-T boundary: the layer of iridium-rich deposits produced by the Ixchulub meteor impact that ended the Cretaceous period.

landslide: the rapid, free-falling downhill movement of soil and rock as a result of failure of the underlying supporting rock structure.

last glacial maximum: the prehistoric period (approximately thirty thousand years ago) in which the glaciers covering Earth were at their thickest.

Legendre polynomials: mathematical functions used to describe conditions such as equipotential surfaces on spheres.

light detection and ranging (LIDAR): a system in which the transmission times of pulsed light signals that are reflected from the ground are used to determine ranges for mapping an area's topography.

limestone: a sedimentary rock composed mostly of calcium carbonate formed from organisms or by chemical precipitation in oceans.

liquefaction: the loss in cohesiveness of water-saturated soil as a result of ground shaking caused by an earthquake.

liquid: the state of any substance at temperatures above its melting point and below its boiling point.

lithology: the description of rocks, such as rock type, mineral makeup, and fluid in rock pores.

lithosphere: Earth's rigid outer layer, composed of the crust and upper mantle, extending 100 to 150 kilometers (60 to 90 miles) below the surface.

long-period comet: a comet originating from the Oort cloud, a spherical shell of comet nuclei and other debris lying between 10,000 and 100,000 AU from the sun.

Love waves: surface-level seismic waves that induce side-to-side motion perpendicular to their direction of propagation.

lower mantle: the seismic region of the earth between 670 and 2,890 kilometers (below the surface, consisting of the DN and DO layers.

low-frequency seismic waves: seismic waves produced during earthquakes that produce slower vibration of the rock structures through which they pass; also called long-period waves.

Luna: a Soviet lunar orbit expedition that culminated in the first impact on the moon's surface in 1959.

lunar reconnaissance orbiter: a 2009 craft that found evidence of large amounts of ice at the moon's south pole.

macroscopic: large enough to be observed by the naked eye.

mafic: descriptive of magma having both high magnesium and iron content and high concentration of dark, heavy, iron-bearing minerals such as olivine and pyroxene; see "Felsic."

magma: a molten material usually composed of silicate material and suspended mineral crystals, that occurs below Earth's crust.

magma chamber: a volcanic reservoir of molten rock that builds within Earth's crust at the site of a volcano.

magma intrusion: the entrance of magma under pressure into available fissures in overlying rock layers.

magma ocean: a deep layer of molten rocks, volatiles, and solids that may have covered large portions of Earth and the moon in their early stages of formation.

magmatic: composed of or originating in the molten rock, or magma, of the earth's mantle.

magnetic anomalies: patterns of reversed polarity in the ferromagnetic minerals present in Earth's crust.

magnetic dipole: a pair of magnetic poles with equal magnitude and opposite signs (generally referred to as "north" and "south").

magnetic domain: a region within a mineral with a single direction of magnetization; mineral grains smaller than about 100 microns contain only one domain, while larger grains can contain several domains.

magnetic field: a field produced by an electric current or magnetic dipole and capable of exerting a localized magnetic force.

magnetic polarity time scale: the geologic history of the changes in the earth's magnetic polarity.

magnetic poles: the two antipodal locations where Earth's magnetic field becomes vertical; not to be confused with geomagnetic poles.

magnetic remanence: the ability of the magnetic minerals in a rock to imprint the magnetic field of Earth prevailing at the time of their formation.

magnetite: a magnetic iron oxide whose molecules are composed of three iron atoms and four oxygen atoms as an isometric mineral, an oxide that is sensitive to magnetic fields.

magnetometer: a device used to detect and measure the strength and direction of magnetic fields.

magnetosphere: the volume around a planetary body where the motion of charged particles is influenced by a planet's magnetic field rather than by the solar wind.

magnitude: a measure of ground motion and energy release of an earthquake; an increase of one magnitude means roughly a thirtyfold increase in energy release.

mantle: the area of molten basaltic rocks separating Earth's crust from its core, estimated to be about 2,900 kilometers (1800 miles) thick.

mantle convection: the thermally driven cyclic movement of molten material in Earth's mantle, believed to be the driving force of plate tectonics.

mantle plume: an upwelling of magmatic material in the mantle within which abnormal amounts of heat are conducted upward to form a hot spot at the crustal surface.

map scale: the ratio that defines the relationship between the measurements of features depicted on the map and those of the actual features.

marker fossil: a species that existed in a wide area but died out in a short time, the presence of which in a stratigraphic layer fixes the date of the strata in which it is found.

mass extinction: a period in which the extinction rate for numerous species rises well above the background extinction rate predicted for species under relatively constant conditions.

mass spectrometer: an apparatus that separates charged particles, including isotopes of a particular element according to their mass difference, based on their movement through a magnetic field in an evacuated chamber.

mass spectrometry: the measurement of the relative quantities of charged particles derived from a single source, including isotope abundances, through the use of a mass spectrometer.

Maunder minimum: the period from 1645 to 1715, when sunspot activity was almost nonexistent.

megaton: a unit of force equivalent to that exerted by the single explosion of one million tons of the high-explosive trinitrotoluene (TNT), used to describe the power of both impact events and nuclear weapons.

mesosphere: the third layer from the surface of Earth's interior when the planet is divided rheologically; a dense, rigid layer that corresponds with most of the mantle.

metamorphic rock: any rock whose mineralogy, mineral chemistry, or texture has been altered by heat, pressure, or changes in composition; it may have igneous, sedimentary, or other, older metamorphic rocks as its precursor.

metamorphosis: a physical and chemical transformation of a type of rock into a different type of rock under the influence of high temperature and pressure.

metasomatism: a process that adds chemicals from surrounding rocks, often carried by water, causing a drastic chemical or physical change in contacted metamorphic rock.

meteoric water: groundwater that has fallen to ground as precipitation as part in the hydrologic cycle before infiltrating the subsurface to become groundwater.

meteorite: a small extraterrestrial body that has struck the surface of Earth; known as a meteor before impact and as a meteoroid before it enters the atmosphere.

methane hydrate: a mineral formed when methane (natural gas) is trapped within the structure of water ice crystals; extensive ocean-floor deposits of methane hydrate might influence climate and could become a major resource.

mid-ocean ridge: a system of undersea rifts in ocean basins, where an underlying magmatic convection process drives seafloor spreading.

milligal: the basic unit of the acceleration of gravity, used by geophysicists in measurement of gravity anomalies; equal to 0.001 centimeter per second squared.

mineral: a natural crystalline substance with a definite range of chemical composition and an ordered internal arrangement of atoms providing well defined physical properties.

Mohorovičić (Moho) discontinuity: the boundary between the crust and the mantle, at which the velocities of S and P waves increase sharply; named after the Yugoslavian seismologist Andrija Mohorovičić, who discovered it in 1909.

mole: the amount of any pure substance that contains as many elementary units as there are atoms in 12 grams of the isotope carbon-12 (^{12}C).

mudstone: sedimentary rock type formed from mixtures of particles ranging from fine clay to coarse sand grains.

multispectral photography: photography generated by utilizing the varying emission wavelengths of different rocks and vegetation.

natural frequency: the frequency at which an object or structure normally vibrates.

natural period: the length of time required for a single vibration of an object or substance when vibrating at its natural frequency.

neap tides: smaller than normal tides that occur when the sun and moon are 90 degrees apart.

near-Earth asteroid: an asteroid in a highly elliptical orbit in the region between Mars and Jupiter, whose orbital trajectory brings it within the orbit of Earth.

neutron: the uncharged particle that is one of the two particles of nearly equal mass forming the nucleus of an atom.

neutron reaction: a nuclear reaction in which a neutron is added to the nucleus, increasing the atomic mass and forming a different isotope of that particular element.

normal polarity: orientation of the earth's magnetic field so that a compass needle points toward the Northern Hemisphere.

normal stress: that component of the stress on a plane that acts in a direction perpendicular to the plane.

north geographic pole: the northernmost region of the earth, located at the northern point of the planet's axis of rotation.

north magnetic pole: a small, nonstationary area in the Arctic Circle toward which a north-seeking compass needle points from any location on the earth.

nuclear reaction: a change in the structure of an atomic nucleus brought about by a collision of the nucleus with another nuclear particle such as a neutron or by fission of the nucleus.

nucleons: the major elementary nuclear particles, consisting of protons and neutrons.

nucleus: the central portion of the atom, which contains all the positive charge and essentially all of the mass of an atom.

nuclide: an isotope; any observable combination of protons and neutrons.

nutation: a periodic wobble or oscillation in the angle of tilt of Earth's rotational axis .

objective lens: the second, magnifying stage of electronic and optical lens focusing, which occurs after the electron beam or incident light passes through the specimen.

oblate spheroid: a spherical body that has become flattened somewhat due to rotational forces acting at the polar regions.

obliquity: the angle of tilt of Earth's rotational axis (about 23.5 degrees) from a perpendicular to its orbital plane (its ecliptic).

oceanic rise: a type of divergent plate boundary that forms long, sinuous mountain chains in the oceans.

ocean wave: a disturbance on the ocean's surface, viewed as an alternate rise and fall of the surface

Oort cloud: a massive spherical region of comets believed to exist beyond Pluto at a distance of between 10,000 and 100,000 AU.

ore: any concentration of economically valuable minerals.

orogeny: mountain building by tectonic forces through the folding and faulting of rock layers.

oscillate: to move back and forth between two states at a steady rate.

outer core: the outer portion of the core, about 2,080 kilometers (1,300 miles) in thickness, which is believed to be composed of molten iron.

paleoceanography: the study of the history of the oceans of the earth, ancient sediment deposition patterns, and ocean current positions compared to ancient climates.

paleomagnetism: the record of the polarity of the planetary magnetic field imprinted in the molecular structure of igneous rock material.

Pangaea: a single supercontinent composed of all of the present continental fragments that existed approximately 300 to 200 million years ago.

particle motion: the motion of a particle in a material volume when it experiences the passage of seismic waves.

partitioning: a physical process by which components of a solution are extracted from one solvent directly into a second solvent immiscible with the first.

pedosphere: the outermost layer of the crust where soil is formed.

pendulum: a suspended mass that can swing freely at its natural frequency.

peridotite: dark-colored iron- and magnesium-rich silicate rock composing much of the earth below the crust, usually containing olivine, pyroxene, and garnet.

perihelion: the closest point of Earth from the sun in its elliptical orbit of the sun, which is located at one focal point of the ellipse.

perturb: to disrupt the stability of a system; to change the path of an orbiting body by the incidental application of a gravitational force.

petrology: the scientific study of the properties and characteristics of rocks.

phase: any part of a system - solid, liquid, or gaseous - that is generally homogeneous and distinct from other entities in the system under investigation; a boundary surface separates adjacent phases.

phase diagrams: graphical devices that show the material phases and stability limits of substances in terms of the variables temperature, pressure, and X, the percent composition.

phase equilibria: the investigation and description of the properties of chemical systems in terms of classical thermodynamics; systems of specified composition are generally investigated as a function of temperature and pressure.

photometer: a device to measure light intensity, using a light meter with a numerical output reading.

photon: a quantum of radiant energy.

photosynthesis: the process of combining atmospheric carbon dioxide and water in the presence of chlorophyll and sunlight to produce glucose and glucose biopolymers in plants, with the concomitant release of free oxygen as a by-product.

physical property contrast: the difference in a characteristic (density, velocity) between the object of interest and its surroundings.

planetary differentiation: a process in which a planetary body separates into distinct layers because of the different chemical and physical characteristics of the substances that make up its composition.

planetesimals: celestial objects that form by the accretion of dust and gravitational attraction and that may eventually combine to form planets.

plastic: descriptive of a solid material with some fluid-like properties that can be deformed by the application of a force.

plastic deformation: a deformation that does not disappear when the deforming stress is removed.

plate boundary: a region where the earth's crustal plates meet, as a converging (subduction zone), diverging (mid-ocean ridge), transform fault, or collisional interaction.

plate tectonics: a geological theory describing the large-scale motions of the earth's lithosphere that account for continental drift and mountain building.

Plinian eruption: a powerful eruption that forces a column of ash and other material directly upward into the sky.

point-source emitters: a power or industrial plant that produces carbon dioxide at its specific location while the plant is operating.

polarity: orientation of Earth's magnetic field relative to the physical orientation of the planet; generally, the relative orientation of electrical or magnetic poles of opposite character.

polarization: a method of filtering light so that only rays aligned with a specific plane are passed.

pore pressure: the pressure of fluid within the pores of a rock structure.

positron: a subatomic particle with the properties of the electron but with a positive electrical charge; the antimatter form of the electron.

postglacial rebound: process by which an area of Earth's crust returns to its original position after a glacier dissipates or moves away and the area is released from the weight of the glacial mass.

precession: the conical movement of the spin axis of a freely rotating body about its center of mass under the influence of an external force such as gravity.

precession of the equinoxes: the change in timing of the equinoxes in the 26,000-year cycle of the precession or wobble of Earth's rotation.

precipitation: the process in which crystals form within a liquid solution and settle out, or precipitate, from the liquid phase of the solution in which they have formed, as from water or magma.

press/pulse model: the theory that extinctions result from the slow and constant buildup of environmental pressures set into motion by major environmental changes.

primary wave (P wave): the primary or fastest wave traveling through rock away from a seismic event and consisting of a series of compressions and expansions within the substance of the rock.

principal vibration directions: directions in a crystal structure in which light vibrates with maximum or minimum indices of refraction.

product: the material that results from a reactant undergoing a chemical process.

propagated: conducted through a medium.

protolith: the original igneous, sedimentary, or metamorphic rock from which a particular metamorphic rock has been formed.

proton: the positively charged particle that is one of the two particles of nearly equal mass forming the nucleus of an atom.

protoplanet: the early stage of a planet's formation, often referred to as a planetary embryo, that may collide with other protoplanets to form a planetary mass or a planet.

pyroclastic flow: the fluidized and superheated mixture of gases and materials that descends from a volcano at extreme velocity.

P wave: the primary or fastest wave traveling through rock away from a seismic event and consisting of a series of compressions and expansions within the substance of the rock.

radioactive decay: a natural process of exponential decay by which an unstable, or radioactive, isotope transforms into a stable, or radiogenic, isotope by the spontaneous emission of ionizing particles from its nucleus. See also "half-life."

radioactive isotope: an isotope of an element that undergoes a radioactive decay process that ultimately forms a stable, non-radioactive isotope.

radioactivity: the effect of the spontaneous radioactive decay of a nucleus into a more stable isotope.

radiogenic isotope: the isotope resulting from radioactive decay of another radioactive isotope.

radiogenic lead: lead formed by the radioactive decay of uranium or thorium.

radioisotope dating: the process of using the decay of radioactive isotopes to determine the date at which nearby geological or archaeological features were formed or deposited in the surrounding rock or soil matrix.

radiometric dating: determination of the numerical age of a rock by measuring the decay of radioactive minerals, such as uranium, rubidium, or potassium.

radionuclide: a radioactive isotope of an element.

radio telescope: an instrument used to detect and track radio waves emanating from other planets and celestial bodies.

rain-out model: the stage of planetary differentiation when the core is formed during the "raining out" of iron from the molten NiFe (nickel and iron) from silicate emulsion toward the center of the planet.

rate-determining step: the particular step in a multistep process that determines the maximum rate at which the overall change is observed to take place.

Rayleigh waves: surface seismic waves that occur in a circular, rolling fashion retrograde to their direction of propagation.

reactant: a starting material in any chemical reaction.

reflected wave: a wave whose direction of propagation has been inverted by encountering the interface between two materials of differing physical properties.

reflection: the redirection of an elastic wave from a boundary or interface between two substances or phases, such as between two rock layers of different seismic velocities.

reflectivity: the ratio of the amplitude of the reflected wave to that of the incident wave.

refracted wave: a wave that is transmitted through the interface between two materials or phases, resulting in a change in the direction of travel of the wave.

refraction: the redirection of an elastic wave as it passes through a boundary or interface between two substances or phases, such as through rock layers of different seismic velocities.

relative velocity: the velocity of one object as measured from the point of reference of another.

remolding: the property of some sensitive clays upon disturbance to reorient their particles, which softens them and allows the material to flow in a liquid-like fashion.

remote sensing: the use of instruments not in contact with an object to gain information about that object from some distance away.

reservoir: a location in which a material quantity or property is retained or stored locally.

resolution: the ability to differentiate between two features that are very close together.

retardation: the progressive lagging of part of a ray vibrating in one direction at a slower rate relative to another part vibrating in another direction at a faster rate.

reverse fault: a thrust fault at which compression at the fault line tends to cause material on one side to rise relative to the material on the other side, with the fault plane typically dipping at more than 90 degrees below the horizontal.

reverse polarity: reversed orientation of Earth's magnetic field relative to the physical orientation of the planet; generally, the reverse of the relative orientation of electrical or magnetic poles of opposite character.

rheology: the study of the flow of matter and particularly of liquids and soft solids having plastic character.

Richter scale: a scale devised to rank the magnitude of seismic disturbances, particularly earthquakes, using the amplitude of seismic waves; named after American seismologist and physicist Charles Richter.

rifting: the process whereby lithospheric plates are drawn apart by tensional forces exerted by the movement of underlying magma in the mantle.

rift valley: a region of extensional deformation in which the central block has dropped down with relation to the two adjacent blocks, corresponding to thinning of the crustal material as opposite sides of the rift separate.

right-slip, or right-lateral strike-slip: sideways motion along a steep fault in which the block of Earth's crust across the fault from the point of reference is displaced to the right, the opposite of left-slip faults.

Ring of Fire: a seismically and volcanically active region paralleling the perimeter of the Pacific Ocean.

rotary drilling: a method of drilling holes to great depths using a rotating drill head.

rotation: the change in orientation of a body about an axis.

sandstone: sedimentary rock type formed from sand-sized silicate grains agglomerated and consolidated usually with carbonate minerals.

satellite: any object or body in a stable orbit; generally, a human-made object put into orbit around a celestial body, typically equipped with instruments for various measurements or observations.

scanning electron microscopy (sem): a technique of electron microscopy, in which surface structure images are obtained as patterns of electron emissions from the surface of the specimen scanned by a beam of electrons.

scarp: a steep cliff or slope created by rapid movement along a normal or reverse fault.

seafloor spreading: the hypothesis that oceanic crust is generated by convective upwelling of magma along mid-ocean ridges, causing the sea floor to spread laterally away from the ridge system.

sea level: the normal position of the surface of the ocean relative to the surface of land.

seamount: an isolated dome formed under the sea by plumes and reaching a height of at least 2,300 feet from the sea floor.

seat earth: a fossil soil layer generally found directly beneath coal beds, often with plant rootlets still in place.

secondary wave (S wave): the secondary wave that travels more slowly through solid rock than the P wave, has vibratory shear motion transverse to its direction of travel, and cannot propagate through a liquid medium.

sedimentary rock: a rock that has formed from the accumulation of sediment from water or air; the sediment might be fragments of rocks, minerals, organisms, or products of chemical reactions.

seismic: pertaining to vibrations and movements of crustal material as a result of the internal dynamics of the planet.

seismic belt: a region of relatively high seismicity, globally distributed, marking regions of plate interactions.

seismicity: the occurrence of earthquakes as a function of location and time.

seismic reflection profiling method: measurements

made of the travel times and amplitudes of events attributed to seismic waves that have been reflected from geological interfaces.

seismic sea wave: an extraordinary wave in the ocean generated by an earthquake under the floor of the ocean or along the seacoast; see "tsunami."

seismic tomography: the science of constructing a cross-sectional image of the subsurface geological structure of Earth from seismic data.

seismic waves: physical waves that travel through the material of a planetary body because of processes such as earthquakes, moving outward in all directions as body waves or along and near the surface as surface waves.

seismogram: a record of the train of seismic waves detected by a seismograph as recorded on paper, photographic film, or electronic display device.

seismograph: a sensitive instrument that detects and makes a record of vibrations at the earth's surface and records their arrival times, amplitudes, and directions of motion, usually by measuring the motion of the ground with respect to a relatively fixed mass.

seismologist: a scientist, often a geophysicist, who specializes in studying the characteristics of earthquakes.

seismology: the study of earthquakes and related seismic activity.

seismometer: an instrument used to record seismic energy; also known as a geophone or a seismic detector.

sensitive (quick): describes fine-grained deposits that are characterized by considerable strength in the undisturbed condition, but upon disturbance their ability to support themselves declines dramatically, a typical example being quicksand.

shale: the most abundant sedimentary rock, composed of very tiny mineral grains that settled out of slowly moving water to form a mud.

shallow-focus earthquakes: earthquakes having a focus less than 60 kilometers below the surface.

shatter cone: a conical fracture zone caused by an impact and marked by distinct lines or ridges radiating outward from the apex.

shear: a stress that forces two contiguous parts of an object laterally in opposite directions parallel to their plane of contact; also called shear stress.

shock metamorphism: permanent physical or chemical changes induced in rocks by a shock wave that is generated either by an impact event or by an explosive force.

siderophile: literally, "iron-loving"; refers to elements, like platinum, palladium, osmium, and iridium, that are readily soluble in molten iron and found commonly in meteorites but only rarely on Earth's surface.

silicate: a compound with a negatively charged complex ion whose central atom is silicon, this type of rock being the main component of Earth's crustal material and continental masses.

silicate rocks: rocks whose main crystal structure is a network of silicon and oxygen atoms as silicate ions

siltstone: sedimentary rock type, typically shale, formed from fine clay and silt particles less than 4 microns in size.

site evaluation: an assessment process whereby a location, or site, is selected or rejected for a particular use such as construction or mining.

sixth extinction: a theoretical extinction predicted to occur sometime in the future and potentially caused at least partially by human activity on Earth's surface.

slickensides: fine lines or grooves along a faulted body that usually indicate the direction of latest movement.

slip: a measure of the amount of offset or displacement across the plane of the fault, relative to either the dip or the strike.

SMART-1: a satellite from the European Space Agency's Small Missions for Advanced Research in Technology program, launched in 2003 and deliberately crashed onto the lunar surface in 2006 as the first European spacecraft on the moon.

snowball Earth: hypothetical condition in Earth's early history that the planet was covered with ice, perhaps extending to the equator.

solar wind: the stream of superheated high-energy charged particles flowing out from the upper atmosphere of the sun, consisting mostly of protons and electrons.

solid: any substance that does not flow and that maintains a definite shape.

solstice: the farthest points on the celestial sphere of the sun's apparent path above and below the celestial equator, where the sun stands at midsummer and midwinter, marking the midpoints between the vernal and autumnal equinoxes.

spectrograph: a spectrometer that produces a photographic record of a spectrum of separated electromagnetic radiation.

spectrometer: an instrument used to separate electromagnetic radiation according to wavelength.

spectrophometric: relating to the measurement of the intensity of absorption of light in a specific wavelength that may or may not be in the visible range, including infrared and ultraviolet wavelengths.

spectrophotometer: a photometer for measuring the relative intensities of the light in different parts of a spectrum.

spectroscopy: the subdiscipline of physics that deals with the theory and interpretation of the interactions of matter and electromagnetic radiation.

spectrum: an ordered array of the components of an emission or wave separated and arranged in the order of some characteristic property such as wavelength, mass, or energy.

spectrum analysis: the determination of the constitution of bodies and substances by interpretation of the characteristic spectra of absorbed, transmitted, and reflected electromagnetic radiation that they produce.

spontaneous fission: the unforced splitting of an unstable atomic nucleus into two smaller nuclei.

spring tides: larger-than-normal tides that occur when the sun and moon move in alignment with each other, by which the lunar and solar tides coincide, occurring about twice each month.

SQUID susceptometer: an extremely sensitive magnetometer capable of detecting and measuring very weak magnetic fields.

stable isotope: an isotope of an element that does not undergo spontaneous nuclear fission.

standard: a material of known composition relative to which all enrichment and depletion may be measured.

strain: a measure of deformation resulting from an applied stress that may involve change in volume, shape, or both.

strain rate: the rate at which deformation occurs, expressed as percent strain per unit time.

strata: defined layers in sedimentary rock, typically separated from each other by identifiable bedding planes.

stratigraphy: the study and interpretation of geologic history from layered, usually sedimentary, rock sequences.

stress: the intensity of forces (force per unit area) acting within a body; may refer to a particular stress acting in a particular direction on a particular plane or to the collection of all stresses acting on all planes at that point.

stress-strain test: a common laboratory test utilized in the study of rock and soil deformation to correlate the strain exhibited by a structure with the applied stress.

strike: the measurement of the northerly orientation of a fault plane at the surface as the angle between the horizontal line on the plane and a horizontal line in the north direction.

strike-slip fault: a fault along which movement is horizontal only, parallel to the strike of the fault.

subduction: the process by which dense lithospheric plate material descends into the mantle beneath another, less dense, plate in a subduction zone.

subduction zone: a region where a plate, generally oceanic lithosphere, sinks beneath a continental plate and into the mantle.

subsidence: the sinking of the earth's surface or a decrease in the distance between the earth's surface and its center, generally corresponding to a thinning of the crustal material at that location.

supercontinents: large ancient landmasses containing two or more of Earth's continental landmasses in a single landmass.

supernova: a massive star that explodes after available energy in the interior is used up and the star collapses under the influence of its own gravity.

surface wave: a seismic wave that propagates parallel to a free surface and whose amplitudes disappear at depth, differentiated as "Rayleigh waves" (first described in 1885) and "Love waves" (first described in 1911).

survey line: a normally straight line along which points are located to mark positions where geophysical measurements will be taken.

S wave: the secondary seismic wave generated at the focus of an earthquake, traveling more slowly than the P waves and consisting of elastic shear vibrations transverse to the direction of travel, incapable of propagating through a liquid medium.

synchronized rotation-revolution: a situation in which the rotation rate of an orbiting body is equal to its rate of revolution about its shepherd body, by which mechanism only one side of the moon is visible from Earth.

system: any part of the universe (for example, a crystal, a given volume of rock, or an entire lithospheric plate) that may be defined for thermodynamic analysis; open systems permit energy and mass to enter and leave, while closed systems do not.

target rocks: existing rocks on the surface of a planet that are smashed during an impact event.

tectonic plate: any one of about seven major and several minor pieces forming Earth's outer layer called the lithosphere, each slowly moving, driven by movement of underlying magma in mantle.

tectonics: a branch of geology that deals with the study of structural or deformational features of Earth's crust.

tektite: a dark, glassy object, typically sphere-shaped, that is formed when molten debris resulting from heat of impact is ejected from an impact crater and subsequently cools in the air.

Tellurometer: a portable electronic device that measures ground distances precisely by determining the velocity of a phase-modulated, continuous microwave radio signal transmitted between two instruments operating alternately as a master station and remote station; it has a range up to 65 kilometers.

terrane: a coastal subsection of a continent, formed by the accretion of a small continental mass or plate to a larger one under the influence of seafloor subduction movement; generally, a fault-ridden rock mass relocated from its point of origin and unrelated to adjacent rock structures.

terrestrial: denoting association with dry land rather than water; the opposite of aquatic; generally, associated with planet Earth.

thermal expansion: an increase in the volume of a material that results from its having become warmer.

thermal remanent magnetization (trm): magnetization imprinted in igneous rock as the magnetic materials in molten magma cool below their Curie temperature.

thermodynamics: the branch of chemical physics that deals with the transformation of energy and the laws that govern these changes; equilibrium thermodynamics is especially concerned with the reversible conversion of heat into other forms of energy.

thermohaline circulation: deep water movement in the oceans driven by differences in density, temperature, and salinity, carrying heated surface water from the tropics to the North Atlantic and cooler water at the ocean floor in the opposite direction.

thermoluminescence: a light-emitting process in which electrons trapped within crystal structures in vacancies normally occupied by negative ions are released when heated.

throw: the vertical displacement of a rock sequence or key horizon measured across a fault.

tidal bulge: the areas of a planet's surface and bodies of water that increase in size or are somewhat deformed toward an external gravitational force.

tidal force: the gravitational attraction exerted by an external astronomical body such as the sun or the moon that causes areas of a planet's surface to deform in the direction of the gravity source.

tidally locked: the condition in which the gravitational pull between two bodies, such as Earth and its moon results in one body orbiting the other at the same rate as it revolves on its axis so that the same hemisphere of the orbiting body always faces the other body.

tides: the rising and falling of water levels, usually twice a day, caused by the moon's gravitational attraction.

time of origin: the time of an earthquake's occurrence in local time, or in Coordinated Universal Time (CUT, or Greenwich Mean Time); its use allows for more convenient analysis of worldwide events on a standard time scale.

topographic map: a map representing the elevations of structural features of a part of a planet's surface, plotted to a definite scale and using lines to connect points of the same elevation in structural features.

topography: the structural features of differing elevations of a planetary surface.

topology: the study of the topography of an area and its changes over time.

total dissolved solids (tds): a quantity of solids, expressed in weight percent, determined from the weight of dry residue left after evaporation of a known weight of water.

trace element: a chemical element, such as iron, copper, or zinc, that is essential in plant and animal nutrition, but only in minute quantities.

transcurrent fault: a fault in which relative motion is parallel to the strike of the fault (that is, horizontal); also known as a strike-slip fault.

transform boundary: the location at which the geological effects of two tectonic plates moving past each other in a side-swiping motion are observed.

transform fault: an elongated region away from a mid-oceanic ridge in which faults are formed by lateral shear stress exerted by the boundary regions of two tectonic plates as they move past each other laterally.

translation: the movement of a body or a phenomenon from one location to another.

transmission electron microscopy (TEM): electron microscopy in which all elements of electron activity involved in contact with the specimen are transferred simultaneously to the image.

transmutation: the conversion of an atom of one element into a corresponding atom of another element by altering the number of protons and neutrons within the nucleus.

travel time: the amount of time it takes seismic energy to travel from the source into subsurface geology and arrive back at a seismic detector.

triple junction: a point on Earth's surface where three different tectonic plate boundaries come together.

troposphere: the part of the atmosphere that is closest to the surface of the earth, containing the biosphere and approximately 90% of the total atmosphere.

tsunami: a sea wave created by a seismic event such as an undersea earthquake, a violent volcanic eruption, or a landslide at sea.

turbidity: the degree to which a fluid scatters or diffuses light, caused by solid microparticles or liquid microglobules suspended in the fluid medium.

ultimate strength: the peak or maximum stress recorded in a stress-strain test.

unreinforced masonry (URM): materials not constructed with reinforcing material such as steel .

uplift: the rising of a planet's surface or the increase in distance between the planet's surface and its center.

upper mantle: the fairly rigid part of the earth's interior below the crust of the earth down to about 700 kilometers, composed of magnesium- and iron-rich rock.

Van Allen belt: one of Earth's two radiation belts located outside of the planetary atmosphere.

van der Waals force: a weak electrostatic attraction that arises because certain atoms and molecules are distorted from a spherical shape so that one side of the structure carries more of the charge than does the other.

vector: a quantity that is defined by both magnitude and direction.

velocity: speed and direction of motion; a vector property.

viscosity: a measure of the internal cohesiveness of a liquid material, which determines the ability of a fluid to flow at a certain rate.

volatile elements: basic chemical elements and compounds, such as nitrogen, helium, water, and methane, that are easily driven into the gas phase or that normally exist as gases.

volcanic island arc: a curving or linear group of volcanic islands associated with the movement of a tectonic plate over a hot spot.

volcanic rock: types of igneous rock that are deposited at the surface of the earth through volcanic action, usually composed of larger crystals inside a fine-grained matrix of very small crystals and glasses.

Wadati-Benioff zone: a narrow zone of earthquake foci that seismically delineate an inclined subduction zone, generally tens of kilometers thick and better known as the Benioff zone.

wavelength: the distance over which a sinusoidal or cyclic phenomenon achieves its maximum positive and negative amplitudes before returning to its neutral value.

wave propagation path: directions in which seismic waves travel following an earthquake.

wave refraction: the process by which the direction of waves moving through shallow water is altered by local submarine conditions.

wave velocity: the rate of propagation of a wave; one factor in seismological research that provides useful information about the structure of Earth's interior, which includes regions in which wave velocity markedly increases or decreases.

X-ray: a photon in the wavelength range from 10^{-8} m to 10^{-10} m, thus having a much higher energy and shorter wavelength than photons of visible light but of the same order of magnitude as the spaces between atoms in a crystal.

zone refining: a process of separation that may occur as mineral ores melt and freeze in a repetitive manner, thought to be one process responsible for the separation of specific elements and minerals in igneous rock formations.

BIBLIOGRAPHY

Abaimov, S. G., et al. "Earthquakes: Recurrence and Interoccurrence Times." *Pure and Applied Geophysics* 165 (2008): 777-795.

Abell, George O. *Exploration of the Universe.* 4th ed. New York: Holt, Rinehart and Winston, 1982.

Adams, Jonathan S. *Species Richness: Patterns in the Diversity of Life.* New York: Springer, 2009.

Adler, I. *X-ray Emission Spectrography in Geology.* New York: Elsevier, 1966.

Adolff, Jean Pierre, and Robert Guillaumont. *Fundamentals of Radiochemistry.* Boca Raton, Fla.: CRC Press, 1993.

Agnew, D., and K. Sieh. "A Documentary Study of the Felt Effects of the Great California Earthquake of 1857." *Bulletin of the Seismological Society of America* 68 (1978): 1717-1729.

Agrawal, D. P., and M. G. Yadava. *Dating the Human Past.* Pune: Indian Society for Prehistoric and Quaternary Studies. 1995.

Aharonov, Einat. *Solid-Fluid Interactions in Porous Media: Processes That Form Rocks.* Woods Hole: Massachusetts Institute of Technology, 1996.

Ahrens, L. H. *Distribution of the Elements in Our Planet.* New York: McGraw-Hill, 1965.

Aki, Keiiti, and Paul G. Richards. *Quantitative Seismology: Theory and Methods.* 2d ed. 2 vols. Sausalito: University Science Books, 2002.

Albarede, Francis. *Geochemistry: An Introduction.* 2d ed. Boston: Cambridge University Press, 2009.

Allison, Ira S., and Donald F. Palmer. *Geology.* 7th ed. New York: McGraw-Hill, 1980.

Almeida, J. Sánchez, and Mari Paz Miralles. *The Sun, the Solar Wind, and the Heliosphere.* New York: Springer, 2011.

Anderson, D. L. "The San Andreas Fault." *Scientific American* 224 (February 1971): 52.

Anderson, Greg M., and David A. Crerar. *Thermodynamics in Geochemistry: The Equilibrium Mode.* New York: Oxford University Press, 1993.

Anderson, Ian. "Drilling Deep for Geothermal Power and Science." *New Scientist* 111 (July 24, 1986): 22-23.

Anderson, Soren T., and Richard G. Newell. *Prospects for Carbon Capture and Storage Technologies.* Washington, D.C.: Resources for the Future, 2003.

Angrist, Stanley W., and Loren G. Hepler. *Order and Chaos: Laws of Energy and Entropy.* New York: Basic Books, 1967.

Armstrong, Amit, Roger Surdahl, and H. Gabriella Armstrong. "Peering into the Unknown." *Public Roads.* 72 (2009): 3.

Arnett, David W. *Supernova and Nucleosynthesis: An Investigation of the History of Matter, from the Big Bang to the Present.* Princeton: Princeton University Press, 1996.

Ashwal, L. D., ed. *Workshop on the Growth of Continental Crust.* Technical Report 88-02.

Asphaug, E., and M. Jutzi. "Forming the Lunar Farside Highlands by Accretion of a Companion Moon." *Nature* 476 (August 3, 2011): 69-72.

Associated Press. *The Associated Press Library of Disasters.* Danbury, Conn.: Grolier, 1998.

Atsuyuki, Inoue, et al. "Applications of Chemical Geothermometry to Low-Temperature Trioctahedral Chlorites." *Clays and Clay Minerals.* 57 (2009): 371-382.

Azaroff, L. V., and M. J. Buerger. *The Powder Method in X-ray Crystallography.* New York: McGraw-Hill, 1958.

Bagenal, Fran, Timothy E. Dowling, and William B. McKinnon, eds. *Jupiter: The Planet, the Satellites, and Magnetosphere.* New York: Cambridge University Press, 2004.

Baird, Colin, and Michael Cann. *Environmental Chemistry.* Toronto, Ont.: W. H. Freeman, 2008.

Bakun, William A., et al. "Seismology." *Reviews of Geophysics* 25 (July 1987): 1131-1214.

Ballard, Robert D. *Exploring Our Living Planet.* Washington, D.C.: National Geographic Society, 1983.

Bar-Cohen, Yoseph, and Kris Zacny, eds. *Drilling in Extreme Environments: Penetration and Sampling on Earth and Other Planets.* Weinheim: Wiley-VCH. 2009.

Bard, E., F. Rostek, and Guillemette Menot-Combes. "A Better Radiocarbon Clock." *Science* 303 (2004): 178-179.

Bard, Edouard, and Wallace S. Broecker, eds. *The Last Deglaciation: Absolute and Radiocarbon Chronologies.* Berlin: Springer-Verlag, 1992.

Barker, James. *Mass Spectrometry.* 2d ed. New York: Wiley, 1999.

Barnothy, Madeleine F., ed. *Biological Effects of Magnetic Fields.* 2 vols. New York: Plenum Press, 1964, 1969.

Basavaiah, Nathani. *Geomagnetism: Solid Earth and Upper Atmosphere Perspectives.* New York: Springer, 2011.

Bascom, Willard. *The Crest of the Wave: Adventures in Oceanography.* New York: Doubleday, 1988.

_____. "Deep Hole Story." *Modern Machine Shop* 54 (March 1982): 92-111.

_____. "Drilling the World's Deepest Hole." *Engineering Digest* 34 (June 1988): 16-24.

_____. "Geothermal Boreholes Make Drilling History." *Machine Design* 53 (October 22, 1981): 8.

_____. *A Hole in the Bottom of the Sea: The Story of the Mohole Project.* Garden City, N.Y.: Doubleday, 1961.

Bashkin, V. N., and Robert W. Howarth. *Modern Biogeochemistry.* Boston: Kluwer Academic Publishers, 2002.

Baskaran, Mark, ed. *Handbook of Environmental Isotope Geochemistry.* Vol. 1. New York: Springer, 2011.

Bath, Markus. *Introduction to Seismology.* New York: John Wiley & Sons, 1981.

Baugher, Joseph F. *The Space-Age Solar System.* New York: John Wiley & Sons. 1988.

Bebout, Gray E. *Subduction Top to Bottom.* Washington, D.C.: American Geophysical Union, 1996.

Beck, A. E. *Physical Principles of Exploration Methods: An Introductory Text for Geology and Geophysics Students.* New York: Halsted Press, 1981.

Bedford, A., and D. S. Drumheller. *Introduction to Elastic Wave Propagation.* New York: John Wiley & Sons, 1996.

Belbruno, E., and J. Richard Gott, III. "Where Did the Moon Come from?" *Astronomical Journal* 129, no. 3 (March 2005): 1724-1745.

Ben-Menachem, Ari, and Sarva Jit Singh. *Seismic Waves and Sources.* Mineola, N.Y.: Dover, 1998.

Benn, Chris R. "The Moon and the Origin of Life." *Earth, Moon, and Planets* 85/86, no. 6 (2001): 61-67.

Benn, Douglas I., and David J. A. Evans. *Glaciers and Glaciation.* 2d ed. London: Hodder Education, 2010.

Bennett, Sean J., and Andrew Simon, eds. *Riparian Vegetation and Fluvial Geomorphology.* Washington, D.C.: American Geophysical Union, 2004.

Berlin, G. Lennis. *Earthquakes and the Urban Environment.* 3 vols. Boca Raton, Fla.: CRC Press, 1980.

Bernard, Burton. *ABC's of Infrared.* New York: Howard W. Sams, 1970.

Berner, Elizabeth K., and Robert A. Berner. *Global Environment: Water, Air, and Geochemical Cycles.* Upper Saddle River, N.J.: Prentice Hall, 1996.

Berner, Robert A., and Antonio C. Lasaga. "Modeling the Geochemical Carbon Cycle." *Scientific American* 260 (March 1989): 74-81.

Beroza, G. C., and T. H. Jordan. "Searching for Slow and Silent Earthquakes Using Free Oscillations." *Journal of Geophysical Research* 95 (1990): 2485-2510.

Bertotti, Bruno, Paolo Farinella, and David Vokrouhlicky. *Physics of the Solar System: Dynamics and Evolution, Space Physics, and Spacetime Structure.* New York: Springer, 2003.

Besse, J., and V. Courtillot. "Apparent and True Polar Wander and the Geometry of the Geomagnetic Field Over the Last 200 Myr." *Journal of Geophysical Research.* 107 (2002): 2300.

Best, Myron G. *Igneous and Metamorphic Petrology.* 2d ed. Malden, Mass.: Blackwell Science, 2003.

Beyer, George L., et al., eds. *Microscopy.* New York: Wiley, 1991.

Billings, Marland P. *Structural Geology.* 3rd ed. Englewood Cliffs, N.J.: Prentice-Hall, 1972.

Biogeochemistry. Dordrecht: Martinus Nijhoff/Dr W. Junk Publishers, 1984.

Birks, L. S. *Electron Probe Microanalysis.* 2d ed. New York: Wiley-Interscience, 1971.

Biryukov, E. "Capture of Comets from the Oort Cloud into Halley-Type and Jupiter-Family Orbits." *Solar System Research* 41, no. 3 (2007): 211-219.

Blakemore, R. P., and R. B. Frankel. "Magnetic Navigation in Bacteria." *Scientific American* 245 (June 1981): 42-49.

Blakey, Ronald C., Wolfgang Frisch, and Martin Meschede. *Plate Tectonics: Continental Drift and Mountain Building.* New York: Springer, 2011.

Blatt, Harvey, and Robert J. Tracy. *Petrology: Igneous, Sedimentary, and Metamorphic.* 3rd ed. New York: W. H. Freeman, 2005.

Boggs, Sam, Jr. *Principles of Sedimentology and Stratigraphy.* Columbus, Ohio: Charles E. Merrill, 1986.

Bohne, Rolf, and Roger Anson, eds. *Inventory of World Topographical Mapping.* 3 vols. New York: Pergamon Press, 1989-1993.

Bolt, Bruce A. *Earthquakes: 2006 Centennial Update.* 5th ed. New York: W. H. Freeman, 2005.

_____. *Earthquakes and Geological Discovery.* New York: Scientific American Library, 1993.

_____. *Inside the Earth.* Fairfax, Va. Techbooks, 1991.

_____. *Inside the Earth: Evidence from Earthquakes.* Rev. ed. San Francisco: W. H. Freeman, 1994.

Bonatti, E., and K. Crane. "Ocean Fracture Zones." *Scientific American* 250 (May 1984): 40.

Boore, D. M. "The Motion of the Ground in Earthquakes." *Scientific American* 237 (December 1977): 68.

Bording, R. P., et al. "Applications of Seismic Travel-Time Tomography." *Geophysics* 90 (1987): 285-303.

Bormann, P., ed. *New Manual of Seismological Observatory Practice.* 2d ed. Potsdam: GFZ. 2002.

Bott, M. H. P. *The Interior of the Earth: Its Structure, Constitution, and Evolution.* London: Edward Arnold, 1982.

Bottke, William F., et al. "An Asteroid Breakup 160 Myr Ago as the Probable Source of the K/T Impactor." *Nature* 449 (September 2007): 48-53.

Bowen, David Keith, and Brian K. Tanner. *High-Resolution X-ray Diffractometry and Topography.* London: Taylor and Francis, 1998.

Bowen, Norman L. *The Evolution of the Igneous Rocks.* Mineola, N. Y.: Dover, 1956.

Bowen, Robert. *Isotopes and Climates.* London: Elsevier, 1991.

Bowers, Richard, and Terry Deeming. *Astrophysics.* Vol. 1, *Stars.* Boston: Jones and Bartlett, 1984.

Boyce, Tammy, and Justin Lewis, eds. *Climate Change and the Media.* New York: Peter Lang, 2009.

Brantley, Susan, James Kubicki, and Art White, eds. *Kinetics of Rock-Water Interactions.* New York: Springer, 2007.

Broad, William J. *The Universe Below: Discovering the Secrets of the Deep Sea.* New York: Simon & Schuster, 1997.

Broecker, Wallace. *The Great Ocean Conveyor: Discovering the Trigger for Abrupt Climate Change.* Princeton, N.J.: Princeton University Press, 2010.

Brookfield, Michael E. *Principles of Stratigraphy.* Hoboken, N.J.: Wiley-Blackwell, 2004.

Brooks, C. E. P. *Climate Through the Ages.* New York: McGraw-Hill, 1949.

Brown, G. C., and A. E. Mussett. *The Inaccessible Earth: An Integrated View to Its Structure and Composition.* 2d ed. New York: Chapman and Hall, 1993.

Brownlow, Arthur H. *Geochemistry.* 2d ed. Englewood Cliffs, N.J.: Prentice-Hall, 1995.

Brush, Stephen G. *Nebulous Earth: The Origin of the Solar System and the Core of the Earth from Laplace to Jeffreys.* Cambridge: Cambridge University Press, 1996.

_____. *Transmuted Past: The Age of the Earth and the Evolution of the Elements from Lyell to Patterson.* New York: Cambridge University Press, 1996.

Buck, Roger, et al., eds. *Faulting and Magmatism at Ocean Ridges.* Washington, D.C.: American Geophysical Union, 1998.

Buck, W. Roger, Luc L. Lavier, and Alexei N. B. Poliakov. "Modes of Faulting at Mid-ocean Ridges." *Nature* 434 (2005): 719-723.

Buhrke, Victor E., Ron Jenkins, and Deane K. Smith, eds. *A Practical Guide for the Preparation of Specimens for X-ray Fluorescence and X-ray Diffraction Analysis.* New York: John Wiley & Sons, 1998.

Bullen, K. E., and B. A. Bolt. *An Introduction to the Theory of Seismology.* 4th ed. New York: Cambridge University Press, 1985.

Bunn, C. W. *Chemical Crystallography.* Oxford, England: Clarendon Press, 1958.

Burch, James L. "The Fury of Solar Storms." *Scientific American* 14, no. 4 (2004): 42-49.

Burchfield, Joe D. *Lord Kelvin and the Age of the Earth.* Chicago: University of Chicago Press, 1990.

Burger, H. Robert, Anne F. Sheehan, and Craig H. Jones. *Introduction to Applied Geophysics: Exploring the Shallow Subsurface.* W. W. Norton & Company, 2006.

Bush, Laura. "The Dynamic World of X-ray Fluorescence." *Spectroscopy* 26 (2011): 40-44.

Butler, Robert F. *Paleomagnetism: Magnetic Domains to Geologic Terranes.* Boston: Blackwell Scientific Publications, 1992.

Calder, Nigel. *The Restless Earth: A Report on the New Geology.* New York: Viking Press, 1972.

Campbell, James B., and Randolph H. Wynne. *Introduction to Remote Sensing.* 5th ed. Guilford Press, 2011.

Campbell, Wallace H. *Earth Magnetism: A Guided Tour through Magnetic Fields.* Burlington, MA: Harcourt/Academic Press, 2001.

Campos, Cristina, et al. "Enhancement of Magma Mixing Efficiency by Chaotic Dynamics: An Experimental Study." *Contributions to Mineralogy & Petrology* 161 (2011): 863-881.

Canup, Robin M., and Kevin Righter, eds. *Origin of the Earth and Moon.* Tucson: University of Arizona Press, 2000.

Cardno, Catherine A. "Reftrofit of Stadium Straddling Active Fault Moves Forward." *Civil Engineering.* 80 (2010): 12-14.

Carey, Van P. *Liquid Vapor Phase Change Phenomena.* 2d ed. Flourence, Ky.: Taylor & Francis, 2007.

Cargo, David N., and Bob F. Mallory. *Man and His Geologic Environment.* 2d ed. Reading, Mass.: Addison-Wesley, 1977.

Carlson, Shawn. "The New Backyard Seismology." *Scientific American* 274 (April 1996).

Carmichael, Ian S. E., Francis J. Turner, and John Verhoogen. *Igneous Petrology.* New York: McGraw-Hill, 1974.

Case, James C., and Green, J. Annette. "How to Make Your Wyoming Home More Earthquake Resistant." *Wyoming State Geological Survey: Information Pamphlet 5* (rev.): 2001.

Castro, Jonathan M., and Donald B. Dingwell. "Rapid Ascent of Rhyolitic Magma at Chaiten Volcano, Chile." *Nature* 461 (2009): 780-783.

Cazes, Jack. *Ewing's Analytical Instrumentation Handbook.* 3rd ed. New York: Marcel Dekker, 2005.

Center for Planning and Development Research. *An Earthquake Advisor's Handbook for Wood Frame Houses.* Berkeley: University of California, 1982.

Chacko, T., D. Cole, and J. Horita. "Equilibrium Oxygen, Hydrogen and Carbon Isotope Fractionation Factors Applicable to Geologic Systems." *Reviews in Mineralogy and Geochemistry* 43 (2001): 1-81.

Chandler, Douglas, and Robert W. Roberson. *Bioimaging: Current Techniques in Light and Electron Microscopy.* Sudbury, Mass.: Jones and Bartlett Publishers, 2009.

Chapis, D. A. "The Theory of the Bouguer Gravity Anomaly: A Tutorial." *The Leading Edge* (May 1996) 361-363.

Chapman, Chris. *Fundamentals of Seismic Wave Propagation.* New York: Cambridge University Press, 2010.

Chapman, John Roberts. *Practical Organic Mass Spectrometry: A Guide for Chemical and Biochemical Analysis.* 2d ed. Chichester, N.Y.: Wiley, 1995.

Chester, Roy. *Furnace of Creation, Cradle of Destruction.* New York: AMACOM Books, 2008.

Chilingar, George V., et al. *Evolution of Earth and Its Climate: Birth, Life, and Death of Earth.* Boston: Elsevier, 2011.

Choppin, Gregory R. *Radiochemistry and Nuclear Chemistry.* 3rd ed. Boston: Butterworth-Heinemann, 2001.

Choppin, Gregory R., Jan Rydberg, and Jan-Olov Liljenzin. *Radiochemistry and Nuclear Chemistry.* 3rd ed. Oxford: Butterworth-Heinemann, 2001.

Chuvieco, Emilio, and Chris Justice. "NASA Earth Observation Satellite Missions for Global Change Research." *Earth Observation of Global Change: The Role of Satellite Remote Sensing in Monitoring the Global Environment,* edited by Emilio Chuvieco. Berlin: Springer, 2008.

Ciccioli, Andrea, and Leslie Glasser. "Complexities of One-Component Phase Diagrams." *Journal of Chemical Education* 88 (2011): 586-591.

Clague, John, Chris Yorath, and Richard Franklin. *At Risk: Earthquakes and Tsunamis on the West Coast.* Tricouni Press, 2006.

Clarke, Thurston. *California Fault: Searching for the Spirit of State Along the San Andreas.* New York: Ballantine Books, 1996.

Clayton, Donald D. *Principles of Stellar Evolution and Nucleosynthesis.* Chicago: University of Chicago Press, 1984.

Cloos, E. "Experimental Analysis of Gulf Coast Fracture Patterns." *American Association of Petroleum Geologists Bulletin* 52 (1968).

Coates, John. "Interpretation of Infrared Spectra: A Practical Approach." In *Encyclopedia of Analytical Chemistry,* edited by R. A. Meyers. Chichester: John Wiley & Sons, 2000.

Coble, Charles R., E. C. Murray, and D. R. Rice. *Earth Science.* Englewood Cliffs, N.J.: Prentice-Hall, 1986.

Coburn, Andrew, and Robin Spence. *Earthquake Protection,* 2d ed. New York: Wiley, 2002.

Coffman, Jerry L., Carl A. Von Hake, and C. W. Stover. *Earthquake History of the United States.* U.S. Department of Commerce Publication 41-1. Washington, D.C.: National Oceanic and Atmospheric Administration and U.S. Geological Survey, 1982.

Collier, Michael. *A Land in Motion: California's San Andreas Fault.* San Francisco: Golden Gates National Parks Association, 1999.

Comins, Neil F. *What If the Moon Didn't Exist? Voyages to Earths That Might Have Been.* New York: HarperCollins, 1993.

Compton, Robert R. *Manual of Field Geology.* New York: John Wiley & Sons, 1962.

Condie, Kent C. *Plate Tectonics and Crustal Evolution.* 4th ed. Oxford: Butterworth Heinemann, 1997.

Cone, Joseph. *Fire Under the Sea: The Discovery of the Most Extraordinary Environment on Earth, Volcanic*

Hot Springs on the Ocean Floor. New York: William Morrow, 1991.

Conn, George Keith. *Infrared Methods, Principles, and Applications*. New York: Academic Press, 1960.

Consortium of Univeristies for Research in Earthquake Engineering. *Earthquake Engineering*. Department of Building Inspection, City & County of San Francisco. 2006

Cook, A. H. *Physics of the Earth and Planets*. New York: Halsted Press, 1973.

Correns, C. W. "Fluid Inclusions with Gas Bubbles as Geothermometers." In *Milestones in Geosciences: Selected Benchmark Papers Published in the Journal "Geologische Rondschau,"* edited by Wolf-Christian Dullo and Geologische Vereinigung e.V. New York: Springer-Verlag, 2010.

Cotner, Sehoya, and Randy Moore. *Arguing for Evolution: An Encyclopedia for Understanding Science*. Santa Barbara, Calif.: Greenwood, 2011.

Cox, A., and B. R. Hart. *Plate Tectonics: How It Works*. Hoboken, N.J.: Wiley-Blackwell, 1986.

Cox, Allan, ed. *Plate Tectonics and Geomagnetic Reversals*. San Francisco: W. H. Freeman, 1973.

Craig, J. R., D. J. Vaughan, and B. J. Skinner. *Resources of the Earth*. 3rd ed. Englewood Cliffs, N.J.: Prentice-Hall, 2001.

Cramp, A., et al., eds. *Geological Evolution of Ocean Basins: Results from the Ocean Drilling Program*. London: The Geological Society, 1998.

Criss, Robert E. *Principles of Stable Isotope Distribution*. New York: Oxford University Press, 1999.

Cromie, William J. *Why the Mohole?* Boston: Little, Brown, 1964.

Crowley, Thomas J. "Causes of Climate Change Over the Past 1,000 Years." *Science* 289, no. 5477 (July, 2000): 270-277.

Cubas, N., et al. "Prediction of Thrusting Sequence Based on Maximum Rock Strength and Sandbox Validation." *Trabajos de Geologia* 29 (2009): 189-195.

Cuff, David J., and Andrew S. Goudie. *The Oxford Companion to Global Change*. New York: Oxford University Press, 2009.

Cullity, B. D., and S. R. Stock. *Elements of X-ray Diffraction*. 3rd ed. Addison-Wesley, 2001.

Cunningham, W. D., and Mann, P. *Tectonics of Strike-Slip Restraining and Releasing Bends,* Special Publication no. 290. Geological Society of London, 2008.

Currie, L. A. "The Remarkable Metrological History of Radiocarbon Dating." *Journal of Research of the National Institute of Standards and Technology* 109 (2004): 185-217.

Dahlen, F. A., and Jeroen Tromp. *Theoretical Global Seismology*. Princeton: Princeton University Press, 1998.

Daily, William, and Abelardo Ramirez. "Electrical Resistance Tomography." *The Leading Edge* 23 (2004): 438-442.

Dalrymple, G. Brent. *Ancient Earth, Ancient Skies: The Age of Earth and Its Cosmic Surroundings*. Stanford, Calif.: Stanford University Press, 2004.

Dalrymple, G. Brent, and Marvin A. Lanphere. *Potassium-Argon Dating: Principles, Techniques, and Applications to Geochronology*. San Francisco: W. H. Freeman, 1969.

Daly, R. A. *Our Mobile Earth*. New York: Scribner's, 1926.

Dartnell, Lewis. *Life in the Universe: A Beginner's Guide*. Oxford, England: One World, 2007.

Das, Braja, M. *Principles of Geotechnical Engineering*. 7th ed. Pacific Grove: CL Engineering, 2009.

Das, Braja M., and G. V. Ramana. *Principles of Soil Dynamics*. Stamford: Cengage Learning, 2011.

Davidson, Jon P., Walter E. Reed, and Paul M. Davis. *Exploring Earth: An Introduction to Physical Geology*. 2d ed. Upper Saddle River, N.J.: Prentice Hall, 2001.

Davies, Geoffrey F. *Mantle Convection for Geologists*. New York: Cambridge University Press, 2011.

Davies, Paul. *The Eerie Silence: Searching for Ourselves in the Universe*. Boston: Houghton Mifflin Harcourt, 2010.

Davies, Thomas A. *Glaciated Continental Margins: An Atlas of Acoustic Images*. London: Chapman and Hall, 1997.

_____. "Scientific Ocean Drilling." *Marine Technology Society Journal* 32 (Fall 1998): 5-16.

Davis, Craig A. "Los Angeles Water Supply Impact from a M7.8 San Andreas Fault Earthquake Scenario." *Journal of Water Supply: Research and Technology* 59.6-7 (2010): 408-417.

Davis, G. H. *Structural Geology of Rocks and Regions*. 2d ed. New York: John Wiley & Sons, 1996.

Davis, George H., and Stephen J. Reynolds. *Structural Geology of Rocks and Regions*. 2d ed. New York: John Wiley & Sons, 1996.

de Hoffmann, Edmond, and Vincent Stroobant. *Mass Spectrometry: Principles and Applications.* 3rd ed. New York: Wiley-Interscience, 2007.

Den Hond, Bas. "Scientists Predict GPS Failures." *Astronomy* 34, no. 4 (2006).

Dennen, W. H., and B. R. Moore. *Geology and Engineering.* Dubuque, Iowa: Wm. C. Brown, 1986.

De Pater, Imke, and Jack J. Lissauer. *Planetary Sciences.* 2d ed. New York: Cambridge University Press, 2010.

De Sitter, L. U. *Structural Geology.* New York: McGraw-Hill, 1959.

Dettmer, R. "Geo-energy Firms Work on Deep-Drilling Breakthrough." *Engineering & Technology* 5 (2010): 12.

Dewey, J. F. "Plate Tectonics." *Scientific American* 226 (May 1972): 56.

_____. "Plate Tectonics." In *Continents Adrift and Continents Aground.* San Francisco: W. H. Freeman, 1976.

Dickey, John S. *On the Rocks: Earth Science for Everyone.* New York: Wiley, 1996.

DiGiacomo, Domenico, et al. "Suitability of Rapid Energy Magnitude Determinations for Emergency Response Purposes." *Geophysical Journal International* 180, no. 1 (2010): 361-374.

Diller, T. E. "Advances in Heat Flux Measurements." In *Advances in Heat Transfer,* Vol. 23, edited by J. P. Hartnett et al. Boston: Academic Press, 1993: 279-368.

Dobrin, M. B., and C. H. Savit. *Introduction to Geophysical Prospecting.* 4th ed. New York: McGraw-Hill, 1988.

Dockal, J. A., R. A. Laws, and T. R. Worsley. "A General Mathematical Model for Balanced Global Isostasy." *Mathematical Geology* 21 (March 1989): 147.

Dodds, Walter K., and Matt R. Whiles. *Freshwater Ecology: Concepts and Environmental Applications of Limnology.* 2d ed. Burlington: Academic Press, 2010.

Doell, Richard R., and Allan Cox. *Paleomagnetism.* Vol. 8, *Advances in Geophysics.* New York: Academic Press, 1961.

Dohr, Gerhard. *Applied Geophysics.* New York: Halsted Press, 1981.

Dolphin, Glenn. "Evolution of the Theory of the Earth." *Science and Education* 18, nos. 3/4 (2009): 1-17.

Donlon, Rachael A., ed. *Haiti: Earthquake and Response.* Nova Science Publishers, 2011.

Dosseto, Anthony, Simon P. Turner, and James A. Van-Orman, eds. *Timescales of Magmatic Processes: From Core to Atmosphere.* Hoboken, N.J.: Wiley-Blackwell, 2010.

Doyle, Hugh A. *Seismology.* New York: John Wiley, 1995.

Drever, J. I. *The Geochemistry of Natural Waters.* 3rd ed. Englewood Cliffs, N.J.: Prentice-Hall, 1997.

Dubrov, A. P. *The Geomagnetic Field and Life.* New York: Plenum Press, 1978.

Duckworth, H. E. *Mass Spectrometry.* Cambridge, England: Cambridge University Press, 1958.

Dudley, Walter, and Min Lee. *Tsunami!* 2d ed. Honolulu: University of Hawaii Press, 1998.

Duffield, Wendell, and John Sas. *Geothermal Energy; Clean Power from the Earth's Heat.* USGS Circular 1249. 2003.

Dunlop, D. J., and O. Ozdemir. *Rock Magnetism.* New York: Cambridge University Press, 2001.

Durrance, E. M. *Radioactivity in Geology.* New York: Halsted Press, 1986.

Edwards, John. *Plate Tectonics and Continental Drift.* North Mankato, Minn.: Smart Apple Media, 2006.

Egerton, R. F. *Physical Principles of Electron Microscopy: An Introduction to TEM, SEM, and AEM.* New York: Springer, 2010.

Egholm, D. L., et al. "Glacial Effects Limiting Mountain Height." *Nature* 460 (2009): 884-887.

Ehlers, Ernest G. *The Interpretation of Geological Phase Diagrams.* San Francisco: W. H. Freeman, 1972.

Eiby, G. A. *Earthquakes.* Auckland, New Zealand: Heineman, 1980.

Eicher, Don L. *Geological Time.* 2d ed. Englewood Cliffs, N.J.: Prentice-Hall, 1976.

Eicher, Don L., A. Lee McAlester, and Marcia L. Rottman. *The History of the Earth's Crust.* Englewood Cliffs, N.J.: Prentice-Hall, 1984.

Ekstrom, Goran, Meredith Nettles, and Victor C. Tsai. "Seasonality and Increasing Frequency of Greenland Glacial Earthquakes." *Science* 311 (2006): 1756-1758

Elion, Herbert A., and D. C. Stewart. *A Handbook of X-Ray and Microprobe Data.* Elmsford, N.Y.: Pergamon Press, 1968

Elkins-Tantan, Linda T., Seth Burgess, and Qing-Zhu Yin. "The Lunar Magma Ocean: Reconciling the Solidification Process with Lunar Petrology and Geochronology." *Earth and Planetary Science Letters*, 304, nos. 3/4 (April 2011): 326-336.

Emergency Management BC. *A Simple Explanation of Earthquake Magnitude and Intensity.* Ministry of Public Safety and Solicitor General: Provincial Emergency Program. 2007.

Emiliani, Cesare. *Planet Earth: Cosmology, Geology, and the Evolution of Life and Environment.* Cambridge, England: Cambridge University Press, 1992.

Emsley, John. *The Elements.* 3rd ed. Oxford: Oxford University Press, 1998.

Erickson, John. *Plate Tectonics: Unraveling the Mysteries of the Earth.* New York: Checkmark, 2001.

_____. *Rock Formations and Unusual Geologic Structures.* Rev. ed. New York: Facts on File, 2001.

Ernst, W. G. *Earth Materials.* Englewood Cliffs, N.J.: Prentice-Hall, 1969.

_____. *Petrologic Phase Equilibria.* San Francisco: W. H. Freeman, 1976.

Evans, D. "True Polar Wander and Supercontinents." *Tectonophysics* 362 (2003): 303-320.

Ewing, G. W., ed. *Chemical Instrumentation.* Easton, Pa.: Chemical Education Publishing, 1971.

Ewing, W. Maurice, Wenceslas S. Jardetzky, and Frank Press. *Elastic Waves in Layered Media.* New York: McGraw-Hill, 1957.

Exon, Neville. "Scientific Drilling Beneath the Oceans Solves Earthly Problems." *Australian Journal of Maritime and Ocean Affairs* 2 (2010): 37-47.

Fagan, Brian M., ed. *The Complete Ice Age: How Climate Change Shaped the World.* London: Thames and Hudson, 2009.

Faris, Stephen. *Forecast: The Surprising and Immediate Consequences of Climate Change.* New York: Henry Holt, 2009.

Farley, John E. *Earthquake Fears, Predictions, and Preparations in Mid-America.* Carbondale: Southern Illinois University Press, 1998.

Faure, Gunter. *Isotopes: Principles and Applications.* 3rd ed. New York: John Wiley & Sons, 2004.

_____. *Origin of Igneous Rocks: The Isotopic Evidence.* New York: Springer, 2010.

_____. *Principles of Isotope Geology.* 2d ed. New York: John Wiley & Sons, 1986.

Faust, Aly. *Chemistry of Natural Waters.* Stoneham, Mass.: Butterworth Publishers, 1981.

Federal Emergency Management Agency. *Designing for Earthquakes: A Manual for Architects.* FEMA 454. 2006.

Fenwick, Peter. "The Inverse Problem: A Medical Perspective." *Physics in Medicine and Biology* 32 (April 1987): 5-10.

Fetter, C. W. *Applied Hydrogeology.* 4th ed. Westerville, Ohio: Charles E. Merrill, 2000.

Field, Ned, et al. "Earthquake Shaking: Finding the 'Hotspots'." *U.S. Geological Survey: FS 001-01.* 2001.

Fite, L. E., et al. "Nuclear Activation Analysis." In *Modern Methods of Geochemical Analysis,* edited by Richard E. Wainerdi and Ernst A. Uken. New York: Plenum Press, 1971.

Fleischer, R. L. *Nuclear Tracks in Solids: Principles and Applications.* 2d ed. Berkeley: University of California Press, 1980.

_____. *Tracks to Innovation: Nuclear Tracks in Science and Technology.* New York: Springer, 1998.

Fleming, Stuart. *Dating in Archaeology: A Guide to Scientific Techniques.* New York: St. Martin's Press, 1976.

Florindo, Fabio, and Martin Siegert, eds. *Antarctic Climate Evolution.* Vol. 8. Miamisburg, Ohio: Elsevier Science, 2008.

Folger, Tim. "Waves of Destruction." *Discover* 15 (May 1994): 66-73.

Font, Robert G. *Engineering Geology of the Slope Instability of Two Overconsolidated North-Central Texas Shales.* Vol. 3, Reviews in Engineering Geology. Washington, D.C.: Geological Society of America, 1977.

Fossen, Haakon. *Structural Geology.* New York: Cambridge University Press, 2010.

Foulger, G. R. *Plates vs. Plumes: A Geological Controversy.* New York: Wiley-Blackwell, 2010.

Fowler, Christine Mary Rutherford. *The Solid Earth: An Introduction to Global Geophysics.* 2d ed. Cambridge: Cambridge University Press, 2004.

Fradkin, Philip L. *Magnitude 8: Earthquakes and Life Along the San Andreas Fault.* Berkeley: University of California Press, 1999.

Francheteau, Jean. "The Oceanic Crust." *Scientific American* 249 (September 1983): 114.

Frankel, Henry. *The Continental Drift Controversy.* New York: Cambridge University Press, 2008.

Frechet, Julien, Mustapha Meghraoui, and Massimiliano Stucchi, eds. *Historical Seismology: Interdisciplinary Studies of Past and Recent Earthquakes.* New York: Springer, 2010

Freymueller, Jeffrey T. "Active Tectonics of Plate Boundary Zones and the Continuity of Plate Boundary Definition from Asia to North America." *Current Science* 99, no. 12 (2010): 1719-1732.

Fridleifsson, Gudmundur O., and Wilfred A. Elders. "The Iceland Deep Drilling Project: A Search for Deep Unconventional Geothermal Resources." *Geothermics* 34 (2005): 269-285.

Frisch, Wolfgang, Martin Meschede, and Ronald C. Blakey. *Plate Tectonics: Continental Drift and Mountain Building.* New York: Springer, 2010.

Fritz, P., and J. C. Fontes, eds. *Handbook of Environmental Isotope Geochemistry.* Vol. 1, *Terrestrial Environment.* New York: Elsevier, 1981.

Fry, Brian. *Stable Isotope Ecology.* New York: Springer Science, 2006.

Fujita, Hiroshi. "The Process of Amorphization Induced by Electron Irradiation in Alloys." *Journal of Electron Microscopy Technique* 3 (1986): 245-256.

Fyfe, W. S., and Mackenzie, W. S. "Some Aspects of Experimental Petrology." *Earth-Science Reviews* 5 (1969): 185-215.

Fyfe, W. S., F. J. Turner, and J. Verhoogan. *Metamorphic Reactions and Metamorphic Facies.* New York: Geological Society of America, 1958.

Galbraith, Rex F. *Statistics for Fission Track Analysis.* New York: Chapman & Hall/CRC, 2005.

Galimov, E. M., and A. M. Krivtsov. "Origin of the Earth-Moon System." *Journal of Earth Systems Science* 114, no. 6 (December 2005): 593-600.

Garfinkel, Simson. "Google Earth: How Google Maps the World." *Technology Review* 110, no. 6 (November/December 2007): 20-21.

Garland, G. D. *Introduction to Geophysics: Mantle, Core, and Crust.* Philadelphia: W. B. Saunders, 1971.

_____. *Introduction to Geophysics: Mantle, Core, and Crust.* 2d ed. Philadelphia, Pa.: W. B. Saunders, 1979.

Garrels, Robert M., and Fred T. Mackenzie. *Evolution of Sedimentary Rocks.* New York: W. W. Norton, 1980.

Garver, J. I. "Fission-Track Dating." In *Encyclopedia of Paleoclimatology and Ancient Environments*, edited by V. Gornitz. New York: Kluwer Academic Press, 2008.

Gere, James M., and Haresh C. Shah. *Terra Non Firma: Understanding and Preparing for Earthquakes.* New York: W. H. Freeman, 1984.

Gibson, Henry Louis. *Photography by Infrared: Its Principles and Applications.* 3rd ed. New York: Wiley, 1978.

Gilmour, Iain, and Neil McBride. *An Introduction to the Solar System.* New York: Cambridge University Press, 2004.

Girard, James E. *Principles of Environmental Chemistry.* 2d ed. Sudbury, Mass.: Jones and Bartlett, 2010.

Glasby, Frank. *Planets, Sunspots, and Earthquakes: Effects on the Sun, Earth, and Its Inhabitants.* Bloomington, Ind.: iUniverse, 2002.

Glen, William. *Continental Drift and Plate Tectonics.* Columbus, Ohio: Charles E. Merrill, 1975.

_____. *The Road to Jaramillo: Critical Years of the Revolution in Earth Science.* Stanford, Calif.: Stanford University Press, 1982.

Gochioco, Lawrence. "Advances in Seismic Reflection Profiling for US Coal Exploration." In *Geophysics: The Leading Edge of Exploration*, 1991.

Goldstein, Joseph, et al. *Scanning Electron Microscopy and X-Ray Microanalysis.* 3rd ed. New York: Springer, 2003.

Gonzalez, Guillermo, and Jay Richards. *The Privileged Planet: How Our Place in the Cosmos Is Designed for Discovery.* Washington, D.C.: Regnery, 2004.

Goswami, J. N., and M. Annadurai. "Chandrayaan-1: India's First Planetary Science Mission to the Moon." *Current Science* 96, no. 4 (February 25, 2009): 486-491.

Gottsman, Joachim, and Joan Marti, eds. *Analysis, Modeling and Response.* Vol. 10, *Caldera Volcanism.* Atlanta: Elsevier Science, 2008.

Graff, Karl F. *Wave Motion in Elastic Solids.* New York: Dover, 1991.

Graham, Bob, et al. *Deep Water: The Gulf Oil Disaster and the Future of Offshore Drilling.* National Commission on the BP Deepwater Horizon Oil Spill and Offshore Drilling, 2011.

Greenberg, Arnold, et al., eds. *Standard Methods for the Examination of Water and Wastewater.* 21st ed. Washington, D.C.: American Public Health Association, 2005.

Greenberg, John L. *The Problem of the Earth's Shape from Newton to Clairaut.* New York: Cambridge University Press, 2010.

Greenwood, Norman Neill, and A. Earnshaw. *Chemistry of Elements.* 2d ed. Oxford: Butterworth-Heinemann, 1997.

Gregor, C. Bryan, et al., eds. *Chemical Cycles in the Evolution of the Earth.* New York: John Wiley & Sons, 1988.

Gregory, Snyder A., Clive R. Neal, and W. Gary Ernst, eds. *Planetary Petrology and Geochemistry*. Columbia, Md.: Geological Society of North America, 1999.

Gribble, C. B., and A. J. Hall. *Optical Mineralogy: Principles and Practice*. New York: Chapman and Hall, 1993.

Grieve, Richard A. F. and Gordon R. Osinski. "Impact Craters on Earth." In *Encyclopedia of Solid Earth Geophysics*, edited by Harsh K. Gupta. Dordrecht, Netherlands: Springer, 2011, 593-599.

Griffiths, D. H., and R. F. King. *Applied Geophysics for Engineers and Geologists*. Elmsford, N.Y.: Pergamon Press, 1976.

Griggs, David T. "Creep of Rocks." *Journal of Geology* 47 (April/May 1939): 225-251.

Griggs, Gary B., and John A. Gilchrist. *Geologic Hazards, Resources, and Environmental Planning*. 2d ed. Belmont, Calif.: Wadsworth, 1983.

Gross, M. Grant. *Oceanography*. 7th ed. Columbus, Ohio: Charles E. Merrill, 1996.

Grotzinger, John, et al. *Understanding Earth*. 5th ed. New York: W. H. Freeman, 2006.

Gubbins, David. *Seismology and Plate Tectonics*. Cambridge: Cambridge University Press, 1990.

Gulrajani, Ramesh M. *Bioelectricity and Biomagnetism*. New York: Wiley, 1998.

Gurnis, Michael, et al., eds. *The Core-Mantle Boundary Region*. Washington, D.C.: American Geophysical Union, 1998.

Halacy, D. S. *Earthquakes: A Natural History*. Indianapolis: Bobbs-Merrill, 1974.

Hall, Anthony. *Igneous Petrology*. 2d ed. Harlow: Longman, 1996.

Hamblin, W. K., and E. H. Christiansen. *Earth's Dynamic Systems*. 10th ed. Upper Saddle River, N.J.: Prentice Hall, 2003

Hammond, Christopher. *The Basics of Crystallography and Diffraction*. 3rd ed. London: Oxford University Press, 2009.

Hanslmeier, Arnold. *The Sun and Space Weather*. 2d ed. New York: Springer, 2010.

Hapgood, Charles H. *Earth's Shifting Crust*. Philadelphia: Chilton, 1958.

Hargraves, R. B., and S. K. Banerjee. "Theory and Nature of Magnetism in Rocks." In *Annual Review of Earth and Planetary Sciences*, Vol. 1, edited by F. Donath. Palo Alto, Calif.: Annual Reviews, 1973.

Harland, David M. *Exploring the Moon: The Apollo Expeditions*. 2d ed. New York: Springer-Praxis, 2008.

_____. *Jupiter Odyssey: The Story of NASA's Galileo Mission*. New York: Springer-Praxis, 2011.

Hart, P. J., ed. *The Earth's Crust and Upper Mantle*. Washington, D.C.: American Geophysical Union, 1969

Hartmann, William, Pamela Lee, and Tom Miller. *Cycles of Fire: Stars, Galaxies, and the Wonder of Deep Space*. New York: Workman Publishing, 1988.

Hatcher, Robert D., Jr. *Structural Geology: Principles, Concepts, and Problems*. 2d ed. Englewood Cliffs, N.J.: Prentice Hall, 1995.

Havskov, Jens, and Lars Ottemoller. *Routine Data Processing in Earthquake Seismology*. New York: Springer, 2010.

Heard, Hugh C. "Effect of Large Changes in Strain Rate in the Experimental Deformation of Yule Marble." *Journal of Geology* 71 (March 1963): 162-195.

Heath, Michael J. "Deep Digging for Nuclear Waste Disposal." *New Scientist* 108 (October 31, 1985): 30.

Hecht, E. *Optics*. 4th ed. Reading, Mass.: Addison-Wesley, 2001.

Heinrich, Kurt F. J., ed. *Quantitative Electron Probe Microanalysis*. U.S. National Bureau of Standards Special Publication 298. Washington, D.C.: Government Printing Office, 1968.

Heirtzler, J. R., "Seafloor Spreading." *Scientific American* 219 (December 1968): 60.

Heirtzler, J. R., and W. B. Bryan. "The Floor and the Mid-Atlantic Rift." *Scientific American* 233 (August 1975): 78.

Heiskanen, Weikko A., and Helmut Moritz. *Physical Geodesy*. San Francisco: W. H. Freeman, 1967.

Hem, J. D. *Study and Interpretation of the Chemical Characteristics of Natural Water*. U.S. Geological Survey Water Supply Paper 2254.

Henbest, Nigel. *The Exploding Universe*. New York: Macmillan, 1979.

Henson, Robert. *The Rough Guide to Climate Change*. New York: Penguin, 2011.

Hibben, Frank C. *The Lost Americans*. New York: Thomas Y. Crowell, 1946.

Hill, M. L. "San Andreas Fault: History of Concepts." *Bulletin of the Geological Society of America* 92 (1981): 112-131.

Hillert, Mats. *Phase Equilibria, Phase Diagrams and Phase Transformations*. 2d ed. New York: Cambridge University Press, 2008.

Hites, Ronald A. *Elements of Environmental Chemistry*. Hoboken, N.J.: John Wiley & Sons, 2007.

Hobbs, Bruce E., Winthrop D. Means, and Paul F. Williams. *An Outline of Structural Geology.* New York: John Wiley & Sons, 1976.

Hodge, Paul. *Meteorite Craters and Impact Structures of the Earth.* Cambridge: Cambridge University Press, 2010.

Hodgson, John H. *Earthquakes and Earth Structure.* Englewood Cliffs, N.J.: Prentice-Hall, 1964.

Hoefs, J. *Stable Isotope Geochemistry.* 6th ed. New York: Springer-Verlag, 2009.

Holland, Heinrich D., and Karl K. Turekian, eds. *Isotope Geochemistry: From the Treatise on Geochemistry.* San Diego, Calif.: Academic Press/Elsevier, 2011.

Hollister, L. S., and M. L. Crawford, eds. *Mineralogical Association of Canada Short Course in Fluid Inclusions: Applications to Petrology.* Mineralogical Association of Canada, 1981.

Holloway, J. R., and B. J. Wood. *Simulating the Earth: Experimental Geochemistry.* Winchester, Mass.: Unwin Hyman, 1988.

Holzer, T. L., T. L. Youd, and T. C. Hanks. "Dynamics of Liquefaction During the Superstition Hills, California, Earthquake." *Science* 244 (April 7, 1989): 56-59.

Hooker, Dolph Earl. *Those Astonishing Ice Ages.* New York: Exposition Press, 1958.

Hough, Susan. *Predicting the Unpredictable: The Tumultuous Science of Earthquake Prediction.* Princeton, N.J.: Princeton University Press. 2010.

Howard, Arthur D., and Irwin Remson. *Geology in Environmental Planning.* New York: McGraw-Hill, 1978.

Howell, Benjamin F. *Introduction to Geophysics.* New York: McGraw-Hill, 1959.

Hsu, Kenneth J. *The Mediterranean Was a Desert.* Princeton, N.J.: Princeton University Press, 1983.

Huey, David P., and Michael A. Storms. "Novel Drilling Equipment Allows Downhole Flexibility." *Oil and Gas Journal* 93 (January 16, 1995): 63-68.

Hulot, G., et al., eds. *Terrestrial Magnetism.* New York: Springer, 2011.

Hunter, Elaine Evelyn. *Practical Electron Microscopy: A Beginner's Illustrated Guide.* 2d ed. New York: Cambridge University Press, 1993.

Hutt, C. Robert, et al. *Albuquerque Seismological Laboratory: 50 Years of Global Seismology.* U.S. Geological Survey: FS 2011-3065, 2011.

Huybers, P. J. "Early Pleistocene Glacial Cycles and the Integrated Summer Insolation Forcing." *Science* 313 (July 2006): 508–511.

Iacopi, Robert, ed. *Earthquake Country: California.* 4th ed. Menlo Park, Calif.: Sunset Books/Lane, 1996.

Iacopi, Robert. *Earthquake Country.* 4th ed. Menlo Park, Calif.: Lane Books, 1996.

Idriss, I. M., and R. W. Boulanger. *Soil Liquefaction During Earthquakes.* Earthquake Engineering Research Institute, 2008.

Illinois State Geological Survey. *How to Read Illinois Topographic Maps.* Champaigne, Ill.: Illinois Department of Natural Resources, 2005.

Imbrie, John, and Katherine Palmer Imbrie. *Ice Ages: Solving the Mystery.* 2d ed. Cambridge, Mass.: Harvard University Press, 1986.

Ioannides, A. A. "Trends in Computational Tools for Biomagnetism: From Procedural Codes to Intelligent Scientific Models." *Physics in Medicine and Biology* 32 (January 1987): 77-84.

Ionescu, Corina, Volker Hoeck, and Lucretia Ghergari. "Electron Microprobe Analysis of Ancient Ceramics: A Case Study from Romania." *Applied Clay Science* 53 (2011): 466-475.

Irving, E. "Pole Positions and Continental Drift Since the Devonian." In *The Earth: Its Origins, Structure, and Evolution,* edited by M. W. McElhinny. New York: Academic Press, 1980.

Iyer, H. M., et al., eds. *Seismic Tomography: Theory and Practice.* London: Chapman and Hall, 1993.

Jablonski, David. "Mass Extinctions and Macroevolution." *Paleobiology* 31 (June 2005): 192-210.

Jackson, I., ed. *The Earth's Mantle: Composition, Structure, and Evolution.* Cambridge: Cambridge University Press, 1998.

Jacobs, John A. *Deep Interior of the Earth.* London: Chapman and Hall, 1992.

Jaeger, John Conrad, Neville George Wood Cook, and Robert Zimmerman. *Fundamentals of Rock Mechanics.* 4th ed. New York: John Wiley & Sons, 2007.

Jambor, J., et al., eds. *Advanced Microscopic Studies of Ore Minerals.* Nepean, Ontario: Mineralogical Association of Canada, 1990.

Jeanloz, R., and B. Romanowicz. "Geophysical Dynamics at the Center of the Earth." *Physics Today* 50 (August 1997): 22.

Jenkins, Ron. *Introduction to X-ray Powder Diffractometry.* 2d ed. New York: John Wiley, 1996.

_____. *X-ray Fluorescence Spectrometry.* 2d ed. New York: Wiley, 1999.

Jiao, Wenjie, et al. "Do Intermediate- and Deep-Focus Earthquakes Occur on Preexisting Weak Zones?

An Examination of the Tonga Subduction Zone." *Journal of Geophysical Research* 105 (2000): 125-128.

Johnson, Arvid M. *Physical Processes in Geology.* San Francisco: Freeman, Cooper, 1984.

Johnson, John E., Jr. "The Electron Microscope: Emerging Technologies." *Journal of Electron Microscopy Technique* 1 (1984): 1-7.

Johnson, Michael, and Simon Harley. *Orogenesis: The Making of Mountains.* New York: Cambridge University Press, 2012.

Johnson, Torrence V. "The Galileo Mission to Jupiter and Its Moons." *Scientific American* 13, no. 3 (2003): 54-63.

Jones, Christopher, et al., eds. *Crsytallographic Methods and Protocol.* Totowa, N.J.: Humana Press, 1996.

Jordan, T. H., and J. Minster. "The Deep Structure of the Continents." *Scientific American* 240 (January 1979): 92-107.

_____. "Measuring Crustal Deformation in the American West." *Scientific American* 259 (August 1988): 48-58.

Judson, Sheldon, and Marvin E. Kauffman. *Physical Geology.* 8th ed. Englewood Cliffs, N.J.: Prentice-Hall, 1990.

Jungreis, Susan A. "Biomagnetism: An Orientation Mechanism in Migrating Insects?" *Florida Entomologist* 70 (1987): 277-283.

Jusoh, Zuriati, M. N. M. Nawawi, and Rosli Saad. "Application of Geophysical Method in Engineering and Environmental Problems." *AIP Conference Proceedings* 1250 (2010): 181-184.

Kanai, Kiyoshi. *Engineering Seismology.* Tokyo: University of Tokyo Press, 1983.

Kearey, Philip, Keith A. Klepeis, and Frederick J. Vine. *Global Tectonics.* 3rd ed. New York: Wiley-Blackwell, 2009.

Keer, Richard A. "Geologists Find Vestige of Early Earth—Maybe World's Oldest Rock." *Science* 321 (2008): 1755-1755

Keller, C. *Radiochemistry.* New York: John Wiley & Sons, 1988.

Keller, Edward A. *Environmental Geology.* 9th ed. Upper Saddle River, N.J.: Prentice Hall, 2010.

Kenneth, J. P., ed. *Magnetic Stratigraphy of Sediments. Benchmark Papers in Geology*, Vol. 54. Stroudsburg, Pa.: Dowden, Hutchinson, and Ross, 1980.

_____. *Marine Geology.* Englewood Cliffs, N.J.: Prentice-Hall, 1982.

Kerr, Paul E. *Optical Mineralogy.* 4th ed. New York: McGraw-Hill, 1977.

Kerr, R. A. "Continental Drilling Heads Deeper." *Science* 224 (June 29, 1984): 1418-1420.

_____. "Take Your Choice: Ice Ages, Quakes, or Impacts." *Science* 243 (January 27, 1989): 479-480.

_____. "Two Geologic Clocks Finally Keeping the Same Time." *Science* 320 (2008): 434-435.

Khatiwala, S., F. Primeau, and T. Hall. "Reconstruction of the History of Anthropogenic CO_2 Concentrations in the Ocean." *Nature* 462, no. 7271 (2009): 346-349.

Kholodov, E. A. *Magnetic Fields of Biological Objects.* Translated by A. N. Taruts. Moscow: Nauka, 1990.

Kile, Daniel E. *The Petrographic Microscope: Evolution of a Mineralogical Research Instrument.* The Mineralogical Record, 2003.

King, Lester C. *Wandering Continents and Spreading Sea Floors on an Expanding Earth.* New York: John Wiley & Sons, 1983.

King-Hele, Desmond. "The Shape of the Earth." *Scientific American* 192 (1976): 1293-1300.

Kious, Jacquelyne W. *This Dynamic Earth: The Story of Plate Tectonics.* Washington, D.C.: U.S. Department of the Interior, U.S. Geological Survey, 1996.

Klein, Cornelis, and Barbara Dutrow. *Manual of Mineral Science.* 23rd ed. New York: John Wiley & Sons, 2008.

Klockenkamper, R. *Total-Reflection X-ray Fluorescence Analysis.* New York: Wiley, 1997.

Klotz, Irving M., and R. Rosenberg. *Chemical Thermodynamics: Basic Theory and Methods.* 6th ed. New York: John Wiley & Sons, 2000.

Klug, H. P., and L. E. Alexander. *X-ray Diffraction Procedures for Polycrystalline and Amorphous Materials.* New York: John Wiley & Sons, 1954.

Kolbert, Elizabeth. *Field Notes from a Catastrophe: Man, Nature, and Climate Change.* New York: Bloomsbury, 2006.

Koutsoukos, Eduardo A. M., ed. *Applied Stratigraphy.* New York: Springer, 2007.

Koyhama, Junji. *The Complex Faulting Process of Earthquakes.* New York: Springer, 2010.

Kozlovsky, Yephrim A. "The World's Deepest Well." *Scientific American* 251 (December, 1984): 98-104.

Kramer, Herbert J., and Arthur P. Cracknell. "An Overview of Small Satellites in Remote Sensing." *International Journal of Remote Sensing* 29, no. 15 (August 10, 2008): 4285-4337.

Kramer, Steven Lawrence. *Geotechnical Earthquake Engineering.* Upper Saddle River, N.J.: Prentice Hall, 1996.

Krauskopf, Konrad B. *Introduction to Geochemistry.* 3rd ed. New York: McGraw-Hill, 2003.

Krebs, Robert E. *The History and Use of Our Earth's Chemical Elements: A Reference Guide.* 2d ed. Westport, Conn.: Greenwood Press, 2006.

Kruger, Paul. *Principles of Activation Analysis.* New York: Wiley-Interscience, 1971.

Kuo, John, ed. *Electron Microscopy Methods and Protocols.* 2d ed. Totowa, N.J.: Humana Press, 2007.

Ladd, J. W., et al. "Caribbean Marine Geology: Active Margins of the Plate Boundary." In *The Caribbean Region.* Geological Society of America, edited by G. Dengo and J. E. Case. *The Geology of North America,* Vol. H (1990): 261-290.

Lade, Poul V., and Jerry A. Yamamuro. "Evaluation of Static Liquefaction Potential of Silty Sand Slopes." *Canadian Geotechnical Journal* 48 (2011): 247-264.

Lahee, Frederic H. *Field Geology.* 6th ed. New York: McGraw-Hill, 1961.

Lalleman, Serge, and Francesca Funicello. *Subduction Zone Geodynamics.* New York: Springer, 2009.

Lam, Buuan, et al. "Major Structural Components in Freshwater Dissolved Organic Matter." *Environmental Science & Technology* 41 (2007): 8240-8247.

Lamb, Simon, and Anthony Watts. "The Origin of Mountains: Implications for the Behaviour of Earth's Lithosphere." *Current Science* 99, no. 12 (2010): 1699-1718.

Lambeck, Kurt. *Geophysical Geodesy.* Oxford, England: Clarendon Press, 1988.

Lambert, D., and the Diagram Group. *Field Guide to Geology.* New York: Facts on File, 1988.

Lambert, David, et al. *The Field Guide to Geology.* 2d ed. New York: Facts on File, 2007.

Lane, N., and G. Eaton. "Seismographic Network Provides Blueprint for Scientific Cooperation." *EOS/Transactions American Geophysical Union* 78 (September 1997): 381.

Langel, R. A. *The Magnetic Field of the Earth's Lithosphere: The Satellite Perspective.* Cambridge, England: Cambridge University Press, 1998.

Langmuir, Donald. *Aqueous Environmental Geochemistry.* Upper Saddle River, N.J.: Prentice Hall, 1997.

Lanza, Roberto, and Antonio Meloni. *The Earth's Magnetism: An Introduction for Geologists.* London: Springer, 2011.

Lapedes, D. N., ed. *McGraw-Hill Encyclopedia of Geological Sciences.* New York: McGraw-Hill, 1978.

Lay, T., and T. C. Wallace. *Modern Global Seismology.* San Diego: Academic Press, 1995.

Lay, T., and Q. Williams. "Dynamics of Earth's Interior." *Geotimes* 43 (November 1998): 26.

Leddra, Michael. *Time Matters: Geology's Legacy to Scientific Thought.* Hoboken, N.J.: Wiley-Blackwell, 2010.

Lee, Der-Chuen, and Alex N. Halliday. "Age and Origin of the Moon." *Science* 278 (November 7, 1997).

Lee, W. E., and K. P. D. Lagerhof. "Structural and Electron Diffraction Data for Sapphire." *Journal of Electron Microscopy Technique* 2 (1985): 247-258.

Legget, Robert F. *Cities and Geology.* New York: McGraw-Hill, 1973.

Leick, Alfred. *GPS Satellite Surveying.* 3rd ed. Hoboken, N.J.: Wiley, 2004.

Lenihan, J. M. A., S. J. Thomson, and V. P. Guinn. *Advances in Activation Analysis.* Vol. 2. New York: Academic Press, 1972.

Lerman, A. "Geochemical Cycles." In *The Oxford Companion to the Earth,* edited by Paul Hancock and Brian J. Skinner. New York: Oxford University Press, 2000.

Levin, Harold L. *The Earth Through Time.* 9th ed. Philadelphia: Saunders College Publishing, 2009.

Levy, Matthys, and Mario Salvadori. *Why the Earth Quakes: The Story of Earthquakes and Volcanoes.* New York: W. W. Norton, 1995.

Lieberman, Bruce M., and Roger Kaesler. *Prehistoric Life: Evolution and the Fossil Record.* Hoboken, N.J.: Wiley-Blackwell, 2010.

Liebhafsky, H. A., and H. G. Pfeiffer. "X-ray Techniques." In *Modern Methods of Geochemical Analysis,* edited by R. E. Wainerdi and E. A. Uken. New York: Plenum Press, 1971.

Lieser, Karl Heinrich. *Nuclear and Radiochemistry: Fundamentals and Applications.* 2d ed. New York: Wiley-VCH, 2001.

Likens, Gene, et al. *Biogeochemistry of a Forested Ecosystem.* New York: Springer-Verlag, 1995.

Lillesand, Thomas, Ralph W. Kiefer, and Jonathan Chapman. *Remote Sensing and Image Interpretation.* 6th ed. Hoboken, N.J.: Wiley, 2008.

Lillie, R. J. *Whole Earth Geophysics.* Upper Saddle River, N.J.: Prentice Hall, 1999.

Lindsay, E. H., et al. "Mammalian Chronology and the Magnetic Polarity Time Scale." In *Cenozoic Mammals of North America*, edited by M. O. Woodburn. Berkeley: University of California Press, 1987.

Lines, L. R. "Cross-Borehole Seismology." *Geotimes* 40 (January 1995): 11.

_____, ed. *The Leading Edge* 17 (July 1998): 925-959.

Lo, Tien-When, and Philip L. Inderwiesen. *Fundamentals of Seismic Tomography*. Tulsa, Okla.: Society of Exploration Geophysicists, 1994.

Lockwood, John P., and Richard W. Hazlett. *Volcanoes: Global Perspectives*. Hoboken, N.J.: Wiley-Blackwell, 2010.

Lomnitz, Cinna. *Fundamentals of Earthquake Prediction*. New York: John Wiley & Sons, 1994.

Longley, Paul A., et al. *Geographic Information Systems and Science*, 3rd ed. Hoboken: John Wiley & Sons, 2010.

Looms, M. C., K. H. Jensen, A. Binley, and L. Nielsen. "Monitoring Unsaturated Flow and Transport Using Cross-Borehole Geophysical Methods." *Vadose Zone Journal* 7 (2008): 227-237.

Lorenz, Klaus, and R. Lal. *Carbon Sequestration in Forest Ecosystems*. Dordrecht: Springer, 2010.

Love, A. E. H. *A Treatise on the Mathematical Theory of Elasticity*. 4th ed. Mineola, N.Y.: Dover, 2011.

Lowe, J. John, ed. *Radiocarbon Dating: Recent Applications and Future Potential*. Chichester, N.Y.: John Wiley and Sons, 1996.

Lowell, Lindsay, et al., eds. *Geology and Geothermal Resources of the Imperial and Mexicali Valleys*. San Diego: San Diego Association of Geologists, 1998.

Lowenstein, Tim K., Brian A. Schubert, Michael N. Timofeeff. "Microbial Communities in Fluid Inclusions and Long-Term Survival in Halite." *GSA Today* 21 (2011): 4-9.

Lowrie, William. *Fundamentals of Geophysics*. 2d ed. New York: Cambridge University Press. 2007.

Lundgren, Lawrence. *Environmental Geology*. 2d ed. Englewood Cliffs, N.J.: Prentice-Hall, 1998.

Lutgens, Frederick K., and Edward J. Tarbuck. *Earth Science*. 13th ed. Upper Saddle River, N.J.: Prentice Hall/Pearson, 2012.

Lutgens, Frederick K., Edward J. Tarbuck, and Dennis Tasa. *Essentials of Geology*, 11th ed. Prentice Hall, 2011.

Macdougall, Doug. *Frozen Earth: The Once and Future Story of Ice Ages*. Berkeley: University of California Press, 2006.

Macdougall, J. D. "Fission-Track Dating." *Scientific American* 235 (December 1976) 114-122.

Mackenzie, Dana. *The Big Splat: Or, How Our Moon Came to Be*. Hoboken, N.J.: John Wiley & Sons, 2003.

MacKenzie, W. S., and C. Guilford. *Atlas of Rock-forming Minerals in Thin Section*. New York: Halsted Press, 1980.

Malmivuo, Jaakko. *Bioelectromagnetism: Principles and Applications of Bioelectric and Biomagnetic Fields*. New York: Oxford University Press, 1995.

Mancktelow, N. S. "Fracture and Flow in Natural Rock Deformation." *Trabajos de Geologia* 29 (2009): 29-35.

Manukin, A. B., et al. "Compact High-Sensitivity Accelerometer-Seismometer." *Cosmic Research* 48 (2010): 346-351.

Mares, S., and M. Tvrdy. *Introduction to Applied Geophysics*. New York: Springer, 2011.

Markl, Hubert. "Geobiophysics: The Effect of Ambient Pressure, Gravity and of the Geomagnetic Field on Organisms." Translated by B. P. Winnewisser in *Biophysics*, edited by Walte Hoppe et al. New York: Springer-Verlag, 1983.

Marsh, B. D. "Island-Arc Volcanism." *American Scientist* 67 (March/April 1979): 161.

Marshak, Stephen, and Gautam Mitra, eds. *Basic Methods of Structural Geology*. Englewood Cliffs, N.J.: Prentice-Hall, 1988.

Martin, Angel, et al. "Compact Integration of a GSM-19 Sensor with High-Precision Positioning Using VRS GNSS Technology." *Sensors* 9, no. 4 (2009): 2944-2950.

Marvin, Ursula B. *Continental Drift*. Washington, D.C.: Smithsonian Institution Press, 1973.

Mason, Brian, and Carleton B. Moore. *Principles of Geochemistry*. 2d ed. New York: John Wiley & Sons, 1982.

Mather, K. F., ed. *A Source Book in Geology, 1900-1950*. Cambridge, Mass.: Harvard University Press, 1967.

Mather, K. F., and S. L. Mason, eds. *A Source Book in Geology, 1400-1900*. Cambridge, Mass.: Harvard University Press, 1970.

Maund, Julian G., and Malcolm Eddleston. *Geohazards in Engineering Geology*. London: Geological Society, 1998.

Maus, S., et al. "Earth's Lithospheric Magnetic Field Determined by Spherical Harmonic Degree 90 from CHAMP Satellite Measurements." *Geophysical Journal International* 164, no. 2 (2006): 319-330.

Maxwell, J. A. *Rock and Mineral Analysis*. New York: John Wiley & Sons, 1968.

Mayaux, Philippe, et al. "Remote Sensing of Land-Cover and Land-Use Dynamics." In *Earth Observation of Global Change: The Role of Satellite Remote Sensing in Monitoring the Global Environment*, edited by Emilio Chuvieco. New York: Springer, 2008.

McBride, Neil, and Iain Gilmour. *An Introduction to the Solar System*. New York: Cambridge University Press, 2004.

McCann, D. M. *Modern Geophysics in Engineering Geology*. London: Geological Society, 1997.

McClay, Kenneth R. *Thrust Tectonics*. London: Chapman and Hall, 1992.

McCully, James Greig. *Beyond the Moon: A Conversational, Common Sense Guide to Understanding the Tides*. Hackensack, N.J.: World Scientific Publishing, 2006.

McElhinny, M. W. *Paleomagnetism and Plate Tectonics*. New York: Cambridge University Press, 1973.

McEvily, A. J., Jr., ed. *Atlas of Stress-Corrosion and Corrosion Fatigue Curves*. Materials Park, Ohio: ASM International, 2000.

McKenzie, D. P. "The Earth's Mantle." *Scientific American* 249 (September 1983): 66-78.

McKinley, Theodore D., Kurt F. J. Heinrich, and D. B. Wittry, eds. *The Electron Microprobe*. New York: Wiley, 1966.

McWilliam, Andrew, and Michael Rauch, eds. *Origin and Evolution of the Elements*. New York: Cambridge University Press, 2010.

Means, W. D. *Stress and Strain*. New York: Springer-Verlag, 1976.

Medhat, M. E., and M. Fayez-Hassan. "Elemental Analysis of Cement Used for Radiation Shielding by Instrumental Neutron Activation Analysis." *Nuclear Engineering & Design* 241 (2011): 2138-2142.

Meissner, Rolf, et al., eds. *Continental Lithosphere: Deep Seismic Reflections*. Washington, D.C.: American Geophysical Union, 1991.

Melbourne, Timothy I., and Frank H. Webb. "Slow but Not Quite Silent." *Science* 300 (2003): 1886-1887.

Melchior, Paul. *The Earth Tides*. Elmsford, N.Y.: Pergamon Press, 1966.

Merrill, Ronald T. *Our Magnetic Earth: The Science of Geomagnetism*. Chicago: University of Chicago Press, 2010.

Merrill, R. T., and M. W. McElhinney. *The Magentic Field of the Earth: Paleomagnetism, the Core, and the Deep Mantle*. San Diego: Academic Press, 1998.

Merril, Ronald T., and Philip L. McFadden. "The Use of Magnetic Field Excursions in Stratigraphy." *Quaternary Research*. 63 (2005): 232-237

Michael, K., A. Golab, V. Shulakova, J. Ennis-King, G. Allinson, S. Sharma, and T. Aiken. "Geological Storage of CO_2 in Saline Aquifers: A Review of the Experience from Existing Storage Operations." *International Journal of Greenhouse Gas Control* 4, no. 4 (2010): 659-667.

Milanković, Milutin. *Canon of Insolation and the Ice-Age Problem*. 1941.

Miller, Russell. *Continents in Collision*. Alexandria, Va.: Time-Life, 1983.

_____. *Planet Earth: Continents in Collision*. Alexandria, Va.: Time-Life Books, 1990.

Miralles, Mari Paz, and Jorge Sanchez Almeida, eds. *The Sun, the Solar Wind, and the Heliosphere*. New York: Springer, 2011.

Mitra, Shankar, et al., eds. *Structural Geology of Fold and Thrust Belts*. Baltimore: Johns Hopkins University Press, 1992.

Mogi, Kiyoo. *Earthquake Prediction*. San Diego, Calif.: Academic Press, 1985.

Mollhoff, M., C. J. Bean, and P. G. Meredith. "Rock Fracture Compliance Derived from Time Delays of Elastic Waves." *Geophysical Prospecting* 58 (2010): 1111-1122.

Molnar, P., and P. Tapponier. "The Collision Between India and Eurasia." *Scientific American* 236 (April 1977): 30.

Monastersky, Richard. "Abandoning Richter." *Science News* 146 (October 1994): 250-252.

_____. "The Flap over Magnetic Flips." *Science News* 143 (June 1993): 378-380.

_____. "Waves of Death." *Science News* 154 (October 3, 1998): 221-223.

Monroe, James S., and Reed Wicander. *The Changing Earth: Exploring Geology and Evolution*. 5th ed. Belmont, Calif.: Brooks/Cole, Cengage Learning, 2009.

Monroe, James S., Reed Wicander, and Richard Hazlett. *Physical Geology: Exploring the Earth*. 6th ed. Belmont, Calif.: Thomson, 2007.

Moore, J. Robert. *Oceanography: Readings from Scientific American*. San Francisco: W. H. Freeman, 1991.

Morris, Simon Conway. *Life's Solution*. New York: Cambridge University Press, 2003.

Mortimer, Charles E. *Chemistry: A Conceptual Approach.* 3rd ed. New York: D. Van Nostrand, 1975.

Motz, Lloyd M., ed. *The Rediscovery of the Earth.* New York: Van Nostrand Reinhold, 1979.

Mozumder, A. *Fundamentals of Radiation Chemistry.* San Diego: Academic Press, 1999.

Muller, Richard A., and Gordon J. MacDonald. *Ice Ages and Astronomical Causes: Data, Spectral Analysis, and Mechanisms.* London: Springer, 2000.

Munyan, Arthur C., ed. *Polar Wandering and Continental Drift.* Tulsa, Okla.: Society of Economic Paleontologists and Mineralogists, 1963.

Murr, Lawrence Eugene. *Electron and Ion Microscopy and Microanalysis: Principles and Applications.* 2d ed. New York: M. Decker, 1991.

Murthy, I. V. *Gravity and Magnetic Interpretation in Exploration Geophysics.* Bangaloree: Geological Society of India, 1998.

Mutter, John C. "Seismic Images of Plate Boundaries." *Scientific American* 254 (February 1986): 66-75.

Myers, Anne B., et al., eds. *Laser Techniques in Chemistry.* New York: Wiley, 1995.

Nabarro, Frank, and F. de Villiers. *The Physics of Creep: Creep and Creep-Resistant Alloys.* London: Taylor and Francis, 1995.

Nance, John, and Howard Cady. *On Shaky Ground.* New York: William Morrow, 1988.

Nash, J. R. *Darkest Hours.* Chicago: Nelson-Hall, 1976.

National Academy of Engineering, National Research Council. *The Carbon Dioxide Dilemma: Promising Technologies and Policies.* Washington, D.C.: National Academies Press, 2003.

National Research Council (U.S.), Panel on the Public Policy Implications of Earthquake Prediction. *Earthquake Prediction and Public Policy.* Washington, D.C.: Government Printing Office, 1975.

Nesse, William D. *Introduction to Optical Mineralogy.* 3rd ed. New York: Oxford University Press, 2004.

Nettleton, Lewis L. *Geophysical Prospecting for Oil.* New York: McGraw-Hill, 1940.

Newmark, Nathan M., and Emilio Rosenblueth. *Fundamentals of Earthquake Engineering.* Englewood Cliffs, N.J.: Prentice-Hall, 1971.

Newton, Robert C. "The Three Partners of Metamorphic Petrology." *American Mineralogist* 96 (2011): 457-469.

Nichols, D. R., and J. M. Buchanan-Banks. *Seismic Hazards and Land-Use Planning.* U.S. Geological Survey Circular 690. Washington, D.C.: Government Printing Office, 1974.

Nicholson, Keith. *Geothermal Fluids: Chemistry and Exploration Techniques.* Berlin: Springer-Verlag, 1993.

Nierenberg, William A. "The Deep Sea Drilling Project After Ten Years." *American Scientist* 66 (January/February 1978): 20-29.

Nimmer, Robin E., et al. "Three-Dimensional Effects Causing Artifacts in Two-Dimensional, Cross-Borehole, Electrical Imaging." *Journal of Hydrology* 359 (2008): 59-70.

Nolet, Guust. *A Breviary of Seismic Tomography.* New York: Cambridge University Press, 2008.

_____. *Seismic Tomography.* Boston: D. Reidel, 1987.

Normile, Dennis, and Richard A. Kerr. "A Sea Change in Ocean Drilling." *Science* 300 (2003): 410.

Norris, Pat. *Watching Earth from Space: How Surveillance Helps Us, and Harms Us.* New York: Springer, 2010.

Norris, R. M., and R. W. Webb. *Geology of California.* 2d ed. New York: John Wiley & Sons, 1990.

Norton, Kevin P., and Andrea Hampel. "Postglacial Rebound Promotes Glacial Re-advances." *Terra Nova* 22, no. 4 (2010): 297-302.

Nuffield, E. W. *X-ray Diffraction Methods.* New York: John Wiley & Sons, 1966.

O'Loughlin, K. F., and James F. Lander. *Caribbean Tsunamis: A 500-Year History from 1498-1998.* Norwell, Mass.: Kluwer Academic Publishers, 2010.

O'Neil, J. R. "Stable Isotope Geochemistry of Rocks and Minerals." In *Lectures in Isotope Geology,* edited by Emilie Jäger and Johannes C. Hunziker. New York: Springer-Verlag, 1979.

O'Reilly, W. *Rock and Mineral Magnetism.* New York: Chapman and Hall, 1984.

Oelkers, Eric H., ed. *Thermodynamics and Kinetics of Water-Rock Interaction: Reviews in Mineralogy and Geochemistry.* Mineralogical Society of America, 2009.

Oerter, Erik J. *Geothermometry of Thermal Springs in the Rico, Dunton, and West Fork Dolores River Areas, Dolores County, Colorado.* Colorado Geological Survey, Department of Natural Resources. 2011.

Ogawa, Yujiro, Ryo Anma, and Yildirim Dilek. *Accretionary Prisms and Convergent Margin Tectonics in the Northwest Pacific Basin.* New York: Springer Science+Business Media, 2011.

Ogg, James G., Gabi Ogg, and Felix M. Gradstein. *The Concise Geologic Time Scale.* New York: Cambridge University Press, 2008.

Okamoto, Shunzo. *Introduction to Earthquake Engineering.* 2d ed. New York: University of Tokyo Press, 1984.

Olsen, Kenneth H., ed. *Continental Rifts: Evolution, Structure, Tectonics.* Amsterdam: Elsevier, 1995.

Oncken, Onno, et al., eds. *The Andes: Active Subduction Orogeny (Frontiers in Earth Sciences).* Berlin: Springer-Verlag, 2006.

Opdyke, Neil D., and J. E. T. Channell. *Magnetic Stratigraphy.* San Diego: Academic Press, 1996.

Ormsby, Tim, et al. *Getting to Know ArcGIS Desktop.* 2d ed. New York: ESRI Press, 2010.

Osterihanskay, Lubor. *The Causes of Lithospheric Plate Movements.* Prague: Charles University, 1997.

Ozima, Minoru. *Geohistory: Global Evolution of the Earth.* New York: Springer-Verlag, 1987.

Pagel, B. E. J. *Nucleosynthesis and Chemical Evolution of Galaxies.* 2d ed. Cambridge: Cambridge University Press, 2009.

Palmer, Donald F., and I. S. Allison. *Geology: The Science of a Changing Earth.* 7th ed. New York: McGraw-Hill, 1980.

Paone, Angelo. "The Geochemical Evolution of the Mt. Somma-Vesuvius Volcano." *Mineralogy and Petrology* 87, nos. 1/2 (2006): 53-80.

Park, R. G. *Foundations of Structural Geology.* 3rd ed. New York: Routledge, 2004.

Parker, Sybil P., ed. *McGraw-Hill Encyclopedia of Geological Sciences.* 2d ed. New York: McGraw-Hill, 1988.

Parry, Susan J. *Activation Spectrometry in Chemical Analysis.* New York: Wiley, 1991.

Pater, Imke de, and Jonathan Lissauer. *Planetary Sciences.* 2d ed. New York: Cambridge University Press, 2011.

Paterson, Mervyn S., and Teng-fong Wong. *Experimental Rock Deformation: The Brittle Field.* 2d ed. New York: Springer, 2005.

Pella, P. A. "X-ray Spectrometry." In *Instrumental Analysis,* edited by G. D. Christian and J. E. O'Reilly. 2d ed. Boston: Allyn and Bacon, 1986.

Peltier, W. R. "Global Glacial Isostatic Adjustment: Paleogeodetic and Space-Geodetic Tests of the ICE-4G." *Journal of Quaternary Science* 17, nos. 5/6 (2002): 491-510.

Penick, J. L., Jr. *The New Madrid Earthquakes.* Rev. ed. Columbia: University of Missouri Press, 1981.

Peters, Shanan E. "Environmental Determinants of Extinction Selectivity in the Fossil Record." *Nature* 454 (2008): 626-629.

Peters, Shanan E., et al. "Large-Scale Glaciation and Deglaciation of Antarctica During the Late Eocene." *Geology* 38, no. 8 (2010): 723-726.

Petersen, James F., Dorothy Sack, and Robert Gabler. *Fundamentals of Physical Geography.* Belmont, Calif.: Cengage Learning, 2011.

Peterson, M. N. A., and F. C. MacTernan. "A Ship for Scientific Drilling." *Oceanus* 25 (Spring 1982): 72-79.

Pinet, Paul R. *Invitation to Oceanography.* 5th ed. Sudbury, Mass.: Jones and Bartlett, 2009.

Pinta, Maurice. *Modern Methods for Trace Element Analysis.* Ann Arbor, Mich.: Ann Arbor Science, 1978.

Pipkin, Bernard W., and Richard J. Proctor. *Engineering Geology Practice in Southern California.* Belmont, Calif.: Star Publications, 1992.

Pitman, Walter C. "Plate Tectonics." In *McGraw-Hill Encyclopedia of the Geological Sciences,* 2d ed. New York: McGraw-Hill, 1988.

Plate, Erich J., et al., eds. *Buoyant Convection in Geophysical Flows.* Boston: Kluwer Academic Publishers, 1998.

Plescan, Costel, and Ancuta Rotaru. "Aspects Concerning the Improvement of Soils Against Liquefaction." *Bulletin of the Polytechnic Institute of Iasi* (2010): 39-45.

Plummer, Charles C., and Diane Carlson. *Physical Geology.* 12th ed. Boston: McGraw-Hill, 2007.

Poblet, J., and Lisle, R. J. *Kinematic Evolution and Structural Styles of Fold-and-Thrust Belts,* Special Publication 349. Geological Society of London, 2011.

Potts, Philip J., and Margaret West, eds. *Portable X-ray Fluorescence Spectrometry: Capabilities for In Situ Analysis.* Royal Society of Chemistry, 2008.

Potts, Phillip J., et al., eds. *Microprobe Techniques in the Earth Sciences.* London: Chapman and Hall, 1995.

Poupinet, Georges. "Seismic Tomography." *Endeavour* 14, no. 2 (1990): 52.

Powell, Robert, et al., eds. *Equilibrium Thermodynamics in Petrology: An Introduction.* New York: Harper & Row, 1978.

_____. *The San Andreas Fault System: Displacement, Palinspastic Reconstruction, and Geological Evolution.* Boulder, Colo.: Geological Society of America, 1993.

Prantzos, N., E. Vangionu-Flam, and M. Cassae. *Origin and Evolution of the Elements*. Cambridge: Cambridge University Press, 1993.

Press, Frank. "Earthquake Prediction." *Scientific American* 232 (May 1975): 14-23.

Press, Frank, and Raymond Siever. *Earth*. 4th ed. New York: W. H. Freeman, 1986.

Prichard, H. M. *Magmatic Processes and Plate Tectonics*. London: Geological Society, 1993.

Prothero, Donald R. *Catastrophes!: Earthquakes, Tsunamis, Tornadoes, and Other Earth-Shattering Disasters*. Baltimore: Johns Hopkins University Press, 2011.

_____. *Interpreting the Stratigraphic Record*. New York: W. H. Freeman, 1989.

_____. "Mammals and Magnetostratigraphy." *Journal of Geological Education* 36 (1988): 227.

Pruitt, Evelyn L. "The Office of Naval Research and Geography." *Annals of the Association of American Geographers* 69, no. 1 (March 1979): 103-108.

Pugin, Andre J.-M., Susan E. Pullman, and James A. Hunter. "Multicomponent High-Resolution Seismic Reflection Profiling." *The Leading Edge* 28 (2009): 1248-1261.

Pujol, Jose. *Elastic Wave Propagation and Generation in Seismology*. New York: Cambridge University Press, 2003.

Rackley, Steve. *Carbon Capture and Storage*. Oxford: Academic, 2009.

Ragan, Donal M. *Structural Geology: An Introduction to Geometrical Techniques*. 4th ed. New York: John Wiley & Sons, 2009.

Rahn, Perry H. *Engineering Geology: An Environmental Approach*. 2d ed. Upper Saddle River, N.J.: Prentice Hall, 1996.

Rakovic, Miloslav. *Activation Analysis*. London: Iliffe Books, 1970.

Ramsay, John G. *Folding and Fracturing of Rocks*. New York: Blackburn Press, 2004.

Rankama, K. *Progress in Isotope Geology*. New York: John Wiley & Sons, 1963.

Raup, David. *Extinction: Bad Genes or Bad Luck?* New York: W. W. Norton, 1992.

Raymo, Maureen E., and Peter Huybers. "Unlocking the Mysteries of the Ice Ages." *Nature* 451 (January 2008): 284-285.

Razin, A. "Excitation of Rayleigh and Stoneley Surface Acoustic Waves by Distributed Seismic Sources." *Radiophysics and Quantum Electronics* 53, no. 2 (2010): 82-99.

Reasenberg, Paul A., et al., eds. *The Loma Prieta, California, Earthquake of October 17, 1989: Aftershocks and Postseismic Effects*. Washington, D.C.: Government Publications Office, 1997.

Reddy, Ramesh, and R. D. DeLaune. "Biogeochemical Characteristics." In *Biogeochemistry of Wetlands*. Boca Raton: Taylor & Francis, 2008.

Redfren, Ron. *The Making of a Continent*. New York: Times Books, 1983.

Reed, S. J. B. *Electron Microprobe Analysis and Scanning Electron Microscopy in Geology*. 2d ed. Cambridge: Cambridge University Press, 2010.

Reimer, Paula J., et al. "IntCal104 Terrestrial Radiocarbon Age Calibration, 0-26 cal kyr BP." *Radiocarbon*. 46 (2004): 1029-1058.

Reimold, W. U. and R. L. Gibson. *Meteorite Impact!: The Danger from Space and South Africa's Mega-Impact; The Vredefort Structure*. Berlin: Springer, 2010.

Reite, M., and J. Zimmerman. "Magnetic Phenomena of the Central Nervous System." *Annual Review of Biophysics and Bioengineering* 7 (1978): 167-188.

Renfrew, Colin. *Before Civilization: The Radiocarbon Revolution and Prehistoric Europe*, 2d ed. London: Penguin Books, 1990.

Reppert, Steven M., Robert J. Gegear, and Christine Merlin. "Navigational Mechanisms of Migrating Monarch Butterflies." *Trends in Neurosciences*, 33 (2010): 399-406.

Reps, William F., and Emil Simiu. *Design, Siting, and Construction of Low-Cost Housing and Community Buildings to Better Withstand Earthquakes and Windstorms*. Washington, D.C.: U.S. Department of Commerce/National Bureau of Standards, 1974.

Rey, Jacques, and Simone Galeotti, eds. *Stratigraphy Terminology and Practice* Paris: Technips, 2008.

Reynolds, John M. *An Introduction to Applied and Environmental Geophysics*. 2d ed. New York: John Wiley, 2011.

_____. *Planet Earth: Earthquake*. Alexandria, Va.: Time-Life Books, 1982.

Rhodes, Richard. *The Making of the Atomic Bomb*. New York: Simon & Schuster, 1986.

Richardson, S. M., H. Y. McSween, Jr., and Maria Uhle. *Geochemistry Pathways and Processes*. 2d ed. Englewood Cliffs, N.J.: Prentice-Hall, 2003.

Richter, Charles F. *Elementary Seismology*. San Francisco: W. H. Freeman, 1958.

Rick, T. C., R. L. Vellanoweth, and J. Erlandson. "Radiocarbon Dating and the 'Old Shell' Problem:

Direct Dating of Artifacts and Cultural Chronologies in Coastal and Other Aquatic Regions." *Journal of Archaeological Sciences* 32 (2005): 1641-1648.

Ridley, Mark. *Evolution*. Hoboken, N.J.: Wiley-Blackwell, 2004.

Rikitake, Tsuneji. *Earthquake Forecasting and Warning*. Norwell, Mass.: Kluwer Academic Publishers, 1982.

Robinson, Edwin S., and Cahit Coruh. *Basic Exploration Geophysics*. New York: John Wiley & Sons, 1988.

Robinson, J. W., Eileen M. Skelly Frame, and George M. Frame III. *Undergraduate Instrumental Analysis*. 6th ed. New York: Marcel Dekker, 2004.

Rochow, T. G., and E. G. Rochow. *An Introduction to Microscopy by Means of Light, Electrons, X Rays, or Ultrasound*. New York: Plenum Press, 1979.

Roe, Gerard. "In Defense of Milankovitch." *Geophysical Research Letters* 33, no. 24 (December 2006).

Roedder, Edwin. "Ancient Fluids in Crystals." *Scientific American* 207 (October 1962) 38-47.

_____. "Fluid Inclusion Studies of Hydrothermal Ore Deposits." In *Geochemistry of Hydrothermal Ore Deposits*, edited by L. B. Barnes. 3rd ed. New York: John Wiley & Sons, 1997.

_____. *Reviews in Mineralogy*. Vol. 12, *Fluid Inclusions*. Washington, D.C.: Mineralogical Society of America, 1984.

Roeder, D. H. "Subduction and Orogeny." *Journal of Geophysical Research* 78 (1973): 5005-5024.

Roeges, Noel P. G. *A Guide to the Complete Interpretation of Infrared Spectra of Organic Structures*. Chichester, N.Y.: Wiley, 1994.

Rolfs, Claus E., and William S. Rodney. *Cauldrons in the Cosmos: Nuclear Astrophysics*. Chicago: University of Chicago Press, 1988. Reprinted in 2005.

Rollinson, Hugh. "When Did Plate Tectonics Begin?" *Geology Today* 23, no. 5 (2007): 186-191.

Rosenzweig, Cynthia, et al. "Attributing Physical and Biological Impacts to Anthropogenic Climate Change." *Nature* 453, no. 7193 (May, 2008): 353-357.

Ruddiman, William F. *Earth's Climate: Past and Future*. 2d ed. Gordonsville, Va.: W. H. Freeman, 2007.

Rundle, John B., Donald L. Turcotte, and William Klein, eds. *Reduction and Predictability of Natural Disasters*. Reading, Mass.: Addison-Wesley, 1996.

Ruska, Ernst. *The Early Development of Electron Lenses and Electron Microscopy*. Translated by Thomas Mulvey. Stuttgart, West Germany: S. Hirzel Verlag, 1980.

Russell, B. H. *Introduction to Seismic Inversion Methods*. Tulsa, Okla.: Society of Exploration Geophysicists, 1988.

Russell, R. D., and R. M. Farquhar. *Lead Isotopes in Geology*. New York: Interscience Publishers, 1960.

Sabins, Floyd F. *Remote Sensing: Principles and Interpretation*. San Francisco: W. H. Freeman, 1978.

Sacher, Hubert, and Rene Schiemann. "When Do Deep Drilling Geothermal Projects Make Good Economic Sense?" *Renewable Energy Focus* 11 (2010): 30-31.

Salisbury, John W., et al., eds. *Infrared (2.1-25 μm) Spectra of Minerals*. Baltimore: John Hopkins University Press, 1992.

Santamaria-Fernandez, Rebeca. "Precise and Traceable Carbon Isotope Ratio Measurements by Multicollector ICP-MS: What Next??" *Analytical & Bioanalytical Chemistry* 397 (2010): 973-978.

Satake, Kenji, and Fumihiko Imamura, eds. *Tsunamis 1992-1994: Their Generation, Dynamics, and Hazards*. Boston: Birkhauser, 1995.

Scales, John Alan. *Theory of Seismic Imaging*. Berlin: Springer-Verlag, 1995

Schellart, W. P. "Overriding Plate Shortening and Extension Above Subduction Zones: A Parametric Study to Explain Formation of the Andes Mountains." *Geological Society of America Bulletin* 120, no. 11 (2008): 1441-1454.

Schenk, Vladimir, ed. *Earthquake Hazard and Risk*. Dordrecht, Netherlands: Kluwer Academic Press, 1996.

Schimel, David Steven. *Theory and Application of Tracers*. San Diego: Academic Press, 1993.

Schlesinger, William H. *Biogeochemistry*. Amsterdam: Elsevier, 2005.

_____. *Biogeochemistry: An Analysis of Global Change*. San Diego: Academic Press, 1997.

Schmatz, Joyce, Oliver Schenk, and Janos Urai. "The Interaction of Migrating Grain Boundaries with Fluid Inclusions in Rock Analogues: The Effect of Wetting Angle and Fluid Inclusion Velocity." *Contributions to Mineralogy & Petrology* 162 (2011): 193-208.

Schminke, Hans-Ulrich. *Volcanism*. New York: Springer, 2005.

Schramm, David, ed. *Supernovae*. Dordrecht, Netherlands: Reidel Press, 1977.

Schrijver, Carolus J., and George L. Siscoe, eds. *Heliophysics: Space Storms and Radiation—Causes and Effects*. New York: Cambridge University Press, 2011.

Schubert, Gerald, Donald L. Turcotte, and Peter Olson. *Mantle Convection in the Earth and Planets.* New York: Cambridge University Press, 2001.

Schulson, Erland M., and Paul Duval. *Creep and Fracture of Ice.* New York: Cambridge University Press, 2009.

Scott, E. M., M. S. Baxter, and T. C. Aitchison. "A Comparison of the Treatment of Errors in Radiocarbon Dating Calibration Methods." *Journal of Archaeological Science* 11 (1984): 455-466.

Seeber, Gahnter. *Satellite Geodesy.* Rev. ed. Berlin: Walter de Gruyter, 2008.

Segar, Douglas. *An Introduction to Ocean Sciences.* 2d ed. New York: Wadsworth, 2007.

Seyfert, Charles K., and L. A. Sirkin. *Earth History and Plate Tectonics: An Introduction to Historical Geology.* New York: Harper & Row, 1973.

Sharma, Vallabh P. *Environmental and Engineering Geophysics.* Cambridge, England: Cambridge University Press, 1997.

Sharpton, Virgil L., and Peter D. Ward, eds. *Global Catastrophes in Earth History.* Boulder, Colo.: Geological Society of America, 1990.

Shea, James H., ed. *Plate Tectonics.* New York: Van Nostrand Reinhold, 1985.

Shearer, Peter M. "Upper Mantle Seismic Discontinuities." In *Earth's Deep Interior: Mineral Physics and Tomography from the Atomic to the Global Scale.* Edited by Shun-Ichiro Karato, et al. American Geophysical Union, 2000.

Shelton, J. S. *Geology Illustrated.* San Francisco: W. H. Freeman, 1966.

Shelton, J. W. "Listric Normal Faults: An Illustrated Summary." *American Association of Petroleum Geologists Bulletin* 68 (1984).

Shepard, Francis P. *Geological Oceanography.* New York: Crane, Russak, 1977.

Sheriff, R. E., ed. *Reservoir Geophysics.* Tulsa, Okla.: Society of Exploration Geophysicists, 1992.

Shklovskii, Iosif S. *Stars: Their Birth, Life, and Death.* San Francisco: W. H. Freeman, 1978.

Shor, Elizabeth Noble. *Scripps Institution of Oceanography: Probing the Oceans, 1936 to 1976.* San Diego, Calif.: Tofua Press, 1978.

Short, Nicholas M., and Robert W. Blair. *Geomorphology from Space: A Global Overview of Regional Landforms.* Washington, D.C.: National Aeronautics and Space Administration, 1986.

Siegel, Frederick R. *Applied Geochemistry.* New York: Wiley, 1975.

Sieh, K. E., M. Stuiver, and D. Brillinger. "A More Precise Chronology of Earthquakes Produced by the San Andreas Fault in Southern California." *Journal of Geophysical Research* 94 (January 10, 1989): 603-623.

Siever, Raymond. "The Dynamic Earth." *Scientific American* 249 (September 1983): 46-55.

Simon, R. B. *Earthquake Interpretation.* Golden: Colorado School of Mines, 1968.

_____. *Earthquake Interpretations: A Manual for Reading Seismographs.* Golden: Colorado School of Mines, 1981.

Sinkankas, John. *Mineralogy.* New York: Van Nostrand Reinhold, 1975.

Skinner, Brian J., and S. C. Porter. *Physical Geology.* New York: John Wiley & Sons, 1987.

Skinner, B. J., et al. *The Dynamic Earth: An Introduction to Physical Geology.* 5th ed. New York: John Wiley & Sons, 2006.

Skinner, Brian J., et al. *Resources of the Earth.* 3rd ed. Englewood Cliffs, N.J.: Prentice-Hall, 2001.

Smith, David G., ed. *The Cambridge Encyclopedia of Earth Sciences.* Cambridge, England: Cambridge University Press, 1981.

Smith, James. *Introduction to Geodesy: The History and Concepts of Mode Geodesy.* New York: Wiley, 1997.

Smith, P. J., ed. *The Earth.* New York: Macmillan, 1986.

_____. *Topics in Geophysics.* Cambridge, Mass.: MIT Press, 1973.

Sorokhtin, O. G., et al. *Evolution of Earth and Its Climate: Birth, Life, and Death of Earth.* Boston: Elsevier, 2011.

Spangler, M. G., and R. L. Handy. *Soil Engineering.* New York: Harper & Row, 1982.

Sparks, Donald L., and Timothy J. Grundl, eds. *Mineral-Water Interfacial Reactions: Kinetics and Mechanisms.* Washington, D.C.: American Chemical Society, 1998.

Spencer, Edgar W. *Dynamics of the Earth.* New York: Thomas Y. Crowell, 1972.

_____. *Introduction to the Structure of the Earth.* 3rd ed. New York: McGraw-Hill, 1988.

Spitaleri, C., C. Rolfs, and R. C. Pizzone, eds. *Fifth European Summer School on Experimental Nuclear Astrophysics.* American Institute of Physics, 2010.

Spradley, Joseph. "Ten Lunar Legacies: Importance of the Moon for Life on Earth." *Perspectives on Science and Christian Faith* 62, no. 4 (December 2010): 267-275.

Stacey, F. D., and Paul M. Davis. *Physics of the Earth*. 4th ed. New York: Cambridge University Press, 2008.

Steger, T. D. *Topographic Maps*. Denver, Colo.: U.S. Geological Survey, n.d.

Stenchikov, Georgiy. "The Role of Volcanic Activity in Climate and Global Change." In *Climate Change: Observed Impacts on Planet Earth*, edited by Trevor M. Letcher. Boston: Elsevier, 2009.

Stewart, I. S., and J. Lynch. *Earth: The Biography*. Washington, D.C.: National Geographic Society, 2007.

Stewart, R. R. *Exploration Seismic Tomography*. Tulsa, Okla.: Society of Exploration Geophysicists, 1991.

Stewart, S. A. "Vertical Exaggeration of Reflection Seismic Data in Geosciences Publications." *Marine & Petroleum Geology* 28 (2011): 959-965.

Stoffer, Philip W. *Where's the San Andreas Fault?: A Guidebook to Tracing the Fault on Public Lands in the San Francisco Bay Region*. University of Michigan Library, 2006.

Stoiber, Richard E., and S. A. Morse. *Microscopic Identification of Crystals*. Reprint. Malabar, Fla.: Robert E. Krieger, 1981.

Strahler, A. N. *Plate Tectonics*. Cambridge, Mass.: GeoBooks, 1998.

Strahler, Alan H., and Arthur N. Strahler. *Environmental Geoscience: Interaction Between Natural Systems and Man*. Santa Barbara: Hamilton, 1973.

Street, Philip. *Animal Migration and Navigation*. New York: Charles Scribner's Sons, 1976.

Stumm, Werner, and James J. Morgan. *Aquatic Chemistry: An Introduction Emphasizing Chemical Equilibria in Natural Waters*. 3rd ed. New York: John Wiley & Sons, 1996.

Sullivan, Walter. *Continents in Motion*, 2d ed. New York: American Institute of Physics, 1993.

Suppe, John. *Principles of Structural Geology*. Englewood Cliffs, N.J.: Prentice-Hall, 1985.

Sutherland, Lin. *The Volcanic Earth: Volcanoes and Plate Tectonics, Past, Present, and Future*. Sydney, Australia: University of New South Wales Press, 1995.

Sutton, Gerard K., and Joseph A. Cassalli, eds. *Catastrophe in Japan: The Earthquake and Tsunami of 2011*. Nova Science Publishers, 2011.

Swift, J. A. *Electron Microscopes*. New York: Barnes & Noble Books, 1970.

Szczepanek, MalGorzata, et al. "Hydrogen, Carbon, and Oxygen Isotopes in Pine and Oak Tree Rings from Southern Poland as Climatic Indicators in Years 1900-2003." *Geochronometria: Journal on Methods & Applications of Absolute Chronology* 25 (2006): 67-76.

Szefer, P., and Grembecka, M. "Chemometric Assessment of Chemical Element Distribution in Bottom Sediments of the Southern Baltic Sea Including Vistula and Szczecin Lagoons: An Overview." *Polish Journal of Environmental Studies* 18 (2009): 25-34.

Tarantola, A. *Inverse Problem Theory: Methods for Data Fitting and Parameter Estimation*. Amsterdam: Elsevier, 1987.

Tarbuck, Edward J., and Frederick K. Lutgens. *Earth Science*. 13th ed. Upper Saddle River, N.J.: Pearson Education, 2012.

Tarbuck, Edward J., Frederick K. Lutgens, and Dennis Tasa. *Earth: An Introduction to Physical Geology*. 10th ed. Upper Saddle River, N.J.: Prentice Hall, 2010.

_____. *Earth Science*. 12th ed. Westerville, Ohio: Charles E. Merrill, 2008.

Tarling, D. H. *Palaeomagnetism: Principles and Applications in Geology, Geophysics, and Archaeology*. London: Chapman and Hall, 1983.

Tarling, D., and M. Tarling. *Continental Drift: A Study of the Earth's Moving Surface*. Garden City, N.J.: Anchor Press, 1971.

Tarling, Donald H., et al., eds. *Paleomagnetism and Diagenesis in Sediments*. London: Geological Society, 1999.

Tatsumi, Y. "The Subduction Factory: How It Operates on Earth." *GSA Today* 15 (2005): 4-10.

Tauxe, Lisa. *Paleomagnetic Principles and Practice*. Norwell, Mass: Kluwer Academic Publishers, 2002.

Taylor, R. E. "Fifty Years of Radiocarbon Dating." *American Scientist* 88 (January/February 2000): 60-67.

Taylor, R. J., ed. *Stellar Astrophysics*. Philadelphia: Institute of Physics, 1992.

Taylor, S. R., and S. M. McLennan. *The Continental Crust: Its Composition and Evolution*. Reissued. Oxford, England: Blackwell Scientific, 1991.

Telford, W. M., L. P. Geldart, and R. E. Sheriff. *Applied Geophysics*. 2d ed. Cambridge, England: Cambridge University Press, 1990.

Tertian, R., and F. Claisse. *Principles of Quantitative X-ray Fluorescence Analysis*. New York: Wiley, 1982.

Terzaghi, K., and R. B. Peck. *Soil Mechanics in Engineering Practice*. New York: John Wiley & Sons, 1948.

Thompson, Graham R. *An Introduction to Physical Geology*. Fort Worth: Saunders College Publishing, 1998.

Thornbury, William D. *Principles of Geomorphology*. 2d ed. New York: John Wiley & Sons, 1968.

Tierney, Kathleen J. *Report of the Coalinga Earthquake of May 2, 1983*. Sacramento, Calif.: Seismic Safety Commission, 1985.

Timofeev, V. E., et al. "Variations of the Interplanetary Magnetic Field and the Electron and Cosmic-Ray Intensities under the Influence of Jupiter." *Astronomy Letters* 33, no. 1 (2007): 63-66.

Tokosoz, M. N. "The Subduction of the Lithosphere." *Scientific American* 233 (November 1975): 88.

Trabalka, J. R., ed. *Atmospheric Carbon Dioxide and the Global Carbon Cycle*. Honolulu: University Press of the Pacific, 2005.

Trimm, Harold H., and William Hunter, III. *Environmental Chemistry: New Techniques and Data* Candor. New York: Apple Academic Press, 2011.

Tsuboi, Chuji. *Gravity*. London: George Allen & Unwin, 1983.

Tsuchlya, Yoshito, and Nobuo Shuto, eds. *Tsunami: Progress in Prediction, Disaster Prevention, and Warning*. Advances in Natural and Technological Hazards Research Series 4. Norwell, Mass.: Kluwer Academic Publishers, 1995.

Tucker, R. H., et al. *Global Geophysics*. New York: Elsevier, 1970.

Tuniz, Claudio, et al., eds. *Accelerator Mass Spectrometry: Ultrasensitive Analysis for Global Science*. Boca Raton, Fla.: CRC Press, 1998.

Turcotte, Donald L., and Gerald Schubert. *Geodynamics*. 2d ed. New York: Cambridge University Press, 2010.

Turner, Gillian M. *North Pole, South Pole: The Epic Quest to Solve the Great Mystery of Earth's Magnetism*. New York: Experiment, 2011.

U.S. Geological Survey Earthquake Website: earthquake.usgs.gov. Contains useful information about earthquakes and seismicity.

Ulmer, G. C. *Research Techniques for High Pressure and High Temperature*. New York: Springer-Verlag, 1971.

US Department of the Interior. *Earthquake Information Bulletin*. Washington, D.C.: Government Printing Office.

U.S. Geological Survey. *COGEOMAP: A New Era in Cooperative Geological Mapping*. Circular No. 1003. Denver, Colo.: Author, 1987.

_____. *Digital Line Graphics from 1:24,000-Scale Maps: Data Users Guide*. Denver, Colo.: Author, 1986.

_____. Earthquake Website: earthquake.usgs.gov.

_____. *Finding Your Way with Map and Compass*. Denver, Colo.: Author, n.d.

_____. *Large-Scale Mapping Guidelines*. Denver, Colo.: Author, 1986.

_____. *National Geographic Mapping Program: Goals, Objectives, and Long-Range Plans*. Denver, Colo.: Author, 1987.

_____. *Topographic Map Symbols*. Denver, Colo.: Author, n.d.

U.S. Geological Survey/National Earthquake Information Center. Website: earthquake.usgs.gov/regional/neic.

Utgard, R. O., and G. D. McKenzie. *Man's Finite Earth*. Minneapolis, Minn.: Burgess, 1974.

Utgard, R. O., G. D. McKenzie, and D. Foley. *Geology in the Urban Environment*. Minneapolis: Burgess, 1978.

Uyeda, Seiya. *The New View of the Earth: Moving Continents and Moving Oceans*. San Francisco: W. H. Freeman, 1971.

_____. *The New View of the Earth: Moving Continents and Moving Oceans*. Translated by Masako Ohnuki. San Francisco: W. H. Freeman, 1978.

Valentine, Andrew P., and John H. Woodhouse. "Reducing Errors in Seismic Tomography: Combined Inversion for Sources and Structure." *Geophysical Journal International* 180 (2009): 847-857.

Valley, J. W., H. P. Taylor, Jr., and J. R. O'Neil, eds. *Reviews in Mineralogy*. Vol. 1b, *Stable Isotopes in Higher Temperature Geologic Processes*. Washington, D.C.: Mineralogical Society of America, 1986.

Van Andel, Tjeerd H. "Deep-Sea Drilling for Scientific Purposes: A Decade of Dreams." *Science* 160 (June 28, 1968): 1419-1424.

Van Burgh, Dana. *How to Teach with Topographical Maps*. Washington, D.C.: International Science Teachers Association, 1994.

van der Leeden, Fritz, Fred L. Troise, and D. K. Todd. *The Water Encyclopedia*. 2d ed. New York: CRC Press, LLC, 1990.

van Hunen, Jeroen, and Arie P. van den Berg. "Plate Tectonics on the Early Earth: Limitations Imposed by Strength and Buoyancy of Subducted Lithosphere." *Lithos.* 103 (June 2008): 217-235.

Van Kranendonk, Martin J., R. Hugh Smithies, and Vickie C. Bennett, eds. *Earth's Oldest Rocks*. Boston: Elsevier, 2007.

Van Loon, Gary W., and Stephen J. Duffy. *Environmental Chemistry: A Global Perspective*. New York: Oxford University Press, 2010.

Van Ness, H. C. *Understanding Thermodynamics*. Dover Publications, 1983.

Vasilopoulou, T., et al. "Large Sample Neutron Activation Analysis of a Reference Inhomogeneous Sample." *Journal of Radioanalytical & Nuclear Chemistry* 289 (2011): 731-737.

Velli, Marco, Roberto Bruno, and Francesco Malara, eds. *Solar Wind Ten: Proceedings of the 10th International Solar Wind Conference*. College Park, Md.: American Institute of Physics, 2003.

Verbyla, David L. *Satellite Remote Sensing of Natural Resources*. Boca Raton, Fla.: Lewis, 1995.

Verhoogen, John, et al. *The Earth*. New York: Holt, Rinehart and Winston, 1970.

Verney, Peter. *The Earthquake Handbook*. New York: Paddington Press, 1979.

Villaverde, Roberto. *Fundamental Concepts of Earthquake Engineering*. CRC Press, 2009.

Vita-Finzi, Claudio. *Recent Earth History*, New York: Halsted Press, 1974.

Vogt, Gregory. *Predicting Earthquakes*. New York: Franklin Watts, 1989.

Wagemans, Cyriel. *The Nuclear Fission Process*. Boca Raton, Fla.: CRC Press, 1991.

Wagner, Geunther A., and Peter Van de Haute. *Fission Track Dating*. Boston: Kluwer, 1992.

Wagner, Gunther A., and S. Schiegl. *Age Determination of Young Rocks and Artifacts: Physical and Chemical Clocks in Quaternary Geology and Archaeology*. New York: Springer, 2010.

Walcott, R. I. "Late Quaternary Vertical Movements in Eastern North America: Quantitative Evidence of Glacio-isostatic Rebound." *Review of Geophysics and Space Physics* 10 (November 1972): 849-884.

Walker, Bryce. *Earthquake*. Alexandria, Va.: Time-Life Books, 1982.

_____. *Geology Today*. 10th ed. Del Mar, Calif.: Ziff-Davis, 1974.

Walker, Hollis N., D. Stephen Lane, and Paul E. Stutzman. *Petrographic Methods of Examining Hardened Concrete: A Petrographic Manual*. Federal Highway Administration, U.S. Department of Transportation, 2006.

Walker, Michael M., et al. "Structure and Function of the Vertebrate Magnetic Sense." *Nature* 390 (November 1997): 371-376.

Walker, Mike. *Quaternary Dating Methods*. New York: Wiley, 2005.

Walker, Sally M. *Earthquakes*. Minneapolis: Carolrhoda, 1996.

Walter, Thomas R. "Structural Architecture of the 1980 Mount St. Helens Collapse: An Analysis of the Rosenquist Photo Sequence Using Digital Image Correlation." *Geology* 39, no. 8 (2011): 767-770.

Walther, John Victor. *Essentials of Geochemistry,* 2d ed. Jones & Bartlett Publishers, 2008.

Ward, Peter, and Donald Brownlee. *Rare Earth: Why Complex Life Is Uncommon in the Universe*. New York: Copernicus, 2000.

Warme, John E., Robert G. Douglas, and Edward L. Winterer, eds. *The Deep Sea Drilling Project: A Decade of Progress*. Tulsa, Okla.: Society of Economic Paleontologists and Mineralogists, 1981.

Watkins, Jim. "Use of Satellite Remote Sensing Tools for the Great Lakes." *Aquatic Ecosystem Health & Management* 13 (2010): 127-134.

Watson, J. Throck, and O. David Sparkman. *Introduction to Mass Spectrometry: Instrumentation, Applications, and Strategies for Data Interpretation*. New York: Wiley, 2007.

Watt, Ian M. *The Principles and Practice of Electron Microscopy*. 2d ed. New York: Cambridge University Press, 1997.

Webb, Stephen. *If the Universe Is Teeming with Aliens . . . Where Is Everybody? Fifty Solutions to Fermi's Paradox and the Problem of Extraterrestrial Life*. New York: Copernicus Books, 2002.

Wedepohl, Karl Hans. *Geochemistry*. Translated by Egon Althaus. New York: Holt, Rinehart and Winston, 1971.

Wegener, Alfred. *The Origin of Continents and Oceans*. Translated by John Biram. Mineola, N.Y.: Dover Publications, 1966.

Weiner, Eugene. *Applications of Environmental Aquatic Chemistry: A Practical Guide*. 2d ed. Boca Raton, Fla.: CRC Press, 2008.

Weiner, Jonathan. *Planet Earth*. New York: Bantam Books, 1986.

Wenk, Hans-Rudolf, and Andrei Bulakh. *Minerals: Their Constitution and Origin*. New York: Cambridge University Press, 2004.

Wesson, R. L., and R. E. Wallace. "Predicting the Next Great Earthquake in California." *Scientific American* 252 (February 1985): 35.

West, Susan. "Diary of a Drilling Ship." *Science News* 119 (January 24, 1981): 60-63.

_____. "DSDP: Ten Years After." *Science News* 113 (June 24, 1978).

_____. "Log of Leg 76." *Science News* 119 (February 21, 1981): 124-127.

Whitley, D. Gath. "The Ivory Islands of the Arctic Ocean." *Journal of the Philosophical Society of Great Britain* 12 (1910).

Wibberley, Christopher A. J., and Shipton, Zoe K. "Fault Zones: A Complex Issue." *Journal of Structural Geology* 32 (2010): 1554-1556.

Wicander, Reed, and James S. Monroe. *The Changing Earth:. Exploring Geology and Evolution.* 5th ed. Belmont, Calif.: Cengage Learning, 2009.

Wiegel, R. L., ed. *Earthquake Engineering.* Englewood Cliffs, N.J.: Prentice-Hall, 1970.

Wilhelm, Helmut, et al., eds. *Tidal Phenomena.* Berlin: Springer, 1997.

Williams, David B., and C. Barry Carter. *Transmission Electron Microscopy: A Textbook for Materials Science.* 2d ed. New York: Plentum Press, 2009.

Williams, Linda D. *Earth Science Demystified.* New York: McGraw-Hill, 2004.

Wilson, David. *The New Archaeology.* New York: New American Library, 1974.

_____. *Continents Adrift.* San Francisco: W. H. Freeman, 1972.

Wilson, J. Tuzo, ed. *Continents Adrift and Continents Aground.* San Francisco: W. H. Freeman, 1976.

Wiltschko, Wolfgang, and Roswitha Wiltschko. "Magnetic Orientation and Magnetoreception in Birds and Other Animals." *Journal of Comparative Physiology.* 191 (2008): 675-693.

Win, David Tin. "Neutron Activation Analysis." *Assumption University Journal of Technology* 8 (2004): 8-14.

Winchester, Simon. *Krakatoa: The Day the World Exploded, August 27, 1883.* New York: HarperCollins, 2003.

_____. *The Map That Changed the World:. William Smith and the Birth of Modern Geology.* New York: HarperCollins, 2001.

Windley, Brian F. *The Evolving Continents.* 3rd ed. New York: John Wiley & Sons, 1995.

Winter, J. D. *Principles of Igneous and Metamorphic Petrology.* 2d ed. Pearson Education, 2010.

Wood, B. J., and D. G. Fraser. *Elementary Thermodynamics for Geologists.* Oxford, England: Oxford University Press, 1976.

Woods, Mary C., et al., eds. *The Northridge, California, Earthquake of January 17, 1994.* Sacramento: California Department of Conservation, Division of Mines and Geology, 1995.

Woodwell, George M. "The Carbon Dioxide Question." *Scientific American* 238 (January 1978): 34-43.

Wright, Karen. "The Silent Type." *Discover* 23 (2002): 26-27.

Wu, C.-M., and G. C. Zhao. "The Applicability of the GRIPS Geobarometry in Metapelitic Assemblages." *Journal of Metamorphic Geology* 24 (2006): 297-307.

Wu, Chun-Chieh. *Solid Earth. Advances in Geosciences,* Vol. 26. London: World Scientific, 2011.

Wu, Jiedi, John A. Hole, and J. Arthur Snoke. "Fault Zone Structure at Depth from Differential Dispersion of Seismic Guide Waves: Evidence for a Deep Waveguide on the San Andreas Fault." *Geophysical Journal International* 182, no. 1 (2010): 343-354.

Wyld, Sandra J., and James E. Wright. "New Evidence for Cretaceous Strike-Slip Faulting in the United States Cordillera and Implications for Terrane-Displacement, Deformation Patterns, and Plutonism." *American Journal of Science* 301 (2001): 150-181.

Wyllie, Peter J. *The Way the Earth Works: An Introduction to the New Global Geology and Its Revolutionary Development.* New York: John Wiley & Sons, 1976.

Xuejing, Xie, et al. "Digital Element Earth." *Acta Geologica Sinica* (English Edition) 85 (2011): 1-16.

Yakolev, O. I., J. Wickert, and V. A. Anufrief. "Effect of the Solar-Wind Shock Wave on the Polar Ionosphere According to the Radio Occultation Data on Satellite-to-Satellite Paths." *Doklady Physics* 54, no. 8 (2009): 363-366.

Yamaguchi, Masuhiro, and Yoshifumi Tanimoto, eds. *Magneto-Science: Magnetic Field Effects on Materials: Fundamentals and Applications.* Berlin: Springer-Verlag, 2010.

Yan, Bokun, et al. "Minerals Mapping of the Lunar Surface with Clementine UVVIS/NIR Data Based on Spectra Unmixing Method and Hapke Model." *Icarus* 208, no. 1 (July 2010): 11-19.

Yeats, Robert S., Kerry Sieh, and Clarence R. Allen. *The Geology of Earthquakes.* New York: Oxford University Press, 1997.

Yoon, Choonhan, et al. "Web-Based Simulating System for Modeling Earthquake Seismic Wavefields on the Grid." *Computers and Geosciences* 34, no. 12 (2008): 1936-1946.

York, Derek, and Ronald M. Farquhar. *The Earth's Age and Geochronology.* Reprint. Oxford, England: Pergamon Press, 1975.

Young, Patrick. *Drifting Continents, Shifting Seas.* New York: Franklin Watts, 1976.

Yount, Lisa. *Alfred Wegener, Creator of the Continental Drift Theory.* New York: Chelsea House, 2009.

Zalasiewicz, Jan. *The Planet in a Pebble: A Journey into Earth's Deep History.* New York: Oxford University Press, 2010.

Zebrowski, Ernest, Jr. *Perils of a Restless Planet: Scientific Perspectives on Natural Disasters.* New York: Cambridge University Press, 1999.

Zeilik, Michael, and Elske Smith. *Introductory Astronomy and Astrophysics.* New York: Saunders College Publishing, 1987.

Zierenberg, R. A., et al. "The Deep Structure of a Sea-Floor Hydrothermal Deposit." *Nature* 392 (April 2, 1998): 485-488.

Zubinaite, Vilma, and George Preiss. "Investigation of the Effects of Specific Solar Storming Events on GNSS Navigation Systems." *Aviation* 15, no. 2 (2011): 44-48.

COMMON EARTH MINERALS

Mineral	Chemical Formulas	Colors	Features	Streak	Diaphaneity
Apatite	$Ca_5(PO_4)_3(F,Cl,OH)$	yellow, green, etc.	fluorapatite, hydroxyapatite	white	trp-trl
Azurite	$Cu_3(CO_3)_2(OH)_2$	vivid blue	copper ore deposit, soft	light blue	trp-subtrl
Barite	$BaSO_4$	colorless	high density, soft	white	trp-opa
Beryl	$Be_3Al_2(SiO_3)_6$	aqua to green	emerald w/Cr, hard	white	trp-subtrl
Biotite	$K(Mg,Fe)_3AlSi_3O_{10}(OH)_2$	black	sheet silicate, "black mica"	gray	trp-opa
Bornite	Cu_5FeS_4	blue to purple	iridescent luster, soft	gray-black	opa
Calcite	$CaCO_3$	clear if pure	doubly refracting, soft	white	trp-opa
Chalcopyrite	$CuFeS_2$	brassy yellow	"yellow copper," not pyrite	green-black	opa
Cinnabar	HgS	red to scarlet	earthy luster, mercury ore	bright red	trp-opa
Corundum	Al_2O_3	colorless if pure	sapphire, ruby w/ Cr, hard	none	trp-trl
Covellite	CuS	indigo-blue	metallic luster, soft	black-gray	opa
Dolomite	$CaMg(CO_3)_2$	white or pink	brittle crystals	white	trp-trl
Epidote	$Ca_2FeAl_2Si_3O_{12}(OH)$	green, gray, etc.	glassy luster, hard	brown-white	trl
Fluorite	CaF_2	green, purple, etc.	may fluoresce in UV light	white	trp-subtrl
Galena	PbS	silver-gray	metallic luster, lead ore	gray-black	opa
Garnet	$(Ca,Mg,Fe,Mn)_3(Fe,Al,Cr)_2Si_3O_{12}$	dark red, variable	glassy luster, hard	white	trp-opa
Graphite	C	dark gray	pencil lead, soft	black	opaque
Gypsum	$CaSO_4 \cdot 2H_2O$	colorless if pure	pearly, waxy, soft	white	trp-trl
Halite	$NaCl$	colorless or white	rock salt, cubic crystals, soft	white	trp
Hematite	Fe_2O_3	red or dark gray	red-brown streaks, iron ore	red-brown	subtrl-opa
Kaolinite	$Al_2Si_2O_5(OH)_4$	white if pure	china clay, earthy luster, soft	white	trp-trl
Kyanite	Al_2SiO_5	blue, green, gray	found in meta-morphic rocks	white	trp-trl

Limonite	$FeO(OH) \cdot nH_2O$	yellow brown	earthy luster, iron ore, soft	yellow-brown	opa
Magnetite	Fe_3O_4	black	strongly magnetic, lodestone	black	opa
Malachite	$Cu_2CO_3(OH)_2$	bright and dark green	forms in concentric rings	light green	trl-opa
Molybdenite	MoS_2	dark gray	like graphite w/ luster, flakes	green-gray	opa
Muscovite	$KAl_2(AlSi_3O_{10})(F,OH)_2$	white, silvery	a mica, forms thin sheets	white	trp-trl
Olivine	$(Mg, Fe)_2SiO_4$	yellow to green	glassy luster, gem peridot	white	trp-trl
Orthoclase	$KAlSi_3O_8$	white, pink aqua	K feldspar, 90° cleavage	white	trp-trl
Phlogopite	$KMg_3AlSi_3O_{10}(F,OH)_2$	shades of brown	"magnesium mica," soft	white	trp-trl
Plagioclase	$(Na,Ca)(Al,Si)_2Si_2O_8$	white to gray	group of 6 Na-Ca feldspars	white	trp-trl
Pyrite	FeS_2	brass-yellow	metallic luster, "fool's gold"	green-black	opa
Pyroxene	$(Na,Ca,Mg,Mn,Li,Cr,Zn,$ $Ti,Sc,V,Fe,Al)_2Si_2O_6$	green or black	includes augite, jadeite	varies	trp-opa
Quartz	SiO_2	colorless, variable	glassy luster, hard	white	trp
Serpentine	$(Mg, Fe)_3Si_2O_5(OH)_4$	green to yellow	waxy luster, fibrous asbestos	white	trl-opa
Sillimanite	Al_2SiO_5	colorless to variable	needlelike crystals, hard	white	trp-trl
Sphalerite	$(Zn,Fe)S$	yellow, brown, black	glassy luster, chief zinc ore	brown-white	trp-trl
Staurolite	$(Fe,Mg)_2Al_9(Si,Al)_4$ $O_{20}(O,OH)_4$	red-brown to black	long crystals, some crosslike	gray	trl-opa
Sylvite	KCl	colorless or white	cubic crystals, salty, bitter	white	trp
Talc	$Mg_3Si_4O_{10}(OH)_2$	white, gray, green	greasy feel, soapstone, soft	white	trl

EARTH FACTS

SURFACE AND SIZE

Average elevation of land above sea level	840 meters
Percent surface area of land	29%
Average depth of oceans	3.8 kilometers
Percent surface area of oceans	71%
Equatorial radius	6,378 kilometers
Polar radius	6,357 kilometers
Radius of sphere with Earth's volume	6,371 kilometers
Surface area	5.1×10^{14} square meters
Volume	1.083×10^{21} cubic meters

MASS, DENSITY, AND GRAVITY

Mass	5.976×10^{24} kilograms
Mass of atmosphere	5.1×10^{18} kilograms
Mass of ice	$25-30 \times 10^{18}$ kilograms
Mass of oceans	1.4×10^{21} kilograms
Acceleration of gravity at equator	9.780 meters/second/second
Acceleration of gravity at North Pole	9.832 meters/second/second
Average density	5,518 kilograms/cubic meter
Density of water	1,000 kilograms/cubic meter
Density of continental crust	2,800 kilograms/cubic meter
Density of oceanic crust	3,000 kilograms/cubic meter

INTERNAL STRUCTURE

Average depth of crust	40 kilometers
Depth of mantle	40–2,890 kilometers
Depth of liquid iron outer core	2,890–5,150 kilometers
Depth of solid iron inner core	5,150–6,370 kilometers
Mass of crust (0.4% of Earth mass)	2.4×10^{22} kilograms
Mass of mantle (67.1% of Earth mass)	4.01×10^{24} kilograms
Mass of outer core (30.8% Earth mass)	1.84×10^{24} kilograms
Mass of inner core (1.7% Earth mass)	1.0×10^{23} kilograms

EARTH, SUN, AND MOON RELATIONS

Mean distance to sun	1.496×10^{8} kilometers
Mean distance to moon	3.844×10^{5} kilometers
Mass of sun/mass of Earth	3.329×10^{5}
Mass of Earth/mass of moon	81.303
Solar constant (sun's power at Earth)	1,366 watts/square meter
Earth's total receipt of solar power	1.74×10^{17} watts
Geothermal power reaching surface	4.7×10^{13} watts

HISTORICAL TIMELINE of GEOPHYSICAL HISTORY

About 600 B.C.E., **Thales** of Miletus described the magnetic properties of the lodestone (naturally occurring magnetite), including its power to attract iron.

About 530 B.C.E., **Pythagoras** taught that Earth is spherical in shape based on the perfect symmetry of a sphere.

About 460 B.C.E., **Empedocles** proposed four elements and explained volcanoes as elemental fire.

About 450 B.C.E., **Anaxagoras** of Athens suggested that celestial objects are not gods and that the moon is made of the element Earth, reflecting light from the sun.

About 440 B.C.E., **Xenophanes** of Colophon studied fossils and speculated on an evolving Earth.

By about 335 B.C.E., **Aristotle** provided empirical arguments for the sphericity of the earth. He also described earthquakes and recorded tides.

About 330 B.C.E., **Heraclides** of Pontus said that the earth rotates daily on its axis.

About 260 B.C.E., **Archimedes** of Syracuse discovered the principle of buoyancy.

About 240 B.C.E., **Eratosthenes** of Cyrene measured the circumference of the earth by measuring the angle of the noontime sun at Alexandria and Syene and the distance between them.

By about 134 B.C.E., **Hipparchus** of Rhodes defined latitude and longitude, and measured the precession of the equinoxes, later explained as due to a wobble of Earth's axis.

In about 55 B.C.E., **Lucretius** claimed, in his *De rerum natura*, that the volcanic action of Mount Etna was driven by underground fire and wind.

In 77 C.E, **Pliny the Elder** described earthquakes in his encyclopedic *Naturalis Historia.*

In 79 C.E, **Pliny the Younger** described the eruption of Mount Vesuvius.

In 529 C.E., **John Philoponus** published *Against Proclus*, which argued against the eternity of the world and demonstrated its finite age. He also showed that all objects fall at the same rate.

About 1000 C.E., **Ibn Sina**, or **Avicenna**, suggested two causes of mountains: upheavals of the crust of the earth, or the eroding action of water on softer strata of the earth.

In 1543 C.E., **Nicolaus Copernicus** published his treatise on the annual revolution of the earth around the sun, its daily rotation, and the 26,000 year wobble of its axis, causing the precession of the equinoxes.

In 1556 C.E., **Georg Bauer**, better known as Georgius **Agricola**, classified minerals and described physical geography in *De re metallica.*

In 1581, **Robert Norman** published *The Newe Attractive*, announcing the discovery of magnetic dip (inclination).

In 1600, **William Gilbert** published *De Magnete* proposing that Earth is a magnet, which causes the compass to point toward the north.

In 1621, **Galilei Galileo** demonstrated that the acceleration of gravity is constant and that all objects fall at the same rate, independent of their mass.

In 1634, **Henry Gellibrand** discovered the magnetic compass declination from true north.

In 1643, **Evangelista Torricelli** invented the mercury barometer and made the first measurement of atmospheric pressure, leading to his "sea of air" hypothesis that Earth is surrounded by a sea of air moving with the earth through space.

In 1648, **Blaise Pascal** demonstrated that atmospheric pressure varies with weather and decreases with altitude, introducing the weather barometer and altimeter.

In 1669, **Nils Steensen**, known as **Steno**, published the basic geophysical principles of superposition, original horizontality, and lateral extension of sedimentary deposits.

In 1671, **Jean Richer** led a scientific expedition to Guiana, finding variations in the acceleration of gravity, decreasing at lower latitudes.

In 1686, **Isaac Newton** published *Principia*, in which he calculated the shape of the earth to be an oblate spheroid and stated the law of universal gravitation, applying it to explain variations in the acceleration of gravity, tides, and the precession of the equinoxes.

In 1692, **Edmond Halley** theorized that anomalous compass readings might be due to inner-Earth shells with differing magnetic poles. Ten years later, he led expeditions to measure variations in Earth magnetism at different locations.

In 1709, **Daniel Fahrenheit** constructed the first alcohol thermometer and, five years later, a mercury thermometer.

In 1722, **George Graham** discovered diurnal variation of magnetic declination from true north.

In 1741, **Graham** in London and **Celsius** in Sweden observed simultaneous magnetic variations due to the polar aurora.

In 1742, **Anders Celsius** developed the centigrade temperature scale that bears his name.

In 1743, **Alexis Clairaut** published his *Théorie de la figure de la terre: tirée des principes de l'hydrostatique*, in which he used **Colin Maclaurin's** hydrostatic model of the earth to give a mathematical relation for the gravity at the surface of a rotating ellipsoid, known as Clairaut's theorem.

In 1748, **James Bradley** explained the nutation (nodding) of the earth's axis, which he had discovered twenty years earlier, as a result of gravitational interaction with the moon.

In 1749, **Pierre Bouguer** published *La figure de la terre* describing his expedition to measure a degree of latitude at the equator, confirming Newton's calculations on the shape of the earth.

In 1778, **Georges-Louis Leclerc, Comte de Buffon** published his estimate that the earth had developed for at least 75,000 years, based on his measurement of the cooling rate of an iron sphere.

In 1787, **Abraham Gottlob Werner** published his *Short Classification and Description of the Various Rocks*, claiming that rocks formed by crystallization in the early Earth's oceans and leading to the obsolete and discredited theory of Neptunism.

In 1795, **James Hutton** published his *Theory of the Earth*, in which he used the uniformitarian principle to explain slow processes that formed the earth's crust, highlighting the role of heat in support of the theory of Vulcanism (or Plutonism).

In 1798, **Henry Cavendish** used a torsion balance to measure the mutual attraction of lead balls, using his data and Newton's law of universal gravitation to calculate the average density and mass of the earth, which later led to an accurate value for the universal gravitation constant.

In 1804, **Alexander von Humboldt** read a memoir to the Paris Institute on his scientific work in South America, including a decrease in Earth's magnetic field from the poles to the equator, linear groupings of volcanoes suggesting a subterranean fissure, and evidence for the igneous origin of rocks. He was one of the first to note the correspondence of Atlantic coastlines.

In 1809, **William Maclure** made the first geological map of the United States.

In 1812, **Georges Cuvier**, in *Discours sur les révolutions de la surface du globe et sur les changemens qu'elles ont produits dans le regne animal*, provided evidence for the extinction of species and supported catastrophism versus uniformitarianism.

In 1815, **William Smith** published the first geological map of Britain and his *Delineation of the Strata of England and Wales, with Part of Scotland*, in which he used fossils to match rock strata across regions.

In 1821, **Pierre Berthier** discovered the mineral bauxite while working in the village of Les Baux-de-Provence, in southern France. He also discovered the mineral berthierite.

In 1829, **Charles Lyell** published *Principles of Geology*, the first synthesis of geology based on the uniformitarian principle, such as gradual building of volcanoes rather than sudden upheavals.

In 1831, **Carl Friedrich Gauss** began a study of Earth's magnetism, establishing a quantitative basis for measuring magnetic intensity, and a mathematical theory for separating the inner (core and crust) and outer (magnetosphere) sources of Earth's magnetic field.

In the 1830s, **Adam Sedgwick** founded the system of classifying Cambrian rocks and helped to work out the order of Carboniferous and Devonian strata. He also studied metamorphism and concretion.

In 1840, **Louis Agassiz** published *Etudes sur les glaciers*, giving evidence for the existence of a glacial epoch in the temperate zones and the basis for our understanding of the Ice Ages.

In 1846, **Robert Mallet** presented his paper "The Dynamics of Earthquakes," helping to found the science of seismology and coining the words "seismology" and "epicenter."

In 1850, **Jean Foucault** demonstrated the rotation of the earth from precession of a pendulum.

In 1855, **Matthew Maury** published his *Physical Geography of the Sea*, the first comprehensive book on oceanography including contributions to charting winds and ocean currents.

In 1861, **Eduard Suess** used fossil evidence to propose that South America, Africa, and India were once connected in a single supercontinent that he called Gondwanaland, and that the Alps were once under an ocean that he later (1893) named the Tethys Ocean.

In 1869, **William Thomson (Lord Kelvin)** presented his paper "Of Geological Dynamics," estimating the earth's age from 20 to 400 million years based on its cooling rate from molten rocks, later revised to 20 to 40 million years based on improved molten rock temperatures.

In 1880, **John Milne** invented the horizontal pendulum seismograph while working in Japan. This instrument can detect different types of earthquake waves and measure their speeds.

In 1884, **Giuseppe Mercalli** revised the Rossi-Forel earthquake scale, leading to the Modified Mercalli Intensity (MMI) scale still in use for describing the surface effects of an earthquake.

In 1887, **Sekiya Seikei** developed a simple visual model for earthquake ground motions based on his study of the Tokyo earthquake of 1887. He also published a classic study of the 1888 eruption of Mount Bandai. His efforts led to the deployment of nearly a thousand seismographs throughout Japan.

In 1896, **Antoine Henri Becquerel** discovered radioactivity in uranium.

In 1897, **Emil Wiechert** presented a model of the earth with an iron core and a mantle. He later improved the seismograph and invented the field of geophysical prospecting with explosives.

In 1898, **Marie Sklodowska Curie** and **Pierre Curie** discovered polonium and radium, and clarified that radiation was an atomic property. M. Curie coined the term "radioactive."

By 1903, **Ernest Rutherford** had identified alpha, beta, and gamma components of radioactivity and suggested that their energy contributes to the earth's heat and a longer cooling age.

In 1906, **Boris Borisovich Galitzine** invented the electromagnetic seismograph.

In 1906, **Richard Dixon Oldham** analyzed seismic waves from several earthquakes and concluded that the earth has a molten core based on the fact that it does not transmit shear waves.

In 1907, **Bertram Boltwood** proposed that the age of the earth could be determined from the half-life of uranium decay and the relative amounts of uranium and lead in rock samples.

In 1909, **Andrija Mohorovičić** identified a discontinuity within the earth that marks the junction between the crust and the mantle from his study of earthquake waves.

In 1910, **Arthur Holmes** performed the first uranium-lead radiometric dating measurement, and three years later published *The Age of the Earth*. He was an early champion of Alfred Wegener's theory of continental drift, proposing in 1931 that convection cells in the earth's mantle dissipated radioactive heat and moved the earth's crust. In the 1940s, he calculated the age of the earth to be 4.5 billion years.

In 1912, **Alfred Wegener** proposed his theory of continental drift based on the shapes of the continents and comparisons of their fossils, suggesting that they began as a single continent he called Pangaea. He opposed the geosyncline theory of rising and sinking continents.

In 1914, **Beno Gutenberg** studied seismic waves with **Emil Wiechert** and confirmed Wiechert's core-mantle theory of Earth's interior, with the core-mantle boundary now called the Gutenberg discontinuity. Later at Caltech, he worked with **Charles Richter** on an earthquake magnitude scale and the energies associated with the Richter scale.

About 1915, **Felix Vening Meinesz** invented a gravimeter that could be used to measure gravity at sea, leading to measurements in the 1920s of gravity anomalies above the ocean floor, which he attributed to continental drift.

In 1924, **Harold Jeffreys** published *The Earth: Its Origin, History and Physical Constitution* and was the first to demonstrate that Earth has a liquid core based on his study of seismic waves. He was an opponent of continental drift due to lack of a driving force.

In 1928, **Kiyoo Wadati** published his study of deep-focus earthquakes, which eventually led to discovery of the Benioff-Wadati subduction zone.

In the 1930s, **Edward Bullard** began undersea seismic measurements and discovered thick wedges under the continental shelf. In the 1940s, he began developing the dynamo theory of Earth magnetism, and in the 1950s, he made measurements of heat flow under the ocean.

In 1935, **Charles Richter** collaborated with **Beno Gutenberg** in the design of a seismograph for measuring the displacements caused by earthquakes. They introduced a logarithmic scale for measuring the intensity of earthquakes, in which magnitudes on the Richter scale differ by multiples of ten. In 1941 Richter published *Seismicity of the Earth*, revised in 1955.

In 1936, **Inge Lehmann** published her study of seismic waves "P'" in Denmark, which interpreted reflected and refracted P waves as a discontinuity between Earth's liquid core and a solid inner core, the rigidity of which was confirmed in 1971. In 1962, at the Lamont Geological Observatory, she discovered a discontinuity at a depth of about 220 km, now called the Lehmann discontinuity.

In 1938, **Alfred O. C. Nier** published his results on the relative abundance of uranium isotopes based on his improvements in mass spectroscopy, supporting the accurate determination of the earth's age by **Arthur Holmes** in 1940.

From 1941, **Walter Elsasser** developed the dynamo theory of Earth magnetism based on electric currents induced in the outer liquid core, publishing his results in 1946-1947. He pioneered the study of magnetic orientation of minerals in rocks to reveal the magnetic history of Earth's magnetic field.

In 1946, **Willard Frank Libby** developed radioactive carbon-14 dating, using the known rate of decay, measured by its half-life, and relative proportion of its decay products.

In 1949, **Maurice Ewing** established the Lamont Geological Observatory and pioneered the study of seismic waves in ocean basins, deep sea coring of the ocean bottom, submarine explosion seismology, and marine gravity surveying. Earlier, he discovered the deep-sea SOFAR sound channel that permits long-distance sound transmission.

In the 1950s, **Bruce Heezen** led Columbia University's Lamont group, of which **Marie Tharp** was a member, which mapped the Mid-Atlantic Ridge, leading to the discovery of the Great Global Rift.

In 1952, **Francis Birch** demonstrated that Earth's mantle is composed mainly of silicate minerals based on his study of Earth-forming minerals at high pressure and temperature. In 1961, he published a linear relation for compressional waves in rocks and minerals now called Birch's law.

In the 1960s, **Hugo Benioff** used sensitive seismometers he had developed over the previous thirty years to observe that deep-focus earthquakes cluster under an overriding tectonic plate, defining a subduction zone now known as a Wadati-Benioff zone.

In 1960, Rear Admiral **Harry Hammond Hess** and **Robert Dietz** published their work on seafloor spreading based on their study of seamounts in the Hawaiian-Midway chain of volcanic islands, providing critical support to plate tectonic theory.

In the 1960s, **Keith Runcorn** and **Edward A. Irving** did research at Cambridge University on the paleomagnetism of Earth's rocks, helping to work out their implications for plate tectonics. Runcorn also did research on heat convection in the earth and the dynamo theory of Earth's magnetism.

In the 1960s, **John Tuzo Wilson** introduced the plume theory of rising heat convection in the mantle as the source of hot spots on the surface, explaining that the increasing age of seamounts away from the current hot spot in the Hawaii-Emperor chain was the result of plate motion over a hot spot. He also introduced the idea of transform faults and suggested the periodic opening and closing of ocean basins, called the Wilson cycle, related to the supercontinent cycle.

In 1963, **Drummond Matthews,** his student **Fred Vine** at Cambridge University, and Canadian **Lawrence Morley** published their work on variations of magnetic rocks on the ocean floor, consistent with reversals of Earth's magnetic field and confirming the seafloor spreading theory of Hess and Dietz in support of plate tectonics theory.

In 1964, **Robert Dietz** identified the Sudbury Basin in Ontario, Canada, as a meteor crater and later discovered several other impact craters.

In 1966, **Dan McKenzie** submitted his PhD thesis at Cambridge and published the first paper on plate tectonics, basing his work on the thermal structure of oceanic plates as they formed and cooled and working out further details in subsequent papers.

In 1968, **Xavier Le Pichon** developed a comprehensive model of plate tectonics at Columbia University before returning to France.

In 1968, **W. Jason Morgan** published a theory of plate tectonics and in the 1970s, applied Tuzo Wilson's plume theory to several ocean hot spots.

In 1979, **Hiroo Kanamori** and his Caltech student **Thomas C. Hanks** published their moment-magnitude earthquake scale to replace the Richter scale for large magnitude earthquakes.

In the 1980s, **Thomas E. Krogh** developed new techniques in uranium-lead radiometric dating of zircons, leading to the highest precision in the dating of Precambrian rocks.

In 1980, **Walter Alvarez** and his Nobel physics laureate father **Luis Alvarez** discovered that the cretaceous-tertiary (K-T) boundary contained high concentrations of iridium, leading them to propose their theory that this was the result of a 10-km meteor impact that was large enough to cause the extinction of dinosaurs 65 million years ago.

In 1990, **Alan R. Hildebrand** and **Glen Penfield** identified the Chicxulub crater in Yucatan, Mexico, which proved to be the largest impact crater (180 km) ever found. It would have required a 10-km meteor impact, matching the calculations of the Alvarez team for their dinosaur extinction theory.

In 1996, **Kevin Pope** led a group that surveyed satellite photos to identify a sinkhole at the center of the Chicxulub crater thought to be due to subsidence of the crater wall.

In 1997, **Walter Alvarez** published his dinosaur extinction theory in his book *T. rex and the Crater of Doom*.

In 1997, **Gary Glatzmaier** and colleagues created a computer simulation of the earth's dynamo that showed its magnetic reversals.

In 2006, **Lianxing Wen** published evidence of continuing changes at the inner core boundary.

In 2010, a group of forty-one international experts, led by **Peter Schulte,** concluded that the impact at Chicxulub 65 million years ago triggered the mass extinction that included the dinosaurs.

SUBJECT INDEX

Note: Page numbers in **bold** *indicate main discussion.*

G

Galapagos Islands, 421

Galilei, Galileo, 297, 306–307

Galileo spacecraft, 297–298, 301

Galitzin, Boris, 487

gamma radiation, 132, 278, 353–354, 436

gamma spectra, 353–354

Gamow, George, 365

gas: anticlines, 26–27; exploration, 50–51, 119; fluid inclusions, bubbles, 210, 211; mixing, 395

gas chromatography-mass spectrometry, 170, 171

gas fields, carbon storage, 11

gas giant planets, Jupiter, 297

gas hydrates (clathrate hydrates), 10

gas-mixing furnaces for laboratory simulation, 174

Geiger counters, 437, 438

Geiger-Müller counters, 437, 438

gemstone authenticity: fluid inclusions, 213; isotopic signatures, 291

geobarometry, **252–257**

geobiomagnetism, **221–225**

geochemical cycle, **226–232**

geochemical phenomena and processes. *See List of Entries by Categories*

geochemistry: infrared analysis, 280; stable isotopes, 386

geochemistry techniques. *See List of Entries by Categories*

geochronology: dating techniques, 135, 476; fossils, 132; time measurement, 334–335. *See also* radiometric dating

geochronology and the age of Earth. *See List of Entries by Categories*

geodesy: defined, 233; geodetic remote sensing satellites, **233–237**; the geoid, 243; glaciation and azzolla event, 258; surveying the San Andreas Fault, 482

geodesy and gravity. *See List of Entries by Categories*

geodetic remote sensing satellites, **233–237**

geodetic surveying, San Andreas Fault, 482

geodynamics, **238–242**

geographic information systems (GIS), 249, 459

geographic North Pole/South Pole, 105, 124–125, 426

geoid, the, **243–247**

geological analysis: electron microprobes, 146; glaciation, 260; periodicity, 133; tracers, 473; X-ray fluorescence, 573

geological life-supporting requirements, 449–450

geological processes: earthquake mapping, 72–73;

phase changes, **394–399**; phase equilibria, **400–406**; site evaluation, 162–163

Geological Society of America, 351

geologic and topographic maps, **248–251**

geomagnetic band mappings, continental drift, 19

geomagnetic North Pole/South Pole, 124–125

geomagnetic storms: causes, 105; effects, 123–124; Mars, 526; satellite monitoring, 527; solar winds, 525; technology outages, 528–529

geomagnetism, **221–225**

geophones, 161, 493

geophysics, engineering, **161–166**

geosynclines, theory, 535–536

geothermal energy: gradient calculations, 271; heating, 270, 545; temperature-depth profiles, 254

geothermometry and geobarometry, **252–257**

geysers, Yellowstone National Park, 421

giant impactor theory, moon formation, 273–275, 276, 306, 307, 309

giant molecular clouds, 107

giant structures, formation, 180

Gibbs, Josiah Willard, 253, 400, 401–402

Gibbs free energy function, 253

Gibbs phase rule, 401–402

GIS (geographic information systems), 249, 459

glacial clays, soil liquefaction, 520

glaciation and azolla event, **258–261**

glaciers: climate change, 15; continental drift evidence, 424–425; glacio-isostatic rebound, 285, 286; ice core analysis, 295; ocean current changes, 17; water, 215

gliding flow, creep, 25

global climate change, 17–18

global components, biogeochemistry, 7

global networks, 486, 488–489

global positioning system (GPS): crust studies, 235; iceberg studies, 234; mapping, 248; outages, 528; plate tectonics studies, 240; satellites, 235, 460–461

Global Seismic Network, 488

global seismology, 499–500

global stratigraphic correlation, 378–379

Glomar Challenger, 377–378, 544

Goldilocks planet (Earth), 449

Golitsyn, Boris, 509

Gondwanaland, 21, 22, 408

goniometer, X-ray powder diffraction, 578

GPS (global positioning system). *See* global positioning system (GPS)

grabens (faults), 184–185

grades, metamorphic rock, 341

granitic rock, 156–157, 175

gravimeter, Earth's vertical deformation, 139

gravitational forces: compression, 108; Earth-moon interactions, 53; Jupiter's effect, 297, 298–299; plate tectonics, 114; satellite orbit, 245; tides, 137

gravity. *See* geodesy and gravity in *List of Entries by Categories*

gravity anomalies, **262–266**

gravity faults, 183

Gravity Field and Steady-State Ocean Circulation Explorer, 240, 246

Gravity Recovery and Climate Experiment satellites, 234, 240, 246, 265

Gravity Recovery and Interior Laboratory, 309–310

gravity surveys, Earth's subsurface, 162, 164

Gray, John, 487

Great East Japan earthquake, 363

greenhouse gases: carbon cycle, 228; climate change, 6, 15, 16; giant-impact, removal, 274–275; industrialization, 9

Greenwich Mean Time (Universal Time), 510

Gribbin, John, 297–298, 299

ground-based sensors, 350–351

ground failure, earthquakes, 70–71

ground-penetrating radar, Earth's subsurface, 161–162, 164

ground shaking: earthquakes, 69–70; structural response, 512. *See also* magnitude

groundwater: aquifer gravity anomalies, 266; pollution, 386

growth (contemporaneous) faults, 184

Guatemala earthquake (1976), 82

Guinier, André, 146

Gujarat, India, earthquake (2001), 361–362

Gutenberg, Beno, 47, 60, 114–115

Gutenberg discontinuity, 47–48, 112, 114–115, 128

guyots, 421

H

Haicheng, China, earthquake warning success, 90

Haiti earthquake (2010), 90, 362

half-lives: about, 436, 456, 553–554; fission track dating, 204; radiometric dating, 290–291, 441–442; radionuclides, 99, 354, 468

Halley, Edmond, 54

hanging walls, 194, 195

Hawaiian Islands: earthquake distribution, 59; Emperor Seamounts, 421; hot spots, 270–271, 408–409, 414, 420; plumes, 421; potassium-argon dating, 431; seismographs, 498; tsunamis, 549

Hayford, John, 284

Hayward fault: earthquake creep, 70; epicenter, 478

hazardous waste storage/disposal, 162–163, 568–569

hazards, earthquake. *See* earthquake hazards

heat flow: calculations, 271; Earth, 108, 158; fluid inclusions, 212; plate margins, 545; sources, **267–272**, 439. *See also* temperature

heat sources and heat flow, **267–272**

heavy isotopes, 383–384

heavy metals, food chain, 170

Heiskanen, Veikko, 284

hematite, 317, 463

hemispheric climate differences, 344

Herschel, William, 278

Hess, Harry Hammond, 19, 201, 416, 541–542, 543

Heyers, Dominik, 223

higher organisms, magnetic properties, 222

higher pressures, phase changes, 397

high-performance liquid chromatography, 171

high tides, 53–54, 138

Hill, Mason L., 481

Hillier, James, 146

Himalayan mountain range, 349, 536

hinge faults, 184

Hipparchus, 345

Hirose, Kei, 105

historical approaches: Earth age estimates, 101; earthquakes, 60, 83–84

Hoffmeister, Cuno, 525

Holmes, Arthur, 132, 334, 416, 468

Holocene epoch, 259–260

homogeneous medium, elastic waves, 142

Honshu, Japan: subduction zones, 44; tsunami, 549–550

Hooke's law, 531

horizontality, stratigraphic, 455

horsts (faults), 184

hot spots, geothermal: about, 270–271, 419–421; Hawaiian Islands, 270–271, 414; mesosphere, 408; motion, 409; volcanoes, 419–420

hot springs, 565

Howardite-Eucrite-Diogenite meteorites, 110

human activities, solid waste, 168–169

human activities impact: carbon release, 9, 11–12; climate change, 15, 16–17; earthquakes, 73; environmental chemistry, 168–169; geochemical cycles, 230; species extinction, 332